Activist Science and Technology Education

Cultural Studies of Science Education
Volume 9

The series is unique in focusing on the publication of scholarly works that employ social and cultural perspectives as foundations for research and other scholarly activities in the three fields implied in its title: science education, education, and social studies of science.

The aim of the series is to establish bridges to related fields, such as those concerned with the social studies of science, public understanding of science, science/technology and human values, or science and literacy. *Cultural Studies of Science Education*, the book series explicitly aims at establishing such bridges and at building new communities at the interface of currently distinct discourses. In this way, the current almost exclusive focus on science education on school learning would be expanded becoming instead a focus on science education as a cultural, cross-age, cross-class, and cross-disciplinary phenomenon.

The book series is conceived as a parallel to the journal *Cultural Studies of Science Education*, opening up avenues for publishing works that do not fit into the limited amount of space and topics that can be covered within the same text.

For further volumes:
http://www.springer.com/series/8286

Larry Bencze • Steve Alsop
Editors

Activist Science and Technology Education

Springer

Editors
John Lawrence (Larry) Bencze
Ontario Institute for
Studies in Education
University of Toronto
Toronto, ON, Canada

Steve Alsop
Faculty of Education
and Department of Science
and Technology Studies
York University
Toronto, ON, Canada

ISSN 1879-7229
ISSN 1879-7237 (electronic)
ISBN 978-94-007-4359-5
ISBN 978-94-007-4360-1 (eBook)
DOI 10.1007/978-94-007-4360-1
Springer Dordrecht Heidelberg New York London

Library of Congress Control Number: 2014941210

Printed on acid-free paper

Springer is part of Springer Science+Business Media (www.springer.com)

Acknowledgements

This project grew out of conversations 'in parallel' with an academic conference. With our backgrounds in science and technology and education and active affiliations with a variety of social and environmental justice groups, we wanted to bring seemingly fractured aspects of our lives closer together. We sat on city busses, walked in fields and forests and lounged in coffee shops and bars, struggling with alternatives to the status quo and prospects of critical scholarship. Out of those conversations came the *Project for Activist Science & Technology Education* [PASTE] (www.wepaste.org) and associated 'community-reviewed' journal – the latter embracing radical self-reflexivity with prospects of 'awakening solidarity, peace, justice, and responsibility to heal and celebrate relationships between people, communities and life on earth'. Contributions to the *Journal for Activist Science & Technology Education* attracted some attention from science education scholars and others and, happily, we were invited – by John Wallace, Editor – to co-edit a special issue of the *Canadian Journal of Science, Mathematics & Technology Education* focusing on activism. Concerns for social justice and environmental wellbeing associated with fields of science and technology emanating from this special issue then, in turn, led to support from Springer and, especially, Bernadette Ohmer, Ken Tobin, Catherine Milne and Chris Siry to publish this edited collection. It is an honour being part of the *Culture Studies of Science Education* (CSSE) book series. Knowing that we are part of a large and growing network of critical science education scholars, teachers and others worldwide, it is difficult to thank everyone contributing to this project. We are, however, extremely grateful to the 40 authors who have contributed to this edited collection. They have, of course, brought great breadth and depth to the PASTE project that we could not, otherwise, have accomplished. Finally, none of this would be possible without the amazing and ongoing support from our respective partners and families.

Larry Bencze
Steve Alsop

Contents

Contributors

Steve Alsop Faculty of Education and Department of Science and Technology Studies, York University, Toronto, ON, Canada

Barbara Bader Chaire de leadership en enseignement des sciences et développement durable, Université Laval, Quebec City, QC, Canada

Jesse Bazzul Department of STEM education and Teacher Development, University of Massachusetts Dartmouth, North Dartmouth, MA, USA

Larry Bencze Ontario Institute for Studies in Education, University of Toronto, Toronto, ON, Canada

David W. Blades Faculty of Education, University of Victoria, Victoria, BC, Canada

Erica N. Blatt School of Education, College of Staten Island, City University of New York, New York, NY, USA

G. Michael Bowen Faculty of Education, Mount Saint Vincent University, Halifax, NS, Canada

Angela Calabrese Barton Department of Teacher Education, Michigan State University, East Lansing, MI, USA

Elena Camino Interdisciplinary Research Institute on Sustainability (IRIS), Department of Life Sciences and Systems Biology, University of Turin, Turin, Italy

Angela V. Carter Department of Political Science, University of Waterloo, Waterloo, ON, Canada

Lyn Carter School of Education, Australian Catholic University, Melbourne, VIC, Australia

Carolina Castano Rodriguez School of Education, Australian Catholic University, Melbourne, VIC, Australia

Laura Colucci-Gray Interdisciplinary Research Institute for Sustainability, School of Education, Aberdeen University, Aberdeen, UK

Leo Elshof School of Education, Acadia University, Wolfville, NS, Canada

Jose Etcheverry Faculty of Environmental Studies, York University, Toronto, ON, Canada

Randolph Haluza-DeLay Sociology, The King's University College, Edmonton, AB, Canada

Derek Hodson Faculty of Education, University of Auckland, Auckland, New Zealand

Darren Hoeg Ontario Institute for Studies in Education, University of Toronto, Toronto, ON, Canada

Rebecca Houwer Faculty of Education, York University, Toronto, ON, Canada

Bernhard Isopp Graduate Program in Science & Technology Studies, York University, Toronto, ON, Canada

Mellita Jones School of Education, Australian Catholic University, Melbourne, VIC, Australia

Ashley S. Kerckhoff Rainbow District School Board, City of Greater Sudbury, ON, Canada

Mirjan Krstovic Erindale Secondary School, Peel District School Board, Mississauga, ON, Canada

Yves Laberge Centre de Recherche en Éducation et Formation Relatives à l'Environnement et l'Écocitoyenneté, Montreal, QC, Canada

Ralph Levinson Institute of Education, University of London, London, UK

Ana Maria Martinez Faculty of Education, York University, Toronto, Canada

Kate Milberry Faculty of Extension, University of Alberta, Edmonton, AB, Canada

Michael P. Mueller College of Education, University of Alaska Anchorage, Anchorage, AK, USA

Janet Newbury Faculty of Human and Social Development, University of Victoria, Victoria, BC, Canada

Chantal Pouliot Département d'études sur l'enseignement et l'apprentissage, Université Laval, Québec, QC, Canada

Giuliano Reis Faculty of Education, University of Ottawa, Ottawa, ON, Canada

Pedro Reis Instituto de Educação, Universidade de Lisboa, Alameda da Universidade, Lisbon, Portugal

Arturo Rodriguez College of Education, Boise State University, Boise, ID, USA

Wolff-Michael Roth Faculty of Education, University of Victoria, Victoria, BC, Canada

David E. Selby Sustainability Frontiers, Adjunct Professor at Mount St Vincent University, Halifax, Nova Scotia, Canada

Laurence Simonneaux Ecole Nationale de Formation Agronomique, Castanet-Tolosan Cedex, France

Erin Sperling Ontario Institute for Studies in Education, University of Toronto, Toronto, ON, Canada

Edna Tan School of Education, University of North Carolina, Greensboro, NC, USA

Kenneth Tobin The Graduate Center, City University of New York, New York, NY, USA

Matthew Weinstein Department of Education Program, University of Washington-Tacoma, Tacoma, WA, USA

Terry Wilkinson Faculty of Education, York University, Toronto, ON, Canada

Brandon Zoras Toronto District School Board (TDSB), Toronto, ON, Canada

Chapter 1
Activism! Toward a More Radical Science and Technology Education

Steve Alsop and Larry Bencze

Abstract What might activism offer science, technology and education? What might science, technology and education offer activism? This chapter provides an introduction to an edited collection exploring these themes. We start by situating assembled responses within contemporary socio-ecological contexts and selected scholarship and practices. We then take up the case for activism as an open question with potentially far-reaching implications for science and technology pedagogies and offer a reading of the following chapters as a more radical complement to existing scholarship in the field. As a basis for greater reflectivity, we then propose four maxims for critical reworking science and technology education praxis; (i) contemporary conditions, (ii) democratic political theory, (iii) subjectivities and agency; and (iv) morals and ethics. The chapter concludes with discussions of partialities and associated tensions, contradictions and limitations, as well as thanking all those involved in bringing this project to fruition.

Keywords Science and technology education • STS • Activism • Radical reflexivity • New social movements • Democracy

It is our pleasure to introduce *Activist Science and Technology Education*. This is, we believe, the first publication of its kind that focuses on exploring activism within science and technology education. With a pioneering spirit, our edited collection has three major goals. The first is to contribute to a growing group of science and

S. Alsop
Faculty of Education and Department of Science and Technology Studies, York University, 4700 Keele Street, Toronto, ON M3J 1P3, Canada
e-mail: salsop@edu.yorku.ca

L. Bencze (✉)
Ontario Institute for Studies in Education, University of Toronto, Bloor St W 252, Toronto, ON M5S 1V6, Canada
e-mail: larry.bencze@utoronto.ca

L. Bencze and S. Alsop (eds.), *Activist Science and Technology Education*, Cultural Studies of Science Education 9, DOI 10.1007/978-94-007-4360-1_1,

technology educators pursuing policies and practices that aim to be more ethically and politically engaged and socially and environmentally responsive. The theories and pedagogies that are outlined in the following pages pay close attention to social, economic and ecological critique and reform. Second, we wish to make the case that this collective work can be viewed as a more radical complement to existing science and technology education. Indeed, we would even go a step further by suggesting that the work assembled herein offers a more progressive orientation to science and technology education than many studies in our field identified by the common acronyms SSI (Socio-Scientific Issues) and STSE (Science Technology Society Environment). Third, as editors, our goal was to bring together people with an openness and courage to share and explore their political and educational commitments such that we might learn together and act more thoughtfully and collaboratively. With these goals in mind, we are delighted that this volume is able to showcase the insightful and inspirational work of so many of our colleagues.[1] It is with a continued disposition of collaboration and an extended openness of plurality and reflexivity that we offer this edited collection in support of on-going and new conversations, movements and actions, within science and technology education and also further afield.

The assembled authors offer the foresight, commitment and care necessary to shape the next round of discussions concerning science and technology education and the public sphere. In the following 33 chapters, authors share ideals, theories and practices that relate to common themes in diverse ways. Our desire is to stimulate conversations by narrating and analysing options, possibilities and challenges associated with considerations of activism, science, technology, and educational praxis. What drives this project is the prospect that education in all of its diverse settings, disguises and narratives might more thoughtfully, more progressively and more effectively, respond to deepening local and global injustices. At this critical contemporary social and environmental juncture, a series of questions unite us:

- What insights might be drawn from our theories and practices as responses to contemporary times?
- How might we take more seriously wider social, political, economic and environmental contexts in which our practices reside and also seek to resist and influence?
- How might we continue to reach out across 'difference' to listen more carefully and proactively to those who remain marginalised by our practices?
- How might we gather collectively to critique and extend our cultural resources as a basis of joining with others within activist transformative democratic politics?

---

[1] Perhaps we might be seen as a gathering of 'critical friends' to use the concept associated with Action Research and Will Carr and Stephen Kemmis (1986).

The audiences that we have in mind for the edited collection are all those in pursuit of reconstituting science, technology and education with aspirations of deep-rooted change. Such discussions may appeal to those critical 'cultural workers' (Giroux 2004) of science, technology and education with desires for radical transformations; those 'destructive daydreamers' with progressive 'political horizons', to use the late Roger Simon's language (1992, p. 9). The following pages imbue science and technology education with moral and ethical commitments and responsibilities. The education that we seek is one more efficaciously engaged with the prospects and hopes of open and normatively situated relational-practices.

Our volume has a particular focus on science and technology in schools. It also explores individuals and organisations that assert themselves within science and technology education in the civic sphere and popular culture. In this regard, it should appeal to school teachers and administrators as well as individuals and groups connected with science and technology in news media, cultural institutions (science centres and museums) and social movements including nongovernmental organisations, civil society organisations, campaign groups, citizen scientists, and various volunteer and hobbyists groups. There is also discussion of post-secondary education, and the text has considerable relevance to scientists and technologists, teacher educators, environmentalists and social scientists within universities, colleges, governments and wider afield.

The following pages brings together a community of researchers and practitioners with shared commitments to inquiries in which knowledge of science and technology, education, political commitments, agency and social and environmental concerns, are considered in relation to one another. Such examinations provide opportunities for complex conversations and actions within various socio-ecological issues. There is a shared desire to explore theory and practices in which activism and education are approached as an open question for examination and mobilization within material, organic and cultural contexts in which we work and play.

We are delighted that contributors have so deeply and comprehensively engaged with the project for an *Activist Science and Technology Education*. The success of the project – of course – is to be found in the way it recognises and celebrates differences, and remains self-critical whilst capturing a common ground in which to fruitfully exchange and critically reimagine our practices whilst joining with others and collaboratively mobilising[2] for progressive change. As the following authors collectively assert, particularly important in such work are critical reconstructive agendas. These embrace on-going analyses of current educational practices and how they are entangled in constituting and representing: academic disciplines (sciences and technologies) and associated subjectivities; wider identities and political agencies; and broader societal and environmental crises and concerns. Such critical stances are simultaneously envisioned within re-constructive agendas

[2] There are a number of adjectives that are applicable here and each carries important meanings: caring, acting, agitating, coalescing, disrupting, resisting, arguing, educating, teaching and learning. We offer the reader an invitation to select an adjective and add to this list if necessary.

that seek to develop, enact and critique new theories, practices and pedagogies. What follows will stimulate critical reflections on so called 'traditional' as well as 'more radical' science and technology education policies and practices.

Navigating the Contemporary

The chapters herein offer diverse pedagogical responses to global and local contemporary contexts of deepening concern. Indeed, many authors situate their work within what might be described as a crumbling project of humanity. David Selby (Chap. 10), for instance, adumbrates a catastrophic future deduced from secondary analyses of climate change models. A future mired by:

> Ubiquitous environmental disaster (including a huge loss of biodiversity), on-going and massive internal and external population displacement in consequence of sea incursions, seasonally recurring wildfire and desertification, and resultant social dislocation, hunger, starvation, internecine strife, violence conflict, tribalism, aggressively defensive localism, as well as the ever-lurking danger of genocide. (Page Needed)

In this and in many of the other chapters what is abundantly clear is that humanity is facing an unprecedented set of challenges and risks. Such challenges include living with: increasing environmental decline, biodiversity loss, climate change and unrelenting and unsustainable resource exploitation; rampant transnational neoliberal markets shamed by unprecedented levels of disparity between the rich and the poor and youth unemployment; weakening state democracies that are restricting possibilities for people and communities and the emergence of global industrial systems as major transnational global political economic players (see Arturo Rodriguez, Chap. 4). Such challenges are compounding and divisive. They are lived, endured and suffered unequally, and throughout the following discussions we are conscious of our privileges, subjectivities and cultural biases as authors located predominantly in the minority wealthy world and educated largely ourselves in traditions of 'spectres of modernity', to use Sandra Harding's term (2008, p. 2).

There seems to be no shortage of academics across of wide range of disciplines seeking to encapsulate the contemporary epoch whether as: a 'shadow of the shopocalypse' (Sandlin and McLaren 2010); a 'reflexive stage of modernity' or 'risk society' (Beck 1992); 'liquid modernity' (Bauman 2008); 'a new era of capitalism' (Hardt and Negri 2009), 'post-normal science' (Funtowicz and Ravertz 1991), or even as a new period in geological history, 'The Anthropocene' (Zalasiewicz et al. 2010). What is abundantly clear in all these accounts is that there are certainly enough reasons to reflect once more on the projects and prospects for science and technology education.

As Kuhnian inspired philosophers remind us, moments of crisis often accompany radical discontinuities. Emerging ruptures in the social sphere make some boundaries and hierarchies fleetingly visible, and zones of resistance and counter cultures might emerge. Naomi Klein's (2007) *The Shock Doctrine* offers a rather distressing interpretation of the crisis as a theatre beckoning in "the rise of disaster capitalism".

But a crisis, as Rebecca Houwer (Chap. 7) more hopefully argues, offers multiple theatres and differing scripts, including sites for rethinking education. In a similar vein, Sheila Jasanoff (2010) writes of global warming and climate change as a 'new climate for society' that necessitates recognising and responding to 'the human and the social' at a time 'that seems to render obsolete important categories of solidarity and experience' (p. 233). Such moments offer science and technology educators opportunities to return to our underpinning assumptions of education in *relation* to the contemporary. There is certainly an abundance of this type of thinking to be admired in what follows.

The narratives within this volume should also be cast against a backdrop of sustained growth of work in Science and Technology Education and Science and Technology Studies [STS] that comport to nurture more normative, critical and reconstructive agendas and perspectives. The work of the science educators Derek Hodson (2003, 2011), Wolff-Michael Roth and Angela Calabrese Barton (2004), Kenneth Tobin and Wolff-Michael Roth (2006), Michael Mueller and Deborah Tippins (in press) and others, of course, serve as inspiration for this work and this volume.[3] During the past few decades there has been sustained interest in examining real-world and local problems within the context of science and technology education as/for social reform. 'Scientific and Technology Literacy', 'The Public Understanding of Science', 'Critical Place Based Science Education' and most recently 'Citizen Science' and 'Ecojustice' have all been topics of extended deliberation and debate. As part of this movement, for the past 5 years we have been coordinating an online project, entitled PASTE[4] [The Project for Activist Science and Technology Education]. PASTE has an associated community-reviewed open access journal, *The Journal for Activist Science & Technology Education (JASTE),* which now has six editions. There is also an associated special issue of the *Canadian Journal of Science, Mathematics and Technology Education* (Alsop and Bencze 2010) – the first of its kind focusing on socio-political activism. This edited book continues this journey and several of the authors in this collection were part of these earlier projects, with their chapters drawing from and extending these experiences.

As might be expected, many of the following authors align their work within different progressive theoretical trajectories. John Dewey, Paulo Freire, Ivan Illich, Herbert Marcuse, Antonio Gramsci, Jürgen Habermas and Michel Foucault are frequently cited. Chapters also make consistent reference to STS scholars including the omnipotent Bruno Latour, Michael Callon, Donna Haraway and Sandra Harding. At an earlier stage in the history of Science and Technology Studies, Ian Hacking's (1999) renowned text, *'The Social Construction of What?'* demarcated 'six grades of constructivist commitment' – historical, ironic, reformist-unmasking, rebellious and revolutionary (pp. 19–21). The chapters in this collection arguably span all these grades but place particular emphasis on the last two: rebellion and

[3] They also feature in this volume.

[4] https://jps.library.utoronto.ca/index.php/jaste/index (website last accessed May 13th, 2014).

revolution. In Hacking's terms, 'rebellious constructivists' are critical of the status quo that they see as not 'inevitable' and needing to be 'radically reformed', whilst 'revolutionary constructivists' go a stage further as 'activist[s] who move beyond the world of ideas and try to change the world' (p. 20). A few years later, Edward Woodhouse, David Hess, Steve Breyan and Brian Martin (2002) mounted a persuasive case for 'reconstructivist agenda' in STS – stemming from a 'rapprochement' between 'more academic and 'activist wings' (p. 297) (also see discussions in Bernhard Isopp in Chap. 17). Woodhouse and colleagues challenge STS scholars to openly embrace their 'normative commitments' with 'thoughtful partisanship'. This edited text builds on these earlier arguments and is written at a time in which there is a noticeable increase in more normative, advocacy based work in the field. Many of the authors make reference to this work in their chapters.

Our edited volume is also open in its support of the increasing number of scientists now identifying as activists. These include Barry Commoner, James Hansen, Michael Mann, David Suzuki, Andrew Weaver and many others whom have sought to publically align themselves with the need for systemic structural changes and continue to endure the fallout that this often entails (see discussions by David Selby, Chap. 10; Bernhard Isopp, Chap. 17; Randoph Haluza-Delay and Angela Carter, Chap. 19). There is, of course, a long history of scientists as critical social commentators, as 'politico-scientists' (McCormick 2009) or as advocates and campaigner spokespersons. During the past few decades, the civic sphere has played an active role in backing a dramatic growth in activist groups, which include the widely known and longstanding *Science for the People* in the United States, as well as the *Science for People* and associated Journal *Radical Science* in the UK. There are now many other organisations which are explicitly linked with science and technology including the *Campaign for Nuclear Disarmament, Doctors and Engineers without Borders, Greenpeace, World Wildlife Fund, EarthRoots, Sierra Club, David Suzuki Foundation, Families Against Radiation Exposure, Sustainability Frontiers, Scientists for the Right to Know, Physicians for Human Rights,* and so many other groups, simply too numerous to mention here. Chapters in this collection give sustained attention to groups including: *The Occupy Movement* (see Lyn Carter, Chap. 2; Kate Milbury, Chap. 15); *Street Medics* (Matthew Weinstein, Chap. 12), *Aboriginal, Religious, Labour and Environmental groups* (Wolff-Michael Roth, Chap. 14; Randy Haluza-Delay and Angela Carter, Chap. 19) and *Corporate and Neoliberal Climate Activists* (see Leo Elshof, Chap. 18).

Throughout these discussions, we are reminded of Chris Jordan's art installation, 'E. Pluribus Unum' (One Among Many).[5] From a distance the electronic image is of a patterned sphere much like Earth, but on closer inspection it reveals a web of thousands of lines comprising names of organisations from around the world that are devoted to peace, environmental stewardship, social justice and the preservation of diverse and indigenous cultures. The dynamic process afforded by the electronic

[5] http://www.chrisjordan.com/gallery/epu/#e-pluribus-unum (website last accessed June 23rd, 2013).

representation of 'zooming in' powerfully captures the silent emergence of thousands of these organisations that together offer the capacity to carry some of the 'weight of the world' (Bourdieu et al. 2003). As James Hawken (2007) powerfully argues they have a collective power to bring about far reaching transformations, although nobody knows the actual numbers of such organisations, whether it is now one or two million, or even more. Authors in this volume remind us, that many of these groups are actively narrating ways in which science and technology and education are implicated in the modern condition but can also be very powerful and important allies in struggles for change. Perhaps each organisation offers multiple sites from which to reconstruct our practices in dynamic and purposeful ways that are more empowering than narratives of loss to neoliberal hegemony. The associated literature on new social movement theories and popular concepts including 'collective action frames', 'resource mobilization' and 'political opportunity processes' (see Benford and Snow 2000, p. 611) have so much to offer our research and practices in the future.

The last few decades has also witnessed the rise of seemingly progressive international policy treatise and rights. United Nations and UNESCO policies have turned to the prospect of 're-orienting' education in pursuit of sustainability and on-going Millennium Development Goals. Such declarations and emerging policies are inspiring. Much evidence suggests they are serving to galvanize and empower locally constituted practices throughout the majority and minority world (see discussions by Laura Colucci-Gray and Elena Camino, Chap. 9; David Blades and Janet Newbury, Chap. 11; Ashley Kerckhoff and Giuliano Reis, Chap. 26).

So, in some ways the chapters herein might fruitfully be framed as re-energized pedagogical 'unrest' as part of a global movement for change. They invite ample opportunities for traditional sites of science and technology education to gaze beyond deterministic and restrictive cultural practices and ingrained assumptions of progress, and embrace an openness to critically learn with others and a resolute willingness to share experiences as part of a growing environmental and social justice movement. Let us not be 'too scientific' (Shrader-Frechette 2001), or too academic and 'High-Church' (Fuller 2007); or for that matter just simply too neutral or too educational to lose sight of an enhanced role in a global movement in which our agencies and expertise might fruitfully and reflexively play a part.[6]

Building a Collection

In building this collection, we enjoyed considerable flexibility in our roles as editors. Early on we adopted an emergent approach, seeking potential authors with diverse scholarship, ideologies and in different locations. Their responses

[6] Whilst of course being ever cognizant and critically aware of the same rhetorical logical within our cultures and reproductive practices.

blend different theories and critical visions of science and technology education within very different educational settings and contexts. Our invitation was for scholarly contributions that explore activist inspired science and technology education. We also offered authors three themes that we thought might be found helpful in organising their contributions – disclosing, mobilizing and celebrating. In terms of:

- *Disclosing*, we envisaged authors writing about various socio-political issues stemming from social hierarchies relating to SSI such as climate change, agriculture and food, energy resources, biotechnologies, information technologies, industrial military complex, material economies and health care;
- *Mobilizing*, we offered authors opportunities to critical explore perspectives and practices in and out of school science that they identify with and promote activism;
- *Celebrating*, we sought narratives that recognise and remember what has been achieved and what needs revisiting and exploring. We sought a context of coming-together to celebrate projects and take-a-breath and garner collective strength.

As we have commented elsewhere (Alsop and Bencze 2009), the label of 'activism' is wrapped up in a series of fractured and imprecise social imaginaries that are themselves not without either potential or concern. At the heart of most of these are desires to act to bring about change. The Oxford English Dictionary describes activism as 'intentional efforts to promote, impeded or direct social, political, economic or environmental change'. Other definitions cast light on the agent, the protester engaged in pursuit of a particular specific cause.

Activism in some ways is a problematic concept. It has, for example, become associated with 'the spectacle' and in so doing runs the risk of being separated from daily-lived actions (see discussion in Angela Calabrese Barton and Edna Tan, Chap. 28). It can become accompanied with a somewhat romanticised notion of activist knowledge and practices as being necessarily, and unquestioningly, superior to other practices. Focusing on activist responses can too readily shift responsibilities to individuals and civic groups whilst seeming to absolve the state and institutions (including schools and universities) of their responsibilities. Furthermore, activism can become erroneously associated with ideologically entrenched ideologies, practices of indoctrination that are intolerant of differences and unable or unwilling to critically self-reflect and learn from and with others. Throughout the following discussions we are certainly cognisant of these tensions, but activism also has some associated benefits. For our purposes here these include:

(i) Activism is a broad contemporary and flexible concept. In Susan Star Lee and James Griesemer's (1989) popular terms, it is a 'boundary object' that a large number of individuals and groups can self-identify with in relatively open and empowering ways;
(ii) It is a concept that brings diverse groups together with sufficiently common but divergent educational and political commitments such that they can share and learn together as 'subjects' rather than 'objects' of educational processes (see Freire 1972);

(iii) It retains controversy in some educational circles and as a consequence invites reflection. A particular strength of the concept of activism – we suggest – is the paradox that it seems to generate concerning the locus of its educational applicability (see Alsop and Bencze 2012).
(iv) It is an action orientated and generative term and thereby offers the prospect of identifying with others and acting with common goals.

A Brief Overview of the Collection

This book comprises both theoretical and more practical discussions that are grouped into four sections. Each section starts with a 'word cloud' and preamble written by the editors. The first section, *Constituting Theories*, includes chapters with an emphasis on theoretically and ideologically informed arguments. Here the authors review a wide array of scholarship that has sought to grapple with science, citizenship and civic participation. They then attend to a series of critical, political and epistemological perspectives. The second section, *The Public Sphere*, takes up our central themes within particular public contexts including street medicine, farming, new social media, news media, and resistance to the Canadian Tar Sands. The third section, *Elementary and Secondary Education*, outlines more practical responses to the challenges of radicalising pedagogies within selected projects set in formal education. Here, the chapters document a series of action-oriented research projects set in school science and environmental clubs. The fourth section, *Post-secondary Education* also has a grounded emphasis, exploring the question of activism within educational contexts of teacher education and university studies. The text concludes with an *Afterword* written by the editors that draws out some emerging themes and signposts future directions for policy, practice and research.

Framing a More Radical Approach to Science and Technology Education

As editors, we opened this introduction with claims of 'radicalism' – that this collection constitutes a basis for a more radical response to science and technology education. Our claims to radicalism are partly inspired by the Radical Science Movement (See Rose and Rose 1979) that was established in the 1970s. The movement became associated with the journals, Undercurrents and Radical Science. It arguably opened up the politics of science, and in a decade of considerable social and environmental instability took-up questions of what science ought we strive toward and what science do we really need? In a similar way, this volume invites radical self-reflection on the aims, methods and purposes of science and technology education. Radicalism, nevertheless, is a term that carries associated tensions of demarcation and elitism; a gesture in which innovations are legitimised through marking their departures from existing practices. In our minds, there is already too much epistemic and pedagogical splintering within educational theorising, associated with a broad array of so-called adjectival educations and emergent turf-wars. Our intent is not to add to this list, nor to inflame sometimes entrenched antagonisms. Instead, we offer this claim as a *provocation* to critically reflect on the augments outlined. These, of course, rightly defy simple party politics or other reductionist and formulaic style categories.

In bringing this introduction to a close we offer four guiding assumptions for future work (two of these have similarities to Ginwright et al.'s (2006: xvi) guiding principles for more democratic youth participation). As we reflect on the assembled chapters, these maxims seem valuable as a basis to frame a more radical science and technology education.

Science and Technology Education Should Be Critically Reworked in Relation to Contemporary Economic, Social, Ecological and Material Conditions

The assembled authors all agree that in order to understand science and technology education better we need to bring contemporary socio-ecological conditions to the foreground. In so doing, we need to question ways in which some of our practices have seemingly become "isolated from the world" and look to more purposefully and critically embed these practices. A host of feminist scholars continue to warn us of the dangers of 'unlocatable knowledges' that are simultaneously 'everywhere' and 'nowhere' all at the same time and thus remaining 'incontestable' (see discussions by Haraway 1991). Such representations serve to render invisible local practices and networks that both constitute and come to represent knowledge, expertise and education. The assembled discussions herein invite future explorations of how our assumptions and local practices of ST education function as part of larger contemporary (and historical) economic, political, social and environmental trends and conditions.

As contemporary educators, our work needs to reflect on the ways in which dominant policies and practices serve to sustain particular social structures and subjectivities. One part of a response to pressing social and environmental issues that we are presently facing is a better understanding of relations of power, and the governmentality with which we constitute ourselves and others in our roles as culture workers with science and technology (see discussions by Lyn Carter in Chap. 2; Jesse Bazzul, Chap. 3).

For decades, sociologists, following Marx, have narrated how formal sites of education (schools and sites of post secondary education) not only serve to teach curricula but they also sustain particular hierarchical structures including labour relations; economic inequalities, cultural and knowledge hierarchies, and state powers (see discussions in Giroux 1998, for example). Environmental education scholars have added to this list by persuasively theorising how education serves to sustain and intensify dialectics of environmental and ecological injustices (Bowers 1997; McLaren and Houston 2005). A pressing task for science and technology educators is to draw on these different theories and methodologies to disclose and animate the multiple and sometimes-contradictory ideological assumptions underpinning dominate modes of ST pedagogy, curriculum and assessment. We need a better sense of the agencies and structures that our practices both propose and also foreclose.

Drawing on ethnographic methods and grounded theories as well as emergent 'methodologies' including Actor Network Theory (and now *After*-ANT), STS theorists have repeatedly revealed the significance of local material and immaterial practices and relationships in which knowledges are continuously and actively re-constructed.[7] Scholars of STS have been extremely diligent in disclosing the genesis and mobility of scientific knowledge in which power, identities and knowledge are continuously entwined. STS scholarship has not, however, paid sustained attention to educational processes and associated institutional practices. Within the context of this volume, this seems especially important.

In contrast, with considerable alacrity and much success, science and technology education continues to focus on exploring complex psychological processes of teaching and learning science. As a consequence it has not always, perhaps, sufficiently situated itself within struggles over its complex, ideological, cultural practices and assumptions. This becomes especially apparent in the case of contemporary scientific and technological issues. As several authors in this collection comment, even more progressive educational practices (associated with SSI and STSE) often favour pedagogies routed in individualised technocratic decision-making (See discussions, for example, by Derek Hodson in Chap. 5). Whilst seemingly uncontroversial (or pleasingly 'cold', see Laurence Simonneaux, Chap. 6) such narrow framing misses the point that contemporary scientific and technological controversies rarely (if ever) emerge solely because of lack of scientific knowledge or sufficient consensus or failing logic, but also because of deeply asymmetric cultural assumptions and power structures. Many SSIs in this respect are not so much discrete problems as they are socio-ecological and socio-technical *conditions*. Climate change, for example, is a lived experience co-produced amongst many actors that comes into being through a collective formation within particular material and immaterial settings (as several of the following authors highlight). A terrain of discourse limited to technocratic "answers" to such a complex socio-ecological and socio-technical phenomena fails to connect these discussions to broader economic, political and ecological contexts and as a consequence questions of power, agency, identity and justice are too easily lost. Several chapters make this point abundantly clear as they highlight

[7] Different theorists bring different metaphors here including coproduction, social construction, representation, performance and enactment.

the ways in which the local effects of SSIs are demonstrably unequal, inequitable and unjust. Those carrying the burden of SSIs are often least responsible for them (see Randy Haluza-Delay and Angela Carter's discussions of climate change in Chap. 19, for instance).

In this regard, pedagogical approaches to SSIs should not be conceived and enacted within an objective tradition of being "out there" as they are also very much "within us". There is, however, this lingering asymmetry associated with too many contemporary educational practices – even those associated with the Nature of Science and Technology [NOST], SSIs and STSE – of "addressing" or "explaining" social and environmental concerns from a position that is somehow invisible, or largely outside and hovering above these concerns. Such approaches escape important questions of how our educational practices are implicated in contributing to the socio-political *conditions* underpinning these very issues. The point is that the existence of SSIs (including climate change) is not simply a technocratic question for science and technology education to answer, but it offers a complex context in which to rethink the very conditions and assumptions underpinning our policies, pedagogies and assessment practices. In so many ways, we are presently teaching climate change (and other SSIs) as though they don't pedagogically exist. The pressing question that the assembled authors bring to our attention is: Where and how are SSIs pedagogical? This question seems to offer a plausible precursor to consider how we might as educators respond to these very complex issues. But as it presently stands this questions seems to have escaped our sustained attention.

The assembled authors cluster their responses around some shared themes including:

- Globalisation, neoliberalism and bio-politics;
- New social movements and localism,
- The democratization of science and technology (including the increasingly politicised, commercial and contested natures of technosciences in the civic sphere);
- Environmental sustainability and epistemic hybrids and relationships;
- Electronic communications and associated new-media literacies.

Within these five distinctive themes, authors offer multiple pathways for rethinking the very terms and conditions of science and technology education in relation to a contemporary period in which science, technology *and* society continue to concurrently change. Each of these themes, we posit, offers promising frames of reference from which to radically rethink science and technology education. In this regard, an open question of activism serves as a provocation for ST educators to look beyond the ways in which science and technology is organised today and to explore education in relation to pressing societal issues *and* underpinning conditions. The more radical science and technology education that we advocate here invites much closer attention to the constitutive conditions of SSIs and the changing contexts of their formations, including, of course, science and technology and education themselves.

Science and Technology Education Should Be Critically Reworked as Political Practice

In many formal educational contexts there is still a real uneasiness associated with the political in science, technology and education. It is therefore perhaps unsurprising that science and technology education in most settings strives for a comforting ideal of apolitical, value-free practices.[8] In contrast, the following arguments situate education in different political contexts and thereby collectively advocate more purposeful opportunities for students to experience science and technology in relation to political theory and identifiable political goals. Such an approach remains unsettling. In part this is because it raises troubling questions of partisanship and indoctrination (Whose and what politics get to count? Where? And how?).

There are, of course, deep-rooted historical origins concerning the separation of politics and pedagogy. A story is widely told of the Athenians accusing Socrates of corrupting the youth and this, in part, led to his execution. In response, his student Plato set the terms for Western education by insisting that rulers (philosopher kings) first remove themselves from society before returning as political leaders ready to deploy their abstract knowledge. Throughout the emergence of science, from the sixteenth to nineteenth centuries and beyond, it is also widely acknowledged that science was conceived as supporting liberal democracies insofar as it remain separated from them. Of course, the relationship between science and politics was never as asymmetrical as was seemingly assumed and this became especially apparent within events concluding the Second World War and the cold-war period (in which, incidentally we both grew up).

A number of the authors in this volume situate their discussions within questions of democracy and prospects of education leading to greater democratic representation (see discussions by Chantal Pouliot, Chap. 29; Ralph Levinson, Chap. 21). Contemporary political representatives of science and technology include specialist groups such as journalists (See discussions by Michael Bowen, Chap. 16; Bernhard Isopp, Chap. 17), opinion polls as well as expect testimonies within the legal system. Increasingly the public understanding and participation in science has emerged as a significant global agenda and 'science-shops', 'lay courts' and 'upstream' engagement in policy and grant applications, are actively engaging interested individuals and groups in science and technology. The rise of new social and environmental justice groups (as previously discussed) might also be interpreted as democratising science and technology.

Sabrina McCormick (2009) documents this shift in two phases – from 'politico-scientists' (of the 1970s and 1980s) to the present day 'democratizing science movements' in which the public have increasingly taken a more central role with the *assistance* of experts (cf. with discussions by Matthew Weinstein, Chap. 12;

---

[8] Despite the logical fallacy of this positioning – see recent discussions by Burns and Norris (2012) and rejoinder by Alsop and Bencze (2012).

Michael Mueller, Chap. 11). Studies of new social movements bring our attention to the ways they assert themselves in the public domain by actively 'framing' (see review by Benford and Snow 2000) or 'spinning' (see Callison 2010) scientific knowledge to suit their particular causes and shared cultural understandings and interests (whether they are medical, farming, climate change, peace, pollution or many others). Their work, in this respect, is not primarily 'epistemological' in the sense of generating new knowledge,[9] but it is more strategic and political. Leo Elshof eloquently illustrates this in Chap. 18 through his analysis of wealthy groups with climate change denial agendas. In response, he suggests formal education needs to offer opportunities for learners to critically analyse now different groups 'politicize' science and technology.

A more radical science and technology education should embrace a critical analysis of political representations in public. It might offer educational opportunities for students to study 'framing' or 'spinning', for instance. Students might take on an active political role by framing science and technology for themselves with particular social and environmental justice agendas in mind. This is the approach broadly advocated, for example, by Derek Hodson (see discussions in Chap. 5). For some time, Hodson has argued for a science and technology education that is not only developmental but is also overtly political. In adopting this stance, Hodson openly recognises the importance of multiple political perspectives and analyses, and draws critical distinctions between approaches that are generative and open, and those that are restrictive, silencing and as such are indoctrinatory.

Underpinning all these discussions are complex questions relating science technology education and democratic theory. Of course democracy is exceedingly complex and in the following discussions different authors might be read in terms of some differing democratic assumptions and preferences. Such a reading might distinguish, for instance, between more: 'structural democratic orientations' (see discussions by Laurence Simonneaux, Chap. 6); 'participatory democratic orientations' (see Jose Etcheverry, Chap. 33; Ana Martinez and Steve Alsop, Chap. 27); and 'standpoint democratic orientations' (Wolff-Michael Roth, Chap. 14; Randy Haluza-Delay & Angela Carter, Chap. 19).[10] This reading is offered here as nothing more than cursory and explorative (at the very least, it needs to be consciously aware of its own structural politics). Nevertheless, it hints at future directions for science and technology education research. A host of questions now seem to emerge concerning relationships between science and technology education and democratic theory and practice. Such discussions seem especially pressing within

[9] Although this does, of course, occur – see inspiring discussions by Corburn (2009).

[10] Dimitris Papadopoulos's (2010) detailed analysis of constituent politics in technoscience guides these distinctions. His analysis identifies five representations of expertise in technoscience; institutional participation, inclusion of non-human others and marginalized experiences and an alternative form of politics, 'constituent politics in technoscience' (proposed by the author). There is insufficient space in this introduction to explore these discussions in any detail, but these discussions seem to offer promising lines of inquiry for future work.

educational responses to SSIs and STSE. However, as Lyn Carter comments in Chap. 2, they have yet to be explored in any depth or detail. A more radical science and technology education should take on the role of more closely responding to democratic political theory in educational policies and practices.

Science and Technology Education Should Be Critically Reworked to Support Learners as Subjects in Change and Not Objects of Change

The concept of a learner can be used to convey a status of 'lacking expertise' or of 'being-in-progress' in some way. Within the context of young people in schools and learners in other venues, the following discussions make abundantly clear that learners (of all ages) have distinctive perspectives and rights. They are not simply 'projects-of-the-future' or 'citizens-to-be', but they are present political subjects and actors. As such, they have rights of enhanced democratic representation and participation (see discussions by Barbara Bader and Yves Laberge, Chap. 23; Brandon Zoras and Larry Bencze, Chap. 24; Erica Blatt, Chap. 25). Education, in this light, shifts from an *a-priori* of political agency to a project of supporting active political engagement and involvement. The conversations in this volume, in this regard, are once more a radical accompaniment to much theorising in the field because they purposefully align themselves with normative prospects of enhancing political empowerment through active participation in debates and controversies within the civic sphere (see discussions by Chantal Pouliot, Chap. 29; Ralph Levinson, Chap. 21).[11] Indeed, in what follows there is much discussion of civic agency and action (see Pedro Reis, Chap. 31; Lyn Carter, Caroline Castono and Mellita Jones Chap. 30). As Derek Hodson suggests, it 'is almost always much easier to *proclaim* that one cares about an issue than to do something about it, and to do it consistently, coherently and effectively' (Chap. 5 – page needed). This should not be interpreted, however, as a case of 'applying knowledge' to everyday contexts (often a term adopted in school curriculum documents), but rather as an invitation to develop education that supports learning *through* political action (see for example the approach adopted by STEPWISE: Erin Sperling, Terry Wilkinson and Larry Bencze, Chap. 20; Mirjan Krstovic, Chap. 22; Darren Hoeg and Larry Bencze, Chap. 32).

[11] Although it seems a distant dream, in this regard the image of a school that we envisage is an institution that seeks to not only nurture democratic participation but also more effectively represents teachers and youth's interests within democracy: a school that is both an internal and external democratic advocate.

Science and Technology Education Should Be Critically Reworked as Moral and Ethical Praxis

Within the context of SSIs and STSE there has been extended consideration of ethics and morals and education (see Sadler et al. 2007). Nevertheless, there are still pressing reasons to reflect on values embedded within our pedagogical assumptions and performances. There is, for example, considerable need to reflect on questions of development and different normative rivals in moral and ethical educational theories (see discussions in Burns and Norris 2012; Alsop and Bencze 2012; Zeidler et al. 2005). There is also need to consider questions of 'ecojustice' and 'ecological ethics' and related philosophies (see discussions by Michael Mueller, Chap. 13; and Michael Mueller and Deborah Tippins [eds.] in press). As editors, we suggest future research in this area would benefit greatly from analyses exploring critical differences between 'retributive', 'restorative' and 'transformative' justice. There are hints of these discussions to be found in this volume that might serve as a basis for future theorising.

Partialities and Possibilities

In any book project there are, of course, limitations. Education in pursuit of answers to deep-rooted change can become lost in 'rhetoric' in which complex educational practices are judged against abstract ideals such as 'democracy', 'justice' and 'equality'. The followings chapters are conscious of social realism in which social factors are awarded the status of fixed explanatory variables – 'social truths'. The discussions herein are all grounded in pedagogical settings and as such are aware of tensions and contradictions navigating 'disruptive daydreaming' (Simon 1992) and very real-world problems and pragmatics.

There are, of course, lots of unexplored questions. Some readers will undoubtedly ask questions of emphasis – the ways in which discussions favour some perspectives more than others. There is a noticeable imbalance, for instance, in terms of traditional identity politics insofar as our deliberations gravitate more toward health, the environment and economic neoliberalism. The women's movement, civil rights movements, the LGBT movement and the peace movement, as well as many other standpoints, continue to profoundly influence work within our field. They do not, however, feature within the following discussions perhaps as much as they should, and we openly acknowledge that this silences essential voices. Similarly, Indigenous groups, practices and knowledges whilst evident in selected chapters are underrepresented in the text as a whole given their importance. There is so much that science and technology education might learn, for instance, from the recent 'Idle No More' protest movement that was started by First Nations peoples in Canada, and also from a vast array of other movements around the World associated with different Indigenous and Aboriginal communities, organisations and groups.

Our discussions, as previously acknowledged, are drawn almost exclusively from writers within Western democracies and this, once more, is a source of limitation and reflection.

Concerning these and other admissions, as editors we partially console ourselves with partiality – that there is only so much that any given text might include and achieve. In this vein, we offer this volume as openly incomplete and tentative; as a work in progress and with an extended invitation to the reader to critically reflect and supplement the arguments and contexts explored.

Before closing this introduction, we also feel a need to openly recognise that science and technology education in all its venues and guises is performed with much love, care and compassion, and with the expertise and commitment of many dedicated students and teachers. It is important within discussions of radical reform not to fall into a trap of misrepreseenting educational institutions and practices as entirely lost to forms of ideological domination. The danger of the project-yet-to-come and the associated politics of the future is that it can all too easily overlook the phenomenal expertise, dedication and possibilities in the past and present. In a reflection on a renowned career in science education spanning 40-years, Ken Tobin (Chap. 8) turns our attention to the power of education as well as 'mindfulness' and 'changing emotional styles' (page needed). This is in many ways perhaps the most radical education of all.

We conclude this introduction by inviting you to explore the assembled collection with hopefulness (Rebecca Houwer, Chap. 7) and prospects of 'remembering forward' (see David Blades and Janet Newbury, Chap. 11). It has been a pleasure editing this collection and working together. We would like to thank all those again who have been involved in bringing this project to fruition. Your work continues to inspire and shape our theorising and practice.

Steve and Larry

References

Alsop, S., & Bencze, L. (2009). Editorial: Beyond the confines of matters of fact. *Journal for Activist Science and Technology Education, 1*(1), i–v.

Alsop, S., & Bencze, L. (2010). Introduction to the special issue on activism: SMT education in the claws of the hegemon. *Canadian Journal of Science, Mathematics, and Technology Education, 10*(3), 177–196.

Alsop, S., & Bencze, J. L. (2012). In search of activist pedagogies in STEM. *Canadian Journal of Science, Mathematics, and Technology Education, 12*(4), 394–408.

Bauman, Z. (2008). *Does ethics have a chance in a world of consumers*. Cambridge: Harvard University Press.

Beck, U. (1992). *Risk society: Towards a new modernity*. London: Sage.

Benford, R., & Snow, D. (2000). Framing processes and social movements: An overview and assessment. *Annual Review of Sociology, 26*, 611–639.

Bourdieu, P., et al. (2003). *The weight of the world: Social suffering in contemporary society*. Stanford: Stanford University Press.

Bowers, C. (1997). *The culture of denial: Why the environmental movement needs a strategy for reforming universities and public schools.* Albany: State University of New York Press.
Burns, D., & Norris, S. (2012). Activist environmental education and moral philosophy. *Canadian Journal of Science, Mathematics, and Technology Education, 12*(4), 380–393.
Callison, C. (2010). *More information is not the problem: Spinning climate change, vernaculars, and emergent forms of life.* Doctoral dissertation, Massachusetts Institute of Technology.
Carr, W., & Kemmis, S. (1986). *Becoming critical: Education, knowledge and action research.* London: Routledge Falmer.
Corburn, J. (2009). *Street science: Community knowledge and environmental health justice.* Cambridge: MIT Press.
Freire, P. (1972). *Pedagogy of the oppressed.* New York: Continuum Books.
Fuller, S. (2007). *Social epistemology.* Bloomington: Indiana University Press.
Funtowicz, S., & Ravertz, J. (1991). A new scientific methodology for global environmental issues. In R. Cosranza (Ed.), *Ecological economics: The science and managements of sustainability.* New York: Columbia University Press.
Ginwight, S., Noguera, P., & Cammarota, J. (Eds.). (2006). *Beyond resistance: Youth activism and community change.* New York: Routledge.
Giroux, H. (1998). Theories of reproduction and resistance in the new sociology of education: A critical review. *Harvard Educational Review, 53*(3), 257–293.
Giroux, H. (2004). Cultural studies, public pedagogy, and the responsibility of intellectuals. *Communication and Critical Cultural Studies, 1*(1), 59–79.
Hacking, I. (1999). *The social construction of what?* Cambridge: Harvard University Press.
Haraway, D. (1991). Situated knowledges: The science question in feminism and the privilege of partial perspective. In D. Haraway (Ed.), *Simians, cyborgs, and women: The reinvention of nature* (pp. 183–201). New York: Routledge.
Harding, S. (2008). *Sciences from below: Feminisms, postcolonialities and modernities.* Durham: Duke University Press.
Hardt, M., & Negri, A. (2009). *Commonwealth.* Cambridge: Belknap Press of Harvard University Press.
Hawken, P. (2007). *The blessed unrest: How the largest social movement in history is restoring, grace, justice and beauty to the World.* London: Penguin.
Hodson, D. (2003). Time for action: Science education for an alternative future. *International Journal of Science Education, 25*(6), 645–670.
Hodson, D. (2011). *Looking to the future: Building a curriculum for social activism.* Rotterdam: Sense Publishers.
Jasanoff, S. (2010). A new climate for society. *Theory, Culture and Society, 27*(2–3), 233–253.
Klein, N. (2007). *The shock doctrine: The rise of disaster capitalism.* Toronto, Canada: Alfred A. Knopf.
McCormick, S. (2009). From "politic-scientists" to democratic science movements: The change climate of citizens and science. *Organisation & Environment, 22*(1), 34–51.
McLaren, P., & Houston, D. (2005). Revolutionary ecologies: Ecosocialism and critical pedagogy. In P. McLaren (Ed.), *Capitalists & conquerors: A critical pedagogy against empire.* Lanham: Rowman & Littlefield.
Mueller, M. P., & Tippins, D. J. (Eds.). (in press). *EcoJustice, citizen science and youth activism: Situated tensions for science education.* Dordrecht: Springer.
Papadopoulos, D. (2010). Alter-ontologies: Towards a constitutive politics in technoscience. *Social Studies of Science, 41*(2), 177–201.
Rose, H., & Rose, S. (1979). Radical science and its enemies. *Socialist Register, 9*, 105–132.
Roth, W.-M., & Barton, A. (2004). *Rethinking scientific literacy.* New York: Routledge Falmer.
Sadler, T. D., Barab, S. A., & Scott, B. (2007). What do students gain by engaging in socioscientific inquiry? *Research in Science Education, 37*(4), 371–391.
Sandlin, J. A., & McLaren, P. (Eds.). (2010). *Critical pedagogies of consumption: Living and learning in the shadow of the "Shopocalypse"*. New York: Routledge.

Shrader-Frechette, K. (2001). *Environmental justice: Creating equality, reclaiming democracy*. New York: Oxford University Press.
Simon, R. (1992). *Teaching against the grain: Texts for a pedagogy of possibility*. New York: Bergin & Garvey.
Star Lee, S., & Griesemer, J. (1989). Institutional ecology, translations and boundary objects: Amateurs and professionals in Berkeley's museum of vertebrate zoology 1907–39. *Social Studies of Science, 19*(3), 387–420.
Tobin, K., & Roth, W.-M. (2006). *Teaching to learn: A view from the field*. Rotterdam: Sense Publishing.
Woodhouse, E., Hess, D., Breyman, S., & Martin, B. (2002). Science studies and activism: Possibilities and problems for reconstructivist agendas. *Social Studies of Science, 32*(2), 297–319.
Zalasiewicz, J., William, M., Steffen, W., & Crutzen, P. (2010). The world of the Anthropocene. *Environmental Science and Technology, 44*(7), 2228–2231.
Zeidler, D., Sadler, T., Simmons, M., & Howes, E. (2005). Beyond STS: A research-based framework for socioscientific issues education. *Science Education, 89*, 357–377.

Part I
Constituting Theories

Steve Alsop and Larry Bencze

Preamble

What forms might an activist response to science and technology education take? In this section, authors draw on a wide range of different social theories to frame their responses. Early chapters question the absence of political analysis in the field and in shape responses routed in Foucauldian discourses of biopolitics and governmentality, as well as Marxist production relations, critical theory and narrative inquiry. Other chapters review extant literature in education and model a variety of student-centered action-orientated pedagogical responses blending social, cognitive, conative and emotional dimensions. The possibilities and challenges of crises as pedagogical and the importance of multiple perspectives and ethically situated research are explored. Several of the authors turn to sustainability as a referent point for advocating hybrid relational epistemic processes, 'contraction' and 'letting go'. As may be apparent from the above word cloud, discussions of education, science, learning and knowledge fuse with economic, social and ecological values and conditions.

The Chapters

2. **The Elephant in the Room: Science Education, Neoliberalism and Resistance**
   Lyn Carter
3. **Science Education as a Site for Biopolitical Engagement and the Reworking of Subjectivities: Theoretical Considerations and Possibilities for Research**
   Jesse Bazzul
4. **A Critical Pedagogy for STEM Education**
   Arturo Rodriguez
5. **Becoming Part of the Solution: Learning About Activism, Learning Through Activism, Learning from Activism**
   Derek Hodson
6. **From Promoting the Techno-sciences to Activism – A Variety of Objectives Involved in the Teaching of SSIs**
   Laurence Simonneaux
7. **Hopeful Practices: Activating and Enacting the Pedagogical and Political Potential in Crisis**
   Rebecca Houwer
8. **Using Collaborative Inquiry to Better Understand Teaching and Learning**
   Kenneth Tobin
9. **From Knowledge to Action? Re-embedding Science Learning Within the Planet's Web**
   Laura Colucci-Gray and Elena Camino
10. **Education for Sustainable Contraction as Appropriate Response to Global Heating**
    David E. Selby
11. **Learning to Let Go of Sustainability**
    David W. Blades and Janet Newbury

Chapter 2
The Elephant in the Room: Science Education, Neoliberalism and Resistance

Lyn Carter

Abstract In this chapter, I examine the place of political discourse in science education, which Erminia Pedretti and Joanne Nazir (Sci Educ 95(4):601–626, 2011) acknowledge has been accorded very "limited" study (p. 618). Specifically, I review the pervasive metadiscourse of neoliberalism, which is now the common-sense way many of us interpret, live in, and understand the world. Exposing and scrutinizing neoliberalism not only enhances the quality of our theorizing about the underdone political in science education, it also facilitates our attempts to develop better science education. I draw a link between neoliberalism and activism by foregrounding two very significant political moments (both as momentary events that were also momentous turning points) that took place some 30 years apart. Firstly, Michel Foucault's lectures to the Collège de France in 1978 and 1979 on biopolitics and governmentality, and secondly, perhaps the better known Occupy Wall Street protests that began during September 2011 in Zuccotti Park in New York City. I finish by drawing out some implications for activism/resistance in science education.

Keywords Science education • Neoliberalism • Occupy • Foucault

---

L. Carter (✉)
School of Education, Australian Catholic University, 115 Victoria Parade, Fitzroy, Melbourne, VIC 3065, Australia
e-mail: lyn.carter@acu.edu.au

L. Bencze and S. Alsop (eds.), *Activist Science and Technology Education*, Cultural Studies of Science Education 9, DOI 10.1007/978-94-007-4360-1_2,

Opening

> Errors, like straws, upon the surface flow;
> He who would search for pearls must dive below.
> John Dryden (1678), *All for Love*.

Researchers such as Camilla Schreiner and Svein Sjøberg (2004) have documented the decline of student interest in school science education within the West. Not surprisingly, many different perspectives have emerged to address this disengagement, some focused on overcoming disadvantages within students themselves (be they cultural, socioeconomic or cognitive), while other approaches look to alter how science and science education are conceptualised. One example of the latter has been the development of the science, technology, society and environment (STSE) education, which generally aims to broaden the remit of science education, making it more relevant to culturally and socially diverse students. Emerging within science education roughly 40 years ago, STSE critically examines the dynamic interfaces between science, technology, society and the environment within a philosophical frame of 'science for all.'

In their cogent overview of STSE, Erminia Pedretti and Joanne Nazir (2011) add a much needed typology to the perplexing array of work that falls within the STSE fold, thereby creating a "heuristic that educators can use for critical analysis of discourse and practice in the field" (p. 603). Beginning with a short historical account that they see as confirmation STSE was "a collective desire for fundamental change in school science" (p. 604), they describe the focus, implementation strategies and critiques of six overlapping ideological trends. Briefly, they identify the *Application/Design*, the *Historical*, the *Value Centred*, the *Socio-cultural*, and finally the *Socio-ecojustice;* the latter which goes beyond critique to problem solving through agency and action. Prominent in this strand is the work of Derek Hodson (2003, 2010) who has been a significant figure in activist science education and is inspirational for this volume.

More relevant to my purpose here in this chapter, Pedretti and Nazir (2011) go on to conclude that "a possible shortcoming of the work we have presented here . . . (is) not hav(ing) adequately explored the effect of political factors" (p. 618), as generally "work on the effect of external political pressure on science education is limited" (p. 618). Indeed, looking back over the long stretch of science education of which he was a part, Peter Fensham (2008) identified politically based policy analysis as one of the great gapping hole in science education research. Interpretations of the term 'politics' vary and beyond referencing educational theorist Michael Apple, Pedretti and Nazir (2011) are spare in their views.

To my mind, David Harvey (2010) provides a useful, if somewhat broad, perspective of politics as our contemporary conditions or landscape, which in this context, shapes and embeds science education. Claimed by the *Times Higher Education* magazine to be amongst the top 20 most cited authors in the humanities, David Harvey wrote the very influential *A Brief History of Neoliberalism* in 2005, and has most recently traced the political and economic nascence of the global financial crisis in *The Enigma of Capital* (2010). In both books, Harvey argues that the political policy programme of neoliberalism has become the pervasive hegemonic discourse to the point it is now the common-sense way many of us interpret, live in, and understand the world. He would probably agree with Jessica Benjamin's (1996) observation that, even if one adopts an oppositional stance, the prior constructs so shape one's starting point and holds on to one's coordinates of thought, that rejecting the postulates doesn't seem to be enough. Thought of this way, neoliberalism has become a type of default position, partly, Harvey (2010) believes, because we have yet to work through any alternative. Harvey (2010) argues that all academics and researchers of whatever domain should be alert to 'the times,' and debate the complexities of the interconnections between political, economic, social and cultural imperatives that form the contemporary condition. He contrasts this with the localism of many scholars, which would include many of those within science education, who incrementally advance their own idiosyncratic fields as if they were enshrouded in a vacuum.

Undoubtedly, politics inhabits all of science education even if it is an under-acknowledged presence, rather like the elephant in the room. Apt here is Brian Joiner's (1981) description of a 'lurking variable.' A lurking variable is one that has an important effect and yet is not included for consideration. Joiner quotes William Hunter and John Crowley (1979, p. 241) who suggest that a lurking variable may be omitted "because its existence is unknown or, if its existence is known, its influence is thought to be negligible or data on it are unavailable." The key question for Joiner (1981) becomes "[h]ow does one even identify the existence of a lurking variable when, by definition, it is not amongst the list of factors identified . . .?" (p. 227). In much the same vein, Annie Coombes and Alison Brah (2000) draw our attention to the notion of the 'unconscious' in research practice. They suggest that what matters most is not so much the intention of researchers/authors, but the nature of the discourses that they produce via the articulation of meanings constituted and disseminated through long-standing practices. Once available, discourses become sedimented into the cultural unconscious of the field of endeavour, and are repeatedly circulated, recited and utilised as unproblematic, shared cultural code that frames further questions of research. Hence, what has always been neglected or excluded as marginal or unimportant like the political in science education can forever be overlooked as kinds of "blind spots" in our research, to quote Jon Wagner (1993, p. 16).

Critical science education research from scholars like Angie Calabrese Barton (2003), Noel Gough (2007) and Matthew Weinstein (2009) are among the most noted for considering political perspectives. Unsurprisingly, many of these researchers are included in Pedretti and Nazir's (2011) typology of STSE. While emphases vary, most critical scholars proactively pursue social, ecological, activist and redistributive justice agendas that may or may not include direct analyses of the contemporary political forces shaping and transforming science education. That said, there is none-the-less very little science education scholarship that explicitly engages neoliberalism as the dominant socio-political project gripping the contemporary condition, and under which we all labour (exceptions here include Carr and Thésée 2008; Carter 2005, 2008, 2011; Bencze 2008, 2010; Tobin 2011).

Hence, in this chapter I examine the 'lurking variable' or 'blind spot' of the political in science education more closely. Specifically, I review the metadiscourse of neoliberalism that is part of the lexicon of the everyday, and thus inhabits Pedretti and Nazir's (2011) currents within STSE even as they were unacknowledged. Exposing and scrutinizing neoliberalism not only enhances the quality of our theorizing about the underdone political, it also facilitates our attempts to develop better science education with a focus on activism.

Rather than reiterate the tenets of neoliberalism[1] that are likely already known to readers of this volume, I want to draw a link between neoliberalism and activism by foregrounding two very significant political moments (both as momentary events that were also momentous turning points), which took place some 30 years apart. The first moment is Michel Foucault's lectures to the Collège de France in 1978 and 1979 on biopolitics and governmentality where he mapped out the polycentric formation neoliberalism governmentality. The second is the Occupy Wall Street protest that began during September 2011 in Zuccotti Park in New York City. For scholars such as Judith Butler (2012) and Bernard Harcourt (2012), Foucault's vision of resistance/activism was enacted during the Occupy encampment. I finish by drawing out some implications of these events for activism in science education.

Foucault's Neoliberalism

My first moment utilises the acute and prescient observations of the French philosopher Michel Foucault on neoliberalism, a little known discourse in France at the time.[2] Collectively known as *The Birth of Biopolitics,* this series of lectures were first published in 2004 and translated into English in 2008. For Lois McNay (2009), Foucault remarkably "predict(ed) crucial aspects of the marketization of social relations" (p. 56) even though his lectures were delivered several years before the emergence of the New Right in the early 1980s. To mark the 25th anniversary of Foucault's death in 2009, the journal *Theory, Culture & Society* produced a special issue that aimed to reengage Foucault with contemporary issues. Philosophers, political theorists, social and cultural commentators including Paul Rabinow, Brian Massumi, Couze Venn, Lois McNay, Tiziana Terranova and Maurizio Lazzarato contributed to the volume furthering Foucault's analyses

[1] Peck (2010) provides a lucid overview of neoliberalism, about the lack of consensus on it as a messy diverse and hybrid hegemon, and about the increasing reams of scholarship it commands particularly after it was declared dead in the wake of the Global Financial Crisis. Rather than gone, it is now viewed as having increased its grip on the world.

[2] Throughout this chapter, I use lengthy quotations from Foucault's work. This is deliberate choice, as Foucault's own words capture the nuances of his meaning better than I could ever hope to achieve. This does mean though, that Foucault's gendered language remains intact.

which, due to his untimely death in 1984, necessarily stopped short of neoliberalism as a "lived phenomenon" (Peck 2010, p. xii).

Capitalism as our historical system is regarded as a contingent product of institutional arrangements and practical rules that enable the conditions under which it can proceed. Hence, different ideological positionings engender different organizing frameworks. In *The Birth of Biopolitics,* Foucault's object was the historical exploration of a particular "framework of political rationality" (p. 317), which Terry Flew (2010), in hindsight, suggests had an intellectual path in common with comparative political economy and 'varieties of capitalism' research (Lane and Wood 2009). This saw Foucault use his 'histories of the present' genealogical approach to trace a selective history of liberalism and capitalism from its eighteenth century classical form, bypassing the nineteenth century and settling on the two cases of twentieth century (neo)liberal thought – the German Freiberg School of *ordoliberalism* and the applied neo-classical economics of the Chicago School in the United States. With antecedents as far back as the 1930s, both schools more fully developed in opposition to the Keynesian form of liberal government that was at its peak between World War II and the 1970s.

Having looked at the *longue durée* of liberal thought, McNay (2009, p. 58) argues that Foucault identified "a catalysing moment" in the shift in capitalism and governance by the Germans as they sought to rethink state, economic and societal relations in response to the Nazism, on the one hand, and the strong interventionism and welfarism of Keynesianism, on the other. They initiated a number of important breaks with traditional liberal understanding of a *laissez-faire* market economy with state intervention to stimulate demand and mitigate market driven social inequalities through programmes of wealth redistribution. As a consequence of their own unique trajectory of history and ideology, the Ordoliberals saw economic problems in terms of an unconstrained state, which Foucault (2008) expresses thus:

> Nothing proves that the market economy is intrinsically defective since everything attributed to it as a defect and as the effect of its defectiveness should really be attributed to the state. So, let's do the opposite and demand even more from the market economy than was demanded from it in the eighteenth century . . . let's ask the market economy itself to be the principle, not of the state's limitation, but of its internal regulation from start to finish of its existence and action. (p. 116)

For the Ordoliberals, central to this constraining of the 'defect' state was the shift towards competition as the organizing plank of the market. Competition replaces exchange, but not as whether markets are competitive or not, but rather that government must implement policies to promote competition. Conditions for competition must be "carefully and artificially furnished. . .. Competition is thus an historical objective of the art of government . . . [T]he market, or rather pure competition . . . cannot emerge unless it is produced, and unless it is produced by an active governmentality" (Foucault 2008, pp. 120–121). In other words, the market can operate as regulatory principle only if competition is made the regulatory principle of society.

Foucault went on to carefully differentiate the neoliberalism of the Chicago School from the Ordoliberals relating a different trajectory unique to the US context, (although the two schools later connected through membership of the Mont Pelerin Society). Foucault's analysis around human capital theory rated the Chicago School a more extreme form of neoliberalism as it sought the extension of the "economic form of the market . . . throughout the social body and including the whole of the social system not usually conducted through or sanctioned by monetary exchange" (Foucault 2008, p. 243). Social policy, the Chicago economists argued, should work to exclusively support economic policy. The focus becomes not supply and demand of goods and services but on the individual as *homo oeconomicus* or *enterprise man*, an active economic subject who "allocates their time and resources between consumption . . . and investment in the self . . . (S)uch an individual is . . . an investor, an innovator, and an entrepreneur" (Flew 2010, p. 29). For Foucault, the required *homo oeconomicus* is not the man of exchange or man the consumer; he is the man of enterprise and production within an enterprise society. The contemporary mission of neoliberal government is that "one must govern for the market" (Foucault 2008, p. 125).

> That is, one must govern according to the rules of the market, by drafting laws, by instituting (fiscal and other) regulatory apparatuses, recalibrating the functions of socio-cultural institutions to bring them into line with the new language and new objectives of the enterprising state, and by constituting appropriate subjectivities, notably *homo oeconomicus* as 'enterprise man.' (Venn 2009, p. 212, italics in original)

Foucault didn't live to see the Chicago School become "the dominant pole in the ideational universe" (Peck 2010, p. xviii), nor to see how fully and completely the hegemons of competition and enterprise would cover the globe. As Lazzarato (2009, p. 113) argues, and Foucault so presciently envisaged, we have managed to produce a "new mode of government (that) substitutes the couple inequality-enterprise in place of the (traditional liberal) couple exchange-equality." In other words, despite the neoliberal discourse that claims the contrary, inequality has become the deliberate progenerator of our social world. We see it in social policy that has reconstructed what it means to be a citizen. We see it in atomisation of everything, in increasing individualisation, and in the organised proliferation of difference absolutely crucial to promoting the inequality that enables the mechanism of competition to work. Where competition and inequality doesn't naturally exist various strategies work to engender it. It occurs when appetites of all types are developed, sharpened, promoted, priced to include/exclude and be met without thought to the limits of nature, when hyper-consumption is normalised, and when the self-managing and promoting entrepreneurial subject hawks his/her talents around a cosmopolitan world that excludes four fifths of humanity. In education, we see it in the reforms that have increased privatisation, decimated public spending, closed schools, blamed teachers and students who must vie for places in the new system. In science education, competition is constructed through the ever-proliferating standards that facilitate the high stakes testing regimes of PISA and TIMSS Foucault's sagacious words that we are now all in a state of 'equal inequality' have certainly come to pass.

Interestingly, rather than Foucault's lectures being designed as an overt criticism of neoliberalism, commentators generally agree that his purpose was to ask if the political left was up to the task of innovating with the same audacity as the right. Neoliberalism, he argued, conceived of different governmental practice and institutional frameworks, and so "define(d) for itself its way of doing things" and formulating a "different capitalism" (Foucault 2008, p. 94). With the passage of time, the answer has become unfortunately obvious!

In their introduction to the special volume of *Theory, Culture & Society*, Venn and Terranova (2009), draw attention to some of Foucault's other work, lest we be left in despair. They are quick to point out that neoliberalism is one of the two possibilities Foucault envisaged. The first possibility, *homo oeconomicus* is:

> [t]he utopian ambition of modernity, namely, the emergence of the calculating, instrumentally-driven 'enterprise man' ... (and) ... the logical outcome of that curious figure ... the Cartesian subject ... who appeared with European modernity and is destined to disappear with the tide of time, as Foucault expressed it at the end of the *Order of Things*. (p. 3)

Foucault's other possibility, suggest Venn and Terranova (2009), is the political subject who 'speaks truth to power,' a subject who is framed by "an ethics and aesthetics of ourselves" (p. 10) and the human as an "essentially collaborative, convivial, spiritual and a historically located social being" (p. 10). This subject is one who understands their times and who acts to "transgresses the limitations placed ... by historically specific conditions" (p. 3). For Foucault (1997), this enlivens the possibility of resistance:

> Even when the power relation is completely out of balance, when it can truly be claimed that one side has "total power" over the other, a power can be exercised over the other only insofar as the other still has the option of killing himself, of leaping out the window, or of killing the other person. This means that in power relations there is necessarily the possibility of resistance because if there were no possibility of resistance (of violent resistance, flight, deception, strategies capable of reversing the situation), there would be no power relations at all. (p. 292)

It is to this other human that I now turn, the one who is not afraid "to 'speak truth to power', in public, at the risk of the consequences" (Venn and Terranova 2009, p. 4) and the one who sees the possibility of resistance, in other words many of the participants of the Occupy movement.

Occupy Wall Street: The Incessancy of Resistance

My second moment in this chapter that is both a momentary event and also a momentous turning point is the Occupy Wall Street protests, where several thousand people in varying and changing configurations occupied Zuccotti Park in New York City from September to November 2011. It was part of a wider movement of protest in the wake of the 2007–2008 global financial crisis, which the *U.S. Senate Permanent Subcommittee on Investigations* has subsequently found was the result of "high risk, complex financial products; undisclosed conflicts of interest; (and a) failure of regulators, credit rating agencies, and the market itself to rein in the excesses of Wall Street" (Levin and Coburn 2011, p. 1). Occupy Wall Street was preceded, and inspired, by similar encampments for example, in Spain (the Indignados) and Britain (the Democracy Village) as well as by the Arab Spring and the occupation of Cairo's Tahrir Square.

Central to its genesis, and crucial to its success, was the *Adbusters Media Foundation,* which first proposed the occupation of Wall Street to protest the growing financial inequality under neoliberal political and corporate governance. Adbusters, an electric group of 'cultural creatives' that promotes information age activism, created the #OCCUPYWALLSTREET hashtag on Twitter™ and registered the OccupyWallStreet.org domain name several months before the actual encampment. They collaborated with *Workhouse,* a Manhattan based public relations firm, to advertise the occupation in its early weeks when few in the mainstream were interested. Before too long though, the Occupy movement had mushroomed across the US and in many major cities internationally animated by the political slogan 'We are the 99 %.' Accredited by *Rolling Stone* magazine (although not all agree), to anthropological scholar, activist and early Occupy organiser David Graeber, this slogan refers to the rampant growth in wealth within the top 1 % reported by the Washington based *Centre on Budget and Policy Priorities* in the following terms:

> Two-thirds of the nation's total income gains in the economic expansion from 2002 to 2007 flowed to the top 1 percent of U.S. households, and that top 1 percent held a larger share of income in 2007 than at any time since 1928. During those years, the real (inflation-adjusted) income of the top 1 percent of households grew more than *ten times faster* than the income of the bottom 90 percent of households. (Shaw and Stone 2010, p. 2, their italics)

While clearly a protest against inequality under neoliberalism, Occupy Wall Street was distinctive for its lack of overt leadership and refusal to articulate specific demands. It used democratic consensus-based decision-making most clearly embodied by its General Assembly, in which participants made both large and small group decisions, anything according to Dan Berret (2011) from articulating the principles of solidarity to deciding how best to stay warm at night.

It is difficult to come to grips with the rhizome (after Deleuze and Guattari 1987, p. 22) that is Occupy. There is so much one could interrogate – from the protester demographics of the mainly highly-educated young White males and the concomitant elision and erasure of the racialised nature of inequality (Kilibarda 2012), to the information-age protest style with its own generator, YouTube™ videos, tweets, blog posts and help from hacktivist group *Anonymous.* Then there are the (neoliberal-style?) public-private partnerships that saw Workhouse win an industry public relation award – the 2012 Platinum PR WOW! Award, part of the citation of which read: "[t]he results, obviously, have been spectacular. There's hardly a newspaper, Internet or broadcast media outlet that hasn't covered OWS." The sheer scope, complexity, and connectedness continue to be immense. Just a casual look at the number of websites spawned is enough to feel overwhelmed and uncertain where to begin.

They include occupiedstudies, weoccupy, occupyhistory, occupyeconomics, interoccupy, occupytogether, occupyresearch, airoccupy, occupytheory, occupy-network, occupyphilanthropy, occupytogether, occupyvideo, occupyeducation, the futureofoccupy and occupy various places like occupyaustralia, occupylondon, occupyseattle – the list goes on!

One part of the rhizome that I find both interesting and relevant to an exploration of the place of political discourse in science education is the engagement with Occupy by public intellectuals and well-known social and cultural academics (many of luminary status). Some like Manuel Castells, Judith Butler, Slavoj Žižek, Angela Davies, David Harvey, Cornell West and Robert Reich lent their support directly by speaking at various encampments. Others such as Saskia Sassen, Gayatri Chakravorty Spivak, George Lakoff, Laurence Lessig, Peter Marcuse, Michael Hardt, Antonio Negri, Craig Calhoun, Wendy Brown, Nancy Fraser, and Bernard Harcourt advocated and/or analysed in essays, Op Ed pieces, blogs, video interviews, Twitter™ posts, magazine articles and so on. While theory in more conventional academic fora is still to come as shape is given to the events, social and mass communication media provided these scholars with opportune avenues and they, in turn, lent a certain intellectual gravitas to the seemingly endless commentary. They have tackled topics such as new trends in social movements, inequality, political representation, strategies of protest, law enforcement and public order, the use of new media, governance, models of democracy including prefigurative democracy, international relations, and so on.

Picking up the thread of the political, Michael Hardt and Antonio Negri (2011) for example, believe that Occupy indicates a failure of political representation. They suggest that not only is our current political system incapable of addressing issues of inequality highlighted by the protesters, but that politicians charged with representing the people's interests "more clearly represent the banks and the creditors" (p. 3). This is hardly a surprise given the international character of large economic processes under globalization that are gradually, but inexorably undermining the legitimacy of Western democracies (Žižek 2011). We no longer have the political representation we have come to expect as our democratic mechanisms are, by definition, limited to nation states. For Slavoj Žižek (2011), this means, "people more and more experience institutional democratic forms (that are) unable to capture their vital interests" (np).

Political representation or its lack is important because much has been made (in terms of word coverage in the media, conventional and social) of the Occupy Movement's diffuse and decentralized nature,[3] as well as its refusal (or inability, depending on one's perspective) to articulate specific grievances and demands that would enable political and corporate governance to respond. Eminent sociologist Judith Butler is most eloquent about this point. In '*Critique, Dissent, Disciplinarity*' written a couple of years before Occupy in 2009, she mobilises Foucault to consider

[3] The decentralized nature of Occupy is readily apparent in the myriad of websites and other online tags that were noted above. It is a true Deleuzian and Guattarian rhizome.

the place of acts of civil or political disobedience. Here, she really prefigures her support for Occupy and its particular repudiation modus operandi that resists neoliberal political representation:

> The impetus for withdrawing one's consent from a given authority consists in trying to establish a limit to governability. And this can, depending on how it is formulated and publicized, result in a more radical inquiry into the legitimacy of the authority in question. . . . The question, how not to be governed? (for Foucault) is always the question of how not to be governed in this or that way. But it is not a question of how not to be governed at all. As a consequence, it is a specific question that emerges in relation to a specific form of government and might well constitute a kind of tactical and provisional anarchy in relation to existing authority—in his words, "how not to be governed like that, by that, in the name of those principles, with such and such an objective in mind and by means of such procedures, not like that, not for that, not by them." (cited in Butler 2009, p. 791)

In post Occupy Wall Street commentary written for the magazine *Tidal-Occupy Theory, Occupy Strategy* in 2012, Butler applies some of her Foucauldian perspectives. She suggests that our traditional notion of the political, or of protest/resistance, is rooted in the belief that we can articulate grievances and specify particular demands. Instead, she argues, the Occupy protesters in drawing attention to the increasing disparity in wealth and structural inequality that crosses most social and economic policies were really critiquing the economic system of global neoliberalism at large. Specifying demands in face of such hegemony would have been reductionist to the point of uselessness. We cannot fix a list of demands, Butler (2012) argues, without understanding the systemic production of inequality. More importantly, it would have conferred legitimacy on the 'authority' (that is, the corporation and neoliberal state). This is because demands come to be framed in the language of the 'authority' that, in turn, reinforces the coordinates of their power, and by disaggregating and meeting one demand or another, neatly circumvents any real changes. After all, how can the very power that has produced such inequality serve as the recipient of demands for change? Better, Butler (2012) argues, that in "withdrawing one's consent from a given authority" (p. 791) through refusal to make demands, the Occupy protesters, experimented with, and helped forge, new forms of social organisation with which to think about and enact the political.

Butler's (2012) view is not dissimilar to that of Bernard Harcourt's (2012). For Harcourt (2012), Occupy Wall Street is best understood "as a new form of *political* as opposed to *civil* disobedience that fundamentally rejects the political and ideological landscape that has dominated our collective imagination" (p. 33, italics in the original). He means here that Occupy has been a rejection of "politics *writ large*... a rejection of conventional political rationality, discourse and strategies" (p. 34, italics in the original) at the same time it is an opening to "new ideas, tactics, and forms of resistance . . . (that) . . . generate possibilities without imposing ideologies" (p. 35). In this notion, he also intones Foucault and suggests that we see in Occupy what Foucault meant by 'voluntary insubordination,' in which we resist the way we are "being governed in this way or that" (Foucault 2002, p. 193). Part of the 'newness,' Harcourt (2012) argues, is in Occupy's leaderless grammar, which means that there is no one to speak for the protesters ensuring multiple voices, perspectives, opinions and grievances can coexist. Moreover, there is also no one or no way to speak to the protesters. For Graeber (2011), this refusal to create internal hierarchy employing instead a consensus based direct democracy, is a prefigurative democracy where the modes of organisation and social relationships embody the future society being sought. In other words, the occupation is first about participation and its decentralised organisation is the protest/resistance or in our terms here, the activism.

Harcourt's (2012) views (as would also be Graeber and Butler's) are at odds with well-known Marxist philosopher Slavoj Žižek who suggests "one should also begin to think seriously about what to propose instead of the predominant economic organization, to imagine and experiment with alternate forms of organization, to search for the germs of the new" (Žižek 2011, np). Harcourt (2012) believes Žižek (2011) has missed the point, as the protesters were about resisting alternative ideological positionings of all types, even if Marxism has something to add, as Žižek believes. It's *"precisely about resisting the old ideologies"* (Harcourt 2012, p. 36, italics in the original). Here, Harcourt (2012) again aligns himself to Foucault:

> When asked whether, after critique, there is "a stage at which we might propose something," Foucault responded: "My position is that it is not up to us to propose. As soon as one 'proposes'—one proposes a vocabulary, an ideology, which can only have effects of domination. . . . These effects of domination will return and we shall have other ideologies, functioning in the same way. It is simply in the struggle itself and through it that positive conditions emerge." It is only by open contestation and struggle that, "in the end," Foucault suggested, "possibilities open up." (p. 39)

Perhaps one thing with which most commentators would agree however, is that whatever else Occupy Wall Street and the rest of the Occupy movement managed to do, it did, according to Noam Chomsky (2012), put the inequality of our times on the agenda.

Science Education, Neoliberalism and Activism/Resistance

I commenced this chapter, arguing the need for more engagement between science education and the political, specifically neoliberalism as the pervasive hegemony of our time. Now with a little help from Foucault, we can envision 'the possibility of resistance.' While resistance in the form of critical pedagogies for example (see Friere 1986; Giroux 2011), are well known within the general education literature, science education has been slow to explore these areas. Notable exceptions have been the Special Issue of the *Canadian Journal of Science, Mathematics and Technology Education* (Bencze and Alsop 2009) on the role of activism in science, mathematics, and technology education, the subsequent response from David Burns and Stephan Norris (2012) and Alsop and Bencze's (2012) rejoinder.

One example of classroom activism/resistance seemingly applicable "like straws, upon the surface flow" (Dryden 1678) to science teachers, is Bree Picower's (2011) study of novice teachers implementing social justice curricula and pedagogies. She described a group of urban primary school teachers in New York City who were able to subvert neoliberal practices by inserting social justice content within the mandated curriculum and standardised testing regimes. Within a literacy unit for instance:

> Jonathan and Nick wanted to provide their students with an opportunity to voice their dissent to the corporations that benefit from child labor on cocoa farms. … the students wrote Valentine cards to the CEO of World's Finest Chocolates. These letters easily fit the criteria of the mandated persuasive essay unit. The cards expressed their anger about child labor and demanded that the company start using fair trade practices. Through this part of the unit, the children engaged in a classic activist strategy, power analysis, to understand which stakeholders were perpetuating and benefiting from this injustice and to decide what could be done. (p. 1126)

Similarly, in Marisa's maths class, the students' used:

> … data to create bar graphs and compared both graphs, contrasting what they thought boys and girls enjoyed with the reality of what they liked to do. By using a traditional math activity … Marissa was able to help her young students develop critical thinking skills and challenge gender stereotypes while never appearing to have strayed from the mandated math curriculum. (p. 1124)

Though admiring the teachers' achievements, Picower (2011) nevertheless laments the small impact such approaches have on the neoliberalisation of education and the proliferation of inequality. "You can decorate the jail cell but you still aren't free" (p. 1130). She goes onto argue that teacher educators must be activists themselves, so as they can be role models and supporters in unreceptive terrains. In considering activism in science education, I found Picower's (2011) study firstly comforting, but ultimately worrying. If individual teachers' subversions of curricula and pedagogy are inadequate, in her terms, to challenging neoliberal forces, then activist teacher educators are no more likely to succeed. Like Dryden's (1678) "pearls" that can only be found at depth, I am driven now to investigate activism in a 'deeper' way than I previously have in order to consider whether 'decorating the jail cell' is okay.[4] To this end, I take from Foucault, his lectures on *Biopolitics,* and his 'voluntary insubordination' from Occupy Wall Street, to be part of 'the struggle itself' in order that 'something positive can emerge'.

Not an Ending but a Beginning

Where usually I would head this section a summary or conclusion or even just an ending, naming it a beginning is more apt for me in my present space. It may well be the end of this chapter but it is only the beginning of my interrogation of activism. My focus now is to work some more within the rhizome of activism so, in the 'becoming,' I may better understand and be a better activist with science education and beyond.

---

[4] What word is appropriate here? I considered 'enough' amongst other like terms, but can activism ever be conceived of as such. I have settled on 'okay' as a compromise.

References

Alsop, S., & Bencze, L. (2012). In search of activist pedagogies in STEM. *Canadian Journal of Science, Mathematics, and Technology Education, 12*(4), 394–408.

Bencze, J. L. (2008). Private profit, science and science education: Critical problems and possibilities for action. *Canadian Journal of Science, Mathematics, and Technology Education, 8*, 297–312.

Bencze, J. L. (2010). Exposing and deposing hyper-economized school science. *Cultural Studies of Science Education, 5*, 293–303.

Bencze, J. L., & Alsop, S. (2009). A critical and creative inquiry into school science inquiry. In W.-M. Roth & K. Tobin (Eds.), *World of science education: North America* (pp. 27–47). Rotterdam: Sense.

Benjamin, J. (1996). *Like subjects, love objects: Essays on recognition and sexual difference*. New Haven: Yale University Press.

Berrett, D. (2011). Intellectual roots of Wall St. protest lie in Academe. *The Chronicle of Higher Education*. http://chronicle.com/article/Intellectual-Roots-of-Wall/129428/. Accessed 2 Mar 2013.

Burns, D., & Norris, S. (2012). Activist environmental education and moral philosophy. *Canadian Journal of Science, Mathematics, and Technology Education, 12*(4), 380–393.

Butler, J. (2009). Critique, dissent, disciplinarity. *Critical Inquiry, 35*(4), 773–795.

Butler, J. (2012). So, what are the demands and where do they go to from here? *Tidal Occupy Theory, Occupy Strategy, 2*, 8–11.

Calabrese Barton, A. (2003). *Teaching science for social justice*. New York: Teachers College Press.

Carr, P. R., & Thésée, G. (2008). The quest for political (il)literacy. In B. Porfilio & C. Malott (Eds.), *The destructive path of neoliberalism* (pp. 233–256). Rotterdam: Sense Publishers.

Carter, L. (2005). Globalisation and science education: Rethinking science education reforms. *Journal of Research in Science Teaching, 42*, 561–580.

Carter, L. (2008). Globalisation and science education: The implications for science in the new economy. *Journal of Research in Science Teaching, 45*(5), 617–633.

Carter, L. (2011). Gathering in threads in the insensible global world: Science education and globalisation. *Cultural Studies of Science Education, 6*(1), 1–12.

Chomsky, N. (2012). *Occupy*. London: Penguin.

Coombes, A. E., & Brah, A. (2000). *Hybridity and its discontents. Politics, science and culture*. London: Routledge.

Deleuze, G., & Guattari, F. (1987). *A thousand plateaus: Capitalism and schizophrenia* (B. Massumi, Trans.). London: Continuum.

Dryden, J. (1678/1998). *All for love*. London: Nick Hern Books Limited.

Fensham, P. (2008). *Science education research and science education policy: A too often overlooked link*. Paper presented at the National Association for Research in Science Teaching, Baltimore.

Flew, T. (2010). *Michel Foucault's the birth of biopolitics and contemporary neoliberalism debates*. Paper presented to the Creative Industries faculty, Queensland University of Technology, Brisbane, Australia.

Foucault, M. (1997). *Michel Foucault: Ethics, subjectivity, and truth* (P. Rabinow, Ed.). New York: The New Press.

Foucault, M. (2002). What is critique? In D. Ingram (Ed.), *The political*. London: Penguin.

Foucault, M. (2008). *The birth of biopolitics: Lectures at the Collège de France 1978–1979*. Basingstoke: Palgrave Macmillan.

Freire, P. (1986). *Pedagogy of the oppressed*. New York: Continuum.

Giroux, H. (2011). *On critical pedagogy*. New York: Continuum.

Gough, N. (2007). Geophilosophy, rhizomes and mosquitoes: Becoming nomadic in global, science education research. In B. Atweh, M. Borba, A. Calabrese Barton, D. Clark, N. Gough, C. Keitel,

C. Vistro-Yu, & R. Vithal (Eds.), *Internationalisation and globalisation in mathematics and science education* (pp. 39–56). Dordrecht: Springer.
Graeber, D. (2011). Occupy Wall Street anarchist's roots. *Aljazeera*. http://www.aljazeera.com/indepth/opinion/2011/11/2011112872835904508.html. Accessed 2 Mar 2013.
Harcourt, B. (2012). Political disobedience: Occupy Wall Street. *Critical Inquiry, 39*(1), 33–55.
Hardt, M., & Negri, A. (2011). The fight for 'real democracy' at the heart of occupy Wall Street. *Foreign Affairs*. http://www.foreignaffairs.com/articles/136399/michael-hardt-and-antonio-negri/the-fight-for-real-democracy-at-the-heart-of-occupy-wall-street. Accessed 2 Mar 2013.
Harvey, D. (2005). *A brief history of neoliberalism*. Oxford: Oxford University Press.
Harvey, D. (2010). *The enigma of capital*. London: Profile Books.
Hodson, D. (2003). Time for action: Science education for an alternative future. *International Journal of Science Education, 25*(6), 645–670.
Hodson, D. (2010). Science education as a call to action. *Canadian Journal of Science, Mathematics, and Technology Education, 10*(3), 197–206.
Hunter, W. G., & Crowley, J. J. (1979). Hazardous substances the environment and public health: A statistical overview. *Environmental Health Perspectives, 32*, 241–254.
Joiner, B. L. (1981). Lurking variables: Some examples. *The American Statistician, 35*(4), 227–233.
Kilibarda, K. (2012). Lessons from #Occupy in Canada: Contesting space, settler consciousness and erasures within the 99 %. *Journal of Critical Globalisation Studies, 5*, 24–41.
Lane, C., & Wood, G. (2009). Capitalist diversity and diversity within capitalism. *Economy and Society, 38*(4), 531–551.
Lazzarato, M. (2009). Neoliberalism in action: Inequality, insecurity and the reconstitution of the social. *Theory Culture & Society, 26*(6), 109–133.
Levin, C., & Coburn, T. (2011). *Wall street and the financial crisis: Anatomy of a financial collapse*. Washington, DC: Permanent Subcommittee on Investigations United States Senate.
McNay, L. (2009). Self as enterprise: Dilemmas of control and resistance in Foucault's the birth of biopolitics. *Theory Culture & Society, 26*(6), 55–77.
Peck, J. (2010). *Constructions of neoliberal reason* (p. 2010). Oxford: Oxford University Press.
Pedretti, E., & Nazir, J. (2011). Currents in STSE education: Mapping a complex field, 40 years on. *Science Education, 95*(4), 601–626.
Picower, B. (2011). Resisting compliance: Learning to teach for social justice in a neoliberal context. *Teachers College Record, 113*(5), 1105–1134.
PRNews. (2012). *Co-winner: OWS with workhouse publicity – Occupy Wall Street: The revolution will be editorialized*. http://www.prnewsonline.com/platinumpr2012/. Accessed 2 Mar 2013.
Schreiner, C., & Sjøberg, S. (2004). *ROSE: The relevance of science education: Sowing the seeds of ROSE*. Oslo: Department of Teacher Education and School Development, University of Oslo.
Shaw, H., & Stone, C. (2010). Tax data show richest 1 percent took a hit in 2008, but income remained highly concentrated at the top. *Centre on Budget and Policy Priorities*. Washington, DC. http://www.cbpp.org/files/10-21-10inc.pdf. Accessed 2 Mar 2013.
Tobin, K. (2011). Global reproduction and transformation of science education. *Cultural Studies of Science Education, 6*, 127–142.
Venn, C. (2009). Neoliberal political economy, biopolitics and colonialism: A transcolonial genealogy if inequality. *Theory Culture & Society, 26*(6), 206–233.
Venn, C., & Terranova, T. (2009). Introduction: Thinking after Michel Foucault. *Theory Culture & Society, 26*(6), 1–11.
Wagner, J. (1993). Ignorance in educational research: Or, how can you *not* know that? *Educational Researcher, 22*(5), 15–23.
Weinstein, M. (2009). Critical science literacy: Identifying inscription in lives of resistance. *Journal for Activist Science & Technology Education, 1*(2), 1–11.
Žižek, S. (2011). Occupy Wall Street: What is to be done next? *The Guardian*. http://www.guardian.co.uk/commentisfree/cifamerica/2012/apr/24/occupy-wall-street-what-is-to-be-done-next?. Accessed 2 Mar 2013.

Chapter 3
Science Education as a Site for Biopolitical Engagement and the Reworking of Subjectivities: Theoretical Considerations and Possibilities for Research

Jesse Bazzul

Maybe the target nowadays is not to discover what we are, but to refuse what we are.

(Foucault 1982, p. 785)

Abstract This chapter advocates viewing science education as a site for biopolitical struggles and the reworking of subjectivities. Such a perspective provides a framework that can unite various struggles for social justice and help educators and students intervene in the forces of what philosopher Michel Foucault calls biopower. This chapter draws from Foucauldian notions of biopolitics and subjectivity and the work of Michael Hardt and Antonio Negri, to provide some notes on biopolitical theory and potential applications in science education. This chapter is just one theoretical articulation of how biopolitics and biopolitical struggle can inform critical, activist work in science education. The full productive 'value' of biopolitical theory has yet to be realized.

Keywords Biopolitics • Biopower • Science education • Foucault • Subjectivities • Politics

One of the primary purposes of the new radio telescope being built in Australasia and Africa is to search the universe for new forms *life* (Amos 2012). It is a reminder that scientific research, even in the coldest and darkest of places (outer space) is always invested in how we think of ourselves in the world. Science's connection with all aspects of life makes it a fertile site for *biopolitical action*; that is, intervening in the way scientific discourses about life, govern and order bodies,

J. Bazzul (✉)
Department of STEM education and Teacher Development,
University of Massachusetts Dartmouth, North Dartmouth, MA, USA
e-mail: jbazzul@gmail.com

L. Bencze and S. Alsop (eds.), *Activist Science and Technology Education*,
Cultural Studies of Science Education 9, DOI 10.1007/978-94-007-4360-1_3,

as well as define acceptable practices and future actions. While science has become arbiter over many aspects of human life, it is simultaneously infused with many social and political agendas. These intertwining contexts provide fertile ground for education communities to view science education as an integral part of what Foucault calls *biopower*, and subsequently intervene *biopolitically*.[1] This chapter attempts to set a theoretical context for viewing science education a site of biopolitical struggle over what constitutes (human) life, and how science education may work to constitute the subjectivities of teachers and students. This involves giving attention to how practices, discourses, and material relations in science education constitute particular kinds of subjectivities; and reworking these subjectivities towards different possibilities for thought and action.

This chapter draws from Foucauldian notions of subjectivity, Paul Rabinow and Nikolas Rose's (2006) elaboration of Biopower, and Michael Hardt and Antonio Negri's (2009) notions of the biopolitical in order to think about how science education can be seen as a site of biopolitical struggle over subjectivities. It is not intended to provide a complete framework for biopolitics in science education as such a framework would be antithetical to the goals of freedom inherent in biopolitical engagement. It is instead meant to be a step towards conceptualizing science education in terms of a struggle over the (re)production of subjectivities, organization of bodies, and ordering of human life. This involves considering science education as a site where subjectivities are produced (in conjunction with other sites) and reworked in dynamic social, cultural, political and economic contexts that neither the metaphor of 'schooling as emancipatory project' nor as 'disciplinary apparatus' can fully encompass. Rethinking science education along biopolitical lines can unite parallel, ongoing, struggles within science education; for example, against White supremacy, Western epistemic hegemony, heteronormitivity, neoliberal restructuring, class/social exclusion, and complicity with environmental destruction. Social change, according to Hardt and Negri (2009), must move forward like a millipede, with many legs (struggles) propelling it forward. Due to its many possible forms, biopolitics remains a necessarily ongoing and unpredictable field of critical thought. Using biopolitical theory as a framework for science education will require teachers and students to engage social theory and political philosophy in creative, interdisciplinary ways.

This chapter contains three interrelated parts that provide a theoretical constellation for thinking about science education as a site of biopolitical struggle. First, the context for science education is addressed, followed by a discussion of biopolitics and subjectivity. Afterwards, I position biopolitics as a productive way to reconceptualize science education as a site of political struggle over subjectivities. Due to limited space, this chapter does not offer a complete framework for thinking about science education, and subsequent political action, from a

[1] I use the terms biopower and biopolitics in the way Hardt and Negri (2009) do. Simply put, biopower represents the control and exploitation of populations, life forces, and human bodies. Biopolitics operates in relation to biopower and consists of interventions and challenges to biopower on the same terrain of bodies and subjectivities toward different subjectivities, social relations and ontologies.

biopolitical perspective. A more comprehensive framework would require substantial input from a wide variety of disciplines such as critical race studies, human biology, political economy, gender studies, and cultural history.

A Context for Science Education

Science education's interstitial locus between the discourses of science (ex. biotechnology and human ecology) and sociopolitical motivations of public and private educational initiatives make it an ideal place to engage biopolitically. While consideration of how subjectivities are constituted[2] through science education discourses and relations of power is important, there also remains the need to describe the broad political, sociocultural and historical contexts this inquiry may take (Bazzul 2012). Creating a more sociopolitically engaged science education will require making commitments to many different social struggles. Like Angela Calabrese-Barton (2003), Larry Bencze and Lyn Carter (2011), I see the desperate need for theoretically divergent ways of highlighting the cultural, political, and social contexts of science and science education. While I agree with Derek Hodson's (2011) call for a less conservative, more socially active science education, there simultaneously needs to be spaces opened for alternative ways of thinking and acting in science education. These spaces will *not* align with what has already been articulated within science education literature and its disciplinary walls, and are needed for educators and students to solve fundamental problems such as climate change and social inequality. Challenging assumptions requires thinking alternatively, especially in a field that is relatively isolated from other disciplines (Carter 2011a). What seems to be a troubling practice in science education scholarship is the reverence given to catch words and acronyms such as NOS, STS, STSE, scientific literacy, that often offer little in the way of political, sociocultural, historical or philosophical transparency. One problem with the prevalence of catch words and acronyms is that the political positions they embody (ex. those aligned with state power) are simultaneously obscured, hidden in science education jargon created by senior researchers and government bureaucrats, yet also exercised 'in the open', as can be seen in the way science educators 'self-reference' a field already well known to be politically conservative (see J. Lemke 2011). It is important that we fundamentally question the focus on science education jargon as well as the apparent pressure of having to use this jargon to encapsulate science education 'correctly.' Packaging science education in field-jargon, while useful for normalizing science education, makes calls for radical

[2] Obviously science education discourses cannot completely constitute subjectivities, but works in conjunction with other discourses, institutions, and material realities. Moreover, as John Fiske (1998) notes of Althusser's work and production of subjects, education systems are not able to tell a different story than the legal system or the family.

action difficult. Sandra Harding's (2006) call for openness regarding the political, social, and cultural contexts of scientific knowledge emphasizes the productive potential of these contexts for democracy.

> What if the goal were not to eliminate all such values and interests, but instead (as indicated above), to evaluate whether there are patterns of productive relations between culture and knowledge, and between prodemocratic culture and knowledge that can advance democratic social relations as well as the growth of knowledge. (p. 51)

Viewing science education as a site for biopolitics can be a way to explore productive relations between the discourses of science and how these work to constitute the subjectivities of students as colonized, sexed, and/or economic subjects.

Clayton Pierce's (2013) book, *Education in the Age of Biocapitalism: Optimizing Educational Life for a Flat World,* advocates for reworking 'scientific literacy' to take into account biocapitalist contexts that are co-extensive with recent education reforms. From a biopolitical perspective, Pierce outlines ways in which neoliberal education reforms are furthering the extraction of (bio)value from student populations, and instilling an ethic (subjectivity) of self-investment that works against sustainable, equitable, community living. Educational restructuring in an age of biocapitalism divides 'haves' and 'have nots,' in terms of those who have the means to recreate themselves as entrepreneurial subjects, in other words along racial and class lines. Pierce recognizes that science education reforms are directly related to the production of subjectivities oriented toward the commodification of life (both biological and social) and the (surplus) labour needed to meet demands of large multinational corporations seeking to influence the aims of science education. Taking a biopolitical perspective can help education communities intervene in these subjectification processes and networks of power.

Biopolitics and Biopower

The term biopolitics has been, and will continue to be, used differently depending on particular analytical necessities. A useful description of these differences can be found in the work of Maarten Simons (2006) and Thomas Lemke (2011). The notion of biopolitics presented here will not be limited to environmentalist conceptions (where ecological concerns become a basis for politics) nor a bioethicist version (where legal or moral questions directly related to biology become the sole basis for political action) (see T. Lemke 2011, Chap. 2). Instead, this chapter will outline and employ Michel Foucault's conception of biopower and biopolitics.

Foucault's historical work (Foucault 1977, 1980; Foucault and Senellart 2010) investigated the shift of sovereign power and its disciplinary practices on the body (public executions, physical brutality) to the management of populations in the seventeenth, eighteenth, and early nineteenth centuries. That is, toward "a power bent on gathering forces, making them grow, and ordering them, rather than one dedicated to impeding them, making them submit, or destroying them" (Foucault 1980, p. 136). As Paul Rabinow and Nicolas Rose (2006) put it, power became "situated and exercised at the level of life" (p. 196). In other words, power exercised in the 'taking' of life diminishes, while the power to foster or stunt life begins to take shape in early modernity, representing a shift towards the development of *biopower* in that "biological existence was reflected in political existence" (Foucault 1980, p. 143). From the late eighteenth century onward, two poles of governance emerge in modern western nation states: the discipline of the body and the management of populations. The discipline of individuals and the regulation of populations are not two extremes but rather two sides of the same "global political technology that simultaneously aims at the control of the human as individual" (T. Lemke 2011, p. 38). Biopolitics focuses on biopower and ways in which human life is controlled through practices such as preventative health, statistical analysis and material infrastructure. While biopower can be seen as the domination of life from above (hierarchal, territorialized), biopolitics can be seen as a response that recognizes the immanent powers of creation, cooperation, and organization inherent in every community toward democratic forms of life (Lewis 2007; Hardt and Negri 2005). Debate may centre on whether biopolitics refers specifically to the intervention and struggle against forms of biopower or whether technologies and discourses of biopower are not also themselves forms of biopolitics. While the latter seems quite reasonable, the former seems to make more definitive sense, and is the general distinction used in this chapter (Hardt and Negri 2000, 2009). It should be acknowledged that practices of biopower and biopolitical interventions into this form of power constitute, and manifest in relation to, each other.

Rabinow and Rose (2006) describe biopolitics as a term used: "[t]o embrace all the specific strategies and contestations over problemitizations of collective human vitality, morbidity and mortality; over the forms of knowledge, regimes of authority and practices of intervention that are desirable, legitimate, and efficacious" (p. 197). They outline spaces for both theorizing and conducting 'molecular' analyses into how biopower operates by describing a plane of actuality encompassing three interrelated elements. These elements and their relation to science education are listed in Table 3.1. Rabinow and Rose (2006) sensibly differentiate between engaging the macro and micro poles of biopower; that is, thinking broadly about the governing and administering of bodies at the population level (macro) and examining individual strategies of biopower (micro). It is in the micro or 'molecular' practices of biopower that science educators can more easily intervene biopolitically. This may involve looking at curriculum materials for discursive modes of subjectification or the structural role of space in science education in maintaining particular political agendas (see Ken Tobin's (2011) article about how school structuring can achieve neoliberal ends). Rose's (2007) theorizing around genomic research and biotechnology offer good starting points for science educators willing to take up difficult questions about the shifting nature of bodies, subjectivities, and their subsequent management and production through science education.

Table 3.1 Elements of Rabinow and Rose's notion of biopower and their relation to science education

Elements of biopower	Relation to science education
1. One or more truth discourses about the 'vital' character of living beings, and an array of authorities considered competent to speak the truth	1. Once biology applies its knowledge(s) to human life there is a situation set up (for better or for worse) whereby an authority speaks
2. Strategies for intervention upon collective existence in the name of life and health, initially addressed to populations. . .sometimes specified in terms of race, ethnicity, gender or religion, as in the emerging forms of genetic or biological citizenship	2. Science education is involved in health and population management; education materials are replete with exercises concerned with problems of collective existence
3. Modes of subjectification, through which individuals are brought to work on themselves, under certain forms of authority, in relation to truth discourses, by means of practices of self. . .or indeed in the name of life or health of the population as a whole	3. How does science education constitute subjects? "Who" can think and act in scientific ways? What relationship to self is required to take action in science?

Hardt and Negri (2009) offer a broad vision of what biopolitics could look like by emphasizing how subjectivities are reproduced through the invasion of capital into all spaces of life where value is produced, and subsequently usurped by mechanisms of biopower. Undergirding Hardt and Negri's version of biopolitics is their conception of *empire*, our current era of capitalism where nothing remains untouched by global capitalism. In this new transnational empire, the (re)production of affects, bodies, and subjectivities maintain the conditions of both material and immaterial production and labour for capital accumulation and the extraction of surplus value (Hardt and Negri 2000). Immaterial labour refers to the production of information and cultural content through the use of media/computer technology, and is a key concept for conducting sociopolitical, cultural and economic analyses into changing patterns of global exploitation. To put it in Rabinow and Rose's (2006) terms, empire is a global regime of *biopower*, regulating social life through 'self-responsible' individuals who continually reactivate this power in their lives. It *both* produces subjects and is, in turn, reproduced by these subjects through immaterial labour – which is why subjectivity needs to be an important focal point for critical educators. Not everything about empire is negative as it is also the very space where struggles for universalizing claims for democracy are fought and won "on the level of the biopolitical" (Lewis 2007, p. 689).

Hardt and Negri's (2000) philosophical framework for biopolitics involves recognizing two forces of modernity, the immanence of human powers (*constituent powers*) and transcendent powers (*constituted powers*), which aims to restore order. Drawing from Baruch Spinoza and Gilles Deleuze, these theorists posit the discovery of immanent creative forces as one of the primary events of modernity, where human beings realize the potential of reason that exists right before them in this world, and not in a transcendent one. At the beginning of modernity, knowledge begins to shift from transcendent ways of knowing towards a practice of transformation: "...the powers of creation that had previously been consigned to the heavens are now brought down to earth" (p. 73). It is through this force of modernity, that of immanence, where singular freedoms and new subjectivities can be attained due to the absence of external mediation. The second force of modernity arises to "wage war against the new forces and establish an overarching power to dominate them" (p. 74). This constituted power tries to bring difference, multiplicity and experience under the control of universalities and hierarchal disciplinary apparatuses. Consequently, modernity can be seen as a constant battle between immanent creative forces and forces of transcendence and control, such as the *rule* of private property, aimed at restoring order and disciplining subjects. The dangers of transcendental control can be seen in the way Hardt and Negri (2009) describe fundamentalisms, namely how they always appeal to transcendental concepts to unify their ontologies and ethics. Fundamentalisms are notorious for their double focus on the body. Employing biopower they simultaneously obsess over bodies (what they are, what they do); and make bodies disappear or refute bodies. In biology, a fundamentalist view of 'human nature' relies on detailed understandings of the body as a particular kind of mammal or set of genomic instructions. Yet, precisely because there is adherence to a transcendental concept, for example natural selection conceived in dogmatic terms, alternate behaviours of the body are made invisible. For instance homosexuality is labeled as a malfunction or anomaly and disappears from legitimized spaces. Modernity as two countervailing forces is best understood as part of an overall narrative for biopolitics. We must *choose* to foster the immanent forces of modernity over the forces that subordinate and bring singularities and difference under 'universal' control. Likewise, we can choose to resist the way structures and discourses of modernity have constituted human life and subjectivities through political action and intercede in the way modes of life have been produced. In science education, this means *asking after* how it has come to be that we find particular ways of thinking about science and human beings natural.

Hardt and Negri (2009) provide broad guidelines biopolitical engagement and reworking how subjectivities are constituted in education.

Free and accessible knowledge – Knowledge must be made free and accessible. Educational efforts towards equity must focus on giving all students the competencies and tools to access this knowledge.

Tearing Down Hierarchies – Seeing science education as a site for biopolitics means tearing down hierarchies and using science to achieve these political ends. Science education might do as Michael Hardt (2010) insists and tear down hierarchies *as* the educational process unfolds.

Trouble the "Expert" – De-privilege the voice of the (education) expert who tells us we have no choice. Wealth comes from below, not through neoliberal efficiencies or extraction of value that could otherwise be put into research and development.

Inclusiveness – Science education must be an inclusive enterprise. Those subjectivities at the "precarious edges" of politics and production have the greatest capacity to challenge how we have been constituted as subjects through mechanisms of biopower.

Preserve the natural commons – Biopolitical action must put the "natural" world to work without consuming it completely. It must put its rationality to the service of life, ecology and social relations between humans and nonhumans.

To conclude these notes on biopolitics it is important to understand that different theoretical orientations for can be employed for different research/political contexts (see Agamben 1998). In the next section, I turn to the production and constitution of subjectivities, a key aspect of engaging biopolitically in science education.

The "Making of Subjects"

The constitution of subjectivities is important when taking a (bio)political perspective in science education. Whether educators and students are interested in challenging privatized biotechnology or science's complicity with 'the North's' geopolitical domination of 'the South,' etc., it is crucial to *ask after* the kinds of subjects constituted by discourses and practices in science education. A focus on subjectivity can help us think differently about how science education plays a role in 'who we think we are, what we find important in research, and how we've come to see many inequitable social conditions as natural.

The constitution of subjectivity needs to be understood not as simply a process by which some exterior power subordinates individuals, though this is certainly involved. Individuals are constituted as subjects in terms of the 'identity' that proceeds from the processes of becoming/being a subject. While a student is shaped by the overt disciplinary structures of school, she is also *always already* a constituted subject when she embodies a particular relation to objects, institutions, others, and herself. As Butler (1997) maintains,

> [s]uch subjection is the kind of power that not only unilaterally acts on a given individual as a form of domination, but also activates and forms the subject....the subject produced and the subject regulated or subordinated are one, and that compulsory production is its own form of regulation. (p. 84)

Subjects are produced through various restrictions, limits, and the ways they are meant to see themselves in relation to objects, meaningful action, and others. Subjectification involves the *perception of freedom* from standpoint of any 'identity', since, as Foucault (1982) insists, power "is exercised only over free subjects, and only insofar as they are free" (p. 790). Complete domination is, to Foucault, not in-line with the relational aspect of power, which means there is always the possibility of resistance and freedom. The freedom inherent in subjectification is both constrained, as the subject has been constituted in one particular way with limited choices for thought and action, and yet very real, since to be free is to *actually have* a range of choices – thus, change is always possible.

Educators should see schooling not just as the *discipline* of bodies (which is the most common usage of Foucault in education), but also as manifestations of power relations, *co-extensive* with the production of subjectivities. Asking after the subjectivities produced in science classrooms/labs/field studies requires examination of the discourses and practices, both macro and micro, that work to constitute them. As Butler (1997) argues, prior to any critical review into what makes us subjects, the law in the abstract, we must also recognize *our vulnerability* to the law. In the context of this chapter, the law in the abstract includes what is indispensible or seems natural in/to science and education. Subjects must understand their *necessary attachment* to how they have already been constituted before critique can continue. In order to engage in critique that challenges the very grounds of the law (what can be said, thought, and acted upon), the subject must also be willing to be *undone* by the critique, since the law is what has 'done-up' the subject in the first place. Butler (2004) stresses that it is not "kicks or thrills" that have people question the very limits of what makes them who they are:

> [O]ne does not drive to the limits for a thrill experience, or because limits are dangerous and sexy, or because it brings us into a titillating proximity with evil. One asks about the limits of ways of knowing because one has already run up against a crisis within the epistemological field in which one lives. (p. 310)

That is to say, the question of subjective foundations is raised because certain subjectivities are *unlivable,* that is the subject is *unable* to reconcile how she has been constituted with lived realities.

Butler (1997) describes three ways the normalizing effects of power, discourse, and the subjectivities they produce may be subverted. First, processes of subjectification and the normative aims of discourses, may exceed the goals for which they are first intended. Second, discursive regimes can merge where one works against the goals of normalization of another. We can see how this works over calls for the freedom of information – at once in line neoliberal appropriation of accountability and the functioning of knowledge economies, yet converse to the privatization of knowledge. Butler, following Louis Althusser, maintains that the making of subjects relies on *repeated* subjectification, and temporal gaps between these moments may allow for anomalies and contradictions to appear. In this way, science educators should consider the repetitive lab activities of science, with their apolitical contexts and cook-book procedures; do they not work to produce a depoliticized subject? Althusser (1998) maintains that there is always the possibility that subjects, in the act of being interpellated, will misrecognize themselves, thereby confounding the repetition necessary for the production of subjects. A key feature about subjectification processes, power and discourse is that possibilities for resistance are *inherent* – they always contain possibilities for their subversion.

Foucault emphasizes the 'doubleness' of power and the agonistic relation between subjection and resistance. It is through the latter that alternative subjectivities are born. Hardt and Negri's (2009) reading of Foucauldian notions of power, freedom and resistance is an encouraging one:

> We should not think of power as primary and resistance as a reaction to it; instead, paradoxical as it may sound, resistance is prior to power. Here we can appreciate the full importance of Foucault's claim that power is exercised only over free subjects. Their freedom is prior to the exercise of power, and their resistance is simply the effort to further, expand, and strengthen that freedom. (pp. 81–82)

New subjectivities can be produced through resistance and new ways of thinking about science and the world. These subjectivities will work against individualism toward collectivity and respect the equality of singularities. In the next Section I will outline, from a biopolitical perspective, some examples of how particular subjectivities may be partially constituted in science education.

Biopolitics and Subjectivities in Science Education

Critique, according to Foucault (1982), must focus on forms of power rather than particular institutions or elite groups.

> This form of power applies itself to immediate everyday life which categorizes the individual, marks him by his own individuality, attaches him to his own identity, imposes a law of truth on him which he must recognize and others must recognize in him, it is form of power that makes individual subjects (Foucault 1982, p. 781).

For example it does not suffice to blame pharmaceutical companies for the corruption of science and education with their commercial interests. Instead, as Rose (2007) and Pierce (2013) demonstrate we must also look to see how power is exercised at the level of how 'healthy', 'sick' or 'responsible' categories are deployed.

Questions related to subjectivity and biopolitics easily fall under the purview of science, for example, what does it mean for a politics 'self,' now that the human genome has been sequenced? There are also the questions of subjectivity related to how the sciences are complicit with oppressions, for example how has science contributed to the colonization of peoples in most parts of the world. There is an added complexity when consider that modes of governing and the production of subjectivity can no longer be isolated to sites such as the factory or the school, but must include networks made available through communication technology and media (Deleuze 1992). This network possibilities made possible by social media makes inquiry into subjectivity more necessary than ever.

The following subsections consider ways science and science education may work to constitute subjectivities related to ethics, neoliberalism, biotechnology, and sex/gender. They should be thought of as examples of how biopolitical theory can inform future inquiry. These categories do not represent 'real' subjectivities, as subjectivities are always multiple and in flux. Rather, they are abstractions that allow us to address questions of subjectivity.

Racisms, Colonialisms and the Power to Make Die

A biopolitical perspective can inform research into the (neo)colonial, Eurocentrist, White supremacist orientations of science and science education. Science students can experience conflict between their multi-dimensional (racialized, sexed) subjectivities and the universal (white male) subject of science. Elizabeth McKinley (2008) highlights this conflict in her study involving the ever-shifting, hybrid subjectivities of Maori women scientists. While colonialisms are part-in-parcel with the controlling forces of modernity (Hardt and Negri 2009), racisms can be understood as a modern biopolitical problem as Ann Stoler (1995) argues in *Race and the Education of Desire: Foucault's History of Sexuality and the Colonial Order of things*. In this work, Stoler explains the discourses of biopower, the development of science, and notions of race and sexuality within a *colonial* frame.

As Stoler points out, for Foucault racism is an incessant social war driven by technologies of purification, and is *internal* to the biopolitical state (rather than just something that happens in the colonies). Racisms solve one problem concerning the transition from a disciplinary to a regulatory society (involving the management of populations) where the 'right to kill' is at odds with the goal of fostering and controlling life – yet still maintains a key role in overall governance. The 'right to kill' serves a disciplinary function in the biopolitical state, through which racism becomes a discourse to designate *who* must die, thus the logic that sustains colonial domination is also used to govern *within* populations. Racism become useful as it not only decides "who" must die, but "establishes a positive relation between the right to kill and the assurance of life" (Stoler 1995, p. 84). The logic becomes, the more subaltern peoples die, the more privileged subjects live. Racisms become a way for the biopolitical state to retain its disciplinary power, justified as an always-incomplete cleansing of the social body (T. Lemke 2011). In this way, scientific cultures may exercise epistemological and ontological forms of the "right to kill," under the formulation of 'what knowledges must die (indigenous/local) for others (scientific) to live'? Other critical questions that be approached from a biopolitical perspective include: Who is "othered" through science education? What sociopolitical functions do racisms and colonialisms serve? How is the health of one person in direct relation to the poverty/ill-health or death of another?

Neoliberal Subjectivity

Matthew Weinstein (2012) argues that neoliberal reforms in science education are inextricably linked to a global capitalist agenda that benefits elites and consequently *must* be a concern for social justice movements. While calls to resist the neoliberalization of science educations have been put forward (see Bencze 2010; Bencze and Carter 2011; Tobin 2011), analyses looking at how neoliberal subjectivity is constituted need to be part of any comprehensive picture. As political theorist Wendy Brown (2005) maintains, it is the sociopolitical aspects of neoliberalism, deployed as governmentality, that claim "the soul" of the citizen subject and "imposes a market rationale for decision making in all spheres" (p. 40). This kind of governmentality produces a "free" subject who makes choices between alternatives, and bears the responsibility for those choices. Educators may look to Foucault's (Foucault and Senellart 2010) lectures concerning the birth of biopolitics, in which he traces the history of liberalism While it may not be intuitive to imagine an entrepreneurial self (see Simons 2006) in science education, one may be surprised by the celebration of private research entrepreneurs like Craig Venter and the ways students are led to invest in their own human capital through discourses of careers. Broad questions that can be asked in relation to neoliberalism are: What is the relationship between the rational, self-responsible subject (Homo economicus) and that of human life as capital – perhaps read through the frame of "careers in science"? How are these relationships forged through normalizing discourses involving health, biotechnology, and ecological knowledge? My own research suggests that economic issues discussed in textbook exercises frame the limits of thought and action around private ownership, the shared values of science, and corporate influences on public and private science research (Bazzul 2012).

Sex/Gender and Sexuality

In the *History of Sexuality Vol 1.*, Foucault (1980) positions sexuality between two modes of power/governance, the disciplinary and population/regulatory. The performance of sex/gender and sexuality, and the power effects that help shape this performance, are both situated on the body and controlled at the level of populations. The erection of norms regarding personal hygiene, as well as concerns about population health make the topic of sex/gender and sexuality ideal for engaging in an *analytics of biopolitics* in science education. For example, my current research with biology textbooks outlines concerns with sexual practices related to aids/population growth as well as issues of individual sexual behavior, such as sexually transmitted infections. The AIDS virus is a biological and social phenomenon that exposes the effects of power at the level of population and the individual.

The philosopher (Irigaray and Oberle 1985) has challenged science's gendered voice – male – under its pretense of "neuterdom". She maintains that "[o]ne of the places most likely to provoke a questioning of the scientific landscape is that of the examination of the subject of science and its psychic and sexed implication in discourse, discoveries and their organization" (p. 79). As Catherine Milne (2011) astutely points out, a sharp dividing practice between what is sociocultural and what counts as science in these materials often works to *exempt* science and science education from engaging tough questions. In this regard, (Irigaray and Oberle 1985) asks a series of pertinent questions:

> Does the alternative become either do science or "be a militant"? Or again to continue to do science and to divide yourself up into different functions, several persons or characters? Should the truth of science and that of life remain separate, at least for the majority of researchers? What science and what life is then under discussion? Especially since life in our time is greatly dominated by science and its techniques (p. 78).

The expanse of sex/sexuality and gender and the way science and science education continue to constrain, define and constitute various subjectivities makes it a rich site for biopolitical action. As Anne Fausto-Sterling demonstrates, the constitution of sex/gendered and sexualized subjectivities in science works to (re)produce heteronormative, male-gendered science research (Fausto-Sterling 2012).

The "Ethical Subject" in Science Education

Science education discourses and practices may also constitute how students and teachers are meant to think and act along ethical lines. What does it mean to be an ethical actor in formalized science education? Engaging questions related to how science students and teachers approach ethical situations from a Foucauldian perspective will involve thinking about "the way a human being turns himself into a subject" (Foucault 1982, p. 778), as well as specific *relations to self* required for ethical actions (Foucault 1986). It will also involve a departure from Foucault's earlier work that focuses on dividing practices, objectification and truth discourses towards his later work that explores how subjectification consists of self-examining ways of being that affect actions towards others and the world (Peters 2004). Butler (2004) insists that examining relations to self (self-making) is essential to the politics of desubjugation that Foucault advocated.

Discourses about ethical issues in science education operate like games of truth carrying both the weight of science and by prescribing particular rational actions (see Bernauer and Rasmussen 1988, Chapter 1). My own research with Ontario biology textbooks, demonstrates that students are meant to engage ethical questions primarily on a legal level; thus rendering ethical issues under the control of state governance; and an ethical actor as someone who attempts to evaluate and amend the law (Bazzul 2013). Results also show that the preservation of personal and population health is a crucial motive for ethical action. While I think the (bio) ethical issues presented in textbooks are all very relevant, we must *ask after* how we've come to view these issues and these actions, delimited by discourses of science education, as important.

Ethics in science (biology) education today may be best understood using a biopolitical framework; that is how they operate along the poles of biopower (disciplinary and regulatory) and can be reformulated through biopolitical intervention. Regarding ethics and biotechnology, we might shift the lines of ethical thought toward questions of how new forms of exploitation and exclusion take place along genetic lines or exposure to insecurity and poverty in global capitalist economies. Thinking again about subjectivity, it is important to ask, 'who' discourses and practices of science education 'expect us to be' when we approach issues of ethical importance (Lather 2012). Biopolitical intervention means that we have a say in what constitutes ethical action, either individually or at the level of populations.

The Biosubject of Biotechnology

Maurizio Lazzarato (2002) credits Foucault as "already pointing out in the seventies what, nowadays, is well on its way to being obvious: 'life' and living being [le vivant] are at the heart of new political battles and new economic strategies" (p. 1). Lazzarato describes how current genomic research, artificial intelligence, and biotechnology are tracing new "cartographies" for biopower. In globalized agriculture, the privatization of seeds and the exploitation of farmers in the global 'south' involves a series of policies related to practices of science research. Shiva and Moser (1995) reminds us that biotechnological progress is not gender or class neutral, and that progress and 'efficiencies', especially when dictated by companies such as Monsanto, often mean different things depending on whether you are from the global 'North' or 'South' (Carter 2011b). Regarding Earth's ever-shrinking biodiversity, Shiva warns that "not until diversity is made the logic of production can diversity be conserved" (p. 207). In biopolitical terms, this involves disrupting and reformulating the rule of private property and the means it gives some bodies to control the material circumstances of others.

As Neil Gerlach et al. (2011) argue in *Becoming Biosubjects*, biotechnology is presenting us with new ways of thinking about governance, human beings and the limits of the body; it is therefore imperative that we consider ourselves as *biosubjects*. The authors stress that biosubjectivity is not divorced from the political contexts of late modernity.

> This biosubjectivity troubles traditional modernist dualisms between natural and artificial, human and animal, private and public, and present and future. The subject is both alienated from and dependent upon a fragmented body. It is a subject outside of humanist ethics and firmly within capitalist relations. . .It is a subject that is always already in conversation with other late modern subjects – the entrepreneurial subject, the prudent subject, the subject under surveillance. (pp. 6–7)

Today, bodies are not only under what Evelyn Fox Keller (1996) calls the "biological gaze", but at a molecular level they can be "informated, sold, killed, manipulated, reproduced, copied, and circulating along networks of exchange and knowledge production" (Gerlach et al. 2011, p. 9). Different forms of identity, as well as normative judgments by authorities such as scientists, pharmaceutical companies, counselors, and education systems, will arise from new biotechnical knowledges and hopefully lead to unforeseeable forms of political activism surrounding science.

Biopolitics as a Path Forward

Viewing science education as a site for biopolitical engagement and reworking subjectivities can help science students and educators confront oppressions linked with scientific practices, and pressing problems such as climate change and social inequality. Since power works as a relation, biopolitical engagement consists not in overthrowing biopower, but working within various networks and institutions to produce alternative subjectivities and forms of life. According to Hardt and Negri (2009), biopolitics can be seen as the flight from power *over* life toward the power *of* life in order to rework how we come to see others, the word, and ourselves. Biopolitics proceeds as a necessarily *queer* endeavor as it challenges norms and puts forward something different. This requires that *all* students have access to science education (Siatras 2012), and that teachers be given opportunities to disrupt the practices, technologies, and discourses of biopower that control bodies and constitute subjectivities and rework them toward the many goals of collective existence and social justice. An important biopolitical goal is to create the freedom for many singularities (subjectivities, fluid identities) as Hardt and Negri (2009) insist they are **essential** to radical change:

> One of the most significant challenges of revolution today, then, which this parallelism of singularities suggests, is that revolutionary action cannot be successfully conducted or even thought in one domain alone. Without its parallel developments any revolutionary struggle will run aground or even fall back on itself. A revolutionary race proposition that ignores or even exacerbates gender hierarchies will inevitably be blocked, as will a class proposition that fails to keep up with its parallels in the racial domain. (p. 343)

As Thomas Lemke (2011) maintains, an *analytics of biopolitics* must include the 'scientific' disciplines that have authority to "tell the truth" about health, populations, what forms of life are socially valuable. Science educators can target processes of subjectification, where students are brought to embrace certain forms of conduct, towards transformative change. Indignation, as suggested by Hardt and Negri (2009) is a good first step to locate what needs changing. For scholars like Clayton Pierce (2013), this means promoting scientific literacies that endeavour "to link learning and teaching science to practices and social movements that are actively resistant to biocapitalist visions of the future, ones that represent cultural practices rooted in communities producing biodemocratic life" (p. 108).

More biopolitical theorization, along with micro-analyses related to subjectification processes in science education, needs to take place. Turning back to the opening quotation from Foucault, biopolitical engagement first means refusing commonsense understandings about 'modes of life' constituted through science education. Science education is perhaps the best field in which to engage biopolitically as it encompasses multiple authoritative, normalizing forces that oscillate between a focus on individual subjects (students) and concern for the population.

References

Agamben, G. (1998). *Homo sacer*. Stanford: Stanford University Press.

Althusser, L. (1998). Ideology and ideological state apparatuses. In J. Rivkin & M. Ryan (Eds.), *Literary theory, an anthology* (pp. 294–304). Malden: Blackwell.

Amos, J. (2012, May 25). *Africa and Australasia to share square kilometre array* [On-line article]. http://www.bbc.co.uk/news/science-environment-18194984. Accessed 25 May 2012.

Barton, A. C. (2003). *Teaching science for social justice*. New York: Teachers College Press.

Bazzul, J. (2012). Neoliberal ideology, global capitalism, and science education: Engaging the question of subjectivity. *Cultural Studies of Science Education, 7*(4), 1001–1020. doi:10.1007/s11422-012-9413-3.

Bazzul, J. (2013, May). *The ethical subject of science education*. Paper presented at the American Educational Research Association international conference, San Francisco, USA.

Bencze, J. L. (2010). Exposing and deposing hyper-economized school science. *Cultural Studies of Science Education, 5*, 293–303.

Bencze, J. L., & Carter, L. (2011). Globalizing students acting for the common good. *Journal of Research in Science Teaching, 48*, 648–669.

Bernauer, J., & Rasmussen, D. (Eds.). (1988). *The final Foucault*. Cambridge: MIT Press (Reprinted from *Philosophy & Social Criticism, 12*(2–3), 109–277).

Brown, W. (2005). *Edgework*. Princeton: Princeton University Press.

Butler, J. (1997). *The psychic life of power: Theories in subjection*. Stanford: Stanford University Press.

Butler, J. (2004). What is critique? An essay on Foucault's virtue (2001). In S. Salih & J. Butler (Eds.), *The Judith Butler reader* (pp. 302–322). Malden: Blackwell.

Carter, L. (2011a). Gathering in threads in the insensible global world: The wicked problem of globalization and science education. *Cultural Studies of Science Education, 6*(1), 1–12.

Carter, L. (Ed.). (2011b). Globalization and science education: Global and local issues. *Cultural Studies of Science Education, 6*, 1–12.

Deleuze, G. (1992, Winter). Post script to societies of control. *October, 59*(1992), 3–7.

Fausto-Sterling, A. (2012). *Sex/gender: Biology in a social world*. New York: Routledge.

Fiske, J. (1998). Culture, ideology and interpellation. In J. Rivkin & M. Ryan (Eds.), *Literary theory, an anthology* (pp. 305–311). Malden: Blackwell.

Foucault, M. (1977). *Discipline and punish: The birth of the prison*. New York: Pantheon Books.

Foucault, M. (1980). *The history of sexuality. Vol. 1: An introduction*. New York: Vintage Books.

Foucault, M. (1982). The subject and power. In H. L. Dreyfus & P. Rabinow (Eds.), *Michel Foucault: Beyond structuralism and hermeneutics* (pp. 208–226). Chicago: University of Chicago Press.

Foucault, M. (1986). *The care of the self* (The history of sexuality, Vol. 3). New York: Vintage Books.

Foucault, M., & Senellart, M. (2010). *The birth of biopolitics: Lectures at the Collège de France, 1978–1979*. New York: Picador.

Gerlach, N., et al. (2011). *Becoming biosubjects: Bodies, systems, technologies*. Toronto: University of Toronto Press.

Harding, S. (2006). *Science and social inequality: Feminist and postcolonial issues*. Urbana: University of Illinois Press.

Hardt, M. (2010). The militancy of theory. *The South Atlantic Quarterly, 110*(1), 19–35.

Hardt, M., & Negri, A. (2000). *Empire*. Cambridge: Harvard University Press.

Hardt, M., & Negri, A. (2005). *Multitude: War and democracy in the age of empire*. New York: Penguin.

Hardt, M., & Negri, A. (2009). *Commonwealth*. Cambridge: Belknap Press of Harvard University Press.

Hodson, D. (2011). *Looking to the future: Building a curriculum for social activism*. Rotterdam: Sense Publishers.

Irigaray, L., & Oberle, E. (1985). Is the subject of science sexed? *Cultural Critique, 1*(1), 73–88.
Keller, E. F. (1996). The biological gaze. In G. Robertson, M. Mash, L. Tickner, J. Bird, B. Curtis, & T. Putnam (Eds.), *Future natural: Nature, science, culture* (pp. 107–121). London: Routledge.
Lather, P. (2012). The ruins of neoliberalism and the construction of a new (scientific) subjectivity. *Cultural Studies of Science Education, 7*, 1021–1025. doi:10.1007/s11422- 012-9465-4.
Lazzarato, M. (2002). From biopower to biopolitics. *Pli: The Warwick Journal of Philosophy, 13*, 112–25. http://cms.gold.ac.uk/media/lazzarato_biopolitics.pdf. Accessed 29 Apr 2012.
Lemke, J. (2011). The secret identity of science education: Masculine and politically conservative? *Cultural Studies of Science Education, 6*(2), 287–292.
Lemke, T. (2011). *Biopolitics: An advanced introduction*. New York: New York University Press.
Lewis, T. (2007). Biopolitical utopianism in educational theory. *Educational Philosophy and Theory, 39*(7), 683–702. doi:10.1111/j.1469-5812.2007.00316.x.
McKinley, E. (2008). From object to subject: Hybrid identities of indigenous women in science. *Cultural Studies of Science Education, 3*, 959–975.
Milne, C. (2011). A convenient dichotomy: Critical eyes on the limits to biological knowledge. *Cultural Studies of Science Education, 6*(2), 305–310.
Peters, M. (2004). Educational research: 'Games of truth' and the ethics of subjectivity. *Journal of Educational Enquiry, 5*(2), 50–63.
Pierce, C. (2013). *Education in the age of biocapitalism: Optimizing educational life for a flat world*. New York: Palgrave Macmillan.
Rabinow, P., & Rose, N. (2006). Biopower today. *Biosocieties, 1*, 195–217.
Rose, N. S. (2007). *Politics of life itself: Biomedicine, power, and subjectivity in the twenty-first century*. Princeton: Princeton University Press.
Shiva, V., & Moser, I. (1995). *Biopolitics: A feminist and ecological reader on biotechnology*. London: Zed Books.
Siatras, A. (2012). *Exploring the role of the science curriculum towards social justice*. Paper presented at the annual meeting of the Canadian Society for the Study of Education [CSSE], Waterloo, ON, Canada.
Simons, M. (2006). Learning as investment: Notes on governmentality and biopolitics. *Educational Philosophy and Theory, 38*(4), 523–540.
Stoler, A. L. (1995). *Race and the education of desire: Foucault's History of sexuality and the colonial order of things*. Durham: Duke University Press.
Tobin, K. (2011). Global reproduction and transformation of science education. *Cultural Studies of Science Education, 6*(1), 127–142.
Weinstein, M. (2012). Science, science education and the politics of neoliberal exceptionality. In J. R. McGinnis, S. J. Lynch, W. C. Kyle, & T. A. Sondergeld (Eds.), *Re-imagining research in 21st century science education for a diverse global community proceedings of the 2012 National Association of Research in Science Teaching international conference*. Indianapolis: NARST e-Publications [CD-ROM].

Chapter 4
A Critical Pedagogy for STEM Education

Arturo Rodriguez

Freedom is acquired by conquest not by gift.

Paulo Freire

Abstract This chapter was first published in JASTE 2.1 2010 as a review of recent work in neo-liberalism and science education as they relate to critical social analysis. At the time I took the position that under neo-liberalism, education in general and science, technology, engineering and math in education in particular press the energies of teachers toward the production of workers, the everyday tools for the expansion of empires, a police state and war machine. A radical departure from the stricture of academic journal writing I wrote the paper to reflect my critical voice as I consider alternative pedagogies for the development of STEM education. A fusion of narrative inquiry, critical social theory and free writing, this chapter provides an update and revision of the original paper.

Keywords Neo-liberalism • Critical pedagogy • Science, technology and activism in education • Neo-liberalism and education

Introduction

This chapter provides a summary of a critical study of an alternative to Science, Technology, Engineering and Mathematics (STEM) education and a fusion of critical theory/pedagogy and political economy.

According to the neoliberal agenda, everything and everyone is for sale. Capitalism is an unfettered tool by which entrepreneurial corporate partners act on the global

A. Rodriguez (✉)
College of Education, Boise State University, Boise, ID, USA
e-mail: arturorodriguez@boisestate.edu

L. Bencze and S. Alsop (eds.), *Activist Science and Technology Education*,
Cultural Studies of Science Education 9, DOI 10.1007/978-94-007-4360-1_4,

market, laying waste to government regulation and state-run businesses while ensuring corporations reap the benefits of patent laws and tax reform (Rodriguez and McLaren 2014). Neoliberalism run-amok is the global poor, the disenfranchised sold on the open market; capitalism is an upwardly mobile neighbor we can rely on to share with us the scraps of their table. While the world's rich parade round the world's beaches, major cities, amusement parks and centers of commerce, they take pride in their purchasing power. Adorned with the latest fashion trends, they hit the golf courses in Thailand, Vietnam and Mexico and, as night falls, descend on the night clubs and brothels exchanging their dollars for a few moments with the young women and men, castoffs of factories that supply the seeds for global wealth.

War, famine and disease have throughout history been the end-run of human experience, yet what science is available to relieve largely-social problems, instead, ensures the dollar is protected while corporations extract the living essence from the global ecology. Capitalism is the quantification of human beings; we are reduced to units of labor that provide surplus value: 15 min of our time mines the world for her resources, produces the printed page, the Gucci™ handbag, Macanudo™ cigars, or smiles on the faces of sex club aficionados.

According to Lacan (2006), we desire all that shines as long as what shines continues to shine: the social meme relating to an original thought, idea or experience. What can we say is an original thought, idea or experience? Since most of our becoming has to do with the people, places and histories with which we are surrounded. Take a person, change the place across space and time; will (s)he be an astronaut, a surgeon, a cleric, a suicide bomber or a freedom fighter? Will (s)he be one of these or all of these? We become the ideas we are given: our ethnicity, social class, gender and sexuality are social tags fixed by the other; "they" determine the course of our living experience and the value of our lives.

The critical experience that is the day-to-day of the classroom, the boardroom, and the playground is you and I sold on the idea our living has exchange value. From the pacifier soothing us to sleep to the car we are awarded at high school graduation, we determine the course of our lives based on what we are willing to sell: your spine, your mind and your sexuality will accept or not accept the world as it is. What was it Marx said? Spaces and places are fields for the production of someone else's knowledge, wealth or phobia (Bowles and Gintis 1976). Karel Kosik (1976) understood, society's eyes of experience, the pseudo-concrete and the pseudo-intellectual, what we know or think we know about our experiences with the world and each other. Does all of this sound esoteric? Sure. But that is part of the design, at least the education made possible under capital.

Global Capitalism

According to Dave Hill (2004), there are five major trends in Global Capitalism: (1) Spread of capitalism both geographically throughout the states of the world and sectorally within those states; (2) Deepening of capitalist social relations with the commodification of everyday life; (3) Increasing use of repressing economic, legal, military, and other state and multi-state apparatuses globally and within states; (4) Increasing use of ideological state apparatuses in the media and education systems; and, (5) Increasing concentration of wealth and power (power to retain and increase wealth) in the hands of the capitalist class.

Echoing this position, Henry Giroux's (2009) analysis of the project of neo-liberalism is pervasiveness evident on its influence globally and on the restructuring of social life. This is restructuring social life to support the interests of capitalists, the cementing of capitalist social relations of production manifested across the known world. The stranglehold on the planet by global industrialists is no-thing and no-body escapes the path of capitalist expansion or consumption. Have a look at the major tourist attractions where children and adults hawk the latest cultural artifacts, artifacts purportedly crafted by local people. Upon closer inspection, their labels prove pieces you can buy at the pyramids in Egypt, Mexico or Guatemala are made in Thailand, India or China. Moreover, children and adults on the streets spend their days collecting, sorting and selling, they experience the capitalist dream they in turn, were sold; the more they sell, the more they sell. Put another way, 100 key-chains sold to tourists will buy a sack of rice, a bag of beans or a bowlful of chickpeas. Fifteen hours on the street or 15 min on your back will ensure you can buy medicines, clothes for your family or food.

Capitalists and capitalism operate under the assumption that if you starve a person enough while offering them a way out of misery, climb the social ladder, produce/sell enough key-chains, while avoiding revolution, you will have arrived at a self-perpetuating system. The minutiae of capital are avoided, the sequence and the outcome are not as necessary as the systems [apparatus] in place to ensure success. It is important to note here the fruits of capital are mere specters; the end goal of capital is the perpetuation and maintenance of the system. People do not continue under ideological slavery because of the rewards set before them: they act upon the idea the reward that is capital buys freedom.

Under such a system, one must wonder what of the price of an education? Can we say it is worth a thousand key-chains, two thousand or ten thousand? And what of the labor power traded along the human chain from the hands that produced the item to the hands that consumed the item. Global Neo-liberalism has many caught on the human chain, from inception of an idea, genesis of a product, to conception of the fruits of capital [surplus value]. How we arrive at the nexus of production is paradoxical. Why do humans continue to believe they can buy their way through life?

STEM and Activism in Education

According to the Congressional Research Service (CRS) report, 2006 and the America Competes Act 2008, US pupils fail to reach adequate levels of proficiency in the STEM disciplines. A lack of performance in the STEM disciplines much as in Literacy education, is reduced to: teacher quality, funding by the federal government and a comparison of data, how the US currently fares compared to other countries. The data disaggregated across K-16 education describes the structure of STEM education in the US and abroad while failing to detail the obvious: the relationship between STEM curricula or programs and student scores.

That students fail to make adequate yearly progress (AYP) is designed into the system of public education. Teachers and administrators are made to adopt a technicist view of science and math curricula. Theory and Method are taught according to the latest and historical school of thought: teachers deposit information, information students then regurgitate during testing at year's end. Diverse methodologies for arriving at solutions or creating/transforming a theory, existing research or paradigm are ignored since the outcome of failing to make AYP is government takeover of public schools and school restructuring. STEM education in the United States is reduced to the memory, and methods fetish (Macedo 2006). Students are taught the appropriate theory, method and test taking strategies, including the correct operations and answers. The preceding must sound familiar, a simple analysis I conducted using personal experience mentoring new and current teachers in US public schools, the latest headlines concerning school takeover and Horace Mann's The Republic and the School, although we could include Dewey's Experience and Education. The basic principles are: create curriculum research supporting policy, legislate the curriculum, tie policy to research and funding, mandate the conduct of schools. The states then toe the line as Capital involves one of two options: take the Billions or be sanctioned.

What is education then in US Public schools? A process of indoctrination and an assumption by the elite that values, principles adopted by the masses will ensure their good conduct. This is what Henry Giroux calls the hidden curriculum or the interplay of ideology and action (1997, 2001) these are the limitations on the unconscious, common sense and critical consciousness meeting agency and the individual. As students are taught in US public schools the years of penal indoctrination and assimilation take hold; we are taught there is a right answer for everything, stray from the path and you will not receive your pot of gold. Educational policy, the America Competes Act, No Child Left Behind or Race to The top, creates institutional thinking among the masses, students, teachers and families; STEM education much as Literacy education becomes a further encroachment by government on the individual (Durkheim 1984). In US public schools, we are taught to accept the status quo for the sake of society, America or as is ever popular we do it "for the children."

STEM Education, Research and Practice

> A land of rigorous abstraction, empty of all familiar landmarks, is certainly not easy to get around in. But it offers compensations in the form of new freedom of movement and offering fresh vistas. The intensified formalization of mathematics emancipated people's minds from the restrictions that the customary interpretation of expressions placed on the construction of novel systems of postulates. New kinds of algebras and geometries were developed which marked significant departures from the mathematics of tradition. (Nagel and Newman 2001, p. 12)

K-12 schools, colleges and universities are more than mere purveyors of information or empirical paradigm (Rodriguez 2008). Students and professors engaged in the practice of teaching and learning do more than argue quanta, the lifespan of a mud-wasp or the credibility of one form of analysis over another in describing the day-to-day experiences of humanity.

We are engaged in our individual practice as we set out to discover/uncover phenomena and processes related to the understanding of all life, objects or ideas. In writing this chapter, I support the existing knowledge in the study of science education, science and technology studies in general and the study of activism in science and technology studies (STS) in education in particular. This is by no means a comprehensive view of the field; instead, it is an interest held by myself in the development of a personal critical revolutionary praxis and, more importantly, standing in solidarity of pedagogies that are progressive, activism-oriented and that seek to promote the critical revolutionary project.

Not a new field nor a new practice, the study of science and technology in education has enjoyed and been conflicted with a similar past as that of language and literacy studies and the urban and social studies in education. The traditional paradigm holds to the Cartesian understanding or engagement of phenomena. What can we surmise about distinctions made among objects, ideas or fields of experience? What can we understand, moreover, to manipulate those fields to the benefit of the individual and society and, more critically put, the employer, the license holder (trademark or copyright) and the individual?

The work of Wolff-Michael Roth provides a progressive look at the traditional forms of the study of science, science education and the study of science and technology in education. In his work from the 1990's, he submits constructivist understandings for the teaching of science and technology. Similarly, Lawrence Bencze, Stephen Alsop and Angela Calabrese Barton add to the existing literature in science in education studies and the constructivist and progressive research and teaching paradigms. In *From Everyday Science to Science Education*, Wolff-Michael Roth submits the notion that science and technology in education can offer the student much more than a clinical understanding of ideas, arguments or phenomena. Education in the pedagogy of sciences operates under the traditional assumption that students are objects to be filled with the diversity of data and methodology that governs the field misunderstood as the scientific method (Roth 1997). Furthermore, he describes what we have seen across other disciplines. Schools, colleges and universities act as gatekeepers; who passes the class, final exam and submits for the degree or diploma are eventually who practices physics, microbiology or engineering in the field.

This understanding of the disciplines, knowledge adopted from the school or the classroom community, enacted and engaged in the field, is contrary to the practice known by philosophers and theorists about being in the world and enacting a chosen profession, vocation or other way of being in the society. Experience, reflection and further practice in a relationship with peers and the natural environment provides an individual with the empirical knowledge to carry out a profession, vocation or other way of being in a society. Knowledge of phenomena in nature or a social experience is affected by history and the social relationships that contribute to the development of knowledge in a given cause or relation among objects or experiences.

That is to say, if we demonstrate the physical properties made by the upward and downward movement of a yo-yo to a student (Roth 1997) but do not push beyond asking students to consider why a yo-yo swings up and down when a person plays with a yo-yo, we can expect that students will provide a pat answer; e.g., 'Yo-yos are affected by gravity or they spin held by a string depending on the amount of friction acting on the central pin as it rotates.' The preceding is perhaps important for the understanding of gravity and friction. But what of the cognitive demands for the creation of string, the plastics and bearings involved for professional yo-yos? Pushing further: why use plastic for the creation of toys instead of wood?

A Critical Pedagogy for STEM Education

The birth of critical pedagogy is the historical fusion of many traditions of inquiry. They include existentialism, Marxism, the critical theories and feminism among others. Critical pedagogy differs radically from traditional forms of education in that pedagogues seek to develop with their students a transformative posture where student and teacher make sense of their lives while challenging the dominant and authoritarian ideologies in schools (Giroux 1997). The focus of critical pedagogues is the auto-emancipation or what Freire calls the re-humanization of their students (1970). What I mean is under Capital education or schooling operates under the construct of human enslavement vis a vis the cementing of social relations of production across the lifespan. The end-game of the dominant class is the classification of students and their acceptance as workers across the strata that are societies.

Constructivism, activism and STS-like critical pedagogy across the disciplines do not assume that students are empty vessels to be filled with factoid about social relations or the natural environment (Freire 1970). Instead, they provide a framework or grounding in knowledge that supports personal critical analyses of distinct phenomena. Why do we want to know about people, or people and their relationships to yo-yos? What I mean is yo-yos, like atom bombs, are interesting subjects, they can be related to a personal experience, but what is the end result of the use of yo-yos or the atom bomb in the world? It is the above idea I wish to continue discussing as analyses of the efficacy of curriculum design in teaching and learning affects teaching and learning outcomes. Teachers, Roth (1997) describes, are more accurately represented in the constructivist paradigm as more knowledgeable old-timers who engage with students in pedagogy where the material and discursive relations are interrogated such that the teacher is an authority figure. The act of teaching similarly related by Freire (1998) is to embark in teaching and learning with students to share information, working with students to uncover the static and dynamic properties of distinct objects or human relationships.

Teachers, instructors and professors then must set the conditions for learning to occur (McLaren 2007). In science education and STS, Roth and McGinn (1998) and

Roth (1997) describe these conditions as the authentic relationships engaged by teachers and students and students and the world situated in ways that reflect the complexity of processes and operations that scientists and other practitioners in the field experience. These are, as Roth and (1997) others describe, authentic learning tasks and authentic environments. Students must not merely mimic experiences from the field but actually undertake figuring, working with and refiguring a problem. They can then reflect on data and experience conducting research to ascertain a series of analyses that might provide a viable solution to the science implemented in the field.

The outcome of such practice is the student-led inquiry process. No longer dependent on their teacher, or codependent as they act out their respective roles, the student with the conditions set for learning embarks on the quest for knowledge by working with ideas from inception of the problem [what do we observe in or about nature?] to developing a research project: choosing the subject to be studied, how it is to be studied and which if any outcomes will influence the field and contribute to the further development of ideas or a research paradigm.

Community and Revolution

I sought a graduate degree as a way of understanding the social relations that overtly and covertly contribute to how and what my students learned. At graduate school I became more familiar with the progressive and critical tradition. The academic experience has been a journey marked by teaching, learning and writing about what Hannah Arendt (1959) refers to as the human condition. That is the complex social relations that happen, at least on this planet, in developing the fullness of humanity. It is my position in engaging the professoriate, teaching profession or instructorship to engage theory beyond dialogue. Sitting at the armchair of freedom, progressive liberals and radicals alike might argue the benefits of liberatory education over constructivist thinking in the schooling experience. They might decry the need for a closer inspection of the functions of US democracies and neo-liberal global capitalism; and yet, still not arrive at the crux of auto-emancipation.

The depth of consciousness necessary to take a critical and collective moral posture in acting on instead of simply knowing about the world: war in the Sudan or Iraq and Afghanistan or the return of the dust bowls to the Sacramento/San Joaquin river valleys and the parchment of the Sub-Saharan African plain.

Theoretical Freestyle

Democracy as currently understood lends itself to a blanket understanding of all a capitalist cabal wishes it to be, history is dead. More aptly put, perhaps, capital wishes history to be killed every day as Wal-Mart profiteering in the form of small town monopolies glutted with family owned business bankruptcies and tent cities spring up in the wake of the mortgage and real estate meltdown. So marks Lenin's understanding of imperialism, the social amnesia that self-converts human lives to human capital (McLaren 2007). Can democracy vanquish capitalism? Can the tip of the spear driven into the heart of global enslavement be marked by the blood of revolutionary liberals? Radicals doing more than sitting at armchairs idling days away with prospective to change the world and turn the tide of human traffic.

Bow-tie and elbow patches, pipe tobacco at the ready, the arm chair philosopher sings out the sound of democracy, crying havoc, ringing the liberty bell of freedom; while, at his white Christian feet, students fawn, grovel and lick up pearls of wisdom as they're spouted in the name of the epistemic tradition, to boldly know what no man has thought before. Truth at the feet of liberty congealing to break out in new ways to light a light bulb, cook a turkey, or impress the latest generation of hyper-consumers every Tuesday and Thursday at 7:00 p.m. sharp.

What are these cool kids singing for in their hipster dens in Brooklyn, Westwood, Palm Beach, and Signal Hill? When will my trust fund recover? Ask CNN, The New York Times or The Wall Street Journal; time and time again, they declare an end to the current state of capitalist affairs on the world stage. Yet what we have here is more akin to market control, the day after the Times and the Journal print a change in the market the market changes. Perhaps Gabriel Garcia Marquez was correct as he submitted his version of magical realism in reflecting social relations, what people believe to be true about foreigners and each other. Magical realism on the world market or a magician's trick, sleight of hand, as what the naked eye sees the mind believes.

The present condition of the United States of America: global capitalist hegemony; we are no longer asked but made to believe and act out in the everyday of our human experience (McLaren 2007). All social relationships are commodified, every chance meeting is an appropriate moment to network; who you know, what you know, and how can you manipulate the present social moment such that all moments from this one forward will be marked by hyper social capital and material consumption. Your success is marked by the Rollei™, Cartier™ and Benz's™ stacked in your drive-way. So the discursive relations of human experience go. The Journal describes an educated workforce as offering 'sub-prime human capital' while the Times signs off on the white house 'push' for an improvement in the teaching of science math and technology in United States schools.

Analytical Freestyle in Science Education

If the above resembles a rant, consider why a string of words that includes political and economic critique and the actual market functions of our global society affect the reader's view of this chapter. The academy turns its nose at work marginally reviewed or constructed as outcry, pedagogy of indignation (Freire 2004) at how people continue to enslave other people while destroying the last useable resources on the planet. Organic and academy taught intellectuals have given the world their lives, their blood sweat and fears chasing the ether, the unifying principle or truth, to solve the worlds mysteries. And how are they repaid? They are distracted from their work by colleagues who scream bloody murder as they find ways to take solace from the everyday right wing never ending barrage.

Is it the argument they are after when they cry foul? Or is it the sign, the symbol of freedom represented by a life's work in the academy sharing the living experience with students, colleagues, all workers alike managing the living, the breathing and the dying. And what is capitalist schooling at its best marked by the alienating principal: fuck the guy that helped you graduate that ensured you made it to the next step, the next position on the research/career ladder. Was it truth we were after as we began our study in the hopes of shedding light on some obscure fact? The mating principles of the mud-wasp or sexuality in the human male, are these black holes in the minds eye as rebellion takes the place of cultural logic and cultural truth? So progressive educators a reflection of the reality that is human destroy the earth and its atmosphere when their pedagogy ensures children learn the science necessary to produce industrial coatings, fertilizer and cyanide without also ensuring they acquire the depth of consciousness necessary to make connections between wearing a gold and diamond ring and the use of cyanide and strip mining for their production.

Pushing still further, why is Marxism such a word of abuse (McLaren and Jaramillo 2009)? Even the right can see the fluidity of accepting the changing condition of the system, what Lacan (2006) refers to as synthome of societies. Radical pedagogy ain't for the timid, it is a critical revolutionary praxis marked by the blood of Zapatistas, Che Guevara, Hugo Chávez and progressive intellectuals that understand a need for change from gripping tight to the cosmic orgone (Reich 1973) that is Capital; it does not allow for any competing principal or ideology. The search for truth is not about finding the source of all energy or a catalyzing principal.

It is the understanding that humans and objects share relationships, principles that adhere to organizational value and metaphysical conception and oscillations. The gangrene of racism, sexism, fascism and homophobia are human made (McLaren and Jaramillo 2009); they are the legacy of the left and of the right. What can be done about them is marked by the way intellectuals enact and participate in their personal and social praxis. A critical reflexivity that draws the kite-string of principal between the market need to produce chemicals for consumption, like Zyklon B, and the necessary day-to-day Socratic discursive practices doing more than shouting out to father capital in the classroom.

Human and environmental devastation are the end result of our social relations (Rodriguez 2009), which includes the needs and whims of markets and of the hyper-complex systems that are societies as they trade in material and human surplus value. The legacy of Marx and critical analyses are not the mere Utopic visions of a few stalwart, yet antiquated, intellectuals (McLaren 2007). They are a cultural critique positioning trade consciousness and social amnesia as the culprits on the market stage of global capitalist domination. Critical social theory does not disclude what is or what the agent knows or has known, like the conglomerate it promulgates all byproduct of human relations bad and good as actors that contribute to the enslavement of the individual and the devastation of the natural environment.

Dissemination, the symbol, the division of units and of labor, the structure of the phenomenon all bear as a derivative of the human and environmental condition of existence markings of each other. All symbols of experience return to the source; that is, we humans police ourselves and each other and we free ourselves and each other.

Closing Remarks

The global market occupies virtually every corner of the struggle for humanity (McLaren and Jaramillo 2007). Children in classrooms are the direct inheritors, as they grow to adulthood of the social and natural environment adults accept. War is class war, as those who reap the benefits, profit margins, on a global scale are never those with most at risk; the soldiers doing the killing in the fields benefit only so far as their use value is justified in controlling the world via the wholesale slaughter of 'enemy combatants.' These are children and adults in the so-called terrorist states who happen to be in the way of cementing capitalist social relations – whether copper, oil, timber or human interests. Furthermore, the human life span is far too short for any one human being to have an effect that significantly impacts the world market. We are far beyond the moment where the Molotov cocktail, the baton or a rock thrown by its self can cause the adoption among the human chain of a worldwide position for revolution. Even when riots occur, the 1960s, 1980s or 1990s, 2000s globally, the market fights individual citizens to a standstill. Hard to throw a rock when you are starving, or when you have to excavate rubble to recover and then bury your children. And yet the US has been successfully fought to a standstill, in the market by Cuba and Venezuela and at war by Afghanistan and Iraq. Why does a military that possesses the sole surviving global Air Force, Navy and Army continue to make war on people that return fire from horseback using muskets and single shot World War I era munitions? The war begun in 2003 was conceived over 10 years prior; in 2009, the US was still at war with, according to Gibson (2009), a military with no long history of defense no internal defense industry of note, no definable supply lines, no clear chain of command or central leadership. Can it be there is more to life and war than production or enslavement?

The classroom, as McLaren and Jaramillo (2009) relate and as Bencze and Alsop (2009) elaborate, were the last truly public domain where students and teachers could engage in a respite from the dominant ideology; to consider the social relations that exist and ways they affect the environment. According to David Hursch (2006), "[n]eo-liberals' desire not to intervene in markets and to focus on economic growth, primarily terms of consumption, has both significantly contributed to the environmental problems that we face and to global warming" (p. 5). The copper canyons in Utah were not put their by meteors, but by mining operations. The depletion of salmon and steelhead in the rivers and streams of California,

Oregon and Washington did not happen as a product of the ravages of time. Human constructed, petrochemicals, positions on the treatment of the environment as things existing solely for the purpose of providing the corporatocracy with surplus value created all of it.

Critical educators in and out of the classroom stand as a measure of change as the onslaught of neoliberalism continues. People cause the ravages of time to negatively affect the planet, surplus accumulation whether it is PCB's in the Hudson, ammonium nitrates at the mouths of the world's major rivers or the debris from surface and subsurface detonations of nuclear material. Yet there is another more insidious form of surplus accumulation with students in classrooms across the globe, the toll of curricula and pedagogies ensuring students leave classrooms functionally illiterate; capable only of reading and acting out the prescribed lives global capitalists have set. Human agency and enslavement result, as people live careless to the effect their actions have on the natural environment and each other. Critical pedagogues in the natural and social sciences do more than share information with their students. They leave a lasting imprint, a seed which may contribute to the production of knowledge. But, more importantly, offer an alternative to the living currently destroying the planet. Our outcome with students is, "a pedagogy, therefore, that can help students reconstruct the objective context of class struggle by examining the capitalist mode of production as a totality" (Allman et al. 2005). Critical pedagogy for STEM education then is working with students to link their human development and human potential with a collective global consciousness occurring outside of capital to resist the further infringement by the corporatocracy on their lives.

References

Allman, P., McLaren, P., & Rikowski, G. (2005). After the box people: The labor-capital relation as class constitution and its consequences for Marxist educational theory and human resistance. In P. McLaren (Ed.), *Capitalists and conquerors: A critical pedagogy against empire*. New York: Rowman & Littlefield.

Arendt, H. (1959). *The human condition*. Chicago: The University of Chicago Press.

Bencze, L., & Alsop, S. (2009). Anticapitalist/pro-communitarian science and technology education. *Journal for Activist Science & Technology Education, 1*(1), 65–84.

Bowles, S., & Gintis, H. (1976). *Schooling in capitalist America*. London: Routledge & Kegan Paul.

Durkheim, E. (1984). *The division of labor in society*. New York: The Free Press.

Freire, P. (1970). *Pedagogy of the oppressed*. New York: Continuum.

Freire, P. (1998). *Teachers as cultural workers: Letters to those who dare teach*. Boulder: Westview Press.

Freire, P. (2004). *Pedagogy of indignation*. Boulder: Paradigm.

Gibson, R. (2009). How shall we live as lambs among wolves? Reason-passion-power and organization. *Radical Notes*.

Giroux, H. A. (1997). *Theory, culture and schooling*. Boulder: Westview.

Giroux, H. A. (2001). *Theory and resistance in education*. London: Bergin & Garvey.

Giroux, H. A. (2009). Neo-liberalism, youth, and the leasing of higher education. In D. Hill & R. Kumar (Eds.), *Global neo-liberalism and education and its consequences* (pp. 30–53). New York: Routledge.
Hill, D. (2004). Books, banks and bullets: Controlling our minds-the global project of imperialists and militaristic neo-liberalism and its effect on education policy. *Policy Futures in Education, 2*(3), 504–522.
Hursch, D. (2006). Neo-liberalism and the high-jacking of globalization in education. *Radical Notes.*
Kosik, K. (1976). *Dialectics of the concrete: A study on problems of man and world.* Boston: Reidel Publishing.
Lacan, J. (2006). *Écrits* (B. Fink, Trans.). New York: W. W. Norton & Company.
Macedo, D. (2006). *Literacies of power*. Boulder: Westview.
McLaren, P. (2007). *Life in schools: An introduction to critical pedagogy in the foundations of education.* Boston: Pearson.
McLaren, P., & Jaramillo, N. E. (2007). *Pedagogy and praxis in the age of empire: Toward a new humanism.* Rotterdam: Sense Publishers.
McLaren, P., & Jaramillo, N. E. (2009). Not neo-Marxist, not post-Marxist, not Marxian: Reflections on a revolutionary critical pedagogy. *Cultural Studies/Critical Methodologies, 10*(3), 251–262. SAGE.
Rodriguez, A., & McLaren, P. (2014). Human right, states' rights, and linguistic apartheid. In P. W. Orelus (Ed.), *Affirming language diversity in schools and society: Beyond linguistic apartheid* (pp. 77–93). Routledge: New York.
Nagel, E., & Newman, J. R. (2001). *Gödel's proof.* New York: New York University Press.
Reich, W. (1973). *Cosmic superimposition & Ether god and devil* (T. Pol, Trans.). Toronto: Doubleday Canada Ltd.
Rodriguez, A. (2008). Toward a transformative teaching practice: Criticity, pedagogy and praxis. *International Journal of Learning, 15*(3), 346–352.
Rodriguez, A. (2009). On democracy and critical citizenship. *International Journal of Learning, 16*(6), 331–343.
Roth, W.-M. (1997). From every day science to science education: How science and technology studies inspired curriculum design and classroom research. *Science and Education, 6*, 373–396.
Roth, W.-M., & McGinn, M. K. (1998). Knowing, researching, and reporting science education: Lessons from science and technology studies. *Journal of Research in Science Teaching, 35*(2), 313–235.

Chapter 5
Becoming Part of the Solution: Learning about Activism, Learning through Activism, Learning from Activism

Derek Hodson

Abstract After making the case for an action-oriented science curriculum as a major component of education for responsible citizenship, the author contends that building such a curriculum has four key elements. First, learning about the issues, that is, focusing on the science and technology aspects of important socioscientific issues (SSI), recognizing the social, cultural and economic contexts in which they are located, developing the nature of science knowledge that builds robust understanding of contemporary scientific practice, and acquiring the media literacy necessary to access and read with critical understanding a wide variety of information sources. Second, learning to care about issues and the people impacted by them, including a focus on dealing with controversy, addressing values and developing concern for the views, needs and interests of others. Third, engaging and managing the powerful emotions often generated by SSI. Fourth, learning about sociopolitical action, taking action and evaluating action. For this key fourth element, the author advocates a 3-stage apprenticeship approach comprising modelling, guided practice and application.

Keywords Individual action • Collective action • Direct action • Indirect action • Apprenticeship

D. Hodson (✉)
Faculty of Education, University of Auckland, Private Bag 92601, Symonds Street, Auckland 1150, New Zealand
e-mail: dandsehodson@xtra.co.nz

L. Bencze and S. Alsop (eds.), *Activist Science and Technology Education*, Cultural Studies of Science Education 9, DOI 10.1007/978-94-007-4360-1_5,

Making the Case for an Action-Oriented Science Curriculum

If you are not part of the solution, you are part of the problem[1]

The past decade has seen a number of calls for a much more radical, politicized form of science and technology education in which students not only address complex and often controversial environmental and socioscientific issues (SSI) and formulate their own position concerning them, but also prepare for, and engage in, sociopolitical actions that they believe will 'make a difference,' asking critical questions about how research priorities in science are determined, who has access to science, how science could (and perhaps should) be conducted differently, how scientific and technological knowledge are deployed, whose voices are heard, whose reading of a situation or interpretation of an issue are considered in formulating policy, and how action can be taken at individual, group and community level in order to influence policy and practice (Roth and Désautels 2002; Hodson 2003, 2011; Alsop and Bencze 2012). This chapter addresses some of the issues relating to the establishment of this particular curriculum emphasis (to use Roberts' (1982) terminology). It looks at ways of enabling young people to be part of the solution to society's problems rather than contributing to them. It can be summed up as a plea for: (i) assisting and supporting students in understanding complex issues, including exploration of the complex sociopolitical context in which the problem/issue is located; (ii) resolving conflicts of interest, considering any moral-ethical dimensions the issue raises and establishing a personal view; and (iii) building a commitment to taking appropriate sociopolitical action, both individually and collectively.

What makes this kind of curriculum unique is its commitment to student action. The simple point is that it is almost always much easier to proclaim that one cares about an issue than to do something about it, and to do it consistently, coherently and effectively. An action-oriented curriculum is predicated on the premise that our opinions and values are worth very little until we live them. Rhetoric and espoused values won't bring about a reappraisal of policy, establish social justice with respect to SSI, or halt environmental degradation. Not only must we change our behaviour, we must take action to change the behaviour of others, and we must ensure that alternative voices and their underlying interests and values, are brought to bear on policy decisions.

There is no doubt that political apathy is increasingly widespread and that many citizens have lost faith and trust in politicians. It is also the case that opportunities to

[1] This quotation is variously attributed to Martin Luther King, Eldridge Cleaver and advertising guru Charles Rosner. Cleaver's exact words, in a speech delivered to the San Francisco Barristers Club in September 1968, were: "there is no more neutrality in the world. You either have to be part of the solution, or you're part of the problem".

participate in key decision-making have declined substantially with the rise of mega-corporations and the increasingly convoluted bureaucracies of local, regional and national governments. Teachers can play a key role in halting this decline in civic participation and firing up citizens to seize opportunities to take control of local matters and to influence national and international decision-making. If this is to happen on any substantial and meaningful scale, students currently in school need opportunities to work together, take responsibility and engage in activities designed to effect change. We need to cultivate a sense of community and develop an awareness of ties to others, obligations and responsibilities and we need to show students how to establish, support and sustain politically active communities. Advocates of STS and STSE education have long argued that it is important for students to learn that scientific/technological activity is influenced by a complex of social, political and economic forces, to formulate their own views on a range of contemporary issues and problems, and to care passionately about them. Erminia Pedretti and Joanne Nazir (2011) have described variations and shifts in the focus of STSE in terms of six "currents": *application/design* (practical problem solving through designing new technology or adapting old technologies); *historical* (understanding the sociocultural embeddedness of science and technology); *logical reasoning* (using a range of perspectives, including many outside science, to understand scientific and technological developments); *value-centred* (addressing the multidimensionality of socioscientific issues, including moral-ethical concerns); *sociocultural* (recognizing and critiquing science and technology as social institutions); *socio-ecojustice* (critiquing and addressing socioscientific issues through direct and indirect action). The position adopted here is that the curriculum needs to focus very overtly on the final two "currents". Students need to learn how to participate, and they need to experience participation. Moreover, they need to encourage others to participate, too: parents, grandparents, friends, relatives, neighbours, local businesses, etc. It is not enough for students to be armchair critics. As Bill Kyle (1996) puts it: "Education must be transformed from the passive, technical, and apolitical orientation that is reflective of most students' school-based experiences to an active, critical, and politicized life-long endeavour that transcends the boundaries of classrooms and schools" (p. 1). In words that would have substantial currency in my native North of England working class community, students need to "put their money where their mouth is!"; that is, they need to engage in action rather than just talk about it Hodson (2009). Indeed, all of us (students, teachers and other citizens) need to "put our money where our mouths are." With that in mind, I have much in sympathy with Mark Elam and Margareta Bertilsson's (2003) notion of the *radical scientific citizen*:

> The radical scientific citizen is fully prepared to participate in demonstrations... street marches, boycotts and sit-ins and other means of publicly confronting those ruling over science and technology... While the scientific citizen as activist may be taking a partisan position in defence of a particular individual or group in society, they are also understood as assuming a moral stance in defence of general ethico-political principles... which are accepted as existing through many different and conflicting interpretations... and subjecting them to continuous contestation. (p. 245)

Building a Curriculum: Learning About the Issues

Before proceeding to the substance of this chapter, it is important to make a number of key points about building an issues-based, action-oriented curriculum. The first concerns the selection of appropriate SSI and their organization into a coherent and theoretically justifiable programme. What should be the criteria of selection? Student interest? Perceived importance in contemporary society? Topicality? Cutting edge science and technology? Lively and public controversy? Ready availability of curriculum resources or conversely the lack of readily available material, thus requiring students to search for further knowledge and information (an important learning goal in itself)? My inclination would be to provide a judicious mix of all these categories, a mix of local, regional/national and global issues, together with a range of idiosyncratic personal interests. For me, coherence would be located in the selection of issues that contribute to rigorous consideration of seven areas of concern: human health; land, water and mineral resources; food and agriculture; energy resources, consumption levels and sustainability; industry (including manufacturing industry, the leisure and service industries, biotechnology, and so on); communications technology and transportation; ethics and social responsibility, including freedom, control and sponsorship in science and technology (Hodson, 1994, 2003).

No matter what the criteria of selection, students need scientific knowledge if they are to get to grips with SSI at any level beyond the merely superficial. Simple common sense tells us that content knowledge is crucial, and that those who know more about the topic/issue under consideration will be better positioned to understand the underlying issues, evaluate different positions, make an informed decision on where they stand in relation to the issue, and argue their point of view. Key questions concern the depth of knowledge required and the manner in which it should be acquired. It seems almost trite to state that the level of scientific knowledge needed is that which enables students to understand the nature of the problem and what might constitute appropriate evidence on which to base their decision-making, and that it will vary substantially from issue to issue, but that is simply the reality of the situation. Whether that scientific knowledge should be acquired through prior instruction or on a ‘need to know’ basis when dealing with a particular issue is best decided on an issue by issue basis. As is so often the case in education, there is no universal answer; different situations demand different approaches and different SSI create widely different knowledge needs. Much depends on whether the entire curriculum is given over to an SSI-oriented approach or SSI are included as occasional add-ons to an otherwise content-oriented curriculum, and on whether that particular science content is likely to be taught and utilized elsewhere in the curriculum.

Of course, no science curriculum can equip students with thorough first-hand knowledge of *all* the science underlying *every* important issue. Indeed, given the pace of scientific and technological development, some of the scientific knowledge students will need to know in order to make important decisions on the many important SSI they will encounter during their lifetimes has yet to be developed. However, we *do* know what knowledge, skills and attitudes are essential for appraising scientific reports, evaluating scientific arguments and moving towards a personal opinion about the science and technology dimensions of real world issues. It includes understanding of the status of scientific knowledge, the ways in which it is generated, communicated and scrutinized by the community of scientists, and the extent to which it can be relied upon to inform critical decisions about SSI. In other words, students need to have a clear understanding of what counts as *good* science, that is, a well-designed inquiry and a well-argued conclusion. They need to be able to interpret reports, make sense of disagreements, evaluate knowledge claims, scrutinize arguments, distinguish among facts, arguments and opinions, make judgements about good science, bad science and non-science, detect error, bias and vested interest, and so on – all the things we have come to know as nature of science (NOS) understanding.

Stein Kolstø (2001) sums up the NOS knowledge and understanding needed for addressing SSI in terms of eight major elements: (i) the ability to distinguish between science-in-the-making, where dispute, disagreement and uncertainty are to be expected, and ready-made science, on which we can rely; (ii) recognizing that sociocultural, political, economic and religious factors can impact on priorities for scientific research and development, and on the knowledge claims that are accepted; (iii) ability to evaluate the quality of scientific and statistical evidence, and to judge the appropriateness of anecdotal and experiential knowledge; (iv) ability to appraise the degree of support for a knowledge claim and the quality of the argument that establishes the warrant for belief; (v) a skeptical approach that includes both a critical, questioning stance and a commitment not to jump to conclusions until compelling evidence and arguments have been assembled; (vi) awareness of the importance of contextual factors when evaluating knowledge claims, including the social status of the actors and their institutional allegiance; (vii) sensitivity to the underlying values, ideologies and potential for bias in the design and reporting of scientific investigations; and (viii) awareness of the constraints that might limit the application of generalized theoretical knowledge to particular real world situations. With regard to reports of specific research studies, a simple checklist of questions can be enormously helpful. For example, who conducted the research and where was it conducted? How was the research funded? Was the research sponsored and, if so, by whom? What is being claimed? What evidence supports the claim? How was the evidence collected? How was the evidence interpreted? What assumptions are made and what theories are used in arguing from evidence to conclusion? Do the authors use well-established theory or do they challenge such theories? Are alternative interpretations and conclusions possible? What additional evidence would help to clarify or resolve issues? Have there been other studies conducted by these scientists or by others?

Because much of the information needed to address SSI is of the science-in-the-making kind, rather than well-established science, and may even be located at or near the cutting edge of research, it is unlikely that students will be able to locate all of it in traditional sources of information like textbooks and reference books. It will need to be accessed from academic journals, magazines, newspapers, TV and radio broadcasts, and Internet sources, thus raising important issues of *media literacy*. Being media literate means being able to access, comprehend, analyze, evaluate, compare and contrast information from a variety of sources and utilize that information judiciously and appropriately to synthesize one's own detailed summary of the topic or issue under consideration. It means recognizing that the deployment of particular language, symbols, images and sound in a multimedia presentation can each play a role in determining a message's impact, and will have a profound influence on its perceived value and credibility. It means being able to ascertain the writer's purpose and intent, determine any sub-text and implicit meaning, detect bias and vested interest. It means being able to distinguish between good, reliable information and poor, unreliable information. It involves the ability to recognize what Nicholas Burbules and Thomas Callister (2000) call *misinformation*, *malinformation*, *messed-up information* and *useless information*. Students who are media literate understand that those skilled in producing printed, graphic and spoken media use particular vocabulary, grammar, syntax, metaphor and referencing to capture our attention, trigger our emotions, persuade us of a point of view and, on occasions, by-pass our critical faculties altogether. They understand that material may be biased and may use a range of journalistic techniques such as emotive language, hyperbole and innuendo, provocative pictures and images, and emotionally manipulative background music, to persuade readers, viewers and listeners of a particular point of view.

Building a Curriculum: Learning to Care

The kind of curriculum being advocated here has a major concern with supporting students in their attempts to formulate their own opinions on important issues and establishing their own value positions, rather than with promoting official or textbook views (the prime motive of what Ralph Levinson (2010) calls the 'deficit view' of citizenship education). It focuses much more overtly than traditional STS or STSE education on values clarification, developing strong feelings about issues, addressing moral-ethical concerns, and actively thinking about what it means to act wisely, justly and rightly in particular social, political and environmental contexts. It is geared towards helping students to become committed to the fight to establish more socially just and environmentally sustainable practices and building the confidence, mindset, insights and skills necessary for effective and responsible change advocacy and change agency. It has much in common with the goals of Peace Education, Multicultural and Antiracist Education, Global Education and Humane Education. It begins with the fostering of self-esteem and personal well-being in each individual, and extends to acceptance of diversity in ideas, opinions, perspectives, practices and values, concern for the welfare of others, respect for the rights of others, building empathy and mutual trust, the pursuit of fairness, equity, justice and freedom, cooperative decision-making, creative resolution of disagreements and conflict between individuals, within and between communities, and throughout the world. It is driven by a deep commitment to *anti-discriminatory* education, that is, exposing the common roots of sexism, racism, homophobia, Eurocentrism and Westism (or Northism) in the tendency to dichotomize and generate a sense of *other*, and working actively to confront the 'us and them' mentality that invariably sees 'us' as the norm, the desirable and the superior. It culminates in a commitment to the belief that alternative voices can and should be heard in order that decisions in science and technology reflect wisdom and justice, rather than powerful sectional interests. Nicholas Maxwell (1984, 1992) defines wisdom as the capacity to realize what is of value in life for oneself and others. He continues as follows:

> In a world in which international affairs are conducted at the intellectual and moral level of gang warfare (as they all too often are), the mere provision of new knowledge and technology, dissociated from a more fundamental concern to help humanity resolve its conflicts and problems of living in more cooperative ways, is an obvious recipe for disaster. It merely increases our power to *act*, without at the same time increasing our power to act humanely, cooperatively and rationally... We urgently need a new, more rational kind of academic inquiry, which gives intellectual priority to the tasks of articulating our problems of living, proposing and critically assessing possible cooperative solutions. (1992, p. 207)

Many SSI are highly controversial: GM crops, governmental DNA banks, gene therapy, cloning, stem cell research, health hazards associated with mobile phones and overhead power lines, toxic waste disposal, euthanasia, abortion, nuclear power generation and nuclear weapons, deep space exploration, xenotransplantation, animal experiments, food irradiation, compulsory MMR vaccination, smart ID cards, priorities for deployment of scarce resources for medical services and for medical research, and ways to deal with ozone depletion, desertification, loss of biodiversity and other environmental crises. Controversy may be *internal* to science, that is, the scientific information required to formulate a judgement about it is incomplete, insufficient, inconclusive or extremely complex and difficult to interpret, or it may be *external* to science, that is, rooted in social, political, economic, cultural, religious, environmental, aesthetic and/or moral-ethical concerns, beliefs, values and feelings. The capacity to address internal controversy depends crucially on students' NOS knowledge and critical reading skills; the capacity and willingness to address external controversy hinges on a consideration of values and feelings, and on the ability to balance rationalistic reasoning with reasoning driven by emotions, feelings, personal experience and sociocultural influences.

Once a decision has been made to include externally controversial issues in the curriculum, teachers have to decide the most appropriate way to do so. Should they take a neutral position, adopt the devil's advocate role or try to present a balanced view? One form of neutrality, *affirmative neutrality*, describes a situation in which teachers present multiple sides of a controversy without revealing which side they support. In *procedural neutrality*, information about the controversy and different points of view are elicited from the students, possibly after opportunity for library-based or Internet-based research. Quite apart from the danger of encouraging relativism, where any idea is accepted as long as it is someone's opinion, neutrality is a position that seriously threatens the teacher's credibility as critic, guide and mentor.

The notion of even-handedness or presenting a 'balanced view' is also extremely problematic. What counts as balance? Whose judgement of balance and selection of perspectives is to count? Who decides what counts as relevant or not relevant, accurate or inaccurate, admissible or inadmissible, important or unimportant? Who decides what should be regarded as facts and what is deemed to be opinion? If all students express similar views, who will provide the alternatives? How should the teacher or the class respond to opinions that seem designed for no other reason than to shock, provoke or 'wind people up'? The key point is that *all* views embody a particular position, and that position needs to be rationalized and justified if indoctrination is to be avoided. Following some supposed notion of even-handedness prevents students from developing the critical skills necessary for judging the worth and validity of different positions, and requires teachers to give equal time, consideration and weight to views and arguments that are clearly not of equal merit.

Understanding the nature of controversy itself entails knowing that views may differ because they are based on different information, different interpretations of the same information or differences in worldviews, values, attitudes, interests, experiences, feelings or emotions. It entails knowing that different value judgements are sometimes a consequence of differences in moral codes or ethical principles deriving from different religious, political or philosophical positions. When addressing a *particular* controversial SSI, students need to ascertain the nature and extent of the disagreement. Is it a consequence of insufficient evidence, evidence of the 'wrong kind', evidence that is conflicting, confusing or inconsistent, or too complex and difficult to interpret? Is the problem of resolution located in the absence of clear criteria for making a judgment? Is it the case that different criteria point to different solutions or actions? And so on. They need to know that individual feelings and emotions or personal experiences can impact the ways in which issues are viewed, data are interpreted, conclusions evaluated and courses of action advocated. This applies to scrutiny of their own views, the views of other students, views expressed in curriculum materials, newspapers, Internet websites and so on, and the teacher's views. With regard to this latter point, it is my experience that confrontation of controversy invariably invites questions about the teacher's view. It is absurd for teachers to pretend that they don't have a view. It is deplorable for teachers to refuse to state that view when requested to do so, especially when they encourage or even require students to state theirs. Even if they choose to remain silent, teachers' views are likely to be evident to the more perceptive students from the questions they ask and the ways in which they respond to (or ignore) student comments, and through tone of voice, maintenance of eye contact (or not), and the ever-potent and revealing classroom body language.

While I acknowledge the right of an individual student to remain silent on a particular issue, I would not extend that privilege to teachers. I believe that students have a right to know their teacher's views on SSI addressed in the curriculum. However, I would not be supportive of teachers who used their own views as justification for excluding opportunities for students to address issues such as abortion, birth control, genetic engineering and cloning. I believe that it is incumbent on teachers to make provision for students to address a wide range of controversial SSI, particularly those in which they express an interest and those with direct impact on their lives. And I believe that it is incumbent on teachers to share their views on these matters with students and to make explicit the ways in which they have arrived at their particular position. It is also incumbent on teachers to adopt the same stance of critical reflection and open-mindedness that they demand of their students, and to be willing to change or modify their views in the light of new evidence, a new way of interpreting evidence, a reappraisal of underlying values, or whatever. Some years ago, Thomas Kelly (1986) proposed the broadly similar approach of "committed impartiality", in which teachers present multiple sides of an issue or argument and, at some stage, share their own views with the class. In my view, it is crucial that teachers identify, clarify and challenge the assumptions of *all* available positions (including their own), acknowledge the influence of sociocultural context, religious beliefs, emotions and feelings, address issues of rationality, equity and social justice, and encourage critical reflection. Kelly (1986) argued that when students are encouraged to debate and challenge their teacher's ideas without fear of sanctions, they not only develop argumentation skills, but also build the courage for social commitment.

It is inevitable that some teachers will lack confidence and expertise in handling unstructured, open-ended discussions on controversial issues, and it is unsurprising that teachers unfamiliar with them often express a concern, bordering on anxiety, that they will be accused of bias, and may possibly lay themselves open to charges of indoctrination. I would make two points in response. First, adoption of the critical approach discussed here (what Ratcliffe and Grace (2003) refer to as the "stated commitment" approach) constitutes a legitimate defence against such charges. As Ivan Snook (1975) reminds us, we are guilty of indoctrination when, and only when, we intend students to believe a proposition (or set of propositions) in the absence of, or despite/regardless of, the evidence. Or when we deliberately suppress or distort evidence to the contrary. What is proposed here is better described as "adopting a critical perspective". Second, the views of students often indicate the exact opposite, with many of the students with whom I have worked expressing the view that confronting socioscientific issues in this critical and collaborative way "opened my eyes to other perspectives", "helped me to sort out my own views" and "enabled me to think more clearly and more carefully" about such matters. Far from feeling that they had been indoctrinated, many students report that the approach provided a stabilizing framework within which their existing views could be accommodated, enriched and used more critically and more effectively.

Almost any discussion of a topical SSI is likely to raise questions not only about what we *can* or *could* do, but also about what is the *right* decision and what *ought* we to do? However, because we live in an increasingly pluralist society, we cannot assume a shared set of moral values and reaching agreement is likely to be difficult. One response is to allow the views of the majority to prevail – a position that necessarily disregards or marginalizes the needs, interests, values and rights of minorities. Even critical discourse between and among all interested parties may fail to bring about consensus, and if consensus were reached there would be no guarantee that it had reached the *right* answer. Ascertaining the right answer (what we *ought* to do) raises questions and concerns about morality (what it is right or wrong to do) and ethics (the reasons and justifications for judging these things to be right or wrong). I am certainly not advocating that students be required to follow a rigorous course in moral philosophy, any more than I would advocate the promotion of a morality based on a particular religion or set of laws and cultural precedents, but I am advocating that they be equipped with some intellectual tools for addressing and resolving contentious issues that cannot be solved solely by scientific considerations – at the very least, some basic understanding of egoism, consequentialist notions (including utilitarianism, deontological ethics, social construct theory (or social contract theory) and virtue ethics).

Patrick Fullick and Mary Ratcliffe (1996) describe a number of strategies that can help to direct student attention to the ethical concerns embedded in SSI and assist them in dealing with ethical dilemmas in a systematic and rational way. Strategies include: consequence mapping or "future wheels" (through which students are asked to consider a range of personal, social, economic, legal, environmental and ethical implications surrounding an issue and the possible responses to it); use of a goals-rights-duties framework (for each player or constituency involved in a controversy, students consider the intentions, rights/expectations and obligations towards others and the environment); and group discussions around carefully focused questions (oral or written questions direct student attention to the nature of the problem, possible solutions, reasons why one solution may be preferred to another, and stimulate reflection on students' own value positions). The New Zealand Biotechnology Learning Hub (www.biotechlearn.org.nz) provides support for students addressing ethical issues in the form of two interactive "thinking tools". The *ethics thinking tool* enables students to structure and evaluate their ideas in relation to four sets of ethical guidelines: benefits and harms; rights and responsibilities; freedom of choice; virtues. The *futures thinking tool* encourages students to consider the existing situation, analyze trends, identify the driving forces and causes of those trends, identify possible and probable futures, and select preferred futures. Use of these tools, together with a wide range of other teaching and learning strategies, is discussed by several authors in the edited collection: *Ethics in the Science and Technology Classroom* (Jones et al. 2010).

Engaging Emotions, Managing Emotions

Personal investment in an issue and commitment to problem solving and action derive, in part, from emotional involvement. The stronger one's emotional involvement, the more likely one is to take positive action – a situation that is well illustrated in students' responses to SSI when they impact directly on their own lives, or those of family members and people in the local community. Reliance on secondary experience, information and knowledge, which is likely to be the case for many students for many SSI, removes them emotionally from the issue and is likely to result in non-involvement and non-action. It easy to react to sudden and catastrophic change brought about by earthquakes and tsunamis, but environmental degradation and climate change are both slow and cumulative. There is a tendency, therefore, to over-estimate the long-term significance of hurricanes and earthquakes and seriously under-estimate the long-term significance of small increases in the mean temperature of the oceans. Large-scale, global environmental problems (such as ozone depletion, loss of habitat and greenhouse gas build-up) are not immediately tangible. We don't see it happen and so it "slips off the radar". The long time-lag between the emission of greenhouse gases and their effects on the climate impedes a proper understanding of the relationship. So, too, the anticipated time lag between any actions taken to reduce emissions and the positive effects they might produce. For many people in the affluent West/North, tangible impact is elsewhere: melting ice caps in Antarctica, rising sea levels in the islands of the South Pacific, pollution of waterways in China. For many people, the fact that the effects are not uniform across different parts of the world seems to be at variance with predictions that associate climate change with mean increase in temperature across the globe. Moreover, predictions by the IPCC and other bodies lose precision at finer geographical scales and so may seem to contradict local experience (González-Gaudiano and Meira-Cartea 2010). Thus, environmental degradation and climate change are seen as distant or future problems, not immediate and local ones. Despite repeated warnings from climate change scientists that the longer we delay measures to reduce greenhouse gas emissions the deeper and more irreversible will be the consequences, significant action at the political level is not forthcoming. Many aspects of SSI relating to health, resource use, industrial and scientific practice may also seem distant to students. Making these issues real means finding ways to stimulate, provoke, irritate, offend, outrage, amuse or delight students as a way of gaining their attention and building involvement and commitment. We need to find ways to make the impact more real, to precipitate feelings of fear, anger, sadness, pain, empathy, compassion and guilt, and link them to positive feelings of agency, control and empowerment. Emotional involvement can be fostered through case studies, drama and role play, literature, art, photographs, movies and music, site visits, interviews with those directly impacted, and so on. Interestingly, Benjamin Lester et al. (2006) have shown that carefully designed writing activities can also play an important role in developing personal investment in an issue and in increasing students' awareness of the need for sociopolitical action, especially when students assume the role of investigative journalist. Site visits (hospitals, factory farms, laboratories, etc.) and guided experiences in areas of ecological significance can play a profound role in raising awareness and engaging emotions. Best of all, of course, is direct engagement with locally based issues, as discussed below. It is important to note that informal learning experiences seem to be much more effective than formal schooling in bringing about awareness of issues, attitudinal shifts, values reorientation and willingness to engage in sociopolitical action.

A sense of wonder and feelings of empathy, respect and compassion towards other living things can also be fostered by such easily organized activities as investigating a rock pool, noting what lives in a wall or hedgerow, taking digital photographs to examine the feathers of birds in a suburban garden, watching a spider spin a web, observing insects through a magnifying lens or pond water under a microscope (see Lindemann-Matthies (2005) for further suggestions along these lines). Nor should we under-estimate the value of caring for pets, growing vegetables, observing activity in an ant colony and watching the dramatic events in the life history of frogs and butterflies. An important part of these experiences is the delight that students experience in becoming absorbed in their observations, the feelings of surprise at seeing the world in new ways, the thrill of encountering previously unfamiliar living organisms and habitats, recognizing new possibilities and seeing new relationships (Liston 2004). Mark Girod, Cheryl Rau and Adele Schepige (2003) refer to this kind of experience as "re-seeing":

> Re-seeing is an attempt to focus our perception on the nuance and detail of the world. Re-seeing requires that we look carefully when we might be tempted to assume we see everything. Re-seeing is also a disposition that causes us to ask questions of what we perceive, such as 'What's really going on here?' 'Why do things look the way they do?' And 'What kinds of things do I need to know more about to really re-see this?' (p. 579)

There is substantial evidence of the power of television, movies, drama, role-play, multi-media materials and language-based activities of various kinds to stimulate interest in an issue, provoke an emotional response, present alternative positions, challenge values and precipitate debate. Stories juxtapose different opinions, voices and perspectives, encouraging the reader (or listener) to deliberate, evaluate and decide on where they stand, or to adopt a different stance. Through stories, and especially through drama, students are stimulated to address issues and events from the perspectives of others, explore and develop understanding, establish new relationships and consolidate existing ones. In other words, engaging with narrative is as much a way of knowing ourselves as it is a way of understanding the views of others. Improvised drama enables students to enrich these explorations with personal experiences, thoughts and linguistic preferences. Poetry is an especially powerful means of generating emotional response and provoking the shift of perspective encapsulated in the notion of "re-seeing". Encouraging students to write poetry and stories creates opportunities for them to explore their ideas, express them in less formal language, manipulate and critique them by placing them in the mouths of others, explore ambiguity and uncertainty, wrestle with dilemmas and, crucially, express the way they feel about their ideas and the ideas of others.

The social context in which the student is located outside school is likely to be a major factor impacting learning. Rejecting knowledge and beliefs that are strongly held within social groups to which the student belongs, or wishes to belong, may be so emotionally stressful that it becomes virtually impossible. Similarly, accepting views that are in opposition to the dominant views within those groups is likely to be a formidable undertaking. The science teacher's job can be seen, in part, as helping students to gain an understanding of what, for many, are alien cultures (the subcultures of science, school and school science) and assisting them in moving freely and painlessly within and between these subcultures and the subcultures of home and community. It is fair to say that many teachers have seriously underestimated the difficulties faced by some students. As Jay Lemke (2001) comments, a student "spends most of every day, before and after science class, in other subject-area classes, in social interactions in school but outside the curriculum, and in life outside school. We have imagined that the few minutes of the science lesson somehow create an isolated and nearly autonomous learning universe, ignoring the sociocultural reality that students' beliefs, attitudes, values, and personal identities – all of which are critical to their achievement in science learning – are formed along trajectories that pass briefly through our classes" (p. 305).

It is also likely that addressing SSI in class will generate strong feelings and emotions, with students' views and assumptions being strongly influenced by personal experiences and the experiences of friends and family, and by socioculturally determined predispositions and worldviews. A student's sense of identity, comprising ethnicity, gender, social class, family and community relationships, economic status and personal experiences extending over many years, will necessarily impact on their values, priorities and preferences, and influence the ways in which they engage in discussion and the conclusions they reach. Teachers introducing SSI into the curriculum need to be sensitive to these influences and will need to assist students in dealing with potentially stressful and disconcerting learning situations. It is here that notions of *emotional intelligence*, *emotional literacy* and *emotional competence* can be helpful.[2] Although these three terms are closely related, Brian Matthews (2005) chooses to draw a distinction between the individualistic nature of emotional intelligence and the strongly social nature of emotional literacy. Thus, he argues, emotional intelligence refers to an individual's ability to perceive, describe, appraise and express emotions, understand emotions and emotional knowledge, access and/or generate appropriate feelings when they facilitate thought, or manage them productively when they might inhibit, while emotional literacy is the capacity to be receptive to a wide range of feelings, empathize with others, and continuously monitor the emotional climate in which one is located. Emotional competence may be seen as an amalgam of the two. In general, the goal of emotional literacy is awareness and management of one's emotions in both joyful and stressful situations, the confidence and self-assurance to understand one's own emotions, and the capacity to deal with them in a positive and intentional way. It is closely related to notions of self-awareness, self-image, self-esteem and sense of identity, and less directly with self-efficacy and agency.

Building a Curriculum: Learning to Act

The most distinctive feature of the issues-based approach advocated here is concern with students findings ways of putting their values and convictions into action, helping them to prepare for and engage in responsible action, and assisting them in developing the skills, attitudes and values that will enable them to take control of their lives, cooperate with others to bring about change, and work towards a more just and sustainable world in which power, wealth and resources are more equitably

---

[2] The following provide a good introduction to the key issues: Goleman (1985, 1996, 1998), Matthews et al. (2004a, b), Saarni (1990, 1999), Salovey and Meyer (1990), Salovey and Shaytor (1997), Steiner (1997), Sharp (2001), Zeidner et al. (2009).

shared. An interesting and thoughtful essay by Alexandra Dimick (2012) discusses a range of issues relating to science education for social justice in terms of three dimensions of student empowerment: social empowerment (provision of a safe, supportive and non-discriminatory environment within the classroom/school); political empowerment (recognition and critical examination of structures and forces that establish and maintain power inequities); and academic empowerment (access to key knowledge and skills, and the capacity to adapt them to specific SSI). Arguments employed in this chapter extend these ideas into the world outside the classroom.

Writing from the perspective of environmental education, Bjarne Jensen (2002) categorizes the knowledge that is likely to inform and promote sociopolitical action and pro-environmental behaviour into four dimensions: (i) scientific and technological knowledge that informs the issue or problem; (ii) knowledge about the underlying social, political and economic issues, conditions and structures, and how they contribute to creating social and environmental problems; (iii) knowledge about how to bring about changes in society through direct or indirect action; and (iv) knowledge about the likely outcome or direction of possible actions, and the desirability of those outcomes. Although formulated as a model for environmental education, Jensen's arguments are readily applicable to the kind of curriculum being advocated here. Little needs to be said about dimensions 1 and 2 in Jensen's framework beyond the discussion earlier in this chapter. With regard to dimension 3, students need knowledge of actions that are likely to have positive impact and knowledge of how to engage in them. It is essential that they gain robust knowledge of the social, legal and political system(s) that prevail in the communities in which they live, and develop a clear understanding of how decisions are made within local, regional and national government, and within industry, commerce, health authorities, environmental agencies and the military. Without knowledge of where and with whom power of decision-making is located, and awareness of the mechanisms by which decisions are reached, effective intervention is not possible. This kind of understanding requires a concurrent programme designed to achieve a measure of *political literacy*, including knowledge of how to engage in collective action with individuals who have different competencies, backgrounds and attitudes, but share a common interest in a particular SSI. Dimension 3 also includes knowledge of likely sympathisers and potential allies, and strategies for encouraging cooperative action and group interventions. What Jensen does not mention, but constitutes a key element of dimension 3 knowledge, is the NOS-oriented knowledge that would enable students to appraise the statements, reports and arguments of scientists, politicians and journalists, and to present their own supporting or opposing arguments in a coherent, robust and convincing way. Jensen's fourth category includes awareness of how (and why) others have sought to bring about change and entails formulation of a vision of the kind of world in which we (and our families and communities) wish to live. It is important for students to explore and develop their ideas, visions, dreams and aspirations for themselves, for their neighbours and families, and for the wider communities at the local, regional, national and global levels – a clear overlap with Futures Studies (Lloyd and Wallace 2004).

The likelihood of students becoming active citizens in later life is increased substantially by encouraging them to take action *now* (in school), by providing opportunities for them to do so, and by providing detailed examples of successful actions and interventions engaged in by others. It is also the case that all who become active at the collective level in later life have, at one time, engaged in individual action. With respect to an environmental focus (by way of illustration), suitable action might include any (or all) of the following: conducting surveys of dump sites, public footpaths and environmentally sensitive areas, monitoring pollution levels in local waterways, disseminating advice to householders, farmers and local industries on safe disposal of toxic waste, generating data for community groups such as birdwatchers and ramblers, establishing neighbourhood "nature watch" initiatives, instituting recycling programmes for glass, paper and aluminium cans, organizing consumer boycotts of environmentally unsafe products and practices, publishing newsletters, lobbying local government officials on policy matters and regulations (for example, traffic conditions and recreational facilities), working on environmental clean-up projects, establishing an "adopt a stream/creek/river/pond/lake" scheme, creating nature trails, conservation ponds and butterfly gardens, planting trees, building a community garden, designing, building and installing nesting boxes for endangered birds or bats, organizing a school "environmental awareness day", setting up a garbage-free lunch programme, assuming responsibility for environmental enhancement of the school grounds (including planting of indigenous species and encouragement of biodiversity), monitoring the school's consumption of energy and material resources in order to formulate more appropriate practices (including use of solar panels, for example), reducing water consumption through recycling schemes, monitoring use and disposal of potentially hazardous materials within the school, setting up a "green purchasing" network, and so on. Suitable actions on other matters might include: making public statements and writing letters, building informative Websites, writing to newspapers, organizing petitions and community meetings, working for local action groups and citizen working groups, making posters, distributing leaflets, demonstrating, making informative multimedia materials for public education, and exerting political pressure through regular involvement in local government affairs.

It is sometimes useful to distinguish between *direct* and *indirect* action. The former includes such things as recycling, cleaning up a stream or a beach, building a compost heap, using a bicycle rather than a car or bus, switching off lights, and using "green bags" at the supermarket; the latter includes compiling petitions, distributing leaflets, writing to newspapers and making submissions to the local council. Bjarne Jensen and Karsten Schnack (1997) characterize these two kinds of action in terms of orientation towards people-environment relations or people-people relations. Oddly, some environmental educators tend to de-value indirect actions as "mere classroom exercises", while extolling the virtues of direct action. Before reaching such a judgement we should look carefully at the likely *effectiveness* and *social significance* of particular actions, both in the short-term and long-term. While direct action can be enormously important and can have some significant impact, it can also divert attention from the root causes of the problem in our social, political and economic activities. It fails to confront the real causes and agents of environmental degradation, avoids critique and questioning, and "deceptively universalizes the different positions individuals have in relation to the distribution of environmental resources, risks, responsibilities, and decision-making power" (Lousley 1999, p. 299). It depoliticizes environmental problems and shifts the burden of responsibility onto individuals and families and away from governments, corporations, the policies that might have long-term and significant impact, and the political negotiations that might lead to change. Cleaning up a beach will have immediate beneficial impact, of course, but without an investigation of the causes and appropriate intervention aimed at those causes there will be no long-lasting solution. While recycling and buying so-called "environmentally friendly" products enable us to feel that we are doing something constructive, they may have no impact whatsoever on the underlying social and economic structures that have created the problems. Setting up a recycling programme may prolong the active life of one or two landfill sites but it doesn't address (and it certainly doesn't change) the unsustainable economy of resource use, production and consumption. Of course, indirect action needs to be *authentic* action: not just a classroom exercise in which a letter to an imaginary newspaper editor is composed, but a real letter to a real newspaper editor, to express real concerns or to make a series of real debating points or policy recommendations, or the preparation of a report for submission to a local government body, or provision of material assistance for an individual or group involved in a local dispute. It is important for students to recognize that individual actions can sometimes be fairly limited in their impact. Much more effective are collective actions that can exert pressure on governments (local, regional and national) to dismantle barriers to change and create alternative, more equitable and ethically and environmentally responsible policies and practices. Ronald Mitchell, Bradley Agle and Donna Wood (1997) remind us that changes at fundamental levels will only result when three key elements of persuasion are in place: *legitimacy* – perception that the action is desirable or morally right; *urgency* – the need for the issue to be addressed quickly; and *power* – the capacity to force another to do something counter to their current practice, using financial means, voting power, etc. It is group action that provides this final element. In other words, collective action is probably the only route to fundamental change in society.

Jensen and Schnack (1997) draw a distinction between *activities* and *actions*. For them, actions must be consciously chosen and focused on solutions to the problem or issue being addressed, or directed towards changing the conditions or circumstances that led to the problem(s). Thus, investigating nitrate and phosphate levels in waterways is classified as an activity; boycotting chemically-based agricultural products and promoting the use of organic fertilisers is classified as an action.[3] Conducting the

[3] In a later publication dealing with the problem-solving nature of actions, Jensen (2004) differentiates between *scientific investigative actions* (for example, student-initiated testing of pollution levels in waterways) and *social investigative actions* (for example, interviewing people in the local community about a socioscientific issue). Morgensen and Schnack (2010) provide further elaboration of these distinctions.

analysis, publicizing the data arising from it, identifying the likely cause of the pollution as run-off from local farms and parks, alerting farmers, ground maintenance staff in sports facilities, park keepers and domestic gardeners to both the causes and the adverse environmental impact of chemically-based products, making them aware of organic alternatives, and encouraging farm suppliers and garden centres to promote those organic alternatives, would be classified as a complex of activity, direct action and indirect action. From a curriculum or pedagogical perspective, some very obvious distinctions can be drawn between simple and quickly achieved actions (building nesting boxes or cleaning up a stream), those that require a sustained commitment over time (establishing and maintaining a fish hatchery or taking responsibility for managing a conservation area) and those that require a substantial level of political literacy (lobbying for policy changes, drafting legislation and filing law suits against those who violate existing codes and regulations). For these reasons, Wolff-Michael Roth (2010) is at some pains to distinguish among actions, activities and activism. In a more elaborate categorization, Paul Stern (2000) distinguishes among *environmental activism* (participation in activities organized by Greenpeace, Friends of the Earth, Sea Shepherd, etc.), *non-activist political behaviours* (voting, joining a community group), *consumer behaviours* (buying "green products", recycling), *ecosystem behaviours* (installing nesting boxes, cleaning up a stream) and *behaviours specific to our expertise or workplace* (reducing both resource consumption and waste generation). Another useful distinction drawn by Stern (2000) is that between "private sphere" actions and "public sphere" actions, a distinction that Susanne Menzel and Susanne Bögeholz (2010) extend into *activism* (e.g., participating in public demonstrations), *non-activist public sphere actions* (e.g., signing petitions), *private sphere actions* (e.g., green purchasing) and *public sphere actions* (e.g., fostering recycling in the work-force).[4] From a school perspective, there is also much value in distinguishing actions that are student initiated from those that are teacher initiated. Adapting the work of Sherry Arnstein (1969), Roger Hart (1992, 2008) outlines a "ladder of student participation", ranging from actions that are assigned by the teacher, through those that are decided by teachers after consultation with students, initiated by teachers but negotiated with students, initiated and directed by students, to those initiated by students and carried out in collaboration with adults outside school.

In light of this discussion of direct and indirect action, activities and action, individual versus collective action and teacher-initiated versus student-initiated activities and actions, I would argue that a key part of preparing for action involves identifying action possibilities, assessing their feasibility and appropriateness, ascertaining constraints and barriers, resolving any disagreements among those who will be involved, looking closely at the actions taken by others (and the extent

[4] Schusler, Krasny, Peters and Decker (2009) identify five forms of (environmental) action: physical environmental improvements (e.g., restoring natural habitats); community education (e.g., organizing festivals and information fairs, producing newsletters and multimedia materials); inquiry (e.g., surveys and mapping, environmental monitoring, etc.); public issue analysis and advocacy for policy change (researching an issue and making recommendations); and products or services (e.g., growing food in community gardens, working in a food bank).

to which they have been successful) and establishing priorities in terms of what actions are most urgently needed (and can be undertaken fairly quickly) and what actions are needed in the longer term. It is essential, too, that all actions taken by students are critically evaluated and committed to an action database for use by others. From a teaching perspective, it is important that care is taken to ensure both the appropriateness of a set of actions for the particular students involved and the communities in which the actions will be situated, and the overall practicality of the project in terms of time and resources. An action-oriented curriculum can generate considerable controversy and may provoke opposition from other teachers, school administrators, parents and members of the local community. While recycling, cleaning up the beach, harvesting rainwater, building nesting boxes or working in the local food bank or shelter for the homeless are safe, benign and non-controversial, challenging local councils, staging demonstrations, conducting vigils and organizing boycotts may raise parental anxiety levels, offend the local community and lead to sustained opposition. Teachers need to be prepared for backlash and they need courage to fly in the face of this opposition. Implementing this kind of curriculum is not "an easy ride".

Learning *about*, *through* and *from* Action

Milton McClaren and Bill Hammond (2005) draw distinctions among learning *about* action, learning *through* action and learning *from* action.

Learning *about* action focuses on learning the skills and strategies of sociopolitical action using movies, biographies and autobiographies, case studies and simulations, role-play and dramatic reconstructions. Providing students with examples of successful action taking, preferably involving other students, fosters the belief that they can change things, too. It is here that an action database can be especially useful, particularly in helping to overcome what Anneleen Kenis and Erik Mathijs (2012) call "strategy skepticism" (doubts about the efficacy of particular interventions). Students can learn from the experiences of others, that is, listening to and/or reading the stories of those who have been intimately involved in such projects. As John Forester (2006) comments:

> In fields of practical activity… we are likely to learn less from recipes or general rules for all times and places, and more from vivid examples of real work, exemplars of sensitive and astute practical-contextual judgement in families of messy and complex cases. Here we need not abstract lists of 'what worked' but specific stories of reconstructive action – not so much experimental results but experimental stories, not so much (or only) abstract rules (or principles alone) about 'what to do' as emotionally rich, morally entangled, contextually specified stories about 'how they really did it.' (p. 573)

What we need are detailed accounts of individual, group and community-based, action-oriented projects of varying degrees of complexity, sophistication and political involvement. We can learn a great deal from what Forester (2006) calls the "friction of actual practice", that is, learning through "the eyes and ears and hopes and dreads and difficulties and surprises of actual people, activists and ordinary – and often extraordinary – people who get up each morning and confront in messy detail the fears and distrust and scheming and self-interest and aggression of others that our abstractions otherwise so thinly render" (p. 569). Tania Schusler and colleagues (2009) provide much helpful advice on how researchers can go about gathering this kind of oral history by asking questions such as: What motivated or inspired you to engage in this kind of work? What were your goals, hopes and expectations? How did this project come about? At whose initiative? Who has been involved? What barriers and problems were encountered? How were they addressed? What successes have there been? What failures have there been? What have you learned? What surprised, delighted or disappointed you? What would you do differently if you were starting again? Would you do it again?

Learning *through* action comprises direct involvement in action-oriented projects outside the classroom that are likely to have tangible outcomes and consequences. While some projects may be chosen and organized by the teacher, especially in the early years, it is important to involve students as quickly as possible in selecting and planning for themselves the actions to be taken. It is important to involve students in local SSI-oriented research activities and support them in participating in community-based organizations that bring citizens together to grapple with serious local issues, particularly those issues often overlooked by government agencies. In confronting real local issues directly, students gain valuable first-hand experience of the ways in which competing social, political and economic interests impact on decision-making. Through participation in community-based activities, they gain access to ideas, experiences, people, institutions and sociopolitical structures that build both individual and collective capacity to address SSI and environmental issues in a responsible, thoughtful, critical and politically effective way, and build the commitment to engage in the struggle for greater freedom, equality and social justice. In other words, engaging in SSI-oriented actions builds a richer and deeper understanding of the issues, assists students in developing and refining their own views about them, and builds the capacity to engage productively in further actions. Sometimes a clear understanding of the scale and complexity of an issue, and clarification of one's own position regarding it, is a consequence of engagement in action rather than an essential precursor to it.

By focusing on the community and the issues and problems that residents confront in their everyday lives, students come to recognize their own experiences as shared, social and political. It is through direct experience of confronting social and environmental problems in the immediate community that public issues acquire personal meaning for young people – for example, working in shelters for the homeless, participating in breakfast programmes, doing volunteer work in hospitals, drug rehabilitation centres, HIV-AIDS support groups and homes for the elderly, involvement in environmental clean-up projects, renovating dilapidated homes, replanting degraded areas, building and maintaining community gardens, creating parks and conservation areas, organizing community festivals and information fairs, producing a local newsletter or community blog. As Paolo Freire

(1973) observed, people learn democracy through the exercise of democracy, or as James Banks (2004) says: "democracy is best learned in a democratic setting where participation is encouraged, where views can be expressed openly and discussed, where there is freedom of expression for pupils and teachers, and where there is fairness and justice" (p. 13). By engaging in public issues at the local level, students see democratic processes in action and learn how to engage in and negotiate them. By working alongside others, they learn about the demands and difficulties of taking action and learn to develop effective coping strategies. Research suggests that participation in these kinds of activities in childhood and adolescence is associated with levels of civic participation, community service and political activism in adulthood up to four times higher than the norm (Chawla and Flanders Cushing 2007). Carlson (2005) reports an interesting venture in Hampton, Virginia, in which the City Council established part-time, paid positions for two high school students to conduct regular surveys of public opinion, facilitate focus group discussions with their peers about local issues of concern, keep other young people informed about opportunities for community engagement, and help to facilitate that engagement. By the time of the next City Council election, some 2 years later, the voting participation rate among eligible young adults was 29 % higher than the national average.

We should make strenuous efforts to involve students in public hearings and town hall meetings, consensus conferences, study circles, focus groups, citizen juries/panels, negotiated rule-making forums, public/citizen advisory committees, and the like. It is through community-based activities that young people gain autonomy, a sense of worth, a sense of personal and civic identity, respect for other people's views, negotiation skills, and so on. When engaged with real problems and issues, students encounter real barriers and obstacles; working with community members to overcome these barriers cultivates students' competency and sense of competency. When people work together, there are opportunities for doing things that individuals would not even contemplate doing alone. By working on a sub-task within a group effort, individuals acquire a level of expertise that wouldn't be achieved alone, at least not so quickly and so painlessly. They also come into contact with perspectives on issues and problems that differ from their own. Sharing experiences, action strategies and success stories, as well as building friendships, can be inspirational and highly motivating, and can lead to lifelong sociopolitical activism. These experiences are immensely valuable because they run counter to the twenty-first century trend of growing social isolation of individuals and individual families, and counter to the values that underpin the pervasive competition and conspicuous consumption of contemporary society.

Learning *from* action occurs when students evaluate the plans, strategies, processes and outcomes of their own action projects and those of others. Debriefing, as some would call it, entails compilation of a record of what happened or what the students perceive to have happened, an attempt to say why (or why not), and reflection by all parties on the significance of the action for themselves and for the community. It almost goes without saying that the process is facilitated by keeping careful logs and journals, consulting with others, sharing experiences and feelings, and communicating with those who were not involved. There is value, too, in recruiting members of the community to act as critical reviewers.

Apprenticeship in Activism

Students can gain experience of action, and thereby learn *through* action and learn *from* action, via the familiar 3-phase apprenticeship approach.

- *Modelling* – the teacher demonstrates and explains the desired behaviour (in this case, social activism) and provides illustrative examples.
- *Guided practice* – students perform specified tasks within an overall action strategy with the help and support of the teacher.
- *Application* – students function independently of the teacher.

In short, it is assumed that students will become more expert in planning, executing and evaluating sociopolitical action by (i) observing teachers or other "experts" as they engage in action, (ii) practising the various sub-skills under controlled and supportive conditions, (iii) taking increasing levels of responsibility for planning and organizing the action, and (iv) engaging with critical evaluative feedback provided by the teacher and generated in inter-group criticism and discussion, and by means of intra-group reflection on the activity, both as it progresses and on completion. Initially, the teacher is responsible for planning the actions and directing the actions of students. However, if students are to achieve intellectual independence (Munby 1980), they must eventually take responsibility for their own learning and for planning, executing and reporting their own projects. In other words, learning as assisted performance must enable students, in time, to go beyond what they have learned and to use their knowledge and skills in creative ways for addressing different issues, solving novel problems and building new understanding. Consequently, alongside the modelled investigations, students should work through a carefully sequenced programme of exercises, during which the teacher's role is to act as learning resource, facilitator, consultant and critic. Complex problems and interventions can sometimes be broken down into a series of smaller problems and suitable interventions, including relatively simple activities in which careful planning by the teacher can almost guarantee that students will succeed, while also creating opportunities for students to act independently of the teacher, thus building confidence and enhancing motivation for assuming greater autonomy. These exercises provide opportunities for students to learn through a cycle of practice and reflection, and to achieve, with the careful assistance and support of the teacher, and of each other, a level of sophistication and performance they could not achieve unaided. In this guided practice phase, teacher and students are *co-activists*, with both parties asking questions, contributing ideas, making criticisms and lending support. Thus, the teacher's role shifts from instructor/demonstrator to director/facilitator. Clearly, such activities will only be productive if teachers and students are able to establish a learning community characterized by respect for diversity, trust, willingness to engage in collaborative learning and eagerness to contribute to the learning of all members of the community. Eventually, as students gain experience and take on increasing control of decision-making, they can proceed independently: choosing their own topics, problems and situations, and approaching them in their own way. From this point on, students are responsible for the whole process, from initial problem identification to final evaluation. Students identify the issue or problem, collect, organize and analyse information, define the problem from a variety of perspectives, formulate and appraise alternative actions, choose which action to take, develop and carry out a plan of action, and evaluate the outcome and the entire undertaking. As a consequence, they experience both "the excitement of successes and the agony that arises from inadequate planning and bad decisions" (Brusic 1992, p. 49). Throughout these activities the teacher's role is crucial: model activist, advisor, learning resource, facilitator, consultant, emotional support and critic. Also, because students are given the opportunity to experience failure as well as success, it is imperative that the class atmosphere is both forgiving and supportive.

Crucial to the notion of apprenticeship is a continuing dialogue about the way the activity is progressing, including frank discussion of problems encountered, avenues that prove fruitless, and barriers to progress that prove insurmountable. Crucial also, if the goal is for students to gain understanding of authentic sociopolitical action, is constant comparison between what students are doing in their project and what others have done (making use of an action database, as discussed above). By engaging in interventions and action-oriented projects alongside a trusted and skilled critic, students increase both their understanding of what constitutes sociopolitical action and their capacity to engage in it successfully. In other words, social activism is a reflexive activity: current knowledge and expertise informs and determines the conduct of the activity and, simultaneously, involvement in actions (and critical reflection on them) refines knowledge and sharpens expertise. In Carole Patemen's (1970) words, "participation develops and fosters the very qualities necessary for it; the more individuals participate the better they become able to do so" (p. 42). Erin Sperling (2009) urges teachers to introduce students to the idea of SMART plans, that is, plans that are **s**pecific, **m**easurable, **a**ttainable, **r**ealistic and **t**imely. Good advice, certainly, but the reality is that the smartest plans in prospect may prove otherwise in practice. And coming to that realization, and seeking to ascertain why the plan proved less than ideal, is a crucial part of the learning experience. So, too, of course, is simply engaging in action. Even though an action may not solve a problem, reach a satisfactory conclusion or have significant environmental impact, it may still have great significance in terms of personal growth, fostering positive attitudes and building commitment.

As well as teaching students the need to be sufficiently resilient and determined to try again, experiences of failure may also impress upon them the need to mobilize others and to engage in collective action. Collective actions are often more effective than individual actions and, in some circumstances, may be the only means of bringing about change. Interestingly, Roth (2009a) reformulates the Vygotskian notion of zone of proximal development to refer to what can be achieved through community-base collaborative efforts compared with what can be achieved by individuals. A key part of preparation for activism, then, is helping students to recognize, mobilize and coordinate the knowledge and skills that are distributed across communities. As Wolff-Michael Roth and Angela Calabrese Barton (2004) state:

> Education needs to focus on the individual as an integral and constitutive part of the collective, and on the distributed nature of knowledge and skill... (and) we have to begin thinking about the modes by which individuals with different expertise coparticipate in resolving the complex problems that their communities, countries, and humanity as a whole face today. (p. 13)

It is highly unlikely that all students will be motivated by the same issues, problems, experiences or situations. Nor will all students be in a position to make substantial changes to their daily behaviour and routines, and more particularly in the context of education at the school level, effect changes in their family's behaviour and routines. Individuals can also vary quite substantially in their disposition to act (that is, in terms of differences in knowledge, self-esteem, values, commitment, emotional involvement, and so on). Clearly, these variations make it difficult to plan an action-oriented curriculum for all. But there is no reason why we should expect different students and groups of students to participate in the same project. Different views and different priorities could (and possibly *should*) lead to involvement in different projects. One final point: it is important that a particular action is not viewed as an end in itself. Students need opportunities to evaluate the action taken, reflect on its nature and impact, and possibly re-formulate the action. The simple point is that an *action orientation* or *action competence* (as Jensen 2004, calls it) are established over time and are rooted in reflective practice.

Further Considerations

It is important to note that young people are more likely to participate in community activities if a parent, some other family member or a close friend is already active and/or expresses approval and gives them lots of support (Pancer and Pratt 1999; Fletcher et al. 2000). The prevalence of references by young people to the influence of parents and other role models in forming their views and attitudes is sufficient testimony to the influence of the old on the young. It is also the case that adults are more likely to join activist groups if their children are already involved or have expressed a desire to be involved. Political power rests with adults, but children can influence the ways in which that power is exercised. Consumer power rests (ultimately) with adults, though children can and frequently do exert considerable influence on family consumption practices. Codes of behaviour, language patterns and tastes in music, fashion and movies adopted by young people frequently act, over time, to shift older people's views and behaviours in a similar direction. On a closely related theme, Roy Ballantyne, John Fien and Jan Packer (1998, 2001a, b) have sought to exploit the ability of students to influence their parents or guardians, especially on environmental issues, by researching the elements in curricula that encourage students to talk with them (usually at mealtimes) about what they have been doing in school environmental education courses. Among the identified features that can easily be incorporated into recommendations for course design are: novel learning experiences, fieldwork, research-oriented homework assignments, discussion of easily-implemented pro-environmental behaviours (walking to school, taking shorter showers, turning off unneeded lights), student presentations at parents' evenings or public meetings, publicizing the programme in the local newspaper, conducting surveys and interviews in the community, and inviting local people to be guest speakers.

In short, effective sociopolitical action requires there to be a mutually supportive relationship between school and surrounding community. Traditional barriers between school and community need to be dissolved or rendered permeable, with community members present and active in the school, and students and teachers active and involved in the community. The difficulty of building such an atmosphere of interest, trust and shared responsibility and commitment should not be under-estimated. It requires strenuous effort on the part of teachers and students. As part of those efforts, we should be encouraging students to use their interest and skills in contemporary communications technology, especially social media such as *Facebook* and *Twitte*r, to establish networks, express concerns, share thoughts and spread messages about the need for action. New forms of ICT enable forms of participation that were not previously possible and may engage significant numbers of people who would previously have been uninvolved. They have the potential to facilitate the building of a more inclusive, participatory, socially just and politically engaged community. Students should be encouraged and enabled to use aspects of youth culture, particularly music, chat rooms and other communications media, to spread a youth-oriented message concerning civic and environmental responsibility. Music, television and the Internet are important sites for identity construction and reinforcement, gaining a better understanding of one's own experiences and the experiences of others, raising political awareness, and building the solidarity and sense of community that can lead to activism. For many urban youth in the United States, the rap music of hip-hop culture can be a particularly powerful vehicle, enabling them to put their feelings, emotions, needs, aspirations, hopes, joys, fears, disappointments and anger into a form that is respectful of their immediate cultural experiences and will be readily understand by their peers.[5] Shawn Ginwright and Julio Cammarota (2007), for example, describe how youth in Oakland (California) organized what they call "guerilla hip-hop" – impromptu mobile concerts with

[5] Christopher Emdin (2010) provides an extended discussion of the ways in which a hip-hop based and hip-hop inspired science curriculum can play a key role in creating opportunities for marginalized and under-served youth to participate successfully in science education. I am proposing an extension to social activism.

music, rapping, distribution of leaflets and other forms of political education in local parks, shopping malls, street corners and other places where young people hang out.

Karim Remtulla (2008) identifies three categories of online political activity: (i) *awareness and advocacy* usage sees the Internet and other forms of ICT as a means of accessing independent and alternative sources of information that may be ignored or suppressed by mainstream media – for example, the Independent Media Center (www.indymedia.org) and Wikinews (en.wikinews.org); (ii) *community-oriented sites* seek to spread awareness, share experiences and ideas and build networks within communities; and (iii) *action groups* endeavour to raise public support for actions related to specific issues (local, regional, national and international). We need to be aware, however, that social inequities and differential access to technological resources can restrict opportunities for those who are already marginalized, unheard or disregarded. They can be further disadvantaged, silenced or excluded from participation in addressing the very problems that most affect them. Massive efforts will be needed to ensure that online spaces, and the communities that use them, are open to everyone. Kelly Garrett (2006) discusses these and related matters in an extensive review of some key literature in sociology, political science and communications studies. Space precludes any further comment here, save to note that Garrett frames the discussion in terms of three interrelated factors: *mobilizing structures* (the mechanisms that enable individuals to organize and engage in collective action), *opportunity structures* (the conditions that facilitate or constrain activist behaviour), and *framing processes* (the ways in which messages are framed, contested or promoted, and disseminated).

In public meetings, ordinary people ("ordinary" in the sense of being non-experts) and students can sometimes feel intimidated or excluded by scientists and engineers (and by politicians and lawyers, too) who use overly technical language and present opinions as fact and options as restricted. This is where Chantal Pouliot's (2008) advice to teach very explicitly about three models of citizen involvement (*deficit*, *public debate* and *citizen involvement*) can be very helpful:

> The purpose of using the deficit, public debate and co-production models is not to augment the consensual character of discussions concerning SSI... it is to encourage citizen participation in the sociotechnical issues confronting society... it is to encourage students to develop a point of view concerning citizens' attitudes, interests and capacities (discursive and interpretative) that moves away from the deficit model; it is to prompt students to articulate representations that accord legitimacy to the statements and experience-based knowledge of citizens and to the collaboration of citizens in the process of producing scientific knowledge. (p. 68)

Even so, strenuous efforts will need to be made if all constituencies are to be represented and all voices heard. In many societies, it is the urban or rural poor, women and members of minority racial, cultural, ethnic and religious groups who are most likely to be excluded from public representation, and to have their needs, interests, views, attitudes, values and aspirations marginalized or ignored. We would do well to heed Gayatri Spivak's (1988) warning that the space for dialogue is invariably structured in exclusionary terms that prescribe who can speak, what they can speak about and how they will be heard. Within any group of participants, however carefully and sensitively recruited, there is unlikely to be a level playing field within which fully autonomous speakers can express their views. There is the ever-present danger that systemic inequalities will be activated and create opportunities for what Lisa Taylor (2008) calls "selective silencing". Even the venue for a public meeting can impact the demographics of the gathering, with location in a church hall, school hall, local RSA,[6] health centre, university lecture theatre or local council debating chamber playing a role in inclusion/exclusion and determining whose voices are heard. For example, on a Māori *marae* gender will be a key determinant of who speaks; in a community hall in Toronto, ethnicity will be influential in positioning the debate; in a village hall in the English countryside, it is likely to be social class that fixes the agenda. Participants need to be constantly vigilant lest activities undertaken in the name of participation result in patronizing tokenism rather than effective representation and participation of diverse groups; lest they reinforce social hierarchies, reflect the dominant hegemonic agenda, and distract attention from key issues of contention by insisting on early consensus. Despite good intentions and efforts to establish open and democratic processes, there is a danger that dominant individuals can (consciously or unconsciously) impose an agenda that supports particular versions of what is appropriate thought, behaviour and action. It is significant that following the large-scale national debate in the United Kingdom about the commercial growing of GM crops, involving a large number of local, regional and national events during the summer of 2003, the establishment of a Website that received 2.9 million hits and the return of 37,000 feedback forms (Irwin 2008), the final report concluded: "It is profoundly regrettable that the open part of the process, far from being a 'public debate', instead became a dialogue mainly restricted to people of a particular social and academic background" (House of Commons Committee 2003, p. 15). It is also the case that community-based groups can fracture around differences in gender, race-ethnicity, sexuality, age and class-based identities. Much skilful and sensitive work is needed to keep diverse groups working well. As Jeppe Laessøe (2010) comments, it is not simply a case of "top-down is bad, bottom-up is good". Rather, it is a case of struggling for the most appropriate and effective balance of experts and non-experts in any particular situation, and for procedures that ensure all views and voices are heard and given consideration.

Common sense tells us that not all community-based SSI-oriented activities will be successful in promoting, developing and sustaining an activist stance. There is an ever-present danger that actions reflect the teacher's agenda rather than the interests and concerns of the students and a danger that students merely "go through the motions" of engaging in action without any real commitment or sense of empowerment, simply to satisfy course requirements or meet the expectations of the

[6] In New Zealand, the Returned Serviceman's Association (RSA) is the equivalent of the RSL (Returned Serviceman's League) in Australia and the British Legion in the UK.

teacher. At the extreme, teachers may be led to compile a list of approved, scripted and "politically safe" actions in which to involve successive groups of students without ever engaging them in the critical debate that should precede and determine action. Cheryl Lousley's (1999) research on the activities of four urban secondary school environment clubs, established to focus attention on such endeavours as naturalizing the school grounds, planting trees, recycling and organizing an Earth Week Festival, shows that students are frequently directed towards uncontroversial issues, guided away from conflict, dissuaded from political debate and censored when their proposals seem likely to challenge school practices, local government policies or the interests of local businesses. In short, she says, "the hidden 'curriculum' of surveillance, regulation, and interrogation which structured the club experience taught the students not to rock the boat and it hints that the liberal-humanist offer of tangible, 'empowering' results – results which do not alter the relations of power and authority within the school and do not take up controversial and challenging issues – amounts to a false perception of 'making a difference' and an education in naïve conformism" (p. 297). In making a similar point, Venka Simovska (2008) distinguishes between *token* participation and *genuine* participation in terms of "focus" (specified content versus knowledge building through critique and reflection), "outcomes" (acceptance of a particular set of beliefs, values and behaviours versus student autonomy, critical consciousness and ability to address novel and complex issues) and "target of change" (individuals and their specific lifestyle versus individuals in context, taking account of inter-personal relations, sociocultural factors, moral-ethical dimensions and existing organizational structures). The same concerns run through Roth's (2009b) urging of teachers not to subordinate experience of activism to the more general aims of schooling.

Thus far, the choice of most teachers seems to have been to reflect (if not actively promote) the values, attitudes, ways of thinking and social structures that have fostered the economic, social and political systems responsible for current social and environmental crises. It is a matter of considerable urgency that we change the way we think, and change the science and technology education that has for too long maintained a particular way of thinking. For example, all the teachers interviewed by Randy McGinnis and Patricia Simmons (1999) felt so intimidated by the prevailing social climate that they expressed support for an SSI orientation but avoided controversial topics, especially those that might challenge religious views of a fundamental nature or the practices of local industries. Similarly and equally regrettably, Ali Sammel and David Zandvliet (2003) note that most approaches to SSI in school are conducted within teachers' perceptions of "politically acceptable limits".

In contrast, the primary thrust of the politicized science education being advocated here entails being critical of industrial, business, military and wider social practices, and where considered necessary, seeking change. Causing surprise, discomfort or offence to one or two parents, school officials, local residents or business interests is simply the price we have to pay in the struggle to create and sustain a better world and a more just, equitable and honourable society. It is imperative that teachers find the courage, enlist the support of others and mobilize the resources to be much more challenging, critical and politicized in their approach. From my point of view, it is enormously encouraging that the Qualifications and Curriculum Authority in the United Kingdom regard teachers as having a *duty* to prepare students to deal with controversial issues.

> Education should not attempt to shelter our nation's children from even the harsher controversies of adult life, but should prepare them to deal with such controversies knowledgeably, sensibly, tolerantly and morally. (QCA 1998, p. 56)

Avoiding controversial issues, especially those with very significant political dimensions, is regarded by many teachers as taking a neutral view. In reality, it is not neutral. Because it fails to confront and challenge the underlying sociopolitical causes of environmental problems, for example, it implicitly supports current social practices, current institutions and current values. Thus, it has to be regarded as education for social reproduction.[7] There is no such thing as political non-involvement. Non-involvement is, in itself, a form of involvement by default and constitutes implicit support for the dominant ideology. Avoiding political matters is, in effect, leaving it for others to decide. There is no doubt that teachers who promote sociopolitical involvement and develop students' action skills and competencies are riding a tiger, but it is a tiger that may well have to be ridden if we really mean what we say about education for civic participation. I do not seek to minimize the difficulties that teachers face in deciding a course of action. All I can do is urge teachers and students to be critical, reflective, robust in argument and sensitive to diverse values and beliefs, and above all to have the courage and strength of will to do what they believe is right and good and just. In the words of Alberto Rodriguez (2001), we need the courage to "expand our gaze… and rise to the challenge of becoming cultural warriors for social change" (p. 290).

References

Alsop, S., & Bencze, L. (2012). In search of activist pedagogies in SMTE. *Canadian Journal of Science, Mathematics, and Technology Education, 12*(4), 394–408.

Arnstein, S. R. (1969). Eight rungs on the ladder of citizen participation. *Journal of the American Institute of Planners, 35*(4), 216–224.

Ballantyne, R., Connell, S., & Fien, J. (1998). Students as catalysts of environmental change: A framework for researching intergenerational influence through environmental education. *Environmental Education Research, 4*(3), 285–298.

Ballantyne, R., Fien, J., & Packer, J. (2001a). Programme effectiveness in facilitating intergenerational influence in environmental education: Lessons from the field. *Journal of Environmental Education, 32*(4), 8–15.

Ballantyne, R., Fien, J., & Packer, J. (2001b). School environmental education program impacts upon family learning: A case study analysis. *Environmental Education Research, 7*(1), 23–37.

[7] There are, of course, many with a vested interest in social reproduction; there are many more who do not recognize that social reproduction is the outcome of most current educational practice.

Banks, J. (2004). *Diversity and citizenship education: Global perspectives*. San Francisco: Jossey-Bass.
Brusic, S. A. (1992). Achieving STS goals through experiential learning. *Theory into Practice, 31* (1), 44–51.
Burbules, N., & Callister, T. (2000). *Watch IT: The risks and promises of information technology*. Boulder: Westview Press.
Carlson, C. (2005). Youth with influence: The youth planner initiative in Hampton, Virginia. *Children, Youth and Environments, 15*(2), 211–216.
Chawla, L., & Flanders Cushing, D. (2007). Education for strategic environmental behavior. *Environmental Education Research, 13*(4), 437–452.
Dimick, A. S. (2012). Student empowerment in an environmental science classroom: Towards a framework for social justice science education. *Science Education, 96*(6), 990–1012.
Elam, M., & Bertilsson, M. (2003). Consuming, engaging and confronting science: The emerging dimensions of scientific citizenship. *European Journal of Social Theory, 6*(2), 233–251.
Emdin, C. (2010). *Urban science education for the hip-hop generation: Essential tools for the urban science educator and researcher*. Rotterdam: Sense Publishers.
Fletcher, A. C., Elder, G., & Mekos, D. (2000). Parental influences on adolescent involvement in community activities. *Journal of Research on Adolescence, 10*(1), 29–48.
Forester, J. (2006). Exploring urban practice in a democratising society: Opportunities, techniques and challenges. *Development Southern Africa, 23*(5), 569–586.
Freire, P. (1973). *Education for critical consciousness*. New York: Continuum.
Fullick, P., & Ratcliffe, M. (Eds.). (1996). *Teaching ethical aspects of science*. Totton: Bassett Press.
Garrett, R. K. (2006). Protest in an information society: A review of literature on social movements and new ICTs. *Information, Communication and Society, 9*(2), 202–204.
Ginwright, S., & Cammarota, J. (2007). Youth activism in the urban community: Learning critical civic praxis within community organizations. *International Journal of Qualitative Studies in Education, 20*(6), 693–710.
Girod, M., Rau, C., & Schepige, A. (2003). Appreciating the beauty of science ideas: Teaching for aesthetic understanding. *Science Education, 87*(4), 574–587.
Goleman, D. (1985). *Emotional intelligence: Why it can matter more than IQ*. New York: Bantam Books.
Goleman, D. (1996). *Emotional intelligence*. London: Bloomsbury Publishing.
Goleman, D. (1998). *Working with emotional intelligence*. New York: Bantam Books.
González-Gaudiano, E., & Meira-Cartea, P. (2010). Climate change education and communication: A critical perspective on obstacles and resistances. In F. Kagawa & D. Selby (Eds.), *Education and climate change: Living and learning in interesting times* (pp. 13–34). London: Routledge.
Hart, R. (1992). *Children's participation: From tokenism to citizenship* (UNICEF Innocenti Essays, No.4). Florence: International Child Development Centre, UNICEF.
Hart, R. A. (2008). Stepping back from 'the ladder': Reflections on a model of participatory work with children. In A. Reid, B. B. Jensen, J. Nikel, & V. Simovsla (Eds.), *Participation and learning: Perspectives on education and the environment, health and sustainability* (pp. 19–31). New York: Springer.
Hodson, D. (1994). Seeking directions for change: The personalisation and politicisation of science education. *Curriculum Studies, 2*, 71–98.
Hodson, D. (2003). Time for action: Science education for an alternative future. *International Journal of Science Education, 25*(6), 645–670.
Hodson, D. (2009). Putting your money where your mouth is: Towards an action-oriented science curriculum. *Journal of Activist Science & Technology Education, 1*(1), 1–15.
Hodson, D. (2011). *Looking to the future: Building a curriculum for social activism*. Rotterdam: Sense Publishers.
House of Commons, Environment, Food and Rural Affairs Committee. (2003). *Conduct of the GM public debate. Eighteenth report of session 2002–2003*. London: HMSO.

Irwin, A. (2008). STS perspectives on scientific governance. In E. J. Hackett, O. Amsterdamska, M. Lynch, & J. Wajcman (Eds.), *The handbook of science and technology studies* (pp. 583–607). Cambridge, MA: MIT Press.
Jensen, B. B. (2002). Knowledge, action and pro-environmental behaviour. *Environmental Education Research, 8*(3), 325–334.
Jensen, B. B. (2004). Environmental and health education viewed from an action-oriented perspective: A case from Denmark. *Journal of Curriculum Studies, 36*(4), 405–425.
Jensen, B. B., & Schnack, K. (1997). The action competence approach in environmental education. *Environmental Education Research, 3*(2), 163–178.
Jones, A., McKim, A., & Reiss, M. (Eds.). (2010). *Ethics in the science and technology classroom: A new approach to teaching and learning*. Rotterdam: Sense Publishers.
Kelly, T. (1986). Discussing controversial issues: Four perspectives on the teacher's role. *Theory and Research in Social Education, 14*, 113–138.
Kenis, A., & Mathijs, E. (2012). Beyond individual behaviour change: The role of power, knowledge and strategy in tackling climate change. *Environmental Education Research, 18*(1), 45–65.
Kolstø, S. D. (2001). Scientific literacy for citizenship: Tools for dealing with the science dimension of controversial socioscientific issues. *Science Education, 85*(3), 291–310.
Kyle, W. C. (1996). Editorial: The importance of investing in human resources. *Journal of Research in Science Teaching, 33*, 1–4.
Laessøe, J. (2010). Education for sustainable development, participation and socio-cultural change. *Environmental Education Research, 16*(1), 39–57.
Lemke, J. L. (2001). Articulating communities: Sociocultural perspectives on science education. *Journal of Research in Science Teaching, 38*(3), 296–316.
Lester, B. T., Li, M., Lee, O., & Lambert, J. (2006). Social activism in elementary science education: A science, technology, and society approach to teach global warming. *International Journal of Science Education, 28*(4), 315–339.
Levinson, R. (2010). Science education and democratic participation: An uneasy congruence? *Studies in Science Education, 46*(1), 69–119.
Lindemann-Matthies, P. (2005). 'Loveable' mammals and 'lifeless' plants: How children's interest in common local organisms can be enhanced through observation of nature. *International Journal of Science Education, 27*(6), 655–677.
Liston, D. (2004). The allure of beauty and the pain of injustice in learning and teaching. In D. Liston & J. Garrison (Eds.), *Teaching, learning and loving: Reclaiming passion in educational practice* (pp. 101–116). New York: RoutledgeFalmer.
Lloyd, D., & Wallace, J. (2004). Imaging the future of science education: The case for making futures studies explicit in student learning. *Studies in Science Education, 39*, 139–177.
Lousley, C. (1999). (De)Politicizing the environment club: Environmental discourses and the culture of schooling. *Environmental Education Research, 5*(3), 293–304.
Matthews, B. (2005). Emotional development, science and co-education. In W. W. Cobern, K. Tobin, H. Brown-Acquay, M. Espinet, G. Irzik, O. Jegede, L. R. Herrara, M. Rollnick, S. Sjøberg, T. H-L, & S. Alsop (Eds.), *Beyond Cartesian Dualism: Encountering affect in the teaching and learning of science* (pp. 173–186). Dordrecht: Springer.
Matthews, G., Zeidner, M., & Roberts, R. D. (2004a). *Emotional intelligence: Science and myth*. Cambridge, MA: MIT Press.
Matthews, G., Roberts, R. D., & Zeidner, M. (2004b). Seven myths about emotional intelligence. *Psychological Inquiry, 15*(3), 179–196.
Maxwell, N. (1984). *From knowledge to wisdom*. Oxford: Basil Blackwell.
Maxwell, N. (1992). What kind of inquiry can best help us create a good world? *Science, Technology & Human Values, 17*(2), 205–227.
McClaren, M., & Hammond, B. (2005). Integrating education and action in environmental education. In E. A. Johnson & M. J. Mappin (Eds.), *Environmental education and advocacy: Changing perspectives of ecology and education* (pp. 267–291). Cambridge: Cambridge University Press.

McGinnis, J. R., & Simmons, P. (1999). Teachers' perspectives on teaching science-technology-society in local cultures: A sociocultural analysis. *Science Education, 83*(2), 179–211.

Menzel, S., & Bögeholz, S. (2010). Values, beliefs and norms that foster Chilean and German pupils' commitment to protect biodiversity. *International Journal of Environmental & Science Education, 5*(1), 31–49.

Mitchell, R. K., Agle, B. R., & Wood, D. J. (1997). Toward a theory of stakeholder identification and salience: Defining the principle of who and what really counts. *Academy of Management Review, 22*(4), 853–886.

Morgensen, F., & Schnack, K. (2010). The action competence approach and the 'new' discourses of education for sustainable development, competence and quality criteria. *Environmental Education Research, 16*(1), 59–74.

Munby, H. (1980). Analyzing teaching for intellectual independence. In H. Munby, G. Orpwood, & T. Russell (Eds.), *Seeing curriculum in a new light: Essays from science education* (pp. 11–33). Toronto: OISE Press.

Pancer, S. M., & Pratt, M. W. (1999). Social and family determinants of community service involvement in Canadian youth. In M. Yates & J. Youniss (Eds.), *Roots of civic identity* (pp. 32–55). Cambridge: Cambridge University Press.

Pateman, C. (1970). *Participation and democratic theory*. Cambridge: Cambridge University Press.

Pedretti, E., & Nazir, J. (2011). Currents in STSE education: Mapping a complex field, 40 years on. *Science Education, 95*(4), 601–626.

Pouliot, C. (2008). Students' inventory of social actors concerned by the controversy surrounding cellular telephones: A case study. *Science Education, 92*(3), 543–559.

Qualifications and Curriculum Authority (QCA). (1998). *Education for citizenship and the teaching of democracy in schools*. London: QCA.

Ratcliffe, M., & Grace, M. (2003). *Science education for citizenship: Teaching socio-scientific issues*. Maidenhead: Open University Press.

Remtulla, K. A. (2008). Democracy or digital divide? The pedagogical paradoxes of online activism. In D. E. Lind & P. R. Carr (Eds.), *Doing democracy: Striving for political literacy and social justice* (pp. 267–280). New York: Peter Lang.

Roberts, D. A. (1982). Developing the concept of "curriculum emphases" in science education. *Science Education, 66*(2), 243–260.

Rodriguez, A. J. (2001). Courage and the researcher's gaze: (Re)defining our roles as cultural warriors for social change. *Journal of Science Teacher Education, 12*(3), 277–294.

Roth, W.-M. (2009a). Activism or science/technology education as a by product of capacity building. *Journal of Activist Science & Technology Education, 1*(1), 16–31.

Roth, W.-M. (2009b). On activism and teaching. *Journal of Activist Science & Technology Education, 1*(2), 33–47.

Roth, W.-M. (2010). Activism: A category for theorizing learning. *Canadian Journal of Science, Mathematics, and Technology Education, 10*(3), 278–291.

Roth, W.-M., & Calabrese Barton, A. (2004). *Rethinking scientific literacy*. New York: RoutledgeFalmer.

Roth, W.-M., & Désautels, J. (2002). *Science education as/for sociopolitical action*. New York: Peter Lang.

Saarni, L. (1990). Emotional competence: How emotions and relationships become integrated. In R. A. Thompson (Ed.), *Socioemotional development: Nebraska symposium on motivation* (Vol. 36, pp. 115–182). Lincoln, NE: University of Nebraska Press.

Saarni, C. (1999). *The development of emotional competence*. New York: Guilford Press.

Salovey, P., & Meyer, M. V. (1990). Emotional intelligence. *Imagination, cognition and personality, 9*, 185–211.

Salovey, P., & Shaytor, D. (Eds.). (1997). *Emotional development and emotional intelligence: Educational implications*. New York: Basic Books.

Sammel, A., & Zandvliet, D. (2003). Science reform or science conform: Problematic epistemological assumptions with/in Canadian science reform efforts. *Canadian Journal of Science, Mathematics, and Technology Education, 3*(4), 513–520.

Schusler, T. M., Krasny, M. E., Peters, S. J., & Decker, D. J. (2009). Developing citizens and communities through youth environmental action. *Environmental Education Research, 15*(1), 111–127.

Sharp, P. (2001). *Nurturing emotional literacy*. London: David Fulton.

Simovska, V. (2008). Learning *in* and *as* participation: A case study from health-promoting schools. In A. Reid, B. B. Jensen, J. Nikel, & V. Simovska (Eds.), *Participation and learning: Perspectives on education and the environment, health and sustainability* (pp. 61–80). New York: Springer.

Snook, I. A. (1975). *Indoctrination and education*. London: Routledge & Kegan Paul.

Sperling, E. (2009). 'More than particle theory': Action-oriented citizenship through science education in a school setting. *Journal of Activist Science & Technology Education, 1*(2), 12–30.

Spivak, G. C. (1988). Can the subaltern speak? In C. Nelson & L. Grossberg (Eds.), *Marxism and the interpretation of culture* (pp. 271–313). Chicago: University of Chicago Press.

Steiner, C. (1997). *Achieving emotional literacy*. London: Bloomsbury.

Stern, P. C. (2000). Toward a coherent theory of environmentally significant behavior. *Journal of Social Issues, 56*(3), 407–424.

Taylor, L. K. (2008). Beyond 'open-mindedness': Cultivating critical, reflexive approaches to democratic dialogue. In D. E. Lind & P. R. Carr (Eds.), *Doing democracy: Striving for political literacy and social justice* (pp. 159–176). New York: Peter Lang.

Zeidner, H., Matthews, G., & Roberts, R. D. (2009). *What we know about emotional intelligence*. Cambridge, MA: MIT Press.

Chapter 6
From Promoting the Techno-sciences to Activism – A Variety of Objectives Involved in the Teaching of SSIs

Laurence Simonneaux

Abstract The educational trend referred to as Socio-Scientific Issues (SSIs) is gradually spreading internationally. It involves decisions in fields of science (and technology) that have controversial impacts on societies (Kolsto SD, Int J Sci Educ 23:877–901, 2001; Sadler TD, Chambers FW, Zeidler DL, Int J Sci Educ 26 (4):387–410, 2004; Zeidler DL, Walker K, Ackett W, Simmons M, Sci Educ 27:771–783, 2002). A parallel can be drawn here with the educational school of thought known as 'Science-Technology-Society' (STS). However, SSIs are not always introduced into schools in the form of an issue or a controversy; nor are they systematically related to current events. In the literature on the teaching of SSIs we can observe very different objectives. There are many different dimensions to the concept of an SSI. Similarly there is a variation in the extent to which teachers 'heat up' or 'cool down' these issues.

Keywords Socially acute questions • Scientific and political education

Variation in Educational Objectives

Referring to Fig. 6.1 (below), at the 'cold end,' an integration of SSIs into a teaching programme is used to motivate students learning science, or even to convince them of the merits of the techno-sciences. At the 'hot end' of the continuum, the teaching focus goes beyond the purpose of developing science conceptual and procedural knowledge to the nurturing of activist commitments amongst learners. Between these two extremities, there is a continuum of educational priorities. At the 'cold end', the chosen context may give the illusion that an SSI is being dealt with: the teacher may use the setting of a fictional environmental problem that he/she expects to be solved

L. Simonneaux (✉)
Ecole Nationale de Formation Agronomique, BP 22687,
31326 Castanet-Tolosan Cedex, France
e-mail: laurence.simonneaux@educagri.fr

L. Bencze and S. Alsop (eds.), *Activist Science and Technology Education*,
Cultural Studies of Science Education 9, DOI 10.1007/978-94-007-4360-1_6,

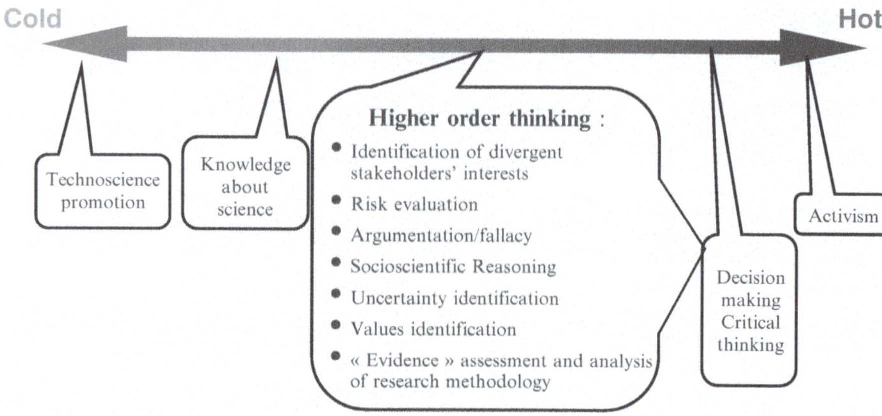

Fig. 6.1 Variation in objectives beyond SSI education

using made-up empirical data and, more particularly, adequate and stabilized concepts. The reasoning only develops at the heart of conventional, 'sedimented,' or 'neutralized' scientific notions and concepts. Introducing SSIs into the classroom may foster student motivation by addressing questions related to 'real life' but, sometimes, the only objective is to teach the underlying stabilized scientific concepts and not to problematize the issues. At times, even, the aim is to reassure the students and convince them of the merits of the techno-sciences.[1]

Midway between the two extremes, the teaching objective may be to develop a better understanding of the nature of the sciences. As we approach the 'hot end', SSIs are dealt with as controversial subjects in research, in the world of work and in society. The educational challenge is to enable students to develop informed opinions on these issues, to be able to debate such issues, to be capable of making choices with respect to preventive measures and to intelligent use of new techno-sciences. Given the increasing importance of many SSIs (biotechnology, nanotechnology, food security, climate change, health and environmental issues linked to agriculture, etc.), each student is already having to or will have to make decisions on such issues and schools should, thus, help them to be informed citizens. In this respect, SSI education also contributes to 'education for'; such as in: sexuality

[1] The concept of technosciences aims to bridge the gap between science and technology due to the need to think scientific discoveries and technological inventions in the same social context in order to account for their strong interactions. The term was introduced in the 1970s by the Belgian philosopher Gilbert Hottois. It has become commonplace in the 1990s. It was used by the philosopher Hans Jonas and the sociologist Bruno Latour.

Two characteristics of technosciences are generally highlighted: operability and circularity between knowledge and instruments. Technosciences not only observe the real, but use it, modify it. Circularity means that sciences produce technologies which in turn produce knowledge. Using the term technosciences can highlight an emerging phenomenon: from modern sciences have gradually emerged technosciences by which our perceptual abilities and possibilities of action on the real were significantly increased.

education, education for health, education for safety, education for the environment. 'Educations for' focus on complex issues involving uncertainties, which inextricably link scientific, economic, political and social questions that challenge values and ethics. These questions are at the heart of the problem of teaching and learning in an uncertain world influenced by the development of techno-sciences, and by environmental and health crises. It is not only the experts who make decisions on SSIs; all citizens are involved (consumers, voters, legislators, etc.). Furthermore, it is impossible to arrive at one single valid and rational decision because conflicting interests produce differing decisions.

The challenge lies in elaborating high level cognitive procedures: identifying the conflicting interests of stakeholders, evaluating the risks and uncertainties, generating debate and pinpointing the fallacies, cultivating socio-scientific reasoning, ethical judgment, identifying the actors' values, assessing the evidence and critically analyzing the research methodology and the expertise, etc. These procedures can contribute to the development of critical thinking and decision making. When critical thinking occurs, the focus moves towards the 'hot end' with a focus on promoting an engaged citizenship. As named by Edgar Morin (1998), specialist of complexity, the issue raised is "*an historical and henceforth crucial problem of cognitive democracy*" (p. 17). SSIs, in Edgar Morin's terms, are 'polydisciplinary', multidimensional, transnational and, in a context of increasing globalisation, planetary in nature. I believe that this didactic approach fits Edgar Morin's analysis well, in that it is education based on "the necessity of reinforcing critical thinking by linking knowledge to doubt, by integrating particular knowledge in a global context and using it in real life, by developing individuals' ability to deal with fundamental problems with which they are confronted in their own historical epoch" (p. 17).

It is possible to identify different objectives that link SSIs to high level thinking (argumentation, epistemic practices, evidence assessment, socioscientific reasoning, critical thinking). In the literature on learning science and, more particularly, on SSIs, it is generally agreed that fundamental importance should be given to debate and notably to the use of evidence to back up the different arguments (Erduran and Jiménez-Aleixandre 2008). A variety of argumentation and reasoning constructs and associated rubrics and scales are used to analyze students' productions about SSIs. Sandoval (2003) specifically studied the epistemic practices used by students when formulating their scientific explanations. More recently, Bravo Torija (2012) analyzed the epistemic levels mobilized by students in order to settle an SSI concerning marine resources management and monitored the discursive moves indicating the passage from one epistemic level to another. However, it is essential to combine these analyses, which are based on mobilizing scientific concepts and empirical data, with the analyses of interdisciplinary or even a-disciplinary reasoning, which are in line with a more authentic approach to the inherent complexity of real-life contexts.

In order to grasp the controversies of SSIs, it is interesting to refer to Habermas's (1987) framework, which distinguishes between *communicative*, *strategic*, *normatively-regulated* and *dramaturgical* action. According to him, communicative action presents itself as an interactive activity aimed at gaining mutual understanding and whose function is to coordinate the actions between participants; this is ideally what is expected in a debate over a controversy. In strategic action, each participant calculates the means and the ends to win the argument by anticipating the decisions of the other participants who are aiming to pursue their own goals. In the case of normatively-regulated action, sometimes called regulated action, the individual orients her/his action according to a common set of values and, in so-doing, conforms to a social norm. In the case of dramaturgical action, the actor tries to influence the other person on an emotional level.

Habermas uses the 'lifeworld' concept (i.e., the shared set of common understandings and values constituted for each individual by his culture and language) to interpret inter-subjective discourse. Habermas distinguishes three different 'worlds' used as reference when claiming the validity of arguments: the *objective* world rooted in the claim to truth (or accuracy) and based on empirical data and 'scientific truth', the *social* world rooted in the claim to rightness (or justice) based on social norms and finally the *subjective* world rooted in the claim to veracity (or sincerity) and based on personal experience. For Habermas, speakers should, ideally, refer to all three worlds – as in the case of communicative action. But, where normative-regulated action is involved, individuals refer to the social world; in the case of strategic action, the objective world is the main reference and, finally, dramaturgical action calls upon the subjective world.

In SSIs discussions, the use of authoritative argument is well known. It becomes a fallacy when it is used to give weight outside the recognized field of expertise. The question of the validity of expertise is complex and central to SSIs and expertise is sometimes questioned. Kolsto (2001) studied students' views on the trustworthiness of claims involved in a local SSI. He found that the students partly sought to evaluate science-related claims and partly took the trustworthiness of these for granted. In addition, he found that the students also focused on the source of information, using evaluation factors like competence and possible interest involved.

Who can claim to be an expert on SSIs? In which field would a person have established his/her reputation? Can we assert, as Isabelle Stengers (2005) does, that "in the case of an open issue challenging the future, a scientist who has been trained according to a discipline is no more of an expert than the concerned or outraged 'man on the street'"? (p. 7). When only non-stabilized knowledge can be taken into account, the status of the scientific evidence should, likely, be challenged. We must not forget that "scientific evidence is only legitimate within the cleansed and prepared environment we call the laboratory. It is not because science has concentrated on the ideology-free 'facts' that it is able to prove something but because it has succeeded in carrying out the ever risky process of deciding what should be taken into account and what should not" (Stengers 2005, p. 5, my translation). In the case of SSIs, can we rely solely on scientific expertise because it is pure, untainted by ideologies and does not complicate the responses "by deploying them in all their complexity and uncertainties" (p. 4)? Is there a risk of neglecting some important aspects outside the field of expertise? Walton (1997) listed six critical questions to evaluate 'the appeal to expert opinion':

1. *Expertise* Question: How credible is E (the expert in question) as an expert source?
2. *Field* Question: Is E an expert in the field A (the given argument) is in?
3. *Opinion* Question: What did E assert that implies A?
4. *Trustworthiness* Question: Is E personally reliable as a source?
5. *Consistency* Question: Is A consistent with what other experts assert?
6. *Back-up Evidence* Question: Is E's assertion based on evidence?

While helpful, this list of questions to evaluate the appeal to expert opinion is limited in our eyes because, for example, the question of consistency implies that the opinion of an expert who is in the minority cannot be taken into account and we know that the same facts can be used to back up different pieces of evidence. In our view, this criticism is even more relevant in the case of controversial SSIs, which are, by their very nature, interdisciplinary and, thus, require opinions of a variety of experts and a complex approach. Besides, expert opinion is sometimes influenced by financial and industrial interests; and, as those who define Post Normal Science argue, it is advisable to open up the expertise to the 'extended peer community' composed of all persons concerned by the issue (Funtowicz and Ravetz 1993). Within the context of the teaching of these questions which are subject to a high level of controversy, it seems essential to train the learners in the methods of constructing the various expert appraisals and also in the identification of fallacious strategies within the discourse of researchers, journalists and even teachers and other learners.

Fostering 'critical thinking' is often advocated in school syllabuses (from primary to higher education) without actually being defined. In the literature on the subject, it is described in terms of skills, procedures, principles and dispositions. In each case, it is essential to describe what we mean by critical thinking. The criteria chosen can be very different, for example: the ability to produce an argument that is backed-up by evidence, to call certain data into question, to problematize, to conduct a socio-epistemological reflection, to identify risks and uncertainties, to think for oneself even in opposition to one's social group. According to Jimenez Aleixandre and Puig (2010), critical thinking is composed of two main elements: (i) rationality; that is, use of evidence and readiness to look for evidence and to question established facts; and, (ii) independent opinion; that is, questioning one's own social group and the critical analysis of discourse that justifies inequality. They equate the first element to argumentation and the second to social emancipation. From an emancipatory perspective, critical thinking can be defined on the basis of the implementation of high level cognitive procedures, but also on the basis of a fundamentally socio-epistemological conception of knowledge building. In line with this conception, the development of critical thinking rests on the critical treatment of the evidence provided by the symbolic producers of knowledge (scientific or not). This implies epistemological reflection (a critical study of the methodology used to produce the evidence, a study of the risks, uncertainties etc.) and socio-epistemological analysis (who are the knowledge producers? What interests do they have in the given question? How does the actor-network function (Callon 1989, 2006)? How do the oppositions and alliances work?).

This is where the French field of '*Questions Socialement Vives*' comes in. Alain Legardez and Laurence Simonneaux (2006) coined the term '*Questions Socialement Vives*' – in English: 'Socially Acute Questions' (SAQs). These questions are 'acute' in society, in research and professional fields and in classrooms and often found in the popular media. The field of SAQs represents a French orientation for the teaching of SSIs. SAQs integrate the (more or less stabilized) knowledge of the 'hard sciences'[2] and knowledge from various disciplines of the humanities and social sciences, but also social and vocational knowledge. I consider that many different actors take part in knowledge production. These include scientists, citizens, philosophers and even whistleblowers. Consequently, Simonneaux (2011) asserts that the knowledge involved in SAQs can be conceived as plural (polyparadigmatic) or/and engaged (analyzing the controversies, uncertainties and risks) or/and contextualized (observing empirical data within a given context), or/and distributed (constructed by different knowledge producers).

Decisions taken on SAQs cannot be based solely upon scientific knowledge (be it knowledge from the area of the social sciences or the hard sciences) but must also take social implications, ideologies and values into account. Unlike the work on SAQs, SSIs approach is mainly based on the didactics of the 'hard sciences.' Although the complexity and uncertainty are recognized in SSIs, the role of interdisciplinarity is rarely studied, nor are the concepts of the humanities and social sciences or those of social or vocational knowledge seriously taken into consideration. The SAQ approach has a common aim with the Science – Technology – Society (Environment) model (Hodson 2003); STSE education has a more holistic nature arguing the need for interdisciplinarity; it aims for students to be committed to make responsible decisions about SAQs. Although SAQs may contribute to scientific literacy, they also have the potential to develop students' political literacy by including such topics as risk analysis, analysis of patterns of political and economic governance, as well as decision-making and action. Even though Dana Zeidler, Troy Sadler, Michael Simmons and Elaine Howes (2005) have claimed that SSI education is a better way than the STS(E) movement to integrate the nature of science, arguments, values and moral judgements, Derek Hodson (2011) has recently critiqued both of these approaches and asserts that STS(E) and SSI education have given too low a priority to the promotion of critical thinking. He asserts that neither STSE nor SSI-oriented teaching go far enough.

---

[2] Some scholars claim that fields like physics and chemistry generate more reliable and valid claims, often because of their reliance on quantitative data, which they consider 'harder' – or more stable – than qualitative data often used in so-called 'softer' fields of investigation, such as sociology.

At the 'hot end' of the spectrum in Fig. 6.1, pioneers of the 'activist' trend have developed a framework called 'STEPWISE' (Science and Technology Education Promoting Wellbeing for Individuals, Societies and Environments) for organizing teaching and learning in science and technology education. It is about encouraging students to take actions to address SSIs/SAQs. For example, in the context of education for sustainable development, the development of eco-gestures, actions implemented by the students, is often at the heart of teaching practices. These eco-gestures correspond, for example, to sorting waste, saving water (taking a shower instead of a bath, turning the tap off when brushing one's teeth) or electricity (switching off the light when leaving a room, not leaving electrical appliances on stand-by. . .). These eco-gestures are not to be sneered at but they should be accompanied by a critical analysis of what determines these choices (e.g., meat consumption generates high energy and water costs which are disproportionate to taking a daily bath) and each eco-gesture should be analyzed within its context and complexity. The STEPWISE program aims at promoting social and environmental justice and tries to foster a desire for change and a sense of responsibility (Bencze et al. 2012). Larry Bencze (2000) suggests that students work on 'student-directed and open-ended' research projects. This involves getting students to work on projects based on their own research for information on socio-scientific issues and encouraging them to make their results public through socio-political actions (for example, by organizing demonstrations and exhibitions, posting activist videos on YouTube™).

Didactic strategies reflect the priorities chosen by the school or by the teachers. Simonneaux (2011) identified four possible didactic strategies for teaching SSIs:

- *A doctrinal* strategy that aims to develop the acceptance of the ideas presented in the high authority of the teacher, who leaves little room for interaction with students;
- *A problematizing* strategy that focuses on students' cognitive activity – here, students take an active part in the construction of an *issue* and develop a line of reasoning rather than finding THE solution;
- *A critical* strategy aims to develop a critical sense – here, the educational purpose is to teach students how to argue and to assess expertise and different stances on complex issues which carry both uncertainties and risks;
- *A pragmatic* strategy is based on involving the students in an activity – here, the challenge is to stimulate student *action*.

Of course, several strategies can be used within the same teaching situation – although, often, one or more strategies may dominate. I should add to the list, however, *an activist* strategy involving a process of problematization and a critical approach. This strategy would not be reduced to action from a pragmatic perspective but would also require taking up a militant and communicative posture. The SSIs may be more or less acute in society, depending on current affairs and media coverage. Activism gains force by defending a minority voice against a dominant one. It implies, as in the case of SAQs, implementing a collaborative approach within the scope of participatory democracy.

The Implications of the Educational Choices on SSIs

There are different ways of teaching SSIs according to teachers' views of the main educational priorities. In an attempt to map out the landscape, these different dimensions are represented as continuums in Fig. 6.2.

At the 'cold end' in Fig. 6.2, knowledge mobilized in the classroom is single-disciplinary science. At the 'hot end,' it is discussed in interdisciplinary sessions in science and humanities. Between the two ends, it is interdisciplinary science (e.g., biology and chemistry). We assert that the study of SAQs forces education to transcend disciplinary divisions, particularly between 'hard' science and 'soft' sciences. When examining (techno)sciences, we realized that many characteristics go beyond the boundaries of the disciplinary divisions and these divisions are as much the result of a social construction as of epistemological specificities. With a French SAQ approach, we argue for real interdisciplinarity – in which science and humanities are integrated in order to account for the complexity of the reality linked to SAQs. This interdisciplinary approach is also advocated by Hodson (2011). Recently,

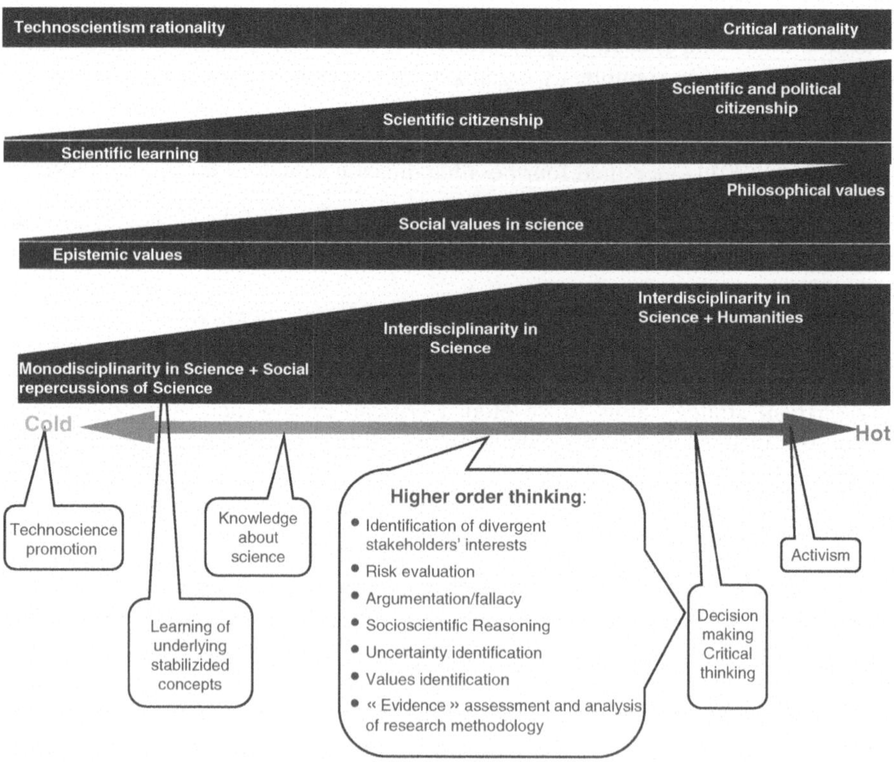

Fig. 6.2 Educational priorities and pedagogies beyond SSI education (Adapted from Simonneaux 2013)

Eastwood, Schlegel and Cook (2011) have described and analyzed an ambitious 4-year interdisciplinary university programme. Here, the interdisciplinary approach goes beyond the single social impact, or the impact of values, or even the impact of culture. The approach requires hybridization of knowledge between the humanities and natural sciences and may often include non-academic knowledge.

I assert that values may be explicit or implicit in the teaching of SSIs. At the 'cold end' in Fig. 6.2, only epistemic values may be mobilized (e.g., validity, reliability, accuracy). At the 'hot end,' philosophical principles underlying the values are explained and discussed. Between the two ends, the social values in science (Longino 2002) are identified and acknowledged. In fact, in the French field of SAQs, values must be clarified, whether they are scientific or social. Such a focus aims to help students to identify the values of different stakeholders, as well as their own, in their decision making.

Beyond science learning, the challenge may be to develop scientific citizenship and even political citizenship so that teaching about SAQs from the activist perspective may lead to the combination of both science education and political education – thus the development of scientific, and even political, literacy. Levinson (2010) has identified a number of democratic participation frameworks that can be used in the teaching of SSIs.

The choice of teaching strategies is influenced by the nature of the teachers' rationality; whether they subscribe to *techno-scientific* rationality (the future techno-sciences will resolve the problems of the current ones) or *critical* rationality involving reflexivity towards the techno-sciences. A teacher who embraces techno-scientific rationality will have the tendency to 'cool down' the issue; whereas, one who adopts a critical rationality will tend to 'heat up' the question. On the other hand, the nature of a teacher's rationality may fluctuate depending on the issue in question.

Research was carried out into the commitment to climate change teaching declared by teachers of different disciplines. It was observed that, depending on their discipline, they engage in three types of pedagogical models; that is, *positivist*, *interventionist* and *critical*. These models range from educating students in accordance with their own opinions and teaching students how to make their own choices (Urgelli 2009; Urgelli et al. 2010). The variety of these engagements may be explained by the teacher's ecological convictions and/or epistemological doubt. Epistemological doubt, which acknowledges that these questions are controversial, that they reflect uncertainty, may be a determining factor in the way these questions are taught. If the teacher accepts the doubt, s/he may choose a critical approach to the question. Sometimes, despite his/her personal doubt, the teacher chooses not to engage students in a critical approach for fear of influencing them on account of her/his institutional position. On the question of climate change, the ecological conviction of the teachers studied by Urgelli (2009) justified an interventionist approach; that is, teachers claimed that students must be convinced of the need to change their habits and patterns of consumption. On issues related to health (e.g., gene therapy, the use of embryonic stem cells...), I assume that ethical convictions may determine ways in which these questions are processed.

I conducted a study on teachers in agricultural education in France. The study focused on SAQs related to animal husbandry (the evolution in meat consumption, the contribution of animal breeding to the greenhouse gas effect, animal welfare). I wanted to discover whether they approached these SAQs on the basis of their ecological or ethical convictions; the latter calling breeding practices into question and/or including a critical analysis of animal husbandry knowledge. This group tended towards a techno-scientific rationality (Simonneaux 2012). Faced with these SAQs, the teachers took sides with the breeders above all else. They empathized with the farmers who are angered by the criticism fired against them and by the measures they are required, by law, to take. These teachers seemed to believe that techno-science will resolve the SAQs. They would like to see more targeted research associated with the development of the techno-sciences in breeding. The majority of these teachers take a positivist approach to the environmental issues offered up for debate. They assimilated the sustainability rhetoric as long as it is associated with productivity. They were confident that techno-scientific progress will resolve the SSIs linked to the environment. But fundamentally, they minimized the responsibility of animal husbandry and the part it plays in the issues rose (e.g., climate change, the food crisis). They were also reticent about the regulations on animal welfare. Some teachers, meanwhile, revealed their critical rationality when dealing with the question of pesticides, denouncing in particular, the environmental problems and to a lesser extent the problems linking the health of consumers and farmers to pesticide use.

Institutional Activism in Agricultural Education in France

The specificity of agricultural education in France is that it is under the supervision of the Ministry of Agriculture and not the Ministry of Education. Consequently, its mission has always been to relay the political choices, in terms of economic strategy, of the Ministry of Agriculture. In 1945, agricultural development was based on a mechanized, motorized and 'chemical' model of agriculture. This model targeted, first and foremost, an increase in productivity, an improvement in the technical aspects, the intensification and integration of farming into the rest of the economy. Scientific, technical/economic and political means were mobilized. These orientations led to a massive increase in standardized production. Parallel to this, we witnessed the advent of the agri-food industry and, in the 1970s and 1980s, the development of mass distribution and large retailing outlets. We entered into a market economy dominated both upstream and downstream by the industries. In agricultural education, the focus became the development of intensive and chemical farming. The aim was the promotion of new farming techniques also encouraged by firms, banks, and organizations working in favor of technical development. An institutional productivist activism developed in agricultural education.

When designing the curricula, techno-scientific and politico-economic choices are made. The popularization of the techno-sciences supported and relayed by schools ensured that farmers embraced the intensive model during the 30-year postwar boom. This, in turn, led to a change in traditional practices. Limits of the productivist farming system (impact on environment, food quality, consumers and farmers' health, employment in the agricultural sector, farmers' dependence vis-à-vis agrochemical companies, etc.) very quickly became apparent. However, the financial stakes were such that the environmental or health risks were minimized in the dominant political discourse. Society began gradually to express its concerns more strongly, particularly following reports of substantial pollution, media coverage of breeding conditions and the break-out of crises such as BSE.[3] The pressure of social demand gave rise to a new kind of institutional activism (European or French, depending on the case) in defense of, for example, animal welfare or a reduction in pesticide use. But, faced with this new institutional activism, teachers expressed different degrees of engagement or resistance that varied according to the issue. The degree of resistance or engagement often was conditioned by the nature of their initial training (environmentalist or productivist); which they reveal to a greater or lesser extent in relation to political correctness and the institutional activism. This political position of the French Ministry of Agriculture can occasionally become schizophrenic because of the necessity to accommodate financial interests. Alongside the new agro-ecological rhetoric, the dominant productivist model is still being taught today.

Scientific, Humanistic and Political Education

Science education that is designed based on a strict disciplinary approach soon reveals its limits. All over the world, educators in science deplore young people's disaffection for scientific studies. At the same time, the selection even for a career in the literary, social, economics, legal sectors etc. is based on the students' performance in math and in the sciences. Science curricula are based on 'cold' stabilized knowledge, which is not hinged on the complex questions of 'real life.' Students who adapt to this type of knowledge, which can be meaningless, are in a sense 'locked in the discipline', poorly trained to tackle complex contextualized questions and to show the critical spirit necessary in the post-modern risk society (Beck 1986). In this post-modern risk society, I advocate a science education rooted in authentic issues implying an interdisciplinary approach, the science concepts being taught using a 'detour-reinvestment' process. In other words, contextualized problematization generates the necessary science detours; and, then, these science

[3] Bovine spongiform encephalopathy (BSE), commonly known as mad cow disease, is a fatal neurodegenerative disease (encephalopathy) in cattle that causes a spongy degeneration in the brain and spinal cord. The disease may be transmitted to human beings by eating food contaminated with the brain, spinal cord or digestive tract of infected carcasses. In humans, it is known as new variant Creutzfeldt–Jakob disease (vCJD or nvCJD), and by October 2009, it had killed 166 people in the United Kingdom. But the statistics are controversial.

concepts and procedures are reinvested critically in resolving the issue. In our eyes, the teaching of SAQs, which are to be grasped in all their socio-economic, political, environmental and ethical acuteness, should be situated along these lines. I consider that this pedagogical approach to SAQs can be a vector for the transformation of school in the post-modern society and encourage democratic participation in environmental and ethical governance. I uphold that, in SAQ education, three areas of education come into play: science education, humanistic[4] education and political education.

References

Beck, U. (1986). *La société du risque, sur la voie d'une autre modernité*. Paris: Flammarion, 2001 French translation.

Bencze, J. L. (2000). Procedural apprenticeship in school science: Constructivist enabling of connoisseurship. *Science Education, 84*(6), 727–739.

Bencze, L., Sperling, E., & Carter, L. (2012). Students' research informed socio-scientific activism: Re/vision for a sustainable future. *Research in Science Education, 42*(1), 129–148.

Bravo Torija, B. (2012). *El desempeno de las competencies cientificas de uso de pruebas y modelizacion en un problema de gestion de recourses marinos*. Tesis doctoral Universidad de Santiago de Compostela.

Callon, M. (1989). *La science et ses réseaux*. Paris: La Découverte.

Callon, M. (2006). (Interview with Michel Ferrary), Les réseaux sociaux à l'aune de la théorie de l'acteur-réseau. *Sociologies Pratiques, 13*(2), 37–44.

Eastwood, J. L., Schlegel, W. M., & Cook, K. L. (2011). Effects of an interdisciplinary program on students' reasoning with socio-scientific issues and perceptions of their learning experiences. In T. D. Sadler (Ed.), *Socio-scientific issues in the classroom – Teaching, learning and research* (pp. 89–126). Dordrecht/Heidelberg/London/New York: Springer.

Erduran, S., & Jimenez-Aleixandre, M.-P. (Eds.). (2008). *Argumentation in science education: Recent developments and future directions*. Dordrecht: Springer.

Funtowicz, S. O., & Ravetz, J. R. (1993). Science for the post-normal age. *Futures, 25*(7), 739–755.

Habermas, J. (1987). *Théorie de l'agir communicationnel, Tome 1: Rationalité de l'agir et rationalisation de la société*. Paris: Fayard.

Hodson, D. (2003). Time for action: Science education for an alternative future. *International Journal of Science Education, 42*(6), 645–670.

Hodson, D. (2011). *Looking to the future – Building a curriculum for social activism*. Rotterdam/Boston/Taipei: Sense Publishers.

Jimenez Aleixandre, M. P., & Puig, B. (2010). Argumentation, evidence evaluation and critical thinking. In B. Fraser, K. Tobin, & C. McRobbie (Eds.), *Second international handbook for science education*. Dordrecht: Springer.

Kolsto, S. D. (2001). To trust or not to trust... Pupils' ways of judging information encountered in a socio-scientific issue. *International Journal of Science Education, 23*, 877–901.

Legardez, A., & Simonneaux, L. (2006). *L'école à l'épreuve de l'actualité*. Issy-les-Moulineaux: ESF.

[4] I refer both to the importance of education in the humanities and to humanistic education in reference to the tradition of the Renaissance. In this tradition, humanism aims for the emergence of educated and fully accountable individuals.

Levinson, R. (2010). Science education and democratic participation: An uneasy congruence? *Studies in Science Education, 46*(1), 69–119.

Longino, H. (2002). *The fate of knowledge*. Princeton: Princeton University Press.

Morin, E. (1998). *Pourquoi et comment articuler les savoirs?* Paris: PUF.

Sadler, T. D., Chambers, F. W., & Zeidler, D. L. (2004). Student conceptualisations of the nature of science in response to a socio-scientific issue. *International Journal of Science Education, 26*(4), 387–410.

Sandoval, W. A. (2003). Conceptual and epistemic aspects of students' scientific explanations. *Journal of the Learning Sciences, 12*(1), 5–51.

Simonneaux, J. (2011). *Les configurations didactiques des questions socialement vives é conomiques et sociales, Habilitation à diriger des recherches*. Université de Provence.

Simonneaux, L. (2012). Les rationalités d'enseignants en productions animales face aux questions socialement vives en elevage. *Revue de Didactique des Sciences et Techniques, 5*, 9–46.

Simonneaux, L. (2013). Questions socialement vives and socioscientific issues: New trends of research to meet the training needs of post-modern society. In C. Bruguière, A. Tiberghien, & P. Clément (Eds.), *9th ESERA conference selected contributions. Topics and trends in current science education* (pp. 37–54). Dordrecht: Springer.

Stengers, I. (2005). L'usage de l'expertise scientifique dans la décision politique. In *Les loyautés du savoir. Positions et responsabilités des sciences et des scientifiques dans l'état de droit dé mocratique*. http://ec.europa.eu/research/conferences/2005/forum2005/docs/progr_stengers_en.pdf

Urgelli, B. (2009). *Les logiques d'engagement d'enseignants face à une question socioscientifique médiatisée, le cas du réchauffement climatique*. Thèse de doctorat de l'Ecole normale supérieure de Lettres et sciences humaines Sciences de l'éducation – Sciences de l'information et de la communication.

Urgelli, B., Simonneaux, L. & Le Marec, J. (2010). Réchauffement anthropique et développement durable – Quelle(s) éthique(s) pour une éducation scientifique citoyenne. *Education au dé veloppement durable et à la biodiversité: concepts, questions vives, outils et pratiques*. Digne les bains.

Walton, D. N. (1997). *Appeal to expert opinion*. University Park: Penn State Press.

Zeidler, D. L., Walker, K., Ackett, W., & Simmons, M. (2002). Tangled up in views: Beliefs in the nature of science and responses to socioscientific dilemmas. *Science Education, 27*, 771–783.

Zeidler, D. L., Sadler, T. D., Simmons, M. L., & Howes, E. V. (2005). Beyond STS: A research-based framework for socio-scientific issues education. *Science Education, 89*, 357–377.

Chapter 7
Hopeful Practices: Activating and Enacting the Pedagogical and Political Potential in Crisis

Rebecca Houwer

> *. . .climate crisis • environmental crisis • AIDS crisis • food crisis • humanitarian crises • refugee crisis • crisis of the subject • crisis of truth • crisis of representation • crisis of evidence • inner-city crisis • rural crisis • personal crisis • currency crisis • constitutional crisis • energy crisis • mid-life crisis • crisis of faith • civilization crisis • urban crisis • financial crisis • intelligibility crisis • housing crisis • cultural crisis • education crisis. . .*
>
> *Ring the bells that still can ring/Forget your perfect offering/There is a crack in everything/That's how the Light gets in. ~ Leonard Cohen*

Abstract This chapter explores the theoretical potential for education to provide learners with structures and processes that mitigate crisis and support "hope grounded in practice" (Freire P, Pedagogy of hope. Continuum, New York, 2008, p. 2). Instead of advocating for a definitive educational response to crisis, I examine the possibilities and challenges crises, when recognized as pedagogical, offer education. With a focus on public rather than private crises, I first situate my analysis within emergent matters of concern and care (Latour B, Crit Inq 30:225–248, 2004; Puig de la Bellacasa M, Soc Stud Sci 41(1):85–106, 2011) to practitioners and theorists of Science, Technology, and Society Education (STSE) studies. Second, I define "crisis" and examine how it is pedagogical. Next, I consider the challenges and barriers to learning through crisis. Fourth, I develop the opportunities crises offer education and finally, I conclude with suggestions that might support the potential for transformative learning from and through crises and considerations of why such learning is desirable.

R. Houwer (✉)
Faculty of Education, York University, 4700 Keele St, Toronto, ON M3J 1P3, Canada
e-mail: rebecca_houwer@edu.yorku.ca

L. Bencze and S. Alsop (eds.), *Activist Science and Technology Education*,
Cultural Studies of Science Education 9, DOI 10.1007/978-94-007-4360-1_7,

Keywords Transformative praxis • Critical pedagogy • Citizenship education • Public participation • Coalition competence • Crisis and freedom • Learning from crisis

In the Age of Reason, many in society invested tremendous hope in promises of modernity that through the rigorous application of scientific rationality humans might emancipate themselves from the shackles of Nature. Ironically, the harder we have tried to flee Nature, the closer to it we have become; inadvertently the scale by which human fates are bound to the fates of natural worlds has expanded exponentially. In our pursuit of independence and disconnection, we have created a proliferation of crises. Still, we place our hope in 'the children', the future, and education. A recent keyword search of my university's library catalogue for 'crisis' returned 9,531 hits. A keyword search for 'crisis AND education' returned 596 hits. Alternatively, a keyword search for 'hope' returned 5,748 hits and 'hope AND education' returned 460 hits. Records of crisis outnumber hope by nearly a 2:1 ratio (9,531:5,748). However, when paired with education, the crisis: hope frequency ratio diminishes to 1.2:1. Paulo Freire (2008) claims that "hope is an ontological need" and that "we need *critical* hope the way a fish needs unpolluted water" (p. 2). Crisis and critical are both etymologically derived from the Greek word krinein which means "able to discern" (Oxford English Dictionary, Web, accessed 21 Mar 2013). In this chapter, I ask if there is something that education provides learners in its structure and processes that mitigates crisis and supports "hope grounded in practice" (Freire 2008, p. 2). I respond to the question, "given compelling evidence of an ecological, sustainability crisis, how might Science, Technology, and Society Education respond?" by asking, more generally, what opportunities and challenges crises, when recognized as pedagogical, offer education. I focus on crisis generally, instead of the "ecological crisis" specifically, because I believe that strategies for learning from and through the latter may be productively applied to former. Moreover, I contend that they are produced within similar power-structures. For this reason, I focus on public as opposed to private crises. The chapter is divided into five sections. First, I situate my analysis within emergent matters of concern and care (Latour 2004; Puig de la Bellacasa 2011) to practitioners and theorists of Science, Technology, and Society Education (STSE) studies. Second, I define "crisis" and examine how it is pedagogical. Next, I consider the challenges and barriers to learning through crisis. Fourth, I develop the opportunities crises offer education and finally, I conclude with suggestions that might support the potential for transformative learning from and through crises and considerations of why such learning is desirable.

Science, Technology, and Society Education (STSE)

In response to the many crises in our many worlds, we need not only all the tools in the toolbox (and those yet imagined, those forgotten, and those presently outside our frame of recognition), but appropriate tools, and competency using them. Since its institutionalization, science education toolboxes have offered students conceptual and material instruments for gaining insights into and understanding about "the natural world". As governments, in particular the Canadian government, set about dismantling world class environmental research facilities, muzzling scientists, and divesting in sciences that are not intimately tethered to private market interests, citizen science for public purposes becomes increasingly important. In order to cultivate an informed, active, and reflexive citizenry, students need opportunities to develop connections between 'the school world' and 'the real world' (Sadler 2009, p. 38). Studies demonstrate that students are more motivated to learn about issues that they can connect to their everyday lives (Sadler 2009; Zeidler et al. 2005; Alsop and Bencze 2010). That said, the sciences have typically not been taught as situated within local contexts or as having more than a functionalist and instrumentalist relationship to students' everyday lives. Instead, the ontological and epistemological resources, and pedagogical technologies, that science education mobilizes are, more often than not, decoupled from their socio-political, cultural, historical, and affective origins (Hodson 2003; Jasanoff 2010). While STSE has taken steps to highlight and trace the intricate interrelatedness of science, technologies and society it "fails to overtly consider the epistemological foundations, moral and ethical development, and emotional aspects of learning science" (Zeidler et al. 2005, p. 371). A reasoned understanding of connective operations is qualitatively different from appreciating the implications of such connections and working to preserve, maintain or restore their integrity. Furthermore, Sheila Jasanoff (2010) contends that whereas "scientific facts arise out of detached observation, meaning emerges from embedded experience" (p. 235). Though a student may have a sophisticated instrumentalist understanding of science, it is very possible that its meaning is divorced from the students' political, or otherwise, experience. Moreover, multiple studies demonstrate that knowledge of a problem does not ensure action will be taken to address the problem. Harold Glasser (2007) portends "awareness of a problem, accessibility of extensive information on its origins and impacts, and, even, stated concern about [a problem] do not guarantee action or imply that, if taken, the action(s) will be appropriate or effective" (p. 42). Furthermore, Dana Zeidler, Troy Sadler, Michael Simmons and Elaine Howes (2005) reiterate that "researchers have confirmed the lack of coherence between the ability to form higher moral judgments and the likelihood of exercising that reasoning in varied contexts" (2005, p. 372). The crises listed at the outset are neither the cause of single individuals nor resolvable by single individuals. They are the products of collective work and they need to be addressed publically rather than privately. Science education, like most other disciplines, currently focuses on individual competency and achievement and not on collaborative work that can be mobilized for the public 'good'. Steve Alsop and Larry Bencze (2010) attest that "education in [socioscientific issues] tends to be an abstract and mostly individualized process – rather than one in which students reach out to communities near and far and act for change" (p. 181). With the proliferation of 'wicked problems' where causes and solutions are multiple, shifting, emergent, embedded in, and the product of, socioscientific enmeshments, we need to reach beyond knowing well to knowing *and* doing well. In recognition that humans are generally not motivated to political action through reason alone, I suggest that we look to the opportunities interdisciplinary, integrative, situated, and participatory pedagogies offer us to *understand and respond* to crisis as offering possibilities for cultivating re-attachment and recommitment to the shared tasks of learning, working and caring for the co-production of a common world.

Understanding and Learning from Crisis

What constitutes a crisis? According to the Oxford English Dictionary (Web, accessed 31 Mar 2013), a crisis [def. 3] is "a vitally important or decisive stage in the progress of anything; a turning-point; also, a state of affairs in which a decisive change for better or worse is imminent" (http://www.oed.com.ezproxy.library.yorku.ca/Entry/44539). Further, Michael Mueller (2009) reminds us that "to have a crisis, there must be humans embedded in a situation to think it that way" (p. 1036). *Humans* experience crisis when our normative "frames of recognition" are disrupted. Judith Butler (2005) explains that frames are the normative conventions and categories "that prepare or establish a subject for recognition" (p. 5). She contends that a frame must circulate "in order to establish its hegemony" and break to "re-install" itself. Through reproduction – at times, in the form of schooling – the frame's fissure "exposes the orchestrating designs of the authority who sought [its control]" (Butler 2005, p. 12). The breakage indicates a "collapsibility of the norm" (ibid.). The disruption of the norm provokes a crisis that is pedagogical in at least two ways: (1) it can teach us that we are subjects; or (2) it can teach us that we are objects. How we respond to the choice crisis offers can be understood as a theory of pedagogy: regardless of how we choose, crisis has made us agents.

The challenges of learning through crisis are many. In the face of catastrophe and the choice crisis puts to us, we are often ambivalent: fight or flee; stasis or progress; desire or fear? We make heroes of those who act in crises. Conversely, those who are seen as unable to act are perceived to be morally corrupt. The firefighters who responded to the Twin Towers on September 11, 2001 are heroes. The then-president moved quickly to action (i.e. war) because to pause would (likely) have been met with accusations of ineffectiveness, or lack of valour and dignity. America's normative frame was ruptured by its own excesses. In order to restore intelligibility to their – and my formative – discursive world, the citizenry frantically sought stasis. The crisis offered a choice (or choices) and 'America' chose to zealously reinforce its normative discursive frame. This observation pertains to the climate crisis and can be generalized to other coexistent crises. These crises offer an opportunity to choose transformative praxis, or to resist learning and change and choose stasis. Schools, by their very design and purpose, preserve the status-quo by mitigating personal crisis to the exclusion of public crisis (Durkheim 1956; Bourdieu 1974). However, the micro and macro, private and public are a tangled together in a complex "assemblage of heterogeneous threads" (Latour 2010, p. 6). William Gaudelli (2008) notes that the question, "Why did I get this grade?" is just a few inquiries and a couple of social connections away from the question "What is the good life?" (p. 80). Due to the ascendance of standardized testing and the importance so assigned to a narrowing definition of academic achievement, students are rarely encouraged to pursue such connections. As Zygmunt Bauman (2008) attests, "the secret of every durable – that is, self-reproducing – social system is the recasting of 'functional prerequisites' into behavioural motives for actors...[in other words it makes] individuals *wish to do* what the system *needs them to* do for it to reproduce itself" (p. 149). One need think only of the desire to get "good grades." Earning or proving self-worth in this way is not natural but naturalized within institutions of schooling. Educations' potential is narrowed by Neoliberal capitalist instrumentalism. Lori Patton (2008) illustrates another way in which schools conserve traditions, perhaps to the detriment of society at large. Interested in how university students make sense of local and global crises such as the genocide in Rwanda, the 2004 Tsunami in the Indian Ocean, the events of September 11, 2001, Hurricane Katrina, the Virginia Tech Massacre, Patton reports that the majority of students that she interviewed knew very little about these events. Students recounted being encouraged not to reflect but to "return to normal" as quickly as possible. Patton offers one students' response as summarizing "the prevailing group sentiment: We have to move on and can't dwell in the past. Life goes on" (p. 11). She argues that the return to normalcy forecloses the potential for transformative learning. However, the purpose of schools has never been social transformation; it has (nearly) always been the conservation of dominant social relations and values. Crisis stimulates learning insofar as it creates a gap between the old and the new. Bridging this gap is schooling's task. It provokes, scaffolds, and manages private micro-crises such that their excesses are seamlessly folded back into the normative public frame. This is called "learning" and students who do this well are rewarded (though differentially depending on class, racialization, gender, sexual orientation, disability, etc.). The primary function in the context of schools is not to connect "the world to the word and the world to the word" (Freire 1970, p. 87) or imagine it anew – as Patton hopes – but to manage the excesses of the conceptual frame through incorporation/absorption (Foucault; Gramsci; Bauman). Mueller, citing Bowers, asserts that learning relies both on metaphors and frames of reference (2008, p. 1035). To make new information intelligible we refer to our conceptual frame to see what it "is like." Similarly, Claudia Ruitenburg (2005) citing Jacques Derrida "contends that there is no such thing as direct unmediated experience...there is nothing outside the text" (p. 214). Crisis provokes deconstruction by providing access to non-normative perspectives that call into question the frame, the integrative hermeneutic that makes our lives intelligible; crises provoke instability and uncertainty.

A second challenge learning from and through crisis encounters is the sheer magnitude of crises (see opening list). Bauman argues that the tide of information, the production of emergencies that demand choice (a moral response), dissolves difference into an "undifferentiated sameness." Eriksen (cited in Bauman) posits: "when growing amounts of information are distributed at growing speed, it becomes increasingly difficult to create narratives, orders, developmental sequences. The fragments threaten to become hegemonic" (p. 164). Gaudelli (2008) argues that in the minority world, crises have become normative due to the daily deluge of media-reported security threats.[1] Citing Sandra Ball-Rokeach, Gaudelli explains that daily reports of crisis produce a condition of "pervasive ambiguity wherein people lack credible information to define a given situation, leading them to attempt to reduce feelings of tension and seek information towards this end" (p. 77). The desire to resolve feelings of tension opens space for "rhetorical agents to define a situation and prescribe actions" (ibid). Bauman contends that the actions prescribed usually entail participating in the market economy as consumers. Instead of inviting agency, the perpetual and incompletely framed "crisis talk" objectifies us. Gaudelli argues that the steady stream of crisis reports "fractures public attention towards episodic focus and cursory awareness" (p. 75). Furthermore, it inculcates a culture of fear and conditioned passivity: we watch and wait and wait to watch (e.g. consume). Adams, Murphy, and Clarke (2009) argue that regimes of security have neocolonial effects in their materialization of anticipatory future risks. In so doing we cede our present, and our accountability to the present, to an imagined and fear-filled future. According to Brian Knowlton (cited in Bauman, p. 192) many are paralyzed and "unsure just how urgently, and fearfully, they should react" to the parade of crises. When crises are habitual ("business as usual") disconnection should come as no surprise. According to Harold Garfinkel's (1967) study decades ago, human trust is the bedrock of our social contracts (cited in Jasanoff 2010). Uncertainty, risk, and fear challenge trust and may erode social cohesion (Jasanoff 2010, p. 244). Feelings of trust are linked to feelings of security. Security can be generated through democratic collaborative practices. Adams, Murphy and Clarke suggest that "instead of ceding to the injunction to anticipate [risk], one might ask what kinds of desirable accountabilities to and kinships with the future might be fostered through [present] work" (2009, p. 260). Likewise, Mueller suggests that because crises induce "shock, confusion, and empty phobias" our time might be more productively spent "conceptualizing how individuals, communities, and environments work together as relational parts of the whole" (2009, p. 1053.). Similar to Bruno Latour's (2004) suggestion that in order to re-gain agency we need to move toward the crisis rather than away from it, Mueller insists "the closer the better." In order to get closer, we must make the implicit explicit; the process of explication reveals the complexity and fragility of our attachments (Latour 2010). Crises do not ask us to "get on with life" by moving away from the social and ecological systems that we take as natural. But rather, crises invite us to reconsider normalcy.

Crisis invites change. Extrapolating from Mario Blaser's (2004) discussion of the characteristics of resilience which "conserve the ability to respond to change" by "embody[ing] the inherently unpredictable and unknown outcomes of interaction between ecosystems and human societies" (p. 39), I suggest that learning from and through crisis cultivates resilience and social fitness, both of which are stated

[1] Whereas in the majority world and marginalized minority world, the crises are not mediated; they are accessed through direct experience, not representation.

aims of schooling. Instead of fleeing crises, it may be more productive to think of them as "arenas in which to gather" (Latour 2004, p. 246). As Latour suggests, in the gathering, the explication of attachments/associations, is the opportunity for renewal.

Learning from and Through Crisis: Opportunities

> I would venture to propose, today, that teaching in itself, teaching as such, takes place precisely only through a crisis: if teaching does not hit upon some sort of crisis, if it does not encounter either the vulnerability or the explosiveness of a (explicit or implicit) critical and unpredictable dimension, it has perhaps not truly taught. (Felman 1992, p. 53)

Felman further argues, as does Freire, that teaching must "make something *happen*, and not just transmit passive knowledge... information that is pre-conceived, substantiated, believed to be known in advance, misguidedly believed, this is, to be (exclusively) a *given*" (ibid). Similarly, Hannah Arendt (2006) asserts that political renewal is only possible through actions which are births in the world (natality). Passive transmission of inert information forecloses politics. For Arendt, politics are necessarily public, meaning in the company – or *gathering* – of others. In other words, without crisis there can be no action. Without action there can be no political renewal and without political renewal there can be no freedom. Both Felman and Arendt's concerns emerge from the events of the Sho'ah where a crisis was ongoing and yet the people of Germany were first unable to recognize it as a crisis and second actively contributed to its perpetuation. Crises, when accessed, familiarize normative frames of reference and provide an opportunity to explore the structural, historical and political roots of the situation. By challenging the normative structure within which our beliefs and actions make sense, crises produce a potentially pedagogical cognitive dissonance (Felman, p. 53). Without the crises, "life goes on".

While inherently pedagogical, crises are only transformative when subjects access their agency, when they retain the capacity to act. Political freedom depends on the ability to act. Freedom is made in the company of others. Again, I quote Bauman:

> Citizens' freedoms are not properties acquired once and for all; such properties are not secure once locked in private safes. They are planted and rooted in the sociopolitical soil, which needs to be fertilized and watered daily and which will dry up and crumble if it is not attended to day in and day out by the informed actions of a knowledgeable and committed public (p. 191).

For Felman, students of crises learn their way through it not by returning to a pre-crisis frame, a nostalgic home to which we can never return (Ruitenburg 2005) but by renewing/remaking the frame through testimony. Testimony does not flee from crisis' aporia but figures the frame's rupture, the excesses, and actively, partially, makes the world anew. Accordingly, Felman argues, the teacher's task is "on the one hand to access but not foreclose the crisis, and on the other hand contain it [such that it is not more] crisis (*sic*) than the class can sustain" (p. 54). Students must be supported by the teacher to "reintegrate the crisis in a transformed frame of meaning" (ibid.). Likewise, Bauman opines that schooling should "resuscitate (sic) the skills of interaction with others – of conducting a dialogue, of negotiating, of gaining mutual understanding, and of managing and resolving the conflicts (e.g. crises) inevitable in every instance of shared life" (p. 190). To that, I believe, Felman would add "witnessing."

Crises are not transformative if they merely evoke feelings. Feeling, in the absence of thought and action, does nothing to challenge or change the conditions that illicit the feeling (Boler 1997). Whereas feeling may be a symptom or outcome of a crisis, transformative learning is only possible through *praxis*. Praxis requires both reflection and action (Freire 1970, p. 60). Freire calls action without reflection "activism" and reflection without action "verbalism" (ibid.). Laura Johnson and Paul Morris (2010) contend that Freirian critical pedagogy supports learning through crisis by "enabling us to both *perceive*, from historical, cultural, economic, personal and political perspectives, and to *act* upon the 'structures of domination'" (p. 83). As noted previously, the words crisis and critical both imply the need to make a decision, a choice. However, as Bauman argues, within liquid modernity the "structures of domination" have dissolved within a proliferation of individual "choices" which masquerade as freedom.

While Felman's treatment of "education and crisis" is insightful, particularly in her recognition that action is required and that the action needs to be supported in order for the crisis to be transformative instead of destructive, her unit of analysis is limited to the individual. Whereas crisis is experienced at the individual level, it also reveals structures external to the individual that produce the conditions for the crisis. Given that the individual did not produce the crisis, it makes sense that the actions required to resolve the crisis be focused on the structures which produced it. Here, I agree with Louise Chawla and Debra Flanders Cushing's (2007) assessment that "the effect of private actions is limited unless it is combined with organizing for collective public change" (p. 438). Drawing on Stern, Chawla and Cushing "distinguish [between] 'private sphere' and 'public sphere' environmentalism" (or "active political citizenship" for our purposes) (ibid.).

Reflecting on the failure of environmental and climate activists to influence negotiations at UN Climate Negotiations in Copenhagen in December 2009, Maryam Adrangi (2010) writes that Toronto activists recognized the need to link with other "struggles for liberation." This awareness emerges from the recognition that "the same power that manifests itself as resource extraction in the countryside, manifests itself as racism, classism, and human exploitation in the city" (p. 13). Each of the theorists I draw on in this chapter advocates, in one way or another, a transformative political pedagogy grounded in, what Mario Blaser calls, a politics of non-dominating partnership, a *gathering*. Mueller (along with others such as Donna Haraway and Sandra Harding) reminds us that knowledge is partial and situated. Moreover, "the climate crisis" is more accurately a crisis in the dominant minority's systems of power. The crises are the blowback from the excesses of capital for capital's sake. Because knowledge is partial, we are always already uncertain. This recognition immediately de-centres Western techno-rational-scientism or any other dominant "ism" for that matter. Uncertainty is mitigated though careful consideration of multiple perspectives. Mueller states, "If people did not have limited perspectives, there would be little need for multiple stakeholders with different perspectives to participate in ecological decisions" (p. 1036). Similarly, Elizabeth Ellsworth (1989) argues, and Bruno Latour would agree, that pedagogy must move away from "oppressive simplifications" (what Cheney and Weston (1999) call "self-validating reduction"). Instead pedagogical praxis should "construct circumstances in which students of difference can thrive" (Ellsworth 1989, p. 324). The crises in educations (Britain, Italy, Montreal, etc.) and the climate emerge precisely because the normative frame of recognition privileges capital and marginalizes other ways of being that are too numerous to name (Mueller lists: other species, non-science derived knowledges, theologians, women, indigenous peoples, "poor" people, families, alternative social arrangements, diverse stakeholders, and there are more).

As a rupture of the normative frame of reference, crisis allows access to marginalized standpoints. Bell Hooks (1990) argues that from marginal standpoints we can theorize counter-hegemonic cultural and political practice. Similar to Felman's notion of "precocious testimony", Hooks posits that coming to voice in the margins, engaging in "the politics of articulation", "the oppressed struggle in language to recover ourselves, to reconcile, to renew" (p. 146). Moreover, and this resonates with Bauman's thoughts on the consequences of hurriedness, Hooks suggests that the politics of articulation from the margins are "a struggle against forgetting" (p. 147). Without crises, without access to the margins, "the past and the future as mental categories are threatened by the tyranny of the moment" (Bauman, p. 159). The anticipatory present is one characterized often by individual and hence collective political paralysis, whereby we cease to be agents or subjects hence producing an atrophied citizenry, an atrophied state. Our inability to participate in the politics of renewal and the creation of political freedom creates objects of us. Hooks argues that the margin is the "space of radical openness" and the normative centre is the foreclosure of possibility. However, as Bauman suggests, our analysis may need to complicate the binaries: centre/margin, freedom/oppression, self/other, past/present, inside/outside, past/present, near/far. Nancy Tuana (2007) suggests that frames are porous and that subjects "are constituted out of relationality" (p. 188). Citing William James, Tuana contends that "what exists are not things made but things in the making" (p. 190). Moreover, things in the making are dependent on both "social practices and natural phenomenon" (p. 193). Crises may alert us to the fact that our subjectivity is relational and emergent; that it is only possible because of interactive associations, gatherings. This has implications for education.

Thus far, I have argued that crises, by rupturing our normative frames of reference, invite the possibility of transforming the conditions that produced the crises. Crisis gives us access to "the margins," shifts our frontiers, and reveals our situatedness. It offers us the chance to critically examine our normative frames of reference and respond by either transforming or re-installing hegemonic structures. Learning through and from crisis requires support, time, space, and the ability to discern and differentiate one crisis from another so as not to succumb to passive ambiguity whereby we are objects of the crisis instead of agents empowered to change the conditions that produced the crisis. As mentioned, the institution of education is designed and intended to resist crisis, to integrate the past, present, and future. Structurally, schools embody some of the supports required to learn from and through crisis. Mitigating the uncertainty crisis provokes requires the ability to explicate, communicate, and associate. These skills can be learned in the various subject areas schooling supports. Schools are adept at provoking and managing micro-crisis at the individual level. Strategies that support learning through and from private crisis might be applied to learning through and from public crisis thereby shifting educational "hopes" from the individual's private success to the pursuit of eco-social justice achieved through active public political participation. By taking a "stubbornly realist" approach to the crisis, Patton (2008) suggests that educators can offer students the opportunity to examine human crises from [at least] two perspectives: they can ask what is there and what is not there (p. 11). Examining "what is there" involves an explication of the frame and its discontents. Examining the omissions gives access to the margins from which we can begin to see the normative frame from a different perspective. Ruitenburg (2005) reminds us that inclusion is only possible through exclusion. Becoming aware of what is excluded by our normative frame opens the possibility of expanding, complicating, and changing the frame. Exclusion is not inherently harmful but when the crisis allows us to hear, for example, the testimonies, made possible through the politics of articulation, of the majority world and their accounts of increased risk and precarity which is produced by the minority world, our normative frame, it charges us with a question: can we choose? Can we create, as Baumann implores us to do, "the conditions that make choice available and within our power" (p. 193)? He argues that consumer societies are not societies of freedom: acting births the new whereas consumption is necrophilic. Crises are produced by the excesses of the normative frame. It signifies the need for change. Change requires resilience and resilience requires response-ability.

Conclusion

> [t]he present crisis calls into question not just the political, economic, social and ecological structures that came into being with the rise of the market economy, but also the actual values that have sustained these structures and particularly the post-Enlightenment meaning of Progress and its partial identification with growth. (Fotopoulos 1997)

I think it is important to remember that children did not cause these many crises. They do, however, live with and through them; they will inherit what remains. The crises ask of education, of schools, "what are your aims, values, and promises"? Furthermore, does education perform the values it is invested with facilitating in others (Bauman, p. 167)? Presently, the normative frame which education reproduces neither integrates private and public crisis nor supports the practice of freedom. Present practices perpetuate consumer society to the detriment not only of the children but to all the associations that make their life possible. It is a time of precarity and uncertainty. According to Butler, precarity is a "politically induced condition" that calls attention to the frames "reproducible as social institutions and relations" that support the conditions within which life can thrive, is livable. However, precarity and uncertainty can be mitigated through gatherings, the politics of partnerships, the community-to-come. Gough and Scott (2006) suggest that education can help individuals to "understand and value their own perspectives while also enabling them to engage with the perspectives of others" (p. 287). By explicating attachments education can support transformative learning and the conditions within which real choices, real actions, real freedom is possible. Freedom must be constantly renewed through collective/public action. Karsten Schnack (1996, in Lundegard and Wickman 2007) advocates that schools create the conditions for students to develop "action competence" which he defines as "the capability – based on critical thinking and incomplete knowledge – to involve yourself as a person with other persons in responsible actions and counter-actions for a more humane world." I advocate for a variation on action competence which I call "coalition competence." Coalition competence links Boler's work on testimonial reading with Ellsworth's reflections on Coalition 101. In the face of crisis, coalition competence allows us to acknowledge and respect the power relations embedded in knowledge's historicity, and to stand in solidarity with diverse Others with the understanding that our knowledge is contingent, but that uncertainty due to its partiality may be mitigated in the gathering.

To conclude, I argue that crises are an opportunity for transformative praxis, for reflexivity. They defamiliarize our normative frames and allow us to see the structures that support their hegemonic reproduction. Crisis need not be debilitating if supported through mindful and cooperative attachments. Like the wild herb that can be both poisonous and medicinal depending on how it is prepared, crisis both provokes uncertainty and invites a gathering of diverse perspectives, earthly associations, the community-to-come. A birth is a crisis, the entrance of the new and the renewal of freedom, of life. Crisis reveals our situatedness, our relationality. Crises offer great teaching material! However, if we fail to respond to crises, we fail to learn; our world stays the same, making objects of us. Crisis is not a danger so much as the (1) failure to recognize crisis[2] and (2) the failure to learn through and from crisis. Crisis opens a crack and shines a light on the complexity and dynamism of our "earthly" attachments. Learning about and from our earthly attachments mitigates crisis. The urgent task for schools is not to respond to "the climate crisis." Rather, the task is to learn how to learn with and through any public crisis. To do so supports the conditions for the renewal of freedom, the renewal of life.

I began this chapter by asking after the relationship between hope and education. I questioned whether there is something about the structure of education that is

[2] At the December 4, 2010 People's Assembly for Climate Justice in Toronto, a participant commented that Canadians don't respond to crises which directly affect others because we are too comfortable. If we do not recognize that the crisis of Other's is also (soon to be) our crisis, we have failed to sufficiently explicate our attachments; this should be a task of schooling.

inherently hopeful. We invest educations with the ambivalent hope for change and for continuity. Learning marks the bridging of a gap between past knowledge and the unknown. Schools provide a structure that supports this transition. Theoretically, schools offer a place for gathering associations and explicating their attachments. In a time of information overload, of ever multiplying choices, education offers a place to focus and possibly to connect. Schools can be a place where we practice acting cooperatively and learn to shift between centres and peripheries. In order to renew the cultural and political purposes of schooling, students cannot be made objects for the markets' use. "Once state politics surrenders to the guidance of the 'economy,' understood as the freeplay of market forces, the balance of the two is switched decisively to the advantage of the first," contends Bauman (p. 189). Standardized testing diminishes our possible futures. It makes students instruments of the markets' needs. The controlling and assessment strategies brought to bear on children fail to prepare them to dynamically and creatively respond to change. Our collective capacity to act is diminished on the flywheel of consumption. Educations have the capacity to slow the flywheel. By making time to explicate attachments and support the integration of private and public crises, education can help us to develop the capacity to live and learn from and through crisis. In order to do so, educations will need to resist the tendency to encourage students to "return to normal." Rather, they must cultivate relationality, cooperative action (coalition competence), and diverse ways of knowing. Crisis shines a light on the hope that educations may yet enact.

References

Adams, V., Murphy, M., & Clarke, A. (2009). Anticipation: Technoscience, life, affect, temporality. *Subjectivity, 28*, 246–265.

Adrangi, M. (2010). Movement ecology: Overcoming the monoculture of mainstream environmentalism. In *Action speaks louder, fall*. Toronto: OPIRG-Toronto.

Alsop, S., & Bencze, L. (2010). Activism in SMT education in the claws of the hegemon (Editorial). *Canadian Journal of Science, Mathematics, and Technology Education, 10*(3), 177–196.

Arendt, H. (2006). *Between past and future*. New York: Penguin Group.

Bauman, Z. (2008). *Does ethics have a chance in a world of consumers?* Cambridge, MA: Harvard University Press.

Blaser, M. (2004). Life projects: Indigenous peoples' agency and development. In M. Blaser (Ed.), *In the way of development*. New York: ZED.

Boler, M. (1997). The risks of empathy: Interrogating multiculturalism's gaze. *Cultural Studies, 11*(2), 253–273.

Bourdieu, P. (1974). The school as a conservative force: Scholastic and cultural inequalities (J. C. Whitehouse, Trans.). In J. Eggleston (Ed.), *Contemporary research in the sociology of education* (pp. 32–46). London: Methuen & Co. Ltd.

Butler, J. (2005). *Frames of war: When is life grievable?* London: Vector.

Chawla, L., & Cushing, D. (2007). Education for strategic environmental behavior. *Environmental Education Research, 13*(4), 437–452.

Cheney, J., & Weston, A. (1999). Environmental ethics as environmental etiquette: Toward an ethics-based epistemology. *Environmental Ethics, 21*, 115–134.

Crisis, n. (OED Online). Oxford University Press, March 2013. Web. 21 March 2013.
Durkheim, E. (1956). *Education and sociology* (S. D. Fox, Trans.). Glencoe: The Free Press.
Ellsworth, E. (1989). Why doesn't this feel empowering? Working through the repressive myths of critical pedagogy. *Harvard Educational Review, 59*(3), 297–325.
Felman, S. (1992). Education and crisis, or the vicissitudes of teaching. In S. Felman & D. Laub (Eds.), *Testimony: Crises of witnessing in literature, psychoanalysis, and history* (pp. 1–56). New York: Routledge.
Fotopoulos, T. (1997). *Towards an inclusive democracy: The crisis of the growth economy and the need for a new liberatory project*. New York: Cassell.
Freire, P. (1970/2000). *Pedagogy of the oppressed*. New York: Continuum.
Freire, P. (2008). *Pedagogy of hope*. New York: Continuum.
Gaudelli, W. (2008). Critical pedagogy as alternative crisis curriculum. In J. Diem & R. Helfenbein (Eds.), *Unsettling beliefs: Teaching theory to teachers* (pp. 75–86). Charlotte: Information Age Publishing.
Garfinkel, H. (1967). *Studies in ethnomethodology*. Englewood Cliffs: Prentice Hall.
Glasser, H. (2007). Minding the gap: The role of social learning in linking our stated desire for a more sustainable world to our everyday actions and policies. In A. E. J. Wals (Ed.), *Social learning toward a sustainable world* (pp. 35–61). Waneningen: Waneningen Academic Publishers.
Gough, S., & Scott, W. (2006). Education and sustainable development: A political analysis. *Educational Review, 58*(3), 273–290.
Hodson, D. (2003). Time for action: Science education for an alternative future. *International Journal of Science Education, 25*(6), 645–670.
Hooks, B. (1990). *Yearning: Race, gender, and cultural politics*. Boston: South End Press.
Jasanoff, S. (2010). A new climate for society. *Theory, Culture & Society, 27*(2–3), 233–253.
Johnson, L., & Morris, P. (2010). Towards a framework for critical citizenship education. *Curriculum Journal, 21*(1), 77–96.
Latour, B. (2004). Why has critique run out of steam? From matters of fact to matters of concern. *Critical Inquiry, 30*, 225–248.
Latour, B. (2010). A plea for earthly sciences. In J. Burnett, S. Jeffers, & G. Thomas (Eds.), *New social connections: Sociology's subjects and objects*. Basingstoke: Palgrave Connect: Palgrave Macmillan.
Lundegard, I., & Wickman, P. (2007). Conflicts of interest: An indispensable element of education for sustainable development. *Environmental Education Research, 13*(1), 1–15.
Mueller, M. (2009). Educational reflections on the "ecological crisis": Ecojustice, environmentalism, and sustainability. *Science and Education, 18*, 1031–1056.
Patton, L. D. (2008). Learning through crisis the educator's role. *About Campus, 12*(6), 10–16.
Puig de la Bellacasa, M. (2011). Matters of care in technoscience: Assembling neglected things. *Social Studies of Science, 41*(1), 85–106.
Ruitenburg, C. (2005). Deconstructing the experience of the local. *Philosophy of Education, 446*, 212–220.
Sadler, T. D. (2009). Situated learning in science education: Socio-scientific issues as contexts for practice. *Studies in Science Education, 45*(1), 1–42.
Schnack, K. (1996). Internationalisation, democracy and environmental education. In S. Breiting & K. Nielsen (Eds.), *Environmental education research in the Nordic countries* (pp. 7–21). Copenhagen: Royal Danish School of Educational Studies.
Tuana, N. (2007). Viscous porosity: Witnessing Katrina. In S. Alaimo & S. Hekman (Eds.), *Material feminisms* (pp. 188–213). Bloomington: Indiana University Press.
Zeidler, D., Sadler, T., Simmons, M. L., & Howes, E. V. (2005). Beyond STS: A research-based framework for socioscientific issues education. *Science Education, 89*(3), 357–377.

Chapter 8
Using Collaborative Inquiry to Better Understand Teaching and Learning

Kenneth Tobin

Abstract In a research career that exceeds 40 years there have been continuous changes in my research methodology on a number of dimensions. The most salient differences involved changes in the theoretical framework that began with logical positivism and gradually changed to embrace sociological and cultural frameworks such as hermeneutic phenomenology, reflexivity, culture, and ethics. A necessity to include multiple voices to obtain participants' perspectives catalyzed ontological issues, including how to deal with difference and embrace polysemia. As well as researching patterns of coherence I adapted methodologies to build understanding based on research on contradictions, which defined events. Thus event-oriented inquiry sought to understand social life through intensive research on spikes in coherence trajectories. Authentic inquiry drew attention to priorities given to theory and improvement of practice on the one hand and multilectical relationships that considered authenticity holistically – recursively considering goals associated with changing ontologies while learning from others, teaching others about personal standpoints and practices, and ensuring that institutions and all individuals benefit from participating in research. I conclude with cautions about the transcendent nature of social inquiry and a reminder of obligations researchers have to participate ethically in research dialogues, listen to learn, and enact right speech to foster social justice for all.

Keywords Collaboration • Reflexive inquiry • Interpretive inquiry • Authentic inquiry • Event oriented inquiry • Multilevel inquiry • Multilogicality

---

K. Tobin (✉)
The Graduate Center, City University of New York, 365 Fifth Avenue,
New York, NY 10016-4309, USA
e-mail: ktobin@gc.cuny.edu

L. Bencze and S. Alsop (eds.), *Activist Science and Technology Education*,
Cultural Studies of Science Education 9, DOI 10.1007/978-94-007-4360-1_8,

Science Curricula as a Central Factor in the Reform of Science Education

In the 1960s there was an energetic debate about the necessity to change the nature of K-12 science education. At the time it appeared to science educators like me that there was agreement on the need for change and what needed to change, as well as healthy disagreement on what to change to and how to get there. John Lake, an influential science teacher educator from my native state of Western Australia, characterized the debate at the elementary level in terms of three nationally funded elementary science curriculum projects from the United States: Science – A Process Approach; the Science Curriculum Improvement Study; and the Elementary Science Study (Lake 1974). These curriculum projects had striking similarities and differences in their approaches to science education – each embracing science inquiry, but emphasizing different outcomes, and prescribing somewhat distinctive roles for teachers and students. Similar investments in curriculum projects having these characteristic orientations, also occurred at middle (e.g., Intermediate Science Curriculum Study) and secondary levels (e.g., Chem Study, Harvard Project Physics, Biological Science Curriculum Study). Differences spanned a variety of orientations (e.g., inquiry, historical, conceptual themes, and psychological foundations). Lake and many others at the time expected that research on the different approaches would somehow identify which approach was preferable and provide a pathway for improving science education. However, this was not to be. Even though there was a great deal of research undertaken on the enactment of different curriculum projects, the macro question of which approach was better was never answered definitively and I maintain that questions like these cannot be decided empirically or decisively by research. Research questions and associated research designs were oversimplified and answers usually failed to take into account participants' voices or quality of enactment. The question of which curriculum is best is macro in that it applies to multiple social fields and does not consider the importance of context, especially issues of implementation fidelity and details concerning the nature and quality of interactions among participants. Furthermore, debates about "which is best" seem to imply that social interaction is irrelevant. Myron Atkin and Paul Black (2003, p. 37) commented: "Both the 'teacher-proof' characterization and the concept of teacher-as-faithful-implementer later came to epitomize what many people saw as the arrogance of this style of curriculum development." It was assumed that teacher training would produce acceptable levels of implementation fidelity, which would then create experiences needed for all individuals to learn. To a large degree it was assumed that adherence to the activities suggested in the curriculum guides, which incorporated psychological learning theories, would enable all students to learn. Equity was considered in terms of opportunities to participate. Atkin and Black (p. 37) remarked succinctly that: "it did not work very well."

Chapter Overview

Although projects like the Elementary Science Study advocated student roles that emphasized autonomy and enjoyment, they did not consider students as research collaborators, curriculum developers, and coteachers. In effect, expanded roles for youth were constrained to peer collaboration – most notably cooperative learning (Johnson and Johnson 1999). In this chapter I describe a wider range of collaborative roles of participants in science education, including doing research for the purpose of improving learning environments, curriculum development, and teacher education. In so doing I illustrate how participants' roles have changed in relation to associated changes in research methodology. Research methodologies I address in the chapter include interpretive, authentic, and event-oriented inquiry. A particular focus concerns the standpoint of difference as a resource and its relationships to polyphonia, polysemia, and multilogicality. The centrality and high value our research squad assigned to collaborative inquiry are illustrated in cogenerative dialogue (hereafter cogen) and coteaching. In a broad treatment of polysemia I show how multilogicality and multilevel research provide complementary windows into social life and combine with other research methodologies to diverse perspectives on science education.

Changing Faces of Research and Science Education

Macro level approaches to framing research questions and the assumptions shared by science educators involved in research, teacher education, curriculum development, and policy, have striking similarities, many of which persist today. For example, a one-size-fits-all approach to theory may derive from adherence to empiricism and models for generalizability that are grounded in inferential statistics. The idea that the results of research applied to a sample and are generalizable to a population provide an underpinning for many graduate level courses in research methods and concerns with internal and external validity of scientific designs for research. Even when arguments were advanced for the use of qualitative data in research, the pressure to apply parallel criteria to interpretive research methods resulted in quality and authenticity criteria being developed and applied that appeared to embrace research that employed experimental and quasi-experimental designs. Accordingly, participants in interpretive research are often considered to be subjects and are referred to as a sample – inadvertently buying into a set of traps that would expose the methodologies and associated methods as deeply flawed and inferior to methodologies that embraced inferential statistics. The labeling of research methodologies as qualitative and quantitative set the stage for a debate that would take for granted many tenets of logical positivism. These included assumptions like the following: measurements and data are objective; the best outcomes from research are parsimonious rather than complex; well-designed research will gradually contribute to discovering a social reality or truth; samples involved in research should be randomly selected to be representative of a population to which outcomes are generalized; the presence of research and researchers does not affect outcomes; random selection of subjects from representative sites in a target area (e.g., city, state, nation) allow differences in individual attributes to cancel out and those that cannot be ignored in a model can be measured and statistically controlled.

Instead of definitive research in the decades that followed the 1960s, mainstream perspectives on the nature of science and axiological commitments of scholars and policymakers framed curricular choices and emphases included in hundreds and perhaps thousands of reports that recommended the reform of science education (Hurd 1997). Furthermore, rather than dramatically changing the faces of science education, reports that advocated reform and associated methods to enact reforms of various persuasions appear to have reproduced forms of science education that have proved to be resilient. Today the cycle continues – there are still calls for reform of science education and what happens in science classrooms bears a family resemblance to what happened in the 1960s when the Sputnik curriculum revolution was in full swing. Of course there were notable exceptions. For example, within the *Science Curriculum Improvement Study*, Mary Budd Rowe researched factors associated with science inquiry, as it was represented in verbal interaction (Rowe 1969). Her seminal work identified wait time, the duration of pauses within utterances, as an important variable associated with the quality of verbal interaction and the presence of pauses between utterances (Tobin 1987). Also, Rowe identified other factors, such as the incidence of verbal rewards, associated with characteristics of verbal interaction that made notable differences to participation levels and the quality of verbal interaction (Rowe 1974). Research like Rowe's addressed an assumption that issues concerning the quality of social interactions are important aspects of learning. Also, her work highlighted the fallibility of the assumption that the curriculum project used was the decisive variable related to the quality of science education and science achievement. There is no guarantee that what is designed and intended will occur during enactment. Certainly curriculum resources, planning, and local school-based factors all contribute to the quality of learning environments. Rowe's research emphasizes that social interactions are paramount when science learning is researched. Of course, the implications are that research about enacted curricula can provide insights into how resources are accessed and appropriated.

What is happening in science classes? A broad question like this would have many answers depending on the context in which science education is embedded. For example, I expect science in a prekindergarten classroom to be quite different than science at a high school level, and for a given grade level science in urban schools might differ from science in rural and suburban schools. Similarly, salient variations in context might include social constructs such as nationality, social class, gender, native language, and religion. The mediating roles of social constructs such as these are almost axiomatic. Perhaps not so obvious is that what happens also depends on how you look and what you can and do see. For example, in 1984 Jim Gallagher and I focused on classroom management, mainly because high school youth in our study were disruptive. Like so many classroom researchers at that time we made sense of learning and doing science education through Piagetian lenses, and adopted a stance that classroom order necessitated teachers establishing and maintaining effective control over students (Tobin and Gallagher 1987).

Knowledge does not exist independently of knowers or structured fields in which knowledge is both represented and enacted. A radical aspect of this assertion is that knowledge is only "known" when it is represented, as Erving Goffman (1983) noted, as a result of an interaction with social artifacts. Alfred Schutz (1967) put it another way; namely, that stocks of knowledge come to hand just in time during social interaction. This is an important idea with many implications for researchers. Social resonance focuses on knowledge as it is produced in the moment as structures unfold. Enactment, that is cultural production, supports fluency when it is timely, anticipatory, and relevant. For this to occur structures are anticipated as they unfold, and the knowledge needed to appropriate them comes to hand at precisely the right time. Since this process is continuous and involves a multifaceted structural flux, most of the process is automatic, beyond awareness, and non agentic. Emmanuel Lévinas (1999) referred to this process as passivity and Wolff-Michael Roth (2007) highlighted the importance of passivity to the agenda of science educators. To tap into passivity it is important to employ methodologies and associated methods that allow participants to become aware of their conduct and interactions that support their practices. Once they become aware they can reveal their ontologies in stories about what is happening and why is it happening. The analysis and interpretation of such stories can be an important thread in research in science education.

Here I argue that appropriate research needs to incorporate multiple methodologies and methods to examine curricular issues in ways that reflect their complexity, yield outcomes that are contingent and nuanced, and acknowledge that decisions about which approach is best will inevitably involve issues associated with axiology, ontology, and epistemology. Furthermore, experienced realities in the social world appear to be mediated by structures that situate individuals in different places in social space. If this is the case then research and science education would necessarily access participants' perspectives and understand similarities and differences in the realities participants perceive in a study. An important part of research methodology concerns ways in which similarities and differences are handled during analysis and interpretation. Theoretical stances concerning polysemia also are salient to ways on which research is designed and conducted.

My approach to research, which began in 1973, involved a gradual shift from quasi experiments and inferential statistics to test hypotheses to interpretive methods using predominantly qualitative data resources, affording emergent and contingent approaches to researching classrooms and schools in ways that were less reductive than our previous research that focused on variables and testing of pre-developed models. Even though it felt at the time that shifts in my methodologies were momentous, in a historical context they appear to be gradual and relatively slow. The most noticeable shifts involved changes from positivistic methodologies grounded in psychology to hermeneutic-phenomenological inquiry related to areas of sociology and anthropology. Increasingly I became aware that I would learn more from research that was multi-voiced and included different robust perspectives among members of our research squad. As I developed a greater understanding of cultural sociology and constructs such as structure and field, I began to understand the power of constructs such as multilogicality, transcendence, and the desirability of using different lenses to study social life. An increase in the complexity of our work necessitated the development of multilevel research methodologies and methods and involvement of teachers and students as researchers.

In my first 20 years of research, theories for teaching and learning were frequently grounded in constructivism and developmental psychology. Social interactions were important, but given a pervasive unruly characteristic of science classrooms the highest priority often was directed to establishing and maintaining control over students. Innovative ways of looking at motivation to learn were incorporated into theories of student agency (e.g., Brophy 1987). As different constructs were used to focus research, the answers to what is happening and why that is happening changed – as did implications for practice, orientating science curriculum, teaching, learning, teacher education, policy, and research.

One noteworthy limitation of our approach, which was beyond our awareness, was the potential impact of the way we considered/dealt with non-confirming data. Frederick Erickson (1986) made it clear that assertions needed to be modified to be consistent with all data – that is, nuance had to be built into the wording of assertions and to some extent non-confirming data had to be explained in the light of a study's assertions. The approach was consistent with a Geertzian model for culture (Geertz 1973) – consisting of thick coherence being enacted in fields contained by strong boundaries. At the time I was most heavily involved in interpretive research and it never occurred to me that culture was central to our research in ways that would deeply relate to my assumptions about epistemology, ontology, and axiology.

Joe Kincheloe and I described how social sciences and associated research and curriculum development have been saturated by pervasive systems of logic that include tenets of positivism, including a tendency to seek simplified causal models that afford prediction, control, and accountability (Kincheloe and Tobin 2009). Lake's idea that answers to macro-level questions such as, "Which approach to curriculum is best?" could be answered definitively (and objectively) by research is flawed – an example of an oversimplified question that implies causal relationships among sets of variables. The idea that curriculum quality can be considered independently of context reflects a reductive view of social life – one that easily could overlook social interactions that make far more meaningful differences than those associated with the type of curriculum used to enact science education. A key point to emphasize is that theoretical frameworks illuminate social life in ways that raise specific issues as salient and at the same time they obscure other ways of framing social life. In science education this point often appears not to have been acknowledged. Possibly due to tenets of positivism, theories are often considered as right or wrong rather than as alternative ways of experiencing, describing and making sense of social life. Different theories highlight patterns and associated contradictions, affording particular ways of construing and learning from research. Furthermore, little research has examined axiology, the values hierarchy that mediates what is considered central and of high priority as distinct from peripheral and of low priority. Often policy decisions are based on either-or thinking about choices.

Although science educators' methodologies and associated methods have changed continuously for the four decades I have been a science education researcher, I am still surprised by policy level pronouncements that are akin to main effects in statistically oriented research which produces assertions that have thick coherence – as if contradictions are not considered in models on which policy tenets are framed. Examples include assertions like the following: inquiry methods enhance science learning; argument strategies improve science achievement, and open-ended questions increase science achievement. One-size-fits-all claims are devoid of nuance and appear to ignore quality – for example, as if inquiry no matter how well, or fully it is enacted is preferable to no inquiry. There are many potential problems associated with research intended to validate best practices. Using a theoretical framework that includes levels of social life (macro, meso, micro): fields that are dynamically structured and unbounded, the enactment of any curriculum project is subject to an ever changing flux of structures that can produce culture that is simultaneously the same and different than what is produced when the "same" curriculum is enacted in another time and place. Rather than viewing enactment like a horse race it makes sense to adopt an approach that embraces phenomenology – learning from researchers' insights into what is happening from the perspectives of the participants and why it is happening. In this way landscapes can be created to reveal possibilities associated with the use of different curriculum projects in the context of them being enacted in different circumstances. Rather than producing simplified models in terms of clearly defined, significant variables, there are advantages in retaining complexity, acknowledging the salience of meanings in use, and recognizing that experiences described by language are underrepresented and always will mean more than can be expressed/represented using language. What is learned from such an approach to research would be grounded in contexts associated with the research (i.e., structures) and any claims about "what works" would be nuanced and considered an integral part of knowledge produced in the study. Users would understand that what is learned is replete with ever-present contradictions and any project involving enactment would necessitate contingent adaptivity that addresses the goals of individuals and collectives, levels of success, and dynamics of the agency|passivity dialectic (here the vertical bar denotes a dialectical relationship). Different theories highlight patterns and associated contradictions, affording particular ways of construing and learning from research.

Dealing with Difference in Research on Teaching and Learning

The relationship between an activity and theoretical frameworks used to experience and describe what happened in an activity are dialectically related. The relationship is synergistic in the sense that applying different theoretical frameworks provides new ways of looking at the activity and characterizing practices and their interrelationships. Theoretical lenses used to shed light on activity are reflected in questions like the following: what to tweak, what to expand, what to truncate, and what to discuss? It is important to realize that as well as shedding light on activity, theoretical lenses obscure other valuable aspects of an activity. This standpoint embraces the relevance of bricolage and polysemia to social inquiry and acknowledges that there are downsides to privileging any one set of frameworks.

A hermeneutic-phenomenological perspective adopts a stance that you can learn about social life by understanding participants' experiences in social life. Furthermore, the approach emphasizes that experiences should be represented by participants' voices. This approach invites possibilities of different accounts of experiences shared by participants who occupy different locations in social space. That is, polyphonia expands possibilities for learning about social life and invites a stance be taken on polysemia. How will researchers handle differences in the process of learning from research?

A revolution in my thinking occurred when I shifted my research to urban schools and included urban youth as researchers and teacher educators (Tobin et al. 2005). The catalyst for reform was that the schools, in inner city Philadelphia, were beyond my experience. Furthermore, when I endeavored to teach in ways that were consistent with how I believed science should be taught in urban schools, my failure to succeed was so pervasive that I needed to take stock of not only my own knowledge but also what was reported as the published "knowledge base of science education." The knowledge needed to teach urban youth had to be enacted. It did not exist independently of the dynamic structures of the urban science classes I had to teach. On the contrary, the knowledge to teach urban youth occurred where the rubber hits the road – in urban classrooms – constituted in dynamic structures as they unfold and are appropriated in chains of interactions. Knowledge of how to teach urban science education could not be separated from all participants' actions – that is, it was in the moment and certainly not something I possessed alone. Furthermore, only some of the knowledge was accessible to language. My experience was a notable example of knowledge being distributed across interaction chains that occurred in a field and that descriptions of research, available in research reports, to positively impact learning had to be enacted appropriately.

An ongoing problem in education generally and science education specifically is an emphasis on individualism. From this perspective learning is regarded as something that individuals do independently of others and elaborate assessment systems are developed based on this premise (Tobin 2012). Aligning with this assumption is a tendency to hold teachers accountable for their students' learning, narrowly construed and assumed to occur primarily at school. That is, science achievement for a particular period of time is a reflection of science teaching at a school during that period of time. At the very least the premises underlying assumptions like these are over-simplifications of very complex processes. The implications of enacting policies based on such assumptions are likely to have profound impacts on education in the near and distant future and have probably been associated with many of the inequities and inadequacies documented in the literature.

How might we think alternatively about individuals and collectives? In our research, for almost two decades, we have considered individual|collective as dialectically related, each recursively associated and presupposing others' existence. From this perspective, as individuals|collectives enact culture in a field, their productions (transformations|reproductions) are interconnected. A recursive relationship between individual and collective implies that changes in one are reflected in the other; the actions of any individual becoming resources for actions of a collective. That is, all individuals in a field are "in action" simultaneously and continuously, enacting culture that has a cascading effect since everybody's actions are resources for everybody else's cultural production. If a teacher acts in ways to expand the learning possibilities of others then, from this perspective, everybody is a teacher for everybody else because acting in a field provides resources to support others' learning. That is, teaching|learning are dialectically related and it is impossible to think of one without the other. Interrelationships between teacher and learner are inextricably linked and whereas learning cannot be separated from teaching, neither can teaching be separated from learning. Learners' actions mediate the possibilities for teaching at every moment enactment occurs in a particular field. Accordingly, it makes no sense to think of teaching in isolation from particular collectives, including students. As most teachers readily acknowledge, the way a person teaches one group of students is often quite different from the manner in which the same person teaches another group. To argue otherwise and assume that teaching can be considered independently of learning and learners is fraught with the potential for failed expectations. For example, accountability systems grounded in assumptions that teaching is a commodity that is transferable across contexts, including schools and students, is suspect at least and damaging at worst.

A current trend among scholars in science education is to consider identity as an outcome (Varelas 2012). There is acceptance of the idea that identities are forged as individuals participate in multiple fields as time unfolds. As individuals think back on what was accomplished in those fields, memory traces reconstruct what happened in much the way that a highlights reel is put together. Events that stand for enactment in a field are reconstructed and it is perhaps in association with these events that individuals construct images of "self" in particular fields. Obviously these constructed images are based on a reduced database and are subject to ongoing revision as an individual returns to a field over time. Whereas most recent studies think of identity as fluid and context dependent very few theoretical models have considered the full implications of an individual|collective relationship. If individuals are considered in relation to collectives in which they practice, then it makes sense for identity to be theorized dialectically rather than as a property of an individual.

Participants Doing Research to Understand and Improve Practice

I began to include high school youth as student researchers in a study I undertook with Stephen Ritchie, in Tallahassee Florida (Ritchie et al. 1997). In that study we utilized a middle school female as a student researcher and, although it did not work out as we envisioned or planned, we both retained our commitment to the idea that youth could provide valuable insights into what was happening and why it was happening. The initial problem we encountered was that the student researcher was not interested in our research and we found it difficult to motivate her to participate as a researcher. In contrast, my research at the University of Pennsylvania was quite different because the students provided their perspectives on the quality of teaching, suggesting ways to make improvements that would suit them. In other words their interests were central (Tobin 2000). Initially our tendency was to privilege their voices because their perspectives were valued. It took time and different frameworks for us to realize that quite likely the greatest benefits of students speaking about teaching and learning involved their participation in the activity. There was value in them speaking with other youth about teaching and learning, and their teachers, who were older and obviously different from them in many social categories. Engaging in dialogue with others who differed markedly in a number of social categories appeared to be a most valuable thing to do.

As director of teacher education at the University of Pennsylvania I inherited a research project proposed by Fred Erickson – largely premised on the idea that students could provide teachers with good ideas on how to be better teachers for kids like them (Tobin et al. 2005). The initial plan called for two youth to serve as advisers to new teachers at least once a week. We instructed the new teachers to select youth from their classes, keeping in mind their differences from one another, often selecting students who were having difficulties in the class. The advantages of the activity were evident almost immediately in that students were not only invited to evaluate the quality of teaching, but also to make specific suggestions about changes to enact. Many of these made an immediate difference and were highly visible, becoming objects for further dialogue in face-to-face meetings. Other benefits were less obvious. For example, in many cases the students involved had not had opportunities to speak with authority and be heard by adults – who were regarded as authority figures (e.g., teachers, school administrators). Not only did the youth make suggestions, but also they received requests for elaboration, clarification and further input. The youth felt respected and demonstrated shared responsibility for the quality of learning environments. During their regular face-to-face meetings the youth and their teachers developed social bonds that, in many cases, transferred into classroom settings. Evidence of such social bonds included cooperative interactions with the teacher and others and efforts to minimize their own and others' disruptive practices.

An unanticipated problem was that the students' voices were privileged in the activity. The youth were regarded as authorities and most of them spoke about exemplary teaching in terms of teachers effectively controlling students. Furthermore, they often considered high quality learning environments in terms of being silent and busy – for example, copying notes from the chalkboard or from a textbook (Tobin et al. 1999). Although youth were sincere, honest, and forthright, a problem resided in their logics about good teaching and learning, including their values concerning what was most important. Frequently students had bad ideas that were oversimplified and included strategies such as corporal punishment, isolation of offenders from others in the class, and exclusion of troublemakers from the class.

Listening to and Learning from Others' Voices

Emerging from the idea of students being mentors for their teachers, Roth and I developed cogen (Tobin and Roth 2006). We highly valued activities in which teachers dialogued with youth, not only sharing the amount and frequency of talk, but also listening and being heard by one another. Accordingly, we decided to undertake research on the nature of the dialogues and change the structure to expand its potential for improving learning environments and schooling more generally (Tobin and Roth 2006). Based on what we learned from youth dialoguing with teachers about "how to better teach kids like me," we expected participants to speak and listen in ways that were focused, in synchrony, and entrained across time and space. Cogen acknowledged that consensus was a goal of an activity in which participants understood one another's perspectives and goals, and endeavored to reach consensus on what was to happen next in class. A valued structure was the right for anyone to have and retain different perspectives while participating fully in the fields of class and cogen.

The research in which Roth and I developed cogen was situated in West Philadelphia. As we developed cogen we also created and researched a coteaching model in which new teachers taught together in urban classrooms for the purpose of better accommodating the needs of urban youth while at the same time learning to teach by teaching at the elbow of another (Tobin and Roth 2006). Cogens were organized to include four or five students together with all participating coteachers, researchers, university supervisors, etc. The requisite for being involved was that all participants in cogen needed to have been substantively and collaboratively involved in the teaching and learning of a lesson. Initially the purpose of cogen was to focus on participation in a dialogue that would identify ways in which the quality of the teaching and learning in the class could be improved in subsequent lessons. Typically cogens at the middle and high school level occurred after school or at lunchtime and occupied 40 min to an hour. Gradually cogen was regarded as an integral part of teaching and learning and teachers and students accepted cogen as part of the ongoing curriculum. The number of participants often included a whole class, and at times one-on-one cogen occurred when a teacher and student met together to resolve classroom-based issues.

Cogen focused on the idea that dialogue had the purpose of converging to produce consensus. Even though individuals may not be in agreement it was essential for participants to reach consensus and then accept responsibility for enacting what had been agreed. This was to change in a number of ways that reflected emergence, contingence, and the synergistic nature of the research in which we engaged. First, we noticed that students who had participated in cogen began to coteach with their teachers. Acceptance of the responsibility for enacting what had been agreed to in cogen resulted in those students assisting the teacher in a variety of ways that included managing the class and most importantly, assisting students with their understandings of what was being taught.

The research drew attention to an important set of dialectical relationships: teacher|learner and teaching|learning to name two. As we reviewed what was happening in classrooms and in cogen it was apparent that there would be times when teachers would be learners with respect to their students and at other times students would be learners with respect to their "official" teachers. These theoretical realizations provided new ways of looking into classrooms and of undertaking research on teaching and learning.

Over approximately 15 years of research and development the purposes of cogen have expanded. For example, because teachers and students differ quite significantly from one another in terms of salient social categories there is an opportunity for participants in cogen to develop adaptive forms of culture for successfully interacting with different others. We regard cogen as a seedbed for cultural production. When it is viewed in this way cogen is an activity that is quite central for new teachers to learn how to successfully teach in urban schools usually characterized by diversity and social categories such as race, ethnicity, native language, English proficiency, religion, and sexual orientation. Through careful selection of participants in cogen it is possible for them to learn how to interact successfully in culturally adaptive ways (Shady 2014). Even though a number of doctoral studies have been undertaken in which cogen has been used to improve the quality of teaching and learning and school level environments (e.g., Bayne 2012), there is obviously much more research that can be done within a sociocultural framework in which collaborative dialogue between individuals who are different from one another can be studied as it evolves in dynamically rich contexts.

Cogen also has been used as a research methodology to afford students and teachers enacting roles of researcher (Tobin and Llena 2011). Within a methodology that involves the enactment of cogen, teachers and students can enact a variety of methods that provide windows into the science of teaching and learning (i.e., the learning sciences). A feature of cogen is that it is an activity structured to foster polyphonia and associated radical listening (i.e., "making an effort to understand others' standpoints without seeking to change them" Hayes et al. 2010, p. xix). That is, everybody is encouraged to participate actively, and as they do so others listen with the explicit purpose of making sense of what is being said and exploring its affordances. Seeking alternatives is done only after a speaker's perspective is understood and its possible affordances have been fully explored. The speaker has a responsibility to "speak for the other" assisting to help others understand what is being proposed and to see its affordances. The speaker has a responsibility to promote interaction with the knowledge that focus will be maintained on the issue that is on the table until there is agreement to move on. At the same time radical listening occurs all participants are encouraged to practice right speech, especially if inequities/injustices are occurring in cogens or the class. When the structural aspects of cogen are enacted the research addresses the authenticity criteria (Tobin 2006) I adapted from Egon Guba and Yvonna Lincoln (1989). That is, participants all get a chance to lay out their ontologies and as a result of objectifying them they can expand and adapt them. Similarly, through radical listening all participants learn about one another's ontologies without seeking to change them. Right speech allows participants to focus on the affordances of all ideas, creating a climate in which consensus can be reached on how to improve the quality of science education institutionally. Similarly, as individuals listen and reflect on their own standpoints, they are well placed to benefit their and others' personal learning.

Initially our work on cogen was broadly theorized within a framework of cultural sociology. We broadened this framework to include the Heideggerian notion of learning by being in with others (Heidegger 1996). This idea was very prominent in our thinking about coteaching and ways in which actors became like the other by being with the other. This theoretical frame was applied also to the ways in which participants in cogen learned from one another. Because of our use of Randall Collins' framework concerning interaction ritual chains our initial concern was with synchrony in speech (Collins 2004). Accordingly, we structured cogen to focus on the distribution of speech, and synchrony and entrainment within and across interactions. For example, when somebody spoke we expected to see a strong focus on the speaker and signs of synchrony involving all or most participants in relation to the speaker. Similarly, at the same time we expected to see synchrony distributed across the entire community i.e., entrainment. Each speaker was expected to act not only for his/her self but also for others; that is, to provide opportunities for social resonance. Other structures also applied to equity in terms of who spoke orally – the number of turns of talk and the duration of talk. Furthermore, we emphasized the obligation of participants to speak for others, meaning that speakers should be attentive to the necessity of others making sense of what was being said and connecting with it in a multitude of ways. Speaking for others embraced a responsibility of each person for learning of the collective.

An initial concern we had in structuring cogen was that we needed a hedge against behaviorism. We did not want to assume that because people were not speaking explicitly that inner speech was not happening. Since we could not access individuals' inner thoughts it was important to emphasize to all participants that activity included inner as well as outer speech. We were explicit concerning legitimate participation including the thinking that occurs as others spoke. We consider this to be salient because the purposes of inner speech can be as varied as the purposes of outer speech (Vygotsky 1962). Obviously, focus, synchrony, and entrainment involve actions on the inside as well as actions on the outside – actions that are not directly accessible to others. Since we had legitimated inner speech we felt it was necessary to address the obligation of each participant to speak out when, and as necessary. This is what I mean by right speech. We considered there was an ethical responsibility for right speech to occur – that is, for individuals to contribute when they could advance collective goals and goals of individuals within a collective. We did not want individuals to sit quietly pursuing their own goals without accepting responsibility to participate equitably, ethically, and responsibly to benefit others in a collective.

Learning to Teach from and with Others

At the time we developed cogen we also were very interested in the development of coteaching models. Initially these models were designed to afford learning to teach for preservice teachers in circumstances where the resident teachers were unwilling to surrender their classes because they themselves were experiencing difficulties that often appeared insurmountable (Tobin et al. 2001). Accepting the advice of a school principal we decided to allow two preservice teachers to teach together without any supervision from a resident teacher. We were able to do this because the school principal was able to obtain emergency certification for the preservice teachers so that the coteaching activity was legally viable. We decided to move forward with this idea on the understanding that we would study it so that we could learn what worked, what we needed to tweak, and what we needed to discard. The initial experiment was so successful that we decided to adopt coteaching as a model for the entire high school teacher education program. At the time we had not fully worked out the characteristics of a heuristic that could be used to guide those who would enact coteaching and it was very much work in progress. In this case collaborative research was a necessity to develop heuristics that could be used to improve the quality of coteaching and broaden its use beyond initial teacher certification to include professional development of practicing teachers.

In order to undertake research on coteaching we opted for a collaborative approach that included new teachers, resident teachers, and high school youth as researchers. It was immediately evident that cogen was a suitable activity for research meetings. Accordingly, we folded coteaching and cogen together for the purpose of improving the quality of teaching and learning. As we did so we developed rules that structured the "talk about praxis" to ensure that power was distributed throughout all participants and that all participants were involved equitably. We had already included most of these ideas into the rule structure and use of the term dialogue was consistent with our theorizing the activity in terms of the work of Lev Vygotsky (1962) and Mikhail Bakhtin (1986).

Searching for and Learning from Spikes in the Curve

How to learn from difference? Having a background in physics and mathematics I am well grounded in statistical analyses in which residuals are calculated and often regarded as error or, having no meaningful consequence. The usual approach is to identify and interpret central tendencies taking them to account for the magnitude and source of variance. However, there are also methodologies that search for outliers and make sense of them. In the context of every voice representing lived experience I had a goal to interpret data resources in terms of central tendencies and contradictions. William Sewell's event-oriented inquiry opened up promising possibilities. He regarded an event as analogous to a rupture of a coherence trajectory – a spike in the curve. For example, if a teacher's average pulse rate while teaching is 98 bpm then a rise to 160 bpm might constitute a spike in the curve. An event would be selected to contain the spike. That is, all salient data would be examined before, during, and after the rapid increase in pulse rate. The selection of an event would be based on all data and would include the spike in pulse rate. Event analysis would then involve a bricolage consisting of methodologies such as multilevel, interpretive, and authentic inquiry.

Event selection begins with the identification of a significant contradiction. After that all data resources I used in the process of identifying and then analyzing an event. A feature of event-oriented inquiry is that we examine what is learned contingently so that the design for subsequent research can be expensive, taking account of what has been learned and continuing to learn more using whatever methodologies make sense in the circumstances. As is the case with other methodologies examined in this chapter event oriented inquiry is considered as a valuable component of a multilogical bricolage that underpins social inquiry that focuses on the science of teaching and learning.

Authentic Inquiry as an Overarching Methodology

Questions about the purposes of research arise from the adoption of models that involve participants as researchers. For example, we pondered the goals of research in terms of models that had privileged theory over practice in the sense that research that produced a new theory was favored over research that improved practices (Arendt 1958). We slowly increased our value for many purposes of research, favoring models in which different goals could be pursued collaboratively by stakeholders who learned from one another, respected the rights of others to hold different understandings and in fact different practices, and actively seek to attain equity, ensuring that all participants benefited from the research. I adapted Guba and Lincoln's models for fourth-generation evaluation (Guba and Lincoln 1989) to embrace polysemia and to accept all stakeholder groups as potential researchers (Tobin 2006) by adapting the four authenticity criteria proposed by Goober and Lincoln, authentic inquiry included two sets of goals related to theory production and to related to improved practices. This approach was consistent with Hannah Arendt's reminder that changes in theory and practice were both valued outcomes from activities such as research.

In order to emphasize authentic inquiry we focused on the creation of models that could be used to educate all participants about the research and what we were learning. Also we designed interventions to afford changes in all participants' understandings, their understandings of one another's different understandings and practices, and changes in conduct for individuals and collectives within the group of research participants.

One form of intervention we designed was quite direct and the other was relatively indirect. For example, a direct intervention involves the use of breathing to ameliorate teachers' and students' expressing high intensity emotions as increases in pulse rate and strength and low levels of oxygen dissolved in the blood. Based on our ongoing research and published literature (Philippot et al. 2002) we designed a breathing meditation intervention which we have now implemented to increase mindfulness at the start of each lesson. That is, the intervention reflects research undertaken by others and what we had learned from our ongoing research.

We also knew from our ongoing research that becoming aware of the unaware can provide participants with things to think about and possibly change. For example, in my research in urban schools, becoming aware that my habitus was breaking down led me to analyze video frame by frame and to get a student researcher to be a mentor for me (Tobin et al. 1999). Recently we have allowed teachers and students to wear finger pulse oximeters in class so that they would become aware of their physiological expression of emotions. Once they were aware of the possible salience of factors like pulse rate, strength of polls, and oxygenation they could use breathing techniques and other practices to gain control over these physical indicators of emotion – when, if, and as necessary.

Finally, we develop sets of characteristics for important constructs that we felt might be improved by allowing participants to become more aware about them. We refer to lists of characteristics for given constructs as heuristics. Two examples that have salience to this chapter are coteaching and cogen. Based on our ongoing research we developed lists of characteristics for coteaching and cogen and asked participants to think carefully about each characteristic in relation to their own conduct. The following are examples of heuristics we developed for cogen: I am respectful to others; I try to get others to contribute to discussions; I try to make sense of what others are saying; Others have opportunities to speak as much as I do; Others try to make sense of what I am saying; and I maintain focus. The following five point scale is provided for each characteristic: 5 = Very often or always, 4 = Often, 3 = Sometimes, 2 = Rarely, 1 = Never or very rarely. In addition, space is provided for participants to comment in regards to their experience with each characteristic. Becoming aware created of the characteristics for a construct like cogen creates a higher potential for participants to make changes on selected characteristics if, when, and as necessary. Importantly, awareness also opens up possibilities for passive change. If a person opens themselves to learning from others then it is possible that changes can occur in characteristics on a heuristic without conscious goals being formulated to make a change.

Heuristics afford change by heightening participants' awareness of characteristics associated with constructs that have emerged from our research as salient – in this case to coteaching and cogen. Heightened awareness creates a context for changing specific characteristics when and as it is deemed desirable to do so. We explicate characteristics of a construct (e.g., mindfulness) as short statements about the construct. The short statements serve the purpose of bringing particular characteristics to the awareness of those who use the heuristic. The inclusion of a Likert scale affords participants connecting each characteristic to their perceptions of its frequency of occurrence in a specific field. We try not to be repetitive, but instead include characteristics to stimulate reflexivity (Bourdieu 1992). As particular uses of a heuristic change in their contextual details we expect the characteristics included in the heuristic to be adapted to better-fit contextual details. We use the metaphor of "shape shifter" to convey the idea that a heuristic can change its characteristics for contexts of interest. Heuristics are used as part of authentic inquiry that employs design studies (Brown 1992) to plan, test, assess and adapt in an ongoing, non-linear cycle, as interventions are planned and validated to afford changes related to characteristics included in heuristics, or characteristics like those included in heuristics. At any moment in time, heuristics reflect our best and are enacted and disseminated to others. Accordingly, the structure of cogen and science teaching and learning in the participant schools will consistently evolve. In terms of emotions, emotional climates and physiological constructs, we will initially create descriptive landscapes. Through dialogues about these data and interrelationships among constructs, participants in the research will become aware of the possibilities for manipulating what happens in class to produce measures and patterns deemed to be desirable.

Brown et al. (2007, p. 212) describe mindfulness as "receptive attention to and awareness of present events and experience," involving nonjudgmental attention to present-moment experiences (e.g., sensations, cognitions, and emotions and sights, sounds and smells in the environment). According to Brown, Ryan, and Creswell, being mindful involves orienting attention toward registering facts observed, shutting down habitual processing, and making efforts to be present in the moment. As well as being less emotional, mindful individuals have greater: control over their thought processes; awareness of experience while being immersed in it; objectivity; tendency to defer judgment; likelihood to act as ecological stewards; levels of cooperation with others; and social attunement. Baer and Sauer (2009) regard mindfulness as a type of attention or awareness that includes qualities such as openness, acceptance, non-judging, non-reactivity, curiosity, and compassion. A concern expressed by Brown and Ryan (2003) is that attachment to emotions can reduce focus, productivity, and physical well-being.

Examples of characteristics developed for the mindfulness heuristic are: I am curious about my feelings as they occur; I easily find words to describe my feelings; I observe my thoughts without being caught up in them; I perceive my emotions without having to react to them; I am compassionate to myself when things go wrong for me; and I quickly recover when things go wrong for me. For each characteristic in the heuristic participants are asked to specify the frequency of occurrence that applies to their enacting the characteristic.

Research suggests that an increase in mindfulness will enhance wellness. For example, Davidson et al. (2003, p. 564) report that mindfulness, involving meditation, produces demonstrable effects on brain and immune function. Davidson identified six emotional styles corresponding with specific locations in the brain (Davidson with Begley 2012). *Resilience* varies from individuals who are slow to recover from adversity through to those who recover quickly when adverse circumstances arise. *Outlook* is an emotional style that pertains to how long a person can sustain positive emotion. *Social intuition* relates to the extent to which a person is adept at picking up social signals from others around him/her. *Self-awareness* concerns how well an individual perceives bodily feelings that reflect emotions (e.g., facial expressions, body temperature, pulse rate). *Sensitivity to context* has to do with an individual being able to regulate emotional conduct to take account of context. Finally, *Attention* concerns the sharpness and clarity of a person's focus. Individuals have a tendency to exhibit characteristic positions along continua associated with these emotional styles – positions that are not set in stone!

Depending on context and life experiences the primary patterns for any of the six emotional styles can vary due to neuroplasticity of the brain. This is a promising scenario as far as education is concerned because individuals might want to change their tendencies as far as some or all of the emotional styles are concerned – if, when, and as necessary. The research by Davidson and colleagues provides micro-level data, associated theories, and empirical validation for the plasticity/adaptability of the brain, raising promising scenarios for education to design and enact curricula that afford the development of tools related to changing emotional styles. Consistent with my involvement in multilevel research (Tobin and Ritchie 2011), our ongoing research is developing interventions that can be used in classrooms and other social institutions to afford individuals changing their emotional styles if, when, and as they choose to do so.

Reflections on the Changing Faces of My Research Methodologies

Doing research and science education is a journey I began more than 40 years ago. In that time I have focused my research on teaching and learning science and learning to teach science. Over time the focus has gradually evolved to building understandings of teaching and learning and learning to teach. Without privileging positivistic definitions of the nature of science, I referred to this evolving research focus as building a science of teaching and learning. My standpoint is that the research is an important part of science education and that more is learned by studying teaching and learning in many different contexts that include, but are not limited to: science, mathematics, music, martial arts, gardening, and everyday activities such as driving a motor vehicle.

As I explained in the chapter, the research methodologies I employed began with positivism and radical behaviorism (Vargas 1972), and gradually evolved to incorporate post-Piagetian constructivism and individual learning (von Glasersfeld 2007), reflexive sociology (Bourdieu 1992), cultural sociology (Sewell 2005), sociology of emotions (Collins 2004), and multilogicality (Kincheloe 2008).

At the present time our methodologies are constantly in flux as improvements in technology provide enhanced tools for multilevel research and we increasingly seek alternative knowledge systems to identify promising frameworks to illuminate our research (e.g., Buddhism, acupuncture, yoga). Our acceptance of methodological bricolage has produced a pastiche of methodologies that include the following forms of inquiry: interpretive, reflexive, multilevel, and authentic. Within this framework we see new knowledge about learning and teaching, embrace theoretical generalizability (Eisenhart 2008), and insist that research produce institutional improvements and equity for all participants. The project on which we have embarked his expansive and there is no logical endpoint, just as there was not a set beginning.

Long before my first formal study of science teaching and learning I was curious about teaching and learning science and gradually developed the tools that allowed me the privilege of joining a conversation that is ongoing. Importantly, the conversation is polyphonic, polysemia, multilevel, and radically continuous. Just as it is my privilege to join and contribute to the dialogue, that is research, the dialogue will continue with fresh voices, hopefully informed by the echoes of earlier conversations. The science is the dialogue that continues, a dynamic flux that moves through time and space, illuminating experience in particular ways while failing to even notice most of what happens. What we know and can learn is radically transcendent, and this thought alone suggests that what we know must be expressed with nuance, humility, and radical doubt – realizing that our knowledge is necessarily incomplete and inadequate. Having said that, we must continue to participate in the dialogue, being open to learn from difference and when the circumstances demand, speak forthrightly about what we know, need to know, and when and how to promote social justice. At the bottom, it is a great privilege to do research with others and the price to pay for the privilege is ethical conduct, compassion for others' well-being, and preparedness to respect and learn from others while maintaining willingness to educate them.

References

Arendt, H. (1958). *The human condition*. Chicago: University of Chicago Press.

Atkin, J. M., & Black, P. J. (2003). *Inside science education reform: A history of curricular and policy change*. New York: Teachers College Press.

Baer, R. A., & Sauer, S. (2009). Mindfulness and cognitive behavioral therapy: A commentary on Harrington and Pickles. *Journal of Cognitive Psychotherapy: An International Quarterly, 23*, 324–332.

Bakhtin, M. M. (1986). *Speech genres & other late essays* (V. W. McGee, Trans., C. Emerson & M. Holquist Eds.). Austin: University of Texas Press.

Bayne, G. (2012). Capturing essential understandings of the urban science learning environment. *Learning Environments Research, 15*, 231–250.

Bourdieu, P. (1992). The practice of reflexive sociology (The Paris workshop). In P. Bourdieu & L. J. D. Wacquant (Eds.), *An invitation to reflexive sociology* (pp. 216–260). Chicago: The University of Chicago Press.

Brophy, J. (1987). Syntheses of research on strategies for motivating students to learn. *Educational Leadership, 45*(2), 40–48.

Brown, A. L. (1992). Design experiments: Theoretical and methodological challenges in creating complex interventions in classroom settings. *Journal of the Learning Sciences, 2*, 141–178.

Brown, K. W., & Ryan, R. M. (2003). The benefits of being present: Mindfulness and its role in psychological well-being. *Journal of Personality and Social Psychology, 84*, 822–848.

Brown, K. W., Ryan, R. M., & Creswell, J. D. (2007). Mindfulness: Theoretical foundations and evidence for its salutary effects. *Psychological Inquiry, 18*(4), 211–237.

Collins, R. (2004). *Interaction ritual chains*. Princeton: Princeton University Press.

Davidson, R. J. with Begley, S. (2012). *The emotional life of your brain*. London: Hodder & Stoughton.

Davidson, R. J., Kabat-Zinn, J., Schumacher, J., Rosenkranz, M., Muller, D., Santorelli, S. F., Urbanowski, F., Harrington, A., Bonus, K., & Sheridan, J. F. (2003). Alterations in brain and immune function produced by mindfulness meditation. *Psychosomatic Medicine, 65*, 564–570.

Eisenhart, M. (2008). Generalized notions from qualitative inquiry. In K. Ercikan & W.-M. Roth (Eds.), *Generalizing from educational research* (pp. 51–66). New York: Routledge.
Erickson, F. (1986). Qualitative research on teaching. In M. C. Wittrock (Ed.), *Handbook of research on teaching* (3rd ed., pp. 119–161). New York: Macmillan.
Geertz, C. (1973). *The interpretation of cultures*. New York: Basic Books.
Goffman, E. (1983). The interaction order: American Sociological Association, 1982 Presidential Address. *American Sociological Review, 48*, 1–17.
Guba, E. G., & Lincoln, Y. (1989). *Fourth generation evaluation*. Thousand Oaks: Sage Publications.
Hayes, K., Steinberg, S., & Tobin, K. (Eds.). (2010). *Key works in critical pedagogy: Joe Kincheloe*. Dordrecht: Sense Publishers.
Heidegger, M. (1996). *Being and time*. Albany: State University of New York Press.
Hurd, P. D. (1997). *Inventing science education for the new millennium*. New York: Teachers College Press.
Johnson, D. W., & Johnson, R. T. (1999). Making cooperative learning work. *Theory into Practice, 38*(2), 67–73.
Kincheloe, J. L. (2008). *Knowledge and critical pedagogy: An introduction*. Dordrecht: Springer.
Kincheloe, J. L., & Tobin, K. (2009). The much exaggerated death of positivism. *Cultural Studies of Science Education, 4*, 513–528. doi:10.1007/s11422-009-9178-5.
Lake, J. H. (1974). The influence of wait-time on the verbal dimension of student inquiry behavior. Doctoral dissertation, Rutgers University, 1973. *Dissertation Abstracts International, 34*, 6476–A.
Lévinas, E. (1999). *Alterity & transcendence* (M. B. Smith, Trans.). New York: Columbia University Press.
Philippot, P., Chapelle, G., & Blairy, S. (2002). Respiratory feedback in the generation of emotion. *Cognition & Emotion, 16*, 605–627.
Ritchie, S. M., Tobin, K., & Hook, K. S. (1997). Viability of mental models in learning chemistry. *Journal of Research in Science Teaching, 34*, 223–238.
Roth, W.-M. (2007). Theorizing passivity. *Cultural Studies of Science Education, 2*, 1–8.
Rowe, M. B. (1969). Science, silence, and sanctions. *Science and Children, 6*(6), 11–13.
Rowe, M. B. (1974). Reflections on wait-time: Some methodological questions. *Journal of Research in Science Teaching, 11*, 263–279.
Schutz, A. (1967). *The phenomenology of the social world* (G. Walsh & F. Lehnert, Trans.). Chicago: Northwestern University Press.
Sewell, W. H., Jr. (2005). *Logics of history: Social theory and social transformation*. Chicago: University of Chicago Press.
Shady, A. (2014). Negotiating cultural differences in urban science education: an overview of teacher's first-hand experience reflection of cogen journey. *Cultural Studies of Science Education, 9*, 31–51. doi:10.1007/s11422-013-9486-7.
Tobin, K. (1987). The role of wait time in higher cognitive level learning. *Review of Educational Research, 57*, 69–95.
Tobin, K. (2000). Becoming an urban science educator. *Research in Science Education, 30*, 89–106.
Tobin, K. (2006). Qualitative research in classrooms: Pushing the boundaries of theory and methodology. In K. Tobin & J. L. Kincheloe (Eds.), *Doing educational research: A handbook* (pp. 15–59). Rotterdam: Sense Publishing.
Tobin, K. (2012). Sociocultural perspectives on science education. In B. J. Fraser, K. Tobin, & C. J. McRobbie (Eds.), *The international handbook of research in science education* (2nd ed., pp. 3–17). Dordrecht: Springer.
Tobin, K., & Gallagher, J. J. (1987). What happens in high school science classrooms? *Journal of Curriculum Studies, 19*, 549–560.
Tobin, K., & Llena, R. (2011). Producing and maintaining culturally adaptive teaching and learning of science in urban schools. In C. Murphy & K. Scantlebury (Eds.), *Coteaching in international contexts: Research and practice* (pp. 79–104). Dordrecht: Springer.

Tobin, K., & Ritchie, S. M. (2011). Multi-method, multi-theoretical, multi-level research in the learning sciences. *The Asia-Pacific Education Researcher, 20*(3), 117–129.
Tobin, K., & Roth, W.-M. (2006). *Teaching to learn: A view from the field*. Rotterdam: Sense Publishing.
Tobin, K., Seiler, G., & Walls, E. (1999). Reproduction of social class in the teaching and learning of science in urban high schools. *Research in Science Education, 29*, 171–187.
Tobin, K., Roth, W.-M., & Zimmermann, A. (2001). Learning to teach in urban schools. *Journal of Research in Science Teaching, 38*, 941–964.
Tobin, K., Elmesky, R., & Seiler, G. (Eds.). (2005). *Improving urban science education: New roles for teachers, students and researchers*. New York: Rowman & Littlefield.
Varelas, V. (Ed.). (2012). *Identity construction and science education research: Learning, teaching, and being in multiple contexts* (pp. 141–156). Dordrecht: Sense Publishers.
Vargas, J. S. (1972). *Writing worthwhile behavioral objectives*. New York: Harper & Row.
von Glasersfeld, E. (2007). *Key works in radical constructivism* (M. Larochelle, Ed.). Rotterdam: Sense Publishers.
Vygotsky, L. (1962). *Thought and language*. Cambridge, MA: MIT Press.

Chapter 9
From Knowledge to Action? Re-embedding Science Learning Within the Planet's Web

Laura Colucci-Gray and Elena Camino

Abstract Global environmental problems are on the rise. If on the one hand a great deal of knowledge is available about the natural systems and their physiological processes; on the other hand, our actions are accompanied by an increasing disorder of the global, ecological patterns regulating the existence of life on the Earth. To deal with such issues, a change of both culture and epistemology is required. The framework of sustainability science calls for a dialogical approach to knowledge production. It values epistemic and reflexive knowledge that is produced in the course of exchanges between disciplines, people and groups, across different sets of experiences, values and methodological frameworks. It is argued that this approach to knowledge production is ethically relevant – bringing forth the values of co-existence and legitimization of the other- and sits at the core of peaceful and sustainable relationships between humanity and the Earth. Hence dealing with complex socio-environmental problems – such as climate change – in education is not simply and solely a matter of content but it involves the redefinition of the process of 'knowing', which is both and at the same time cognitive and relational, emotional and ethical. In this view, learning in science will involve a multiplicity of knowledge competences – linguistic, social, logical, practical and creative – to value reflexivity and collective engagements in a global context.

Keywords Nonviolence • Participatory methodologies • Nature of science • Ecology • Equity

---

L. Colucci-Gray (✉)
Interdisciplinary Research Institute for Sustainability, School of Education, Aberdeen University, MacRobert Building, King's College Campus, Aberdeen AB245UA, UK
e-mail: l.t.gray@abdn.ac.uk

E. Camino
Interdisciplinary Research Institute on Sustainability (IRIS), Department of Life Sciences and Systems Biology, University of Turin, Via Accademia Albertina 13, 10123 Turin, Italy
e-mail: elenacamino1946@gmail.com

L. Bencze and S. Alsop (eds.), *Activist Science and Technology Education*,
Cultural Studies of Science Education 9, DOI 10.1007/978-94-007-4360-1_9,

Introduction

Back in the early 1990s, we began experimenting with participatory methodologies – namely debates to deal with complex and controversial socio-environmental issues in the science classroom. We shared the concerns raised by many educators around the world that an increasing separation was occurring between science teaching (and more generally, school life) and the lives of the children; at a broader level, we were witnessing the alienation of young people from civic debate particularly with respect to important issues that were affecting them both personally and as members of a community.

Not only the methods of teaching appeared to fail to motivate students, but more importantly, school science education as a whole appeared to be silent about the competences required by citizens to respond to the complex and problematic issues arising from scientific and technological developments. The problems of waste, resource depletion, risk and threats to human health and the environment were making their appearance on the news undermining the promise of a continuous, and increasingly prosperous economic growth; yet school science education remained unchanged and unchallenged – transmitting disciplinary notions and reinforcing the image of a science as a body of objective, consolidated and representational knowledge of the world.

Complex Problems and the Role of Science and Technology

Socio-environmental problems are on the rise and they are becoming increasingly more urgent. After more than two millions of years of evolution, in the past 200 years human beings have acquired the capability of inducing the largest and most profound transformations of the natural systems. Will Steffen et al. (2007) refer to the current 100 years as the Anthropocene, "a geological epoch in which human beings and their societies have become a global geophysical force, affecting global level changes in:

- The biological fabric of the Earth;
- The stocks and flows of major elements in the planetary machinery (such as nitrogen, carbon, phosphorus, and silicon);
- The energy balance at the Earth's surface" (p. 614).

While humanity accounts for about 0.5 % of the heterotrophic biomass on the Planet, it extracts about 32 % of the total, available Net Primary Production (Imhoff et al. 2004). NPP is the net amount of solar energy converted to plant organic matter through photosynthesis. It can be measured in units of elemental carbon and it represents the primary food energy source for the world's ecosystems. Human appropriation of net primary production (also known as HANPP) occurs for example, through the production and consumption of food (e.g. agriculture), the extraction of wood and fibres (e.g. in the production of paper) and mineral extractions; as a result, HANPP alters the composition of the atmosphere, the levels of biodiversity, the energy flows within food webs and the provision of important ecosystem services.

Hence global environmental problems range from the micro to the macro-scale and are located on multiple levels – from the level of molecules and compounds through to the higher levels of organisms and communities' life. The explosion of environmental 'issues' that we currently perceive can be effectively understood a global, environmental imbalances of the energy and matter flows through the global ecosystems, affecting patterns of food distribution, soil use and access to water services (Tilman et al. 2009).

Science and Technology are identified as key players in current analyses and discussion of the environmental crisis but their role is ambiguous. In relation to large-scale problems such as global climate change, scientific research grapples with numerical descriptions and estimates:

> We can now say with some confidence that the increased rainfall intensity in the latter half of the twentieth century cannot be explained by our estimates of internal climate variability. (Schiermeier 2011, p. 316)

The most rated academic journals embrace and augment the view of a science that will continue to bring solutions and innovation to overcome the current crisis:

> Scientists and engineers [...] share a belief that increased fundamental knowledge about the natural world will lead to human progress, because they see this happen in their own fields. (Alberts 2008, p. 1435)

From this it follows the encouragement of citizens, public administrations and policy-makers to confide in science in order to make private and public decisions, in full alignment with the well-known perspective described by Michael Polanyi (1962) and Thomas Merton (1968, 1973) who characterized Science as an open-minded, universalist, disinterested, and communal activity.

On the contrary, by rebutting to Alberts, David Guston, Daniel Sarewitz and Clark Miller (2009) underline the problematic implications of an idea of science, which seems to ignore the wider social issues that are emerging from the transformation and manipulation of ecological systems:

> Science and technology bring not only wonderful benefits, but also challenges and risks, from threats to personal and national security, to skewed distribution of wealth and social capital, to environmental and cultural degradation. (Guston et al. 2009, p. 582)

Science and its technological applications provide the essential know-how for undertaking increasingly extensive transformations of the natural systems, but also of the network of social and ecological interrelations, leading to unforeseen, unpredictable outcomes.

A Critical Role for Science Education

Science educators from around the world have recognised the importance of teaching students about the nature of science; extensive research from around the world is available documenting students' images of science (Kolstø 2001). Various strands of pedagogical research and innovation have followed, with the teaching of the history and philosophy of science in secondary schools as one of the examples (Matthews 1994).

With respect to the teaching of socio-environmental issues we find an interesting dialectics between the field of sociology of science and the field of education. At the outset of the first episode that hit the public's imagination – namely the nuclear explosion at Chernobyl in 1987 – sociologists of science have grappled with the emerging profile of a scientific and technological enterprise that was inextricably interwoven within the actions and values of a society. Citizens could no longer be simple spectators of technological and scientific innovations but the impacts of those innovations were to be effectively incorporated within a broader and more diffused form of 'citizen science' (Irwin 1995). Citizens would hold a type of knowledge which would not only allow them to make use of science but also to develop tools for empirical research to be applied on matters affecting them in everyday contexts. In this view, Melissa Leach and Ian Scoones (2007) further extended the discussion on citizens' science by highlighting the recognition of the global impacts of the production and use of science and technology; citizen science was not simply a local dimension of citizenship but needed to be understood as a form of wider awareness of the broader scenario of global inequities, resource management and associated environmental impacts.

It is in this context that curricular innovations such as Science, Technology and Society [STS] and Science Technology Society Environment Education [STSE] (i.e. Solomon 1990; Pedretti 1996) which occurred in the early 1990s can be understood as an important point of departure from traditional science education. The STS curriculum focused on developing citizens' awareness of the interconnections between facts and values in any scientific development and the necessity to develop abilities for active participation in democratic decision-making processes. This educational trend was recently revisited under the Public Understanding of Science initiative (Millar 2002), encouraging pedagogical implementations focusing more specifically on the acquisition of scientific language and skills to enable students (or the public) to produce logical, scientific arguments based on evidence (Lewis and Leach 2006).

All such approaches are clearly recognising the importance of preparing citizens to understand the world of science and technology and acquire practical knowledge for action. In face of the growing recognition of the global scale of socio-environmental problems however, another stream of epistemological reflection examining the role of science and technology in the global economy is placing stronger emphasis on the values and beliefs underpinning the fabric of scientific knowledge production (Jasanoff 2003). In particular, the recognition and acceptance of the limits of scientific knowledge along with the acknowledgement of the impacts of our actions in an increasingly complex world, should open the way towards forms of knowledge production that are inclusive of the voices of other people, communities, language and traditions (Aikenhead 2006).

Such change of epistemology leads to important changes of aims for education. The emphasis will shift from the acquisition of sophisticated scientific knowledge and technical skills to the development of relational, dialogical and reflective competences; the view is to build awareness and understanding of the biological, cultural and evolutionary webs of relationships that sustain our life on the Planet.

The Outline of a Science for Sustainability

Our experiences with the use of participatory methods alongside other experiences of science and environmental education have led us in recent years to reflect more deeply on the nature of scientific knowledge in the current globalised world (Colucci-Gray et al. 2006, 2013).

We are now living at a time in which Western science has taken a dominant role and it has become 'global'. International science journals feature contributions of scholars from all over the world that by means of the English language can communicate and collaborate in the production of new knowledge. Albeit its inevitable contradictions and revisions, modern science is considered both by the scientific community and the global society at large to be a 'universal' type of knowledge, and by implication, superior to other existing forms of knowledge.

Its relevant features are objectivity, generalizability, forecasting power and an increasing capacity for quantification and modeling of reality. In other words, Western science is perceived to be the most suitable and promising instrument for knowing the world and orienting future choices.

However, like in most situations, the positive aspects are always accompanied by the negative ones: the dominance of the Western view of science (and the use of a single language) has suppressed – and in some cases it has swept away – modes of interpreting and of 'being in the world' which have been elaborated in the course of thousands of years by other cultures and other populations. Hence an important element for ensuring adaptability and resilience of a community – its diversity – is going missing (Maffi 2005).

Acquiring Consciousness of Limits

A variety of critical voices can now be more frequently heard casting light on many aspects of modern Western science, with a view to understand the nature of its heuristic power but also the nature of its limitations:

- Objectivity: any description of the world cannot exclude the choice of a language which – by its very nature – is the result of a particular culture and a particular historical time. Hence every language automatically conveys a world-view (e.g. Dodman et al. 2008; Camino and Dodman 2009).
- Universality: contributions from post-colonial theory and gender studies in science have described the dominance of Western science as being the result

of historical domination rather than pure and simple superiority of interpretive power. Such considerations have acquired further relevance nowadays in the light of the increasing disorder of the ecosystems caused by particular modes of production and uses of scientific knowledge.

- Predictive power: the possibility to make predictions about the course of natural events appears increasingly challenged as man's techno-scientific power continues to impact at increasingly deeper levels in the network of life processes (for example, it is very difficult to predict the consequences of oceans' acidifications, or the increasing turbulence in the atmosphere, or even, the unknown effects of chemical compounds of different origin in the water systems etc.). As it is increasingly the case with studies on climate change for example, the consequences of our actions appear to be known only a posteriori (Schiermeier 2011).
- Heuristic features: because of inherent reductionist nature, certain events are potentially known by scientists but they are not measured because they are considered to be irrelevant, expensive or unlikely (e.g. the additive effects of pesticides in the aquifers, on bees populations, or on respiratory conditions…). Some variables or processes that fall outside our mental schemata do not acquire the status of variables (e.g. the biological components of the biosphere). Finally, the complexity and interdependency of the natural systems are ignored (Capra 2002).
- Neutrality: the growing gap between rich and poor, the inequities in resource distribution, the different levels of attention paid to the problems of restricted élites when compared to the problems of those who are deprived of essential means, are not simply deriving from bad governance but they are also related to the choices and responsibilities of the scientific community.

While many people – by virtue of what is mainly a traditional idea of science – identify scientific and technological progress as the most secure way for resolving the global environmental crisis (e.g. Keith et al. 2010; Lovelock and Rapley 2007), there are many others (Orr 1992; Panikkar 2005; Sachs 2002; Sterling 2002) who are asking whether the problem does not lie so much with the environment but with a particular way of thinking, the worldview that our society has elaborated and which is translated into political choices, environmental practices, individual behaviours and more specifically in the ways we produce and make use of scientific knowledge about the world in which we live.

In this perspective, the sustainability of human presence on the planet is dependent upon a radical shift of both culture (Worldwatch Institute 2010) and epistemology (Gallopin et al. 2001). As indicated by Bruno Latour (2007) we need to recognise our nature of 'Earthlings' and move away from the idea of modernization and emancipation from Nature to a concrete scenario of explicit recognition of our identity as beings dependent on Nature:

> Everything that earlier was merely "given" becomes "explicit". Air, water, land, all of those was present before in the background: now they are made explicit because we slowly come to realize that they might disappear – and we with them. (p. 3)

The recognition of our dependency from the natural systems calls for a re-thinking of current views of science for "we are still trapped in conceptualizing it from our old bases, rather than achieving any real shift of consciousness and of being" (Sterling 2002, p. 2). In this perspective, knowledge can be acquired along with a reflective and critical attitude about how, why and when our own knowledge and other ways of knowing (epistemological reflection about learning) have been developed.

Opening Up Towards Dialogue

Scholars working within the field of 'sustainability science' (Kates et al. 2001; Gallopin et al. 2001; Clark et al. 2005) have called for the importance of continuous dialogue amongst experts of different disciplines in order to allow for integration of knowledge and the enrichment of perspectives. Another strand of reflection in sustainability science, namely the concept of post-normal science advanced by Silvio Funtowicz and Jerome Ravetz (1999) argued for the inclusion of all relevant stakeholders in any debate surrounding science in society. To deal with the complex and controversial problems that we are continuously facing, a multiplicity of legitimate perspectives is needed. This mode of knowledge production recognises that qualitative experiences, personal experiences of a place and familiarity with a situation or context are all equally legitimate contributions and can have relevance alongside expert scientific knowledge when addressing open questions.

Other scholars (natural scientists as well as anthropologists, philosophers and educators) stress the need to question the traditional boundaries of academic science in order to enter into dialogue with different ways of seeing the world and create together new research paths. According to Tim Ingold (2010) Western thought built upon the Aristotelian premises of form (morphé) and matter (hyle) – has progressively shifted towards a conceptualisation of the world that is made prevalently of well distinct and separate objects while losing sight of the 'process' that is associated with all things. In recomposing our culture – a process of healing as indicated by Brian Goodwin (2007) – we can transform the boundaries of traditional scientific practice towards a richer and more significant way of practicing science, by means of re-educating and re-equipping ourselves for a new life style, that includes cooperation with other cultures and with nature (Goodwin 2007).

The Thinking of the 'Others'

Recently a special issue of the international journal '*Cultural Studies of Science Education*' has been entirely devoted to reflecting on the relationships between scientific knowledge and indigenous knowledge. In the introduction,

Glen Aikenhead (2008) underlines the substantial difference that has characterised the two forms of knowledge for a long time: one is directed towards the description of how the world functions (episteme) and the other one is directed towards undertaking actions in the world (phronesis). Aikenhead also notes that phronesis does not simply mean 'practical thought' but 'practical wisdom': it is prudent, necessary, moral and appropriate and it is a way of knowing the world recognised by the majority of indigenous populations.

On the contrary, the equivalent of episteme is not recognised by indigenous populations. This concept presupposes an idea of knowledge that is separate from the subject: this idea is alien to the indigenous perspective that recognises an intimate interconnection between the knowing subject and its ways of living (Aikenhead and Ogawa 2007). By means of this dialogue with other forms of knowing we may be able to recognise not simply the differences but more specifically the limitations characterizing our own form of knowledge; with the idea of limitations we refer to those aspects that modern science had indeed neglected or suppressed (Berkes et al. 2000).

Glen Aikenhead and Masakata Ogawa (2007) listed differences between aboriginal knowledge and western scientific knowledge at a number of different levels, involving methodologies, values and objectives. For example, the two forms of knowledge differ for their social goals; the wisdom-in-action aimed at survival differs from the idea of individual scientific credibility, corporate profits and knowledge for its own sake. The two systems also differ for their intellectual goals: co-existence with the mystery of nature through the maintenance of a host of relationships is contrasted by the Western idea of eradication of mystery by means of description. Other important differences also exist at the level of the relationship between knowledge and action: the intimate, subjective, moral and ethical wisdom stretching to account seven generations to come is compared with the formally and objectively decontextualized knowledge of western science. Finally they also differ at the level of cosmology: the holistic perspective of aboriginal people with their accommodating, spiritual, intuitive wisdom differs from the collection of concepts, principles, and techniques that are mainly dualist, reductionist, anthropocentric, and that aspire to universality. They even diverge for their concept of time: circular for the aboriginal people and linear for western science.

While often presented in counter-opposition, Aikenhead and Ogawa (2007) maintain that the two different knowledge systems should enter some form of dialogue; yet this is different from the accommodation of one perspective into the other (Stephens 2000). Rather, such dialogue may begin from a re-thinking of the idea of knowledge in context, which is not simply applied knowledge or traditional, local knowledge. Such expressions – as indicated by Aikenhead (2006) – are not suitable to convey a new idea of knowledge but remain encapsulated within a Eurocentric framework according to which only particular ways of knowing are given legitimacy and authenticity.

Ways of thinking that diverge from the dominant one can be found in modern times, both in the East and West. For example, in current modern India, a group of scholars has recently published a "Manifesto on Science and Technology" (KICS 2009) in which it is pointed out – in the first instance – that in the new knowledge society, universities and research labs are no longer the sole producers of knowledge. The extraordinary development of electronic means of communication that are accessible to the public obliges those places that are traditionally invested with the production and transmission of knowledge to enter into contact with other sources and ways of knowing, and which are conveyed by a variety of new languages, symbols and images. After this premise, the authors introduce the concept of 'lokavidya' a perspective on knowledge which recognises that everyday life is a centre of knowledge production; not simply a place to apply knowledge built or constructed elsewhere. So for example, the knowledge held by a group of craftsmen working with wood, clay, metal and other materials used to produce objects of common use (clothes, tools, toys and so on...) is a dynamic form of knowledge, which is continuously changing and adapting in relation to the availability of raw resources, market forces and technical advancement. Hence it is both traditional and modern knowledge, practical and reflective because it has dynamically responded and adapted to new conditions (Gupta 2000). Within this perspective, the authors of the manifesto found a means for achieving goals of social sustainability and a nonviolent approach to dealing with nature:

> By reworking the idea of the citizen as possessing a repertoire of knowledge, the secluded spaces into which modern knowledge has condemned the nomad, the tribal and the informal economy are opened up (...). To pluralise time is to pluralise the possibilities of life and living for cultures that do not follow modernity calendars. If time becomes unilinear and historical, the tribe might remain only as oral memory and the craft may only survive as an archive. The challenge here is mutual and reciprocal. The poetics of modern science lies in the multiplicity of time that it offers. (KICS, p. 13)

Enhancing Life Through Cultural Diversity

On the importance of biodiversity there is widespread agreement, both in science as well as in social research and education. The United Nations proclaimed the year 2010 as the year of biodiversity and called for a series of concerted efforts at the International level to support research and action aimed at promoting and safeguarding the Earth's biodiversity.

Since the 1980s however, there were calls for the recognition of cultural diversity as a good to protect and respect: "cultural capital includes the large variety of ways in which societies interact with their environments: thus it includes cultural diversity" (Gadgil 1987). Some scholars proposed to overcome the dichotomy that exists between 'biological diversity' and 'cultural diversity' (and which is particularly evident since the first documents produced by the Rio Convention in 1992), pointing to the recognition that such separation is typical of Western thought, as it was already observed by Gregory Bateson: "the continuum of nature is constantly fragmented in a discontinuity of variables in the act of description" (cited by Berkes and Berkes 2009, p. 12).

In this scenario we can find numerous scholars who have argued about the risks and dangers of monocultures – also those ones of the mind (Shiva 1998). A group of scholars (Pretty et al. 2009) has recently published a literature review on the topic summarising that:

- The diversity of life includes both biological diversity and cultural diversity which is expressed in beliefs, values, views of the world and cosmologies;
- Every natural environment provides the basis for the development of cultural processes, actions, beliefs; the interaction between nature and culture gives rise to a variety of language systems and modes of sociality which are intimately interconnected with their respective ecosystems.
- Many cultures and particularly those elaborated by indigenous and non-industrialised populations – have developed worldviews in which the human component is strictly interdependent with nature; they have put in place practices, norms and institutions aimed at maintaining a relationship that will allow for indefinite access to natural system services, source of life, food and well-being.
- The loss of cultural diversity which is happening by means of extinction of the language systems, cultures and knowledge will inevitably bring a loss of biodiversity which for thousands of years had been safeguarded by a complex set of practices based on local knowledge.

These scholars discuss how the mechanisms of cultural assimilation associated with the domination of western science are depriving us of a wealth of approaches, experiences and ways of being that could – instead – be a means for enrichment of the Western view. Without a dialogue between cultures we are deprived of that "other" who can allow one (as a process of mirroring) to acquire awareness of oneself and that by means of dialogue could produce new forms of creativity, adaptation and resilience.

Redefining Science Education?

To question the universality and dominance of modern western science is equivalent to re-discussing the whole of the epistemological and methodological implantation of science teaching (e.g. Aikenhead 2006, 2008).

If we share the idea that a multiplicity of views and ways of interpreting the world is healthy and necessary, then it is important to critically assess one's own educational role: for example, the implications of the hidden curriculum and the posture of the teacher in the class; but more generally an important role is played by the school, with its structures, timetables, textbooks, assessment criteria and the ways in which they exclude and silence alternative voices, languages and knowledge systems.

In light of these reflections, our task has been devoted to carrying out research on participatory learning contexts where attention was paid to both the learning process and its contextual conditions, namely looking at personal involvement, self-expression, development of relationships and ethos. In this scenario, role-plays alongside other methodologies of interdisciplinary and interactive teaching acquire new educational value. They become instruments through which students can reflect on the problematic features of disciplinary integration (for example, by uncovering the methodological basis underpinning the definition and meaning of particular concepts); at thc methodological level, students can examine the working of political and social structures and recognise the value and necessity to hear other people's voices and views (i.e. the non-experts, the citizens, people of different cultural extraction ...) in order to elaborate ways of living which are both biologically and socially more sustainable.

From Objective and Objectifying Knowledge to a Science of Relationships

A possible itinerary (one of many...) that could help shape one's own way of teaching science in a way that is a little different from the traditional approach could be that of always keeping in mind the *necessities of the other*.

- Other... is the colleague next door that is also involved in teaching my students in the realms of language, art or history. Interdisciplinary collaboration is crucial for helping students to re-compose the fragmented mosaic of knowledge. Even small-scale experiences may open the way to other initiatives that may become established over time.
- Other... are my students, each one with a different personality, history, talents, sometimes even language and culture. Each one of them has a treasure of experiences and knowledge that could – in appropriate conditions – be shared with one's peers and the teacher as part of an educating community, operating under principles of socio-constructivism (e.g. Roth and Lee 2004) and practices of Participatory Action-Research (Wicks and Rearson 2009).
- Other... is my body, that I have been taught to approach as if it was separated from me, a shell, a passive instrument to the services of my will. Recent research in the neurosciences (e.g. Gallese 2005, 2010) is showing the inseparability of motor and sensorial structures on the one hand and the environment on the other. Both the body and the environment interact in giving shape to the person in its entirety and provide the ability to adapt and respond to the world. Adolescents' bodies trapped at their desks cannot be doing a good service to cognitive structures and neither can they produce motivation for learning. A more dynamic management of the class and the use of forms of communication that include the body can promote and strengthen the embodied cognition.
- Other... is nature, from which we are progressively more separated. It is no longer a question of not knowing habitats and animals; behaviors but it is the sense of not feeling at home in nature, not knowing how to move on foot on a

mountain path or feeling lost in a woodland, or not being able to perceive the difference between a material object (a 'product') and a living entity (a 'process' in continuous becoming). Getting out of the enclosures of the classroom and the school to immerse oneself in the woods or climb up a mountain could (or maybe should?) be considered a necessary aspect of science education.
- Other... is my internal I, which is often silenced and left aside but with whom I can get in contact through the experience of silence. Experiences of active silence training proposed to children and young people have produced evidence of the importance of a period of silence as a means for re-encountering oneself, restore attention and develop biophilia (Barbiero 2007).
- Other... are those people that are different from me, carrying other cultures and other sciences, who look at the world with different eyes and tell the world with words that I do not know. Those are the people that continuously make me aware that there are infinite possibilities for existence and for achieving a meaningful life and together they allow me to know and transform myself by means of comparison. Educational exchanges between schools from different parts of the world can be a source of extraordinary discoveries about one's culture and one's environment (Ferrero et al. 2005).

The need for all these 'others' can be fulfilled by means of a relationship that is dialogue. Science can thus be re-thought as a modality of dialogical relationship between living things, people and the natural systems. The premises of such change are both epistemological and methodological:

- Not ONE science but many different ways of knowing and interpreting the natural systems, which are in dialogue with one another (epistemological pluralism);
- Involvement of all subjects (participatory democracy);
- Recognition of the 'embodied nature' of all levels of life: from the cells, the body, the mind to the socio-ecosystems that evolve to produce the holarchy of Gaia (Sahtouris 2000).

Dealing with Conflict

Any dialogical relationship does not exclude conflict. Rather, every relationship contains an intrinsic element of conflict (Galtung 1996). For such reason it is important to develop competences for action that aim to develop non-destructive relationships; life on Earth accounts for a multiplicity of such relationships: from the mitochondria in symbiosis with the primordial cells, to the male and female gametes allowing for the birth of a new individual; from the microscopic guests living on and inside our bodies to the different forms of life populating the same ecosystem, and finally, to the strategies of indigenous populations adapting to their natural environments. However, acquiring a positive attitude towards conflict and more generally, towards other people who may have different intentions, interests and goals, is not something that occurs spontaneously. It is not simply a matter of tolerance for other views. It requires attentive listening to oneself with respect for other people's views: this is a psychological and relational attitude requiring development and practice.

According to the framework of sustainability science, in order to deal with the pressing problems facing us at the local and global levels we need to gather not only a variety of scientific information and knowledge but also a multiplicity of points of view, which are all equally legitimate. However, the legitimisation of the different voices is possible only if Western culture – and its most powerful instrument – techno-science – learn to step back from the ambition of supremacy and universality that have characterised its development so far.

Already a few years ago, Robert Chambers (1997) put this concept very clearly: in order to establish genuine dialogue between two participants, simply valuing the other is not sufficient...but one needs to revisit the impact of one's presence. With the words of Chambers:

> The first step: empowerment. Putting the last first: altruism, generous behaviour towards poor, weak, isolated, vulnerable... but the first remains first.
>
> To go further: disempowerment... putting the first last: the uppers give up something, make themselves vulnerable... and gain effectiveness, liberation, fulfillment. (p. 138)

This perspective challenges established criteria of policy-making in science and the industry which are often based on an economics of efficiency and utilitarian ethics. If the last is to be first, or, according to Gandhi's Talisman – if the activity should ameliorate the condition of the weakest and the poorest individual – or, according to Christ's pronouncement, that any actions should be assessed with regard to the smallest person, it would follow a change also at the cognitive level to account for issues in science and technology that have been traditionally associated with a non-cultural view of scientific knowledge. So for example, if science and technology are preoccupied with the large electricity systems, fuels and efficiency of production and distribution of power, a non-violent and pluralist view would look at everybody's opportunities to exist and affirm one's life within the biophysical limits of the biosphere. Vaclav Smil (2008) for example expressed this concept in numerical terms:

> [a]t the beginning of the twenty-first century a purposeful society could guarantee a decent level of physical well-being and longevity, varied nutrition, basic educational opportunities and respect for individual freedom with annual TPES[1] of 50–70 GJ per capita. (p. 387)

So in a perspective of democratic pluralism based on equity, the analysis of energy flows and matter transformations is no longer and not exclusively pertaining to the realm of technocratic expertise or political analysis; rather, it is process of collective concern and engagement which is founded upon the observance of humans' basic needs, cooperation and multiple space-time perspectives.

---

[1] TPES (Total Primary Energy Supply): in 003 the absolute range was from 1 GJ pro-capita for the poorest African countries to the 450 GJ pro-capita of Canada.

Conclusions

Currently, with the many changes that have occurred in society, with the multiplication of environmental problems, the growth of techno-scientific power, and the rising of inequities within society and amongst populations (Day et al. 2009), it is possible to give new value to participatory and interdisciplinary activities. The aim is that of re-composing the fractures, contextualising situations and allowing young people to practice with direct forms of dialogue: in the first instance, by putting into relation the body with emotions, images and languages – and so to give meaning to the many scientific concepts that by means of such activities can be acquired and integrated. In addition, we need to develop specific competences for nonviolent transformation of conflict (Galtung 1996). Constructive transformation is at the heart of nonviolence and arguably, the most important acquisition of modern political culture for entering a sustainable phase of global socio-environmental change.

References

Aikenhead, G. S. (2006). *Science education for everyday life. Evidence-based practice.* New York: Teachers College, Columbia University.

Aikenhead, G. (2008). Objectivity: The opiate of the academic? *Cultural Studies of Science Education, 3*, 581–585.

Aikenhead, G. S., & Ogawa, M. (2007). Indigenous knowledge and science revisited. *Cultural Studies of Science Education, 2*, 539–591.

Alberts, B. (2008). A scientific approach to policy. *Science, 322*(5907), 1435.

Barbiero, G. (2007). Una mente silenziosa immersa nella natura. In E. Camino (Ed.), *Di silenzio in silenzio.* Cesena: Anima Mundi.

Berkes, F., & Berkes, M. K. (2009). Ecological complexity, fuzzy logic and holism in indigenous knowledge. *Futures, 41*, 6–12.

Berkes, F., Colding, J., & Folke, C. (2000). Rediscovery of traditional knowledge as adaptive management. *Ecological Applications, 10*(5), 1251–1262.

Camino, E., & Dodman, M. (2009). Language and science. In D. Gray, L. Colucci-Gray, & E. Camino (Eds.), *Science, society and sustainability education and empowerment for an uncertain world.* New York: Routledge.

Capra, F. (2002). *The hidden connections.* New York: Doubleday.

Chambers, R. (1997). *Whose reality counts? Putting the first last.* London: ITDG Publishing.

Clark, W. C., Crutzen, P. J., & Schellnhuber, H. J. (2005). *Science for global sustainability: Toward a new paradigm* (CID Working Paper No. 120). Cambridge, MA: Science, Environment and Development Group, Center for International Development, Harvard University.

Colucci-Gray, L., Camino, E., Barbiero, G., & Gray, D. (2006). From scientific literacy to sustainability literacy: An ecological framework for education. *Science Education, 90*(2), 227–252.

Colucci-Gray, L., Perazzone, A., Dodman, M., & Camino, E. (2013). Science education for sustainability, epistemological reflections and educational practices: From natural sciences to trans-disciplinarity. *Cultural Studies of Science Education, 8*(1), 127–183.

Day, J. W., Jr., Hall, C. A., Yáñez-Arancibia, A., Pimentel, D., Ibáñez Martí, C., & Mitsch, W. J. (2009). Ecology in times of scarcity. *Bio-Science, 59*(4), 321–331.

Dodman, M., Camino, E., & Barbiero, G. (2008, September). Language and science: Products and processes of signification in the educational dialogue. *Journal of Science Communication, 7*(3), 1–8

Ferrero, E., Cappa, A., Lovesion, P., Prando, R., Molinaro, E., & Mortara, G. (2005, October 2–6). Proposte di educazione alla sostenibilita' attraverso un progetto di gemellaggio e cooperazione. In *Proceedings of the third environmental education world congress,* Torino (Italy) (pp. 121–125).

Funtowicz, S., & Ravetz, J. (1999). Post-normal science: An insight now maturing. *Futures, 31*(7), 641–646.

Gadgil, M. (1987). Diversity: Cultural and ecological. *Trends in Ecology and Evolution, 2*, 369–373.

Gallese, V. (2005). Embodied simulation: From neurons to phenomenal experience. *Phenomenology and the Cognitive Sciences, 4*, 23–48.

Gallese, V. (2010). Embodied simulation and its role in intersubjectivity. In T. Fuchs, H. C. Sattel, & P. Henningsen (Eds.), *The embodied self. Dimensions, coherence and disorders.* Stuttgart: Schattauer.

Gallopin, G., Funtowicz, S., O'Connor, M., & Ravetz, J. (2001). Science for the twenty-first century: From social contract to the scientific core. *International Journal of Social Science, 168*, 219–229.

Galtung, J. (1996). *Peace by peaceful means.* London: Sage.

Goodwin, B. (2007). *Nature's due. Healing our fragmented culture.* London: Floris Books.

Gupta, A. K. (2000). Shifting cultivation and conservation of biological diversity in Tripura, northeast India. *Human Ecology, 28*, 605–629.

Guston, D. H., Sarewitz, D., & Miller, C. (2009). Scientists not immune to partisanship. *Science, 323*(5914), 582.

Imhoff, M., Bounoua, L., Ricketts, T., Loucks, C., Harriss, R., & Lawrence, W. T. (2004, June 24). Global patterns in human consumption of net primary production. *Nature, 429*, 870–873.

Ingold, T. (2010). Bringing things to life: Creative entanglements in a world of materials. *Realities* Working Paper, No. 15.

Irwin, A. (1995). *Citizen science: A study of people, expertise and sustainable development.* London: Routledge.

Jasanoff, S. (2003). Citizens' participation in governing science. *Minerva, 41*, 233–244.

Kates, R. W., Clark, W. C., Corell, R., Hall, J. M., ... (2001). Sustainability science. *Science, 292*(5517), 641–642.

Keith, D. W., Parson, E., & Morgan, M. G. (2010). Research on global sun block needed now. *Nature, 463*(28), 426–427.

KICS (Knowledge in Civil Society). (2009). *Knowledge Swaraj: An Indian Manfesto on science and technology.* Accessed at http://kicsforum.net/kics/kicsmatters/Knowledge-swaraj-an-Indian-S&T-manifesto.pdf

Kolstø, S. D. (2001). Scientific literacy for citizenship: Tools for dealing with the science dimension of controversial socio-scientific issues. *Science Education, 85*, 291–310.

Latour, B. A. (2007, April). *Plea for earthly sciences keynote lecture for the annual meeting of the British Sociological Association, East London.* Accessed at http://www.bruno-latour.fr/articles/article/102-BSA-GB.pdf

Leach, M., & Scoones, I. (2007). *Mobilising citizens: Social movements and the politics of knowledge* (IDS Working Paper 276). Brighton: IDS.

Lewis, J., & Leach, J. (2006). Discussion of socio-scientific issues: The role of science knowledge. *International Journal of Science Education, 28*(11), 1267–1287.

Lovelock, J. E., & Rapley, C. G. (2007). Ocean pipes could help the earth to cure itself. *Nature, 449*, 403.

Maffi, L. (2005). Linguistic, cultural and biological diversity. *Annual Review of Anthropology, 29*, 599–617.

Matthews, M. R. (1994). *Science teaching: The role of history and philosophy of science*. London: Routledge.
Merton, R. K. (1968). *Science and democratic social structure, social theory and social structure*. New York: Free Press.
Merton, R. (1973). *The sociology of science*. Chicago: University of Chicago Press.
Millar, R. (2002). Towards a science curriculum for public understanding. In S. Amos & R. Boohan (Eds.), *Teaching science in secondary schools: A reader* (pp. 112–128). London: Routledge.
Orr, D. (1992). *Ecological literacy*. Albany: State University of New York Press.
Panikkar, R. (2005). *La porta stretta della conoscenza*. Milano: Rizzoli.
Pedretti, E. (1996). Facilitating action research in science, technology and society (STS) education: An experience in reflective practice. *Educational Action Research, 4*, 307–327.
Polanyi, M. (1962). The republic of science. *Minerva, 1*, 54–73.
Pretty, J., Adams, B., Berkes, F., Ferreira de Athayde, S., ... (2009). The intersections of biological diversity and cultural diversity: Towards integration. *Conservation and Society, 7*(2), 100–112.
Roth, W.-M., & Lee, S. (2004). Science education as/for participation in the community. *Science Education, 88*, 263–291.
Sachs, W. (2002). Fairness in a fragile world: The Johannesburg agenda. *Development, 45*(3), 12–17.
Sahtouris, E. (2000). *Earthdance: Living systems in evolution*. New York: iUniverse.
Schiermeier, Q. (2011, February 16). Increased flood risk linked to global warming. *Nature, 470*, 316. Published online 16 February, 2011.
Shiva, V. (1998). *Monocultures of the mind*. London: Zed Books.
Smil, V. (2008). *Energy at the crossroads: Global perspectives and uncertainties*. Cambridge, MA: MIT Press.
Solomon, J. (1990). The discussion of social issues in the science classroom. *Studies in Science Education, 18*, 105–126.
Steffen, W., Crutzen, P. J., & McNeill, J. R. (2007). The anthropocene: Are humans now overwhelming the great forces of nature. *AMBIO: A Journal of the Human Environment, 36*(8), 614–621.
Stephens, S. (2000). *Handbook for culturally responsive science curriculum*. Fairbanks: Alaska Native Knowledge Network.
Sterling, S. (2002). A Baker's Dozen-towards changing our "loaf". *The Trumpeter, 18*(1), 1–15.
Tilman, D., Socolow, R., Foley, J., Hill, J., Larson, E., Lynd, L., Pacala, S., Reilly, J., Searchinger, T., Somerville, T., & Williams, R. (2009). Beneficial biofuels – The food, energy, and environment trilemma. *Science, 325*(5938), 270–271.
Wicks, P. G., & Rearson, P. (2009). Initiating action research: Challenges and paradoxes of opening communicative spaces. *Action Research, 7*(3), 243–262.
WorldWatch Institute. (2010). *The state of the world 2010. Transforming cultures: From consumerism to sustainability*. Washington, DC: Worldwatch Institute.

Chapter 10
Education for Sustainable Contraction as Appropriate Response to Global Heating

David E. Selby

Abstract Human-induced climate change is happening, opinions differing as to what window of opportunity remains to mitigate its direst effects. Responses to the climate change threat are characterized by denial and cognitive dissonance, the cultural pathology extending to those situated on the reform to transformation spectrum, including proponents of education for sustainable development. The climate crisis brings into question the usefulness and appropriateness of a lexicon of development. An alternative to sustainable development – sustainable contraction – is proposed. Nine propositions are nailed to the laboratory door to mark out what an education for sustainable contraction would entail. They call for an educational approach that: confronts denial by engendering disequilibrium in learning spaces; addresses despair, pain, grief and loss; combats consumerism and offers alternative conceptions of the "good life"; endows learning with a deep ecological paradigm; embraces intimacy and cultivates the poetic; folds marginalized "educations" such as anti-discriminatory, peace and media literacy education into sustainability learning; addresses emergency and disaster risk reduction learning; localizes and brings "denizenship" to prevailing "citizenship" discourse and practice; discards mechanistic thinking in favor of holistic and systemic ways of seeing, and acting in, the world. These propositions, it is suggested, constitute an appropriate agenda – that STEM is well-placed to help effect – for addressing the profound crisis in human ethics, values and worldview laid bare by potentially runaway climate change.

Keywords Climate change • Global heating • Climate change denial • Cognitive dissonance • Education for sustainable development • Education for sustainable contraction

---

D.E. Selby (✉)
Sustainability Frontiers, Adjunct Professor at Mount St Vincent University, Halifax, Nova Scotia, Canada
e-mail: dselby@sustainabilityfrontiers.org

L. Bencze and S. Alsop (eds.), *Activist Science and Technology Education*, Cultural Studies of Science Education 9, DOI 10.1007/978-94-007-4360-1_10,

The Heating Is Happening. . .

In a summary for policy makers, the international collectivity of scientists making up the physical science working group of the UN Intergovernmental Panel on Climate Change (IPCC) asserts that: "Warming of the climate system is unequivocal, as is now evident from observations of increases in global average air and ocean temperatures, widespread melting of snow and ice, and rising global average sea level" (IPCC 2007, p. 1). Confirming the anthropogenic nature of climate change and the likelihood of some "abrupt and irreversible" impacts (Ibid, p. 13), the scientists project a rise in surface air temperature of between 1.8 and 4.0 °C during the twenty-first century relative to the pre-industrial period and a sea level rise of between 0.18 and 0.59 m, the latter projection not taking into account any future "rapid and dynamical" Arctic and Antarctic ice conversion events (Ibid, p. 7).

Future histories, each informed by a meta-analysis of scientific papers (Lynas 2007; Romm 2007), offer scenarios of a twenty-first century marked by ubiquitous environmental disaster (including a huge loss of biodiversity), ongoing and massive internal and external population displacement in consequence of sea incursions, seasonally recurring wildfire and desertification, and resultant social dislocation, hunger, starvation, internecine strife, violent conflict, tribalism, aggressively defensive localism, as well as the ever-lurking danger of genocide. Elizabeth Kolbert's empirical study, *Field Notes from a Catastrophe* (2007), forewarns of a similarly dire future.

Not that the present is short on trauma and tragedy. A report on the human impact of climate change from the Global Humanitarian Forum (2009) describes the "silent crisis" of climate change already upon us that, on yearly average, is causing over 300,000 deaths, seriously affecting 325 million people, and bringing about economic losses of $US 125 billion every year: "4 billion people are vulnerable, and 500 million people are at extreme risk," the report adds (2009, p. 1).

Scientific opinion varies as to whether we still have latitude to prevent runaway climate change. "We have a short period – a very short period – in which to prevent the planet from shaking us off," writes George Monbiot (2006a, p. 15), a view largely endorsed by the UN Intergovernmental Panel on Climate Change (IPCC 2007, p. 20). Early in 2009 world-renowned climatologist James Hansen of NASA advised incoming US President, Barack Obama that "we have only 4 years left to act on climate change" (McKie 2009, p. 44). Others are far less sanguine. "The time," says another eminent scientist "is already 5 min past midnight" (Kolbert 2007, p. 58). "Our future," writes another (Lovelock 2006, p. 6), "is like that of passengers on a small pleasure boat above Niagara Falls, not knowing that the engines are about to fail." For Joseph Romm (2007), any perceived window of opportunity to mitigate climate change is closing fast: "Climate change is coming faster and rougher than scientists have expected," he warns (p. 231). There is mounting evidence that surface temperature rise cannot be held at the 2.0 °C rise relative to pre-industrial levels that governments and the UN are taking as livable with (Adam 2009, p. 14). For James Hansen (2009, p. 42) a 2.0° rise is nothing short of a "disaster scenario" anyway. "Business as usual" is not an option.

Denial and Cognitive Dissonance in Response to Global Heating

Arguably, the greatest hindrance to shaking off a "business as usual" mindset are responses to global heating marked by a presenting acceptance, often fulsome, of the severity of the looming crisis coupled with an ill-preparedness to follow through in terms of embracing and promoting the radical personal and societal change needed to stave off the worst effects of climate change. As such, they constitute a form of self-deceptive or furtive denial characterized by fully conscious or threshold of consciousness dissonance between perception of problem and identified acted upon (or not acted upon) remedies, with profoundly unhealthy ramifications for both the individual concerned and society at large. Responses of this kind are captured by Sandra Postel's (1992) prescient words of over 20 years ago:

> Psychology as much as science will determine the planet's fate . . .denial, among the most paralyzing of human responses, . . .can be as dangerous to society and the natural environment as an alcoholic's denial is to his or her own family. Because they fail to see the addiction as the principal threat to their well-being, alcoholics often end up by destroying their own lives. Rather than facing the truth, denial's victims choose slow suicide. In a similar way, by pursuing lifestyles and economic goals that ravage the environment, we sacrifice long-term health and well-being for immediate gratification – a trade-off that cannot yield a happy ending. (p. 4)

To Monbiot (2006a), there is an unspoken and barely acknowledged collusion of denial between citizenry and leadership, electorate and elected:

> But the thought that worries me most is this. As people in rich countries. . .begin to wake up to what science is saying, climate-change denial will look as stupid as Holocaust denial, or the insistence that AIDS can be cured with beetroot. But our response will be to demand that the government acts, while hoping that it doesn't. We will wish our governments to pretend to act. We get the moral satisfaction of saying what we know to be right, without the discomfort of doing it. My fear is that the political parties in most rich nations have already recognized this. They know that we want tough targets, but that we also want those targets to be missed. They know that we will grumble about their failure to curb climate change, but that we will not take to the streets. They know that nobody ever rioted for austerity. (pp. 41–2)

For Diarmuid O'Murchu (2004), the central feature of our "addictive trap" is "an illusion of power and control that has become progressively compulsive, acquisitive, manipulative and destructive. . . .In our addictive commitment to power, we ourselves have become quite powerless, but like all addicts we vehemently deny and disown that fact." Joanna Macy and Molly Young Brown (1998) call the source of our addiction the "Industrial Growth Society," a society that cannot last in that "it is inexorably and exponentially destroying itself" (p. 23). There is ubiquitous evidence of systemic "runaway," they maintain, that should "rivet our attention, summon up the blood, and bond us in collective action" but the evidence before our eyes tends to have the opposite effect making us "want to pull down the blinds and busy ourselves with other things" (Ibid, p. 26). Reminding us of the etymology of the word "apathy", the Greek *apatheia*, literally the inability or refusal to experience pain, Macy and Young Brown identify a range of forms of Western cultural conditioning through which we repress deep concern about the planetary circumstance:

- *Fear of pain* – Seeing pain as dysfunctional and as evidence of an inability to cope, rather than as opportunity for re-empowerment and renewal.
- *Fear of despair* – Fearing that to admit to despair about the state of the world will undermine all we believe in and bring paralysis rather than resolve.
- *Fear of appearing morbid* – Believing that only sanguinity and optimism are culturally appropriate indicators of and keys to successfulness and that dystopia anguish is an indication of lack of confidence, even incompetence.
- *Fear of guilt* – Fearing to expose the moral pain of individual and societal complicity in the exploitation of peoples and other-than-human life forms around the globe and of the planet itself.
- *Fear of causing distress* – Believing it is compassionate not to distress others, especially the young, about the state of the world rather than seeing disclosure as a healthy connecting of people to the world.
- *Fear of being unpatriotic* – Holding that to speak things as they are will somehow harm the national fibre and interest.
- *Fear of appearing weak and emotional* – Falling for the objectivist fallacy that emotional-tinged responses are weak while impassivity is evidence of strength.
- *Belief in the separate self* – Fearing that expressing concern about the world is simply a reflection of unprocessed inner turmoil and believing that the discrete self is the only locus of empowerment and transformation.
- *Fear of powerlessness* – Believing that global threats are so huge and intractable that the individual can do nothing of significance. (After Macy and Young Brown 1998, pp. 27–32)

The consequences of such processes of repression are what Robert Gifford (2007, p. 209) calls "environmental numbness" and Robert Lifton (1967) terms "psychic numbing". We immure ourselves from the way the world is going by divorcing our personal trajectory from the global trajectory. We immure ourselves, too, through forms of displacement or self-delusion on a spectrum from quick fix hedonism to cozy reformism. "We live in a dark age," concludes O'Murchu (2004, p. 140), "but, alas, nobody wishes to entertain that notion. We are unable to befriend the darkness because our addictiveness and compulsiveness keep us firmly rooted in *denial*. The whole thing is too painful to look at, so we choose to befriend our pathology rather than befriend its deeper truth."

Denial and Cognitive Dissonance in the Field of Education for Sustainable Development

A befriending of our pathology afflicts those positioned along the reform to transformation continuum in their responses to global heating, including many proponents of education for sustainable development.

The Learning and Skills Council, a body responsible for the UK further education sector, argues in its *Strategy for Sustainable Development* (2005, p. 3) that "we are living in an unsustainable world" not least because "global temperatures are rising faster than previously recorded". Alongside this, it identifies "the maintenance of high and stable levels of economic growth and employment" as a key sustainable development objective and in language that resonates with global marketplace-speak, it continues:

> Experience shows there is a strong business case for sustainable development. Businesses, companies, colleges and learning providers that adopt environmental management systems can make significant financial savings. They can also enhance their reputation, gain access to new markets and better motivate their staff. (p. 4)

The conception of the humanity/nature relationship within the *Strategy* is one of nature as resource or "natural capital" or "ecosystem service" (Porritt 2006; Orr 2009, pp. 21–2) to be managed, and having instrumental rather than intrinsic value.

The same is the case in the pronouncements of Forum for the Future, the influential UK sustainable development charity with a significant educational arm to its work. Noting in the presumptuously titled report on its 2003 activities, *Sustainable Development – the only game in town*, that "evidence of unsustainable development kept piling up" (Forum for the Future 2004, p. 1), Forum rehearses its "Five Capitals" framework for responding to the global environmental crisis in which nature, human beings and human communities and forms of social organization are viewed as capital assets alongside financial and manufactured capital (Ibid, pp. 3, 5, 8, 9, 12). Nature is conceived of as resource – "the stock or flow of energy and material" (p. 3) – underpinning a system of capitalistic development that needs to be husbanded properly to safeguard its upward trajectory. The response to the global crisis is better care of assets for status quo maintenance. Following from this, Forum's emphasis on skills-based "sustainability literacy" in its Higher Education Partnership for Sustainability with 18 UK universities comes as no surprise:

> A sustainability literate person will be equipped with a number of intellectual and practical tools that enable them to make decisions and act in a way that is likely to contribute positively to sustainable development. They will be able to make decisions on specific matters, such as advising on financial investment, buying food or writing new policy for prisons, by applying the 'at the same time' rule – that is, taking environmental, social and economic considerations into account simultaneously, not separately. (Parkin et al. 2004, p. 9)

While recommending action, the prescription for educational change is apathetic in the sense used earlier of inability or refusal to confront and experience pain. The concept of "sustainability literacy," much vaunted across the field of education for sustainable development (see, for instance, DfES 2003; John Foster Associates 2006; Stibbe 2009) is itself objectivist in its explicit or implicit emotion avoidance – even in skills terms, the skills and capacities for handling despair, distress, pain, guilt and grief are not addressed – and in its failure to position the transformative dispositions and capabilities of the individual within a conscious reconnecting with the flow of life.

This phenomenon is not restricted to the United Kingdom. In their report to Macquarie University and the Australian Department of the Environment and Heritage, *Change in Curricula and Graduate Skills Towards Sustainability*, Daniella Tilbury and colleagues (2004, p. 3) write: "Education for Sustainability involves students and educators in a process of active learning and futures thinking, and addresses the generic skill needs of business and industry". The skills list offered recites critical, creative and future-thinking skills, needs assessment and action-oriented skills, interpersonal and intercultural skills, the skills of dealing with uncertainty, and problem solving skills (Tilbury et al. 2004). Important as they are, they are set within a frame of the "generic skills needs of business and industry," eschewing alternative frames and dispositions crucial to a context of looming or actual civilizational threat. The emphasis on skills, as with the Learning and Skills Council and Forum for the Future, also tends to obfuscate the centrality of a values and socio-affective response to a threatened world. None of the education for sustainable development proposals reviewed here call for the curricular treatment of themes and issues that might reasonably be seen as imperative for actively addressing the deepening multiple crisis syndrome of global heating, a point to be returned to later.

A fundamental issue for proponents of education for sustainable development is the relevance of continuing to talk about development. As James Lovelock (2006) so powerfully puts it:

> (W)hen change was slow or non-existent, we might have had time to establish sustainable development, or even have continued for a while with business as usual, but now is much too late; the damage has already been done. To expect sustainable development or a trust in business as usual to be viable policies is like expecting a lung cancer victim to be cured by stopping smoking; both measures deny the existence of the Earth's disease, the fever brought on by a plague of people. (p. 3)

For Lovelock, "what we need is a sustainable retreat" (2006, p. 7). Lauding the orderly quality of the Napoleonic 1812 retreat from Moscow and exhorting a1940 Dunkirk spirit, he adds: "We need people of the world to sense the real and present danger so that they will spontaneously mobilize and unstintingly bring about an orderly and sustainable withdrawal to a world where we try to live in harmony with Gaia" (p. 150).

Preferring the softer and more ecological concept of "contraction," a concept devoid of militaristic connotations and tending to infer the systemic rather than the systematic, the organic rather than the lockstep, let me examine what "education for sustainable contraction" in the face of runaway climate change might entail.

Education for Sustainable Contraction (ESC): Nailing Nine Propositions to the Laboratory Door

On October 31, 1517, Martin Luther nailed 95 propositions to the door of the Castle Church in Wittenberg, Germany. The propositions fulminated against the widespread practice of the Catholic Church of selling indulgences, paper certificates guaranteeing relief from punishment in Purgatory, to those who had committed sins. This is regarded as the seminal moment in the Reformation of the Western Christian Church (Davies 1996, pp. 484–5). It has not gone unnoticed by climate change commentators that the hypocrisy and cognitive dissonance of the pre-Reformation period finds its echo in the thinking and practices of those accepting – but not following through on the consequences of accepting – the climate change threat. A notable example is carbon offsetting. "Just as in the fifteenth and sixteenth centuries you could sleep with your sister, kill and lie without eternal damnation," suggests Monbiot (2006a, p. 210), "today you can leave your windows open while the heating is on, drive and fly without endangering the climate, as long as you give your ducats to one of the companies selling indulgences." Inspired by this nifty connect (see also, Monbiot 2006b), I offer nine propositions for Education for Sustainable Contraction that we might nail to the doors of our STEM learning and teaching places and spaces.

But, before exploring the propositions, it is important to identify what frequently recurring features of education for sustainable development would be markedly absent from or significantly less ubiquitous within an ESC landscape:

- The uncritical or tacit embrace of the neo-liberal economic growth and global marketplace model, and of rampant consumerism.
- An instrumentalist and utilitarian view of nature, emphasizing the "desirability of sustaining those natural systems that are conducive to *human* flourishing" (Bonnett 1999, p. 315), with its correlative denial of the intrinsic value of the natural world and of human embeddedness in nature.
- A managerialist and policy orientation to sustainability in which "natural resources" and the world are looked upon as what Bonnett (Ibid, p. 317), citing Mitchum (1997), calls "a spaceship in need of an operating manual".
- Absorption with technical fixes to remedy unsustainability, with skills development of the learner prioritized and values issues left on the rhetorical shelf.
- A conception of change potential as fundamentally *individual* as against *individuated*, i.e. the person acting from a sense of being largely alone even if working in tandem with others, rather than from a sense of their orchestrated and holographic enfoldment within the social and environmental whole (O'Murchu 2004, pp. 91–3).
- An exteriority of focus (issues in the world out there) as against a dynamical interplay between interiority and exteriority (the learner's inner and outer worlds) in processes of personal and social transformation.

Uprooting features such as these, so prejudicial to transformation, would clear space for an Education for Sustainable Contraction.

ESC: Proposition 1. A concerted effort is needed in the light of looming runaway climate change to confront denial by moving learner assumptions, understandings and responses towards disequilibrium (fomenting dissipative structures).

ESC: Proposition 2. Given the likely impending severity of global heating, Education for Sustainable Contraction needs to address despair, pain, grief and loss.

Global heating is beginning to turn the world on its business-as-usual head, exposing the fragility of the normal and the vulnerability of the taken-for-granted. As Monbiot (2005) puts it:

> Everything we thought was good turns out to be bad. It is an act of kindness to travel to your cousin's wedding. Now it is also an act of cruelty. It is a good thing to light the streets at night. Climate change tells us it kills more people than it saves. (...) Climate change demands a reversal of our moral compass, for which we are plainly unprepared (p. 23).

STEM classrooms at various levels are marked by the comfortable equilibrium of the agitated pendulum where any movement stimulated by the learning dynamic often tends to reduce to minimal swing followed by a more or less settled state. What, asks Ilya Prigogine (1989, p. 396), if we turn the pendulum on its head? It is difficult to predict what will happen. The notion of the upturned pendulum, Prigogine avers, has been "ideologically suppressed" in that its message is inconvenient for a culture that seeks to dominate and suppress nature (Prigogine 1989, pp. 396–7). For confronting a world that threatens to make castles built of sand of our assumptions, the notion acquires huge consequence, as does Prigogine's concept of "dissipative structures" within self-organizing systems. Prigogine distinguishes between systems at equilibrium or near-equilibrium where huge disturbances would be required to effect radical change, and hence where creativity is low, and far-from-equilibrium systems. In the case of the latter, a fluctuation can induce movement to disequilibrium – dissipation – at which point the system responds by bringing to bear on a situation as wide and coherent a range of forces as is necessary to effect a new complexified condition of equilibrium. It is at the far-from-equilibrium where the potential for deep creativity lies (Capra 1996, pp. 180–3) and, within learning community dynamics, where reversals of the moral compass, held back by denial, are more likely to happen.

Science and technology curricula have tended to reinforce the myths we tell ourselves: of unending economic growth; of ever upward progress; of technological fix to preserve "business as usual"; of human separation from, and dominance over, nature. There is a storehouse of sound science, none better than the work of James Hansen (2009), that can be deployed to bring unsettling but creative far-from-equilibrium thinking to STEM classrooms.

Confronting despair, pain, grief and loss within communities of learners by giving space for both cognitive and affective response to scientific data as well as case studies and personal narratives of climate change (e.g. Kolbert 2007; Selby and Kagawa 2013) is a likely harbinger of dissipative structures. At a conference several years ago, I found myself in a small minority arguing for a confronting of the pain of "gloom and doom" as a vital rite of passage in any transformative learning process. Most of those in the room regarded "gloom and doom" as disabling and disempowering for the learner. For me, their position smacked of denial, and also of reluctance to recognize that, from within an ecological or holistic worldview life and death are marked by dynamical unity, "the cycle of birth-death-rebirth" (O'Murchu 2004, p. 190).

Recognizing that present and future generations need hope, we have to ask what the hope is grounded in and what kind of hope it is. Is it spurious optimism, a comfortable fiction based on what we would prefer to see happen while keeping our "eyes wide shut" (Hilman et al. 2007)? Or, is it a pared down and realistically straitened optimism born of confronting the present and future earth condition? Is it cozy but inauthentic hope or a hard-edged but more authentic hope forged out of what Martin Seligman (1992, p. 292) calls "the courage to endure pessimism"? Are STEM classrooms helping students keep their "eyes wide open" as to the global climate change condition?

From within a quantum theological frame, O'Murchu writes of the importance of the "Calvary moment" (Ibid.), encapsulating the idea that transformation entails a conscious and thoroughgoing progression by groups and individuals through despair towards empowerment, healing and renewal. The "Great Turning," as Joanna Macy calls it, involves breaking through denial to confront the pain of the world, heroic holding actions to stop things getting worse, analysis of the structural causes of the damage wreaked by the Great Industrial Society allied to the nurturing of alternative institutions and, most fundamentally, a cognitive, spiritual and perceptual reawakening to the wholeness of everything (Macy and Young Brown 1998, pp. 17–22). Macy's despair and empowerment work provides a powerful canon of activities and exercises for breaking out of denial to connect with the state of the world (Macy 1983, 1991; Macy and Young Brown 1998). Such exercises would be the food and drink of education for sustainable contraction. They are recommended to STEM practitioners.

ESC: Proposition 3. Given the "powerful wave of neo-liberalism rolling over the planet" (Jickling and Wals 2008, *p. 2), destructive of ecosphere and ethnosphere, climate change education needs to offer alternative conceptions of the "good life", combat consumerism, and help learners explore and experience alternatives to a growth economy.*

For the peoples of the metaphorical North and elites in the South who have taken on the western worldview, it is important that an education in "voluntary simplicity" (Elgin 1981) is made available, the term connoting frugal consumption, ecological awareness, connectedness and community, and personal growth based upon evolving material and spiritual aspects of life in harmony. A countercultural idea amongst such populations, the transition to "voluntary simplicity", its originator argues, is more than made up for through the quality of revitalized experience and the cultivation of "conscious watchfulness", the ability to see the close-at-hand world through an intimate eye (pp. 149–51). STEM curricula have a key part to play in fostering alternative conceptions of the "good life" by helping learners envision and then concretize appropriate technologies for societies and communities relearning how to live simply and frugally as they adapt to and seek to mitigate the drivers and effects of climate change.

Dovetailed with the promotion of "voluntary simplicity" among such populations should be anti-consumerism education. Defined as "consumption beyond the level of dignified sufficiency" (McIntosh 2008, p. 180), consumerism not only violates the indentured slave, the sweatshop worker and the natural environment but also enslaves the consumer herself (McGregor 2003, p. 3). Consumerism, Sue McGregor avers, "is an acceptance of consumption as a way of self-development, self-realization and self-fulfillment. In a consumer society, an individual's identity is tied to what he or she consumes" (2). Anti-consumerism education, then, has the twin goal of protecting the ecosphere and ethnosphere while liberating the individual from the thrall of consumerism for a journey of self-discovery and self-growth. It is to be distinguished from "consumer awareness education" with its subliminal agenda that consumerism can be made benign just as, given the exigencies of structural racism, "race awareness education" had to give way to "anti-racist education".

As a backcloth to this agenda item, it is vital that Education for Sustainable Contraction in STEM and other disciplines provides age-appropriate windows for engaging with ideas for transition to a global slow-growth or no-growth economy (Victor 2008), concretizing those ideas through learning-in-community experimentation and practice.

ESC: Proposition 4. The view of the human<>nature relationship needs to shift from the dominstic, the instrumental and the exploitative to one of embeddedness and intrinsic valuing; from a shallow ecological to a deep ecological paradigm.

ESC: Proposition 5. The embrace of intimacy with nature calls for the cultivation of the poetic.

As has been noted, common to articulations of education for sustainable development is representation of nature as resource. This is indicative of collusion with the dominant corporate paradigm but also suggests that the precocious and headstrong infant that was ESD in the 1990s paid insufficient heed to a heritage of eco-philosophical responses to the question of humanity's relationship to the environment, each response offering its own insights on how to live ethically and responsively on the planet. Had the infant listened, in particular, to deep ecology, ESD might have thought more deeply about the human<>nature relationship and divested itself of some of its anthropocentrism, at least giving space in its debates to the biospherical egalitarian position (Selby 2006, p. 359). Key principles of deep ecology include:

- The well-being and flourishing of both human and other-than-human life have value in themselves
- Richness and diversity of life forms (and cultures) are valuable in themselves
- Human interference with the other-than-human world is excessive
- Quality of life matters more than standard of living (after Devall and Sessions 1998, p. 147).

Re-bonding with nature would be a key goal of ESC and would involve the cultivation of the poetic dimension of human awareness, thus marrying sense and sensibility (Selby 2006, p. 361). It was in the time of Galileo Galilei, says the poet T.S. Eliot that "a dissociation of sensibility" set in from which the West never recovered (cited in McIntosh 2008, p. 154). This "breaking up of the ability to feel and relate to life", according to Alastair McIntosh (p. Ibid), lies behind the "mindlessness that underlies anthropogenic climate change" (Ibid, p.112). Following from such an insight, it would seem evident that an education responsive to climate change should also help learners cultivate a sense of oneness and interconnectedness with nature through poetic and spiritual ways of knowing such as attunement, awe, celebration, enchantment, intuition, reverence, wonder and an oceanic sense of the oneness of being. Education for sustainable development has given barely any space to the poetic and the numinous in its reliance on scientific rationality. There are questions to be asked about rationality "in resolving issues as complex, subtle and multidimensional . . .as environmental concern", especially given how rationality has proved so effective a tool in the exploitation of the environment (Bonnett 1999, p. 321). There are correlative questions to be asked about the "deadening language", emphasizing "the terminology of science and policy over that of ethics and philosophy" and relying upon '"practical" utilitarian arguments', as employed by environmental educators and public awareness raisers (Goodstein 2007, p. 76). Effective political communication on climate change, writes Eban Goodstein (Ibid, p. 77), "comes from the heart, and the heart of concern about the impacts of humanity's climate destabilization is a spiritual connection to nature".

If this sounds "unscientific" then it does not have to be so. Intimacy with nature can be about walking the interface between the scientific and the numinous and thereby cultivating resistance to forces destroying cultural and natural environments. In a time of violation of flora and peoples occasioned by the English land enclosures and agrarian "modernization" of the 1820s, the laborer poet John Clare conveyed a sense of loss through finely-detailed sketches and descriptions of flower species under threat, images that in their detail and exactitude also betokened a sense of oneness between flowers and laborers as "fellow members of the great commonwealth of the fields" now sharing a common fate in their eviction (Mabey 2010, pp. 115–26). His radicalism and expansiveness were bred of nature intimacy in which were folded together science, spirituality and social justice. In a time of present and looming runaway climate change eroding environments, social relations and livelihoods, it is profoundly important to cultivate a sense of enfoldment in nature and correlative disposition to hold on to what is being lost by fostering scientific intimacy alongside poetic and spiritual ways of knowing and relating. UNESCO (2010) has identified bio-diversity loss as one of three "key action themes" for the second half of the 2005–14 Decade of Education for Sustainable Development. STEM curricula can make a signal contribution to the advance of biodiversity education by adopting an activist science approach placing due emphasis on emotional intelligence.

The cultivation of a consciously watchful intimacy with nature suggests a return to something akin to local nature study programs based on close observation of local fauna and flora through the seasons that characterized earlier educational practice, programs long discarded as environmental educators of the 1970s immersed themselves in ecosystemic thinking and global environmental problematics. Diane Pruneau et al. (2001, p. 135) make precisely this point when they recommend "more in-depth educational work focusing on developing knowledge and appreciation of regional fauna and flora, and various seasonal weather patterns". "The idea of going outside and perceiving ambient elements creates a link with these elements," they continue. 'A reflection exercise could follow this direct contact: "Do we really want to lose the piping plover, the Clintonia borealis [sic], snow, the return of spring, and so on?"'.

ESC: Proposition 6. "Educations" that have been marginalized within education for sustainable development are of pivotal importance.

ESC: Proposition 7. With global heating under way, sustainability education and emergency/disaster risk reduction education need to fold together.

It is perhaps indicative of the "business as usual" mindset pervading the field of education for sustainable development that, while holistic and integrative in original intention (UNCED 1992), the insights of certain key social, political and moral educations are virtually ignored.

Confronted with all the societal ramifications of potential, some would say inevitable, degree-by-degree global heating, giving peace education a central place within the panoply of sustainability "educations" would seem essential. A field concerned with non-violence, conflict avoidance and conflict resolution, confronting and unpacking negative and enemy images of the "other," and processes and outcomes of direct and structural violence (Smith and Carson 1998) would have the potential to bring wisdom and insight to learners facing the looming prospect or immediacy of what is being predicted (especially massive population displacements and the tensions they will bring). For similar reasons, anti-discriminatory education, concerned with confronting all the negative and hegemonic "centrisms" that foment societal and inter-human injustice, and with dissecting inner and outwardly-manifesting processes of "othering" (Plumwood 1993, 1996; Selby 2001), needs to be brought to an ESC agenda. In helping learners confront ubiquitous social, political and media global heating denial, the "crap detecting" skills and insights of media literacy education would also be given prominence (Duncan et al. 2000).[1]

The gulf that has so far characterized the relationship between education for sustainable development and emergency and disaster risk reduction education will urgently need to be bridged as the heating happens. Emergency education, that is, education in crisis or disaster contexts occasioned by armed strife and/or environmental cataclysm, has achieved increasing prominence since the end of the Cold

[1] I am, as ever, indebted to Neil Postman and Charles Weingartner (1969) for the their delightfully incisive term, "crap detection".

War period (Kagawa 2005; Kagawa and Selby 2006). As Fumiyo Kagawa (2007) explains, with the world moving ever more inexorably into multiple crisis syndrome, the theory and practice of sustainability and emergency need to coalesce. Disaster risk reduction is the younger cousin of emergency education and is a response to the mounting incidence and severity of natural disaster globally. It has been described as involving a "combination of actions, processes and attitudes necessary for minimizing underlying factors of vulnerability, improving preparedness and building resilience" (Global Education Cluster et al. 2011, p. 2). Key focuses, then, are: understanding the science and mechanisms of natural disasters; learning and practicing safety measures and procedures; understanding social, economic and environmental risk drivers that exacerbate vulnerability and turn a hazard into a disaster; resilience building; building school and community cultures of safety and resilience (Kagawa and Selby 2012). The STEM subjects have the potential to make a huge scientific and technological contribution to this agenda. Clearly, too, there are major implications for any sustainability education agenda in that the mounting incidence of natural disaster coupled with rampant climate change presents potentially insuperable obstacles to the realization of a sustainable future (Ibid.). As events such as Hurricane Sandy illustrate, developed countries can no longer maintain an "out there but not here" attitude to emergency and disaster.

ESC: Proposition 8. Cozy assumptions about the relationship between education for sustainability and education for citizenship need unpacking and formal and informal learning programs need to offer alternative and localized conceptions of "good citizenship" (or "good denizenship").

In *Earth Democracy: Justice, Sustainability and Peace* (2005), Vandana Shiva makes a powerful case for localism in response to the global crisis of unsustainability. "Conservation of the earth's resources and creation of sustainable livelihoods," she writes (p. 10), "are most caringly, creatively, efficiently and equitably achieved at the local level. Localization of economies is a social and ecological imperative". For Shiva, localism allows for "living democracy" integrated with a "sustenance economy" within which "people can influence the decisions over the food we eat, the water we drink, and the health care and education we have" (Shiva 2005).

There has been an all-too-cozy connection between education for sustainable development, on the one hand, and citizenship education (including what is called "global citizenship education") and education for democracy, on the other. The respective "educations" are, more often than not, assumed to enjoy a dovetailed relationship (see, for instance, Bourne 2005; National Assembly of Wales 2005; QCAA Wales 2002). A thorny, but largely untouched, problem concerns how representative democracy drawing upon an electorate immured in and, on that account, not readily teased from, a pervasive consumerist ethic can be squared with an environmental narrative predicated on the finiteness of the Earth. If we embrace the notion of finiteness and the ecological imperatives deriving from it, "certain policies are proscribed," writes Michael Bonnett (1999, p. 315). He elaborates:

> They are in effect not only removed from the area of democratic debate, but set the parameters within which democratic debate can be allowed to function. . . .Insofar as such enframing is broad in scope, it is tantamount to defining a conception of the good life to which citizens need to be brought to conform and thus both runs counter to the assumption of democracy of valuing diversity of view and holds the danger of peripheralising democracy as a contingent value, instrumental to achieving the public acceptance of these imperatives. (Bonnett 1999)

Shiva's "living democracy" provides a means of negotiating this seeming impasse. While it is almost certainly the case that citizenship education focused upon (consumerism-fuelled) representative democracy will never sit easily with the sustained and draconian intervention by government regarded as essential by such as Romm (2007) if the worst global heating scenarios are to avoided, "living democracy" offers the potential for a reinvented citizenship ethic and education. For Shiva, "it is essential to dispel the notion that globalization is natural and inevitable" (2005, p. 106) and that we see it as a political and profiteering process that continues to encroach on the commons, i.e. that held to be common property or of shared accessibility, through appropriation, privatization, exclusion and "the enclosure of minds and imagination" whereby the global market is portrayed by its adherents as the only way forward (2005, p. 20). The alternative path Shiva advocates is the actual lived experience of two-thirds of humankind, in "which humans produce in balance with nature and reproduce society through partnerships, mutuality, and reciprocity" (2005, p. 17). Turning globalization on its head, Shiva envisages a future in which the "most intense relationships are at the local level and the thinnest interactions at the international level" with decisions being taken "at the level closest to where the impact is felt" (2005, p. 64). Localization would not only offer a more fertile arena for participatory democracy to flourish but, based on a keener, immediately lived, appreciation of the "interdependence between nature and culture, humans and other species" (Ibid, p. 82), would open the way for a more biocentric and less consumerist and exploitative democracy. Following the principle of subsidiarity, the centralized draconian approach to preventing perilous levels of global heating would stand in negative correlation with "living democracy", that is, like a thermostat, only being triggered if the climate change determinations of localities fell short.

In educational terms, subsidiarity would also apply to curriculum, with thinnest input into curriculum framing emanating from the national level. Learning would involve a rebalancing of the mind<>hand interface through local craft learning and craft apprenticeships. Generally within localized living democracies and sustenance economies, there would be a move away from learning as expert induction to a livelihood, communitarian orientation fostering new "tools of conviviality" (Illich 1973).

Within such a scenario, the weighting within citizenship education would shift towards local participatory democracy with commensurately reduced emphasis on national citizenship. In treating the global level, education would be responsive to the need to globalize "compassion, not greed" (Shiva 2005, p. 115). The notion of "citizenship" might give way to that of "denizenship," a denizen being an inhabitant or occupant of a particular place, the term connoting primacy of immediate context while also neatly sidestepping the built-in anthropocentrism of citizenship in that a denizen can be either human or other-than-human.

What would STEM curricula look like if set within a localized "living democracy" frame of reference with an accent on "denizenship"?

ESC: Proposition 9. Everyone has to understand and come to terms with the fact that we are threatening our own existence. To confront this requires a Copernican revolution in our view of the world and in the aims, structures and processes of education and, perhaps, in the loci of learning.

While some sustainability educators have emphasized that social transformation towards sustainability calls for a relinquishing of the pervasive mechanistic and reductionist way of seeing the world and a radical shift to holistic and systemic perception (Selby 2006; Sterling 2007), the field of education for sustainable development remains by and large wedded to mechanism (Selby 2007), an argument that has been implicitly and explicitly present throughout this chapter. It will not be further elaborated here. As the heating happens, institutional and in-community learning cannot afford the self-indulgence of being other than holistic and systemic.

If habituation to mechanism/reductionism and a "business as usual" mindset afflict those embracing change, as suggested here, how much more so is that the case within schools and universities. Mechanism, writes Robin Richardson (1990, p. 54), is "institutionalized in all sorts of structures and career patterns". It is certain that, as the heating happens, learning programs and educational institutions as we presently know them will be faced with deep challenge and disruption and, if unresponsive to the need for transformation, will disintegrate as people go to find other, more relevant, loci for learning what they have to learn.

I leave the last word to George Monbiot:

> For the campaign against climate change is an odd one. Unlike almost all other public protests which have preceded it, it is a campaign not for abundance but for austerity. It is a campaign not for more freedom but for less. Strangest of all, it is a campaign not just against other people, but also against ourselves. (2006a, p. 215).

References

Adam, D. (2009, March 11). Sea level could rise more than a metre by 2100. *Guardian, 14*.

Bonnett, M. (1999). Education for sustainable development: A coherent philosophy for environmental education? *Cambridge Journal of Education, 29*(3), 313–324.

Bourne, D. (2005). Interconnectedness versus interdependence: Reflections in response to David Selby. *Zeitschrift fuer internationale Bildungsforschung und Entwicklungspaedagogic* [Journal for International Educational Research and Development Education], *28*(1), 29–34.

Capra, F. (1996). *The web of life: A new scientific understanding of living systems.* New York: Anchor/Doubleday.

Davies, N. (1996). *Europe: A history*. Oxford: Oxford University Press.
Department for Education and Skills (DfES). (2003). *Sustainable development action plan for education and skills*. London: DfES.
Devall, B., & Sessions, G. (1998). Deep ecology. In L. P. Pojman (Ed.), *Environmental ethics: Readings in theory and application* (pp. 144–148). Belmont: Wadsworth.
Duncan, B., Pungente, J., & Shepherd, R. (2000). Media education in Canada. In T. Goldstein & D. Selby (Eds.), *Weaving connections: Educating for peace, social and environmental justice* (pp. 323–341). Toronto: Sumach.
Elgin, D. (1981). *Voluntary simplicity: Toward a life that is outwardly simple and inwardly rich*. New York: William Morrow.
Forum for the Future. (2004). *Sustainable development – The only game in town: Annual report 2004*. London: Forum for the Future.
Gifford, R. (2007). Environmental psychology and sustainable development: Expansion, maturation and challenge. *Journal of Social Issues, 63*(1), 199–212.
Global Humanitarian Forum. (2009). *The anatomy of a silent crisis* (Human impact report). Geneva: Global Humanitarian Forum.
Global Education Cluster, UNICEF, Plan International & Save the Children. (2011). *Disaster risk reduction in education in emergencies: A guidance note for education clusters and sector coordination groups*. http://education.humanitarianresponse.info/document/disaster-risk-reduction-emergencies-guidance-note-education-clusters-and-sector. Accessed 16 May 2014.
Goodstein, E. (2007). *Fighting for love in the century of extinction: How passion and politics can stop global warming*. London: University of New England Press.
Hansen, J. (2009). *Storms of my grandchildren: The truth about the coming climate catastrophe and our last chance to save humanity*. London: Bloomsbury.
Hilman, M., Fawcett, T., & Rajan, S. C. (2007). *The suicidal planet: How to prevent global climate catastrophe*. New York: Thomas Dunne.
Illich, I. (1973). *Tools for conviviality*. New York: Harper & Row.
IPCC. (2007). *Summary for policymakers of the synthesis report of the IPCC fourth assessment report*. Geneva: IPCC Secretariat.
Jickling, B., & Wals, A. (2008). Globalization and environmental education: Looking beyond sustainable development. *Journal of Curriculum Studies, 40*(1), 1–21.
John Foster Associates. (2006). *Sustainability literacy: Embedding sustainability into the curriculum of Scotland's universities and colleges*. Edinburgh: Scottish Further and Higher Education Council.
Kagawa, F. (2005). Emergency education: A critical review of the field. *Comparative Education, 41*(4), 487–503.
Kagawa, F. (2007). Whose emergencies and who decides? Insights from emergency education for a more anticipatory Education for Sustainable Development. *International Journal of Innovation and Sustainable Development, 2*(3/4), 395–413.
Kagawa, F. & Selby, D. (2006). 'Now she is singing!' Emergency education in theory and practice. *Escalate Newsletter* (Newsletter of the higher education academy subject centre for education), *6*, 13–14. http://escalate.ac.uk/2936
Kagawa, F., & Selby, D. (2012, September). Ready for the storm: Education for disaster risk reduction and climate change adaptation and mitigation. *Journal of Education for Sustainable Development, 6*(2), 207–217.
Kolbert, E. (2007). *Field notes from a catastrophe: A frontline report on climate change*. London: Bloomsbury.
Learning and Skills Council. (2005). *From here to sustainability: The Learning and Skills Council's strategy for sustainable development*. Coventry: LSC.
Lifton, R. J. (1967). *Death in life: Survivors of Hiroshima*. New York: Random House.
Lovelock, J. (2006). *The revenge of Gaia: Why the Earth is fighting back – And how we can still save humanity*. London: Allen Lane.
Lynas, M. (2007). *Six degrees: Our future on a hotter planet*. London: Fourth Estate.

Mabey, R. (2010). *Weeds: How vagabond plants gatecrashed civilization and changed the way we think about nature*. London: Profile.
Macy, J. (1983). *Despair and personal power in the nuclear age*. Philadelphia: New Society.
Macy, J. (1991). *World as lover, world as self*. Berkeley: Parallax.
Macy, J., & Young Brown, M. (1998). *Coming back to life: Practices to reconnect our lives, our world*. Gabriola Island: New Society.
McGregor, S. (2003).*Consumerism as a source of structural violence*. http://www.kon.org/hswp/archive/consumersim.html. Accessed 15 Mar 2009.
McIntosh, A. (2008). *Hell and high water: Climate change, hope and the human condition*. Edinburgh: Birlinn.
McKie, R. (2009). We have only four years left to act on climate change: America has to lead. *Observer, 44*, 18.
Mitchum, C. (1997). The sustainability question. In R. Gottleib (Ed.), *The ecological community*. London: Routledge.
Monbiot, G. (2005, May 24). A restraint of liberty. *The Guardian, 23*.
Monbiot, G. (2006a). *Heat: How to stop the planet from burning*. Toronto: Doubleday Canada.
Monbiot, G. (2006b). *Selling indulgences*. http://www.monbiot.com/archives/2006/10/19/selling-indulgences/. Accessed 9 Aug 2007.
National Assembly of Wales. (2005, May). *Global citizenship secondary school map – Education for sustainable development and global citizenship*. Cardiff: National Assembly of Wales Circular no. 11/2005.
O'Murchu, D. (2004). *Quantum theology: Spiritual implications of the new physics*. New York: Crossroad.
Orr, D. (2009). *Down to the wire: Confronting climate change collapse*. Oxford: Oxford University Press.
Parkin, S., Johnson, A., Buckland, H., Brookes, F., & White, E. (2004). *Learning and skills for sustainable development: Developing a sustainability literate society*. London: Forum for the Future/Higher Education Partnership for Sustainability.
Plumwood, V. (1993). *Feminism and the mastery of nature*. New York: Routledge.
Plumwood, V. (1996). Androcentrism and anthropocentrism: Parallels and politics. *Ethics and the Environment, 1*(2 Fall), 119–132.
Porritt, J. (2006). *Capitalism as if the world matters*. London: Earthscan.
Postel, S. (1992). Denial in the decisive decade. In L. R. Brown & L. Starke (Eds.), *State of the world 1992: A Worldwatch Institute report on progress towards a sustainable society* (pp. 3–8). New York: W. W. Norton.
Postman, N., & Weingartner, C. (1969). *Teaching as a subversive activity*. New York: Delacorte.
Prigogine, I. (1989). The philosophy of instability. *Futures, 21*(4), 396–400.
Pruneau, D., Liboiron, L., Vrain, E., Gravel, H., Bourque, W., & Langis, J. (2001). People's ideas about climate change: A source of inspiration for the creation of educational programs'. *Canadian Journal of Environmental Education, 6*, 121–138.
Qualifications, Curriculum and Assessment Authority for Wales. (2002). *Education for sustainable development and global citizenship*. Cardiff: QCAA, Wales.
Richardson, R. (1990). *Daring to be a teacher*. Stoke-on-Trent: Trentham.
Romm, J. J. (2007). *Hell and high water: Global warming – The solution and the politics – And what we should do*. New York: William Morrow.
Selby, D. (2001). The signature of the whole: Radical interconnectedness and its implications for global and environmental education. *Encounter: Education for Meaning and Social Justice, 14* (1 Winter), 5–16.
Selby, D. (2006). The firm and shaky ground of education for sustainable development. *Journal of Geography in Higher Education, 30*(2), 351–365.
Selby, D. (2007). Reaching into the holomovement: A Bohmian perspective on social learning for sustainability. In A. E. G. Wals (Ed.), *Social learning towards a sustainable world: Principles, perspectives, and praxis* (pp. 165–180). Wageningen: Wageningen Academic.

Selby, D. & Kagawa, F. (2013). *Climate change in the classroom: UNESCO course for secondary teachers on climate change education for sustainable development.* Paris: UNESCO. http://www.unesco.org/new/ccesd. Accessed 15 May 2013.
Seligman, M. (1992). *Learned optimism.* New York: Knopf.
Shiva, V. (2005). *Earth democracy: Justice, sustainability and peace.* London: Zed.
Smith, D. C., & Carson, T. R. (1998). *Educating for a peaceful future.* Toronto: Kagan & Woo.
Sterling, S. (2007). Riding the storm: Towards a connective cultural consciousness. In A. E. G. Wals (Ed.), *Social learning towards a sustainable world: Principles, perspectives, and praxis* (pp. 63–82). Wageningen: Wageningen Academic.
Stibbe, A. (Ed.). (2009). *The handbook of sustainability literacy: Skills for a changing world.* Totnes: Green Books.
Tilbury, D., Podger, D., & Reid, A. (2004). *Change in curricula and graduate skills towards sustainability: Final report.* Sidney: Australian Government Department of the Environment and Heritage/Macquarie University.
UNESCO. (2010). *UNESCO strategies for the second half of the United Nations decade of education for sustainable development.* Paris: UNESCO.
United Nations Conference on Environment and Development (UNCED). (1992). *Agenda 21/Rio declaration on environment and development.* New York: United Nations Department of Public Information.
Victor, P. (2008). *Managing without growth: Slower by design not disaster.* Cheltenham: Edward Elgar.

Chapter 11
Learning to Let Go of Sustainability

David W. Blades and Janet Newbury

Abstract By critically engaging with current sustainability discourses and practices, this chapter strives to open space for more/different possibilities from the illusion of restoration or the seduction of neo-romanticism. Recognizing the global interconnectedness of humans and nonhumans alike by tracing (some of) the journey of a glass jar, the authors consider the roles of economic development, gender dynamics, political realities, and our relationships with the material world that may perpetuate unsustainable practices—even in the name of sustainability.

D.W. Blades (✉)
Faculty of Education, University of Victoria, PO Box 3010 STN CSC,
Victoria, BC, V8W 3N4, Canada
e-mail: dblades@uvic.ca

J. Newbury
Faculty of Human and Social Development, University of Victoria,
PO Box 1700 STN CSC, Victoria, BC, V8W 2Y2, Canada
e-mail: janet.neubury@gmail.com

L. Bencze and S. Alsop (eds.), *Activist Science and Technology Education*,
Cultural Studies of Science Education 9, DOI 10.1007/978-94-007-4360-1_11,

This chapter thus complicates what can otherwise be dangerously simplified notion of restoration or desire for a 'return' to more sustainable days gone. Deconstructing the discourses of sustainability begin to reveal opportunities for other ways forward in the move from technical-rational fixes in favour of ontological approaches to change. In this chapter we examine how ontological shifts can substantially alter power relations, inviting us to recognize multiple and simultaneous possibilities for change through a hermeneutics of sustainability that strives to make space for emergent, democratic, and responsive actions when charting more equitable and 'sustainable' ways to live.

Keywords Sustainability • Education • Deconstruction • Ethics • Globalization • Restoration • Neo-romanticism • Remembering forward • Social justice

Introduction

Plastics are not sustainable products; items made of plastic remain essentially unchanged in landfills (Rathje and Murphy 2001); when heated or exposed to microwaves almost all plastics produce estrogen-mimicking compounds (Yang et al. 2011), and plastics are generally made from petroleum, itself a non-renewable resource. Inspired by students and family members who are attempting plastic-free lifestyles, we decided to substitute glass containers—the container in the photo below is one—for the used yoghurt, cottage cheese and other plastic containers in our collection for storing leftovers.

Glass is made from one of the most abundant compounds on the Earth, silicon dioxide or common sand. When heated and cooled properly, silicon dioxide forms into the substance we call, 'glass,' thus glass containers can also be easily recycled into new glass products by first grinding the old glass and then heating it. Glass is also non-toxic for food storage, hardy when thick, and relatively inexpensive. In any move to 'act more sustainably' switching from plastic to glass containers seems a good first step.

Upon closer examination, however, these new glass containers reveal that they were made in Italy, which means that the containers we purchased locally were shipped at least 9,000 km from their point of manufacture. Shipping glass this vast distance requires packaging that ironically could be plastics-based and, of course, includes the environmental costs of the fuel and associated materials to move such products. While the glass container hopefully will last a long time, it nevertheless is clearly not as sustainable an option as it first seems. This, in fact, is true of *every* attempt to 'act sustainably'; each action we take is interconnected with a host of related acts and discourses. In our chapter, we examine, as participants within the discourse of 'sustainability,' the illusion of restoration or a neo-romanticism of past practices. We argue that both approaches to sustainability are counterproductive to finding ways of living that have less impact on the environmental conditions supporting the existence of humans and those species with similar needs. But neither are we hopeless, for the very word 'sustain' indicates possibilities for action.

The artefact of the glass jar provides a useful entry point into discussions of sustainability and potentially fruitful alternatives. Considering the journey of a single glass jar opens up a multitude of questions about what sustainability asks of us, what we ask of each other in its name, what we are in fact striving to sustain, and what the unacknowledged implications of our actions might be. Our concern rests partly in the prescriptive (and in turn arbitrarily limited) nature of current discourses of sustainability. What avenues for action remain unexplored when norms around ethical social action become prescriptive? What assumptions and divisions among us develop? And, what aspects of our supposedly 'sustainable' choices remain unexamined?

By critically engaging with notions of sustainability and how they are currently taken up, we do not wish to throw our hands in the air and suggest there is no hope of ethical ways to proceed. Rather, we wish to *open up* the range of potential ethical actions by not only critically engaging with those we currently embrace, but by widening our gaze to also include those we do not (Caputo 2000).

Sustainability as Restoring

Buying a glass jar is an immediate investment in a set of complex, interconnected discourses all linked by a commitment to 'sustainable development.' The jar uses a non-magnetic metal frame to secure the lid. Since the jar can be used for canning, the metal also is non-rusting and is, thus, likely aluminium. Mining aluminium ore, extracting the pure metal and then shaping the metal into wire requires enormous amounts of energy and produces toxic tailings. The Rio Tinto Alcan™ plant in northern British Columbia is the third largest producer of aluminium in the world (after Russia and China) and, according to their *Sustainable Development Report* (2011), "our commitment to sustainable development is integrated into everything we do" (Rio Tinto Alcan, p. 8). This company adopts a wide, somewhat liberal, view of 'sustainable development,' which includes attempts to balance environmental, economic, social and governance concerns. Reading closer, however, we find that Rio Tinto Alcan is working *towards* sustainability since, as they admit in their report, the plant still requires "environmental permit non-compliance" for pollution beyond federal regulations (p. 11). As well, the company produces sulphur dioxide, a major atmospheric pollutant and, even though this production has decreased over the past 4 years, it has only just reached levels slightly less than in 2002. When I pick up a glass jar for purchase, then, I participate as a consumer in the activities and by-products of metal mining, smelting and production and the associated energy costs of production and shipping the aluminum wire for the jar, whether from Canada, Russian or China, to Italy.

The jar also has a 'rubber' gasket to help seal the lid. There is very little natural rubber used in the world since rubber from trees tends to break down quickly, especially when heated. The gasket is thus likely a synthetic rubber, probably a variation of neoprene. The basic chemical platform for the production of neoprene comes from the fractionalization of crude oil. Therefore, the glass jar has a double connection to oil extraction, both in the production of the gasket and also in the manufacture and use of the fuel to transport the jars from Italy to Canada.

The tar sands in Alberta comprise the third largest proven source of oil in the world (Alberta Energy 2013a). Extracting this oil is very much complicated by the fact that the oil is mixed with sand in a gooey, semi-sold, state called, "bitumen" or "tar." The oil can be extracted, but this requires huge amounts of energy to heat the large volumes of water, resulting in the release of tonnes (metric tons) of carbon dioxide pollution from the natural gas used to heat the water and significant toxic pollution of the water used in the extraction process. To remove the bitumen, almost prehistoric-sized shovels are used in an open-pit approach. Despite the clear wreckage to the tundra ecosystem of the tar sands, this oil reserve is described as a form of "sustainable development" by the industries involved in the extraction and movement of the recovered oil. In his book, *Tar Sands,* Andrew Nikiforuk (2010) notes that companies working in this region claim that upon completion of the extraction of the tar sands, they will "replicate the stability and robustness of the original natural systems" (Imperial Oil, cited in Nikiforuk 2010, p. 107), or will return the land "to a stable, biologically self-sustaining state" (Syncrude, cited in Nikiforuk 2010, p. 107).

The Paradox of Sustainable Development

When companies involved in the mining of oil and metal speak of "sustainable development," they refer to the belief in minimal environmental impact during extraction and development and *restoration* as a possible, even inevitable, consequence of when the resource is depleted. This optimistic view of sustainability is the official policy of the Government of Canada with the enactment in 2008 of the *Federal Sustainable Development Act.* According to the Sustainable Development Office of Environment Canada, the Federal agency responsible for the health of the environment in Canada, this act legally defines the approach of the Government of Canada as "sustainable development"; that is, "a commitment to minimize the environmental impacts of its policies and operations as well as maximize the efficient use of natural resources and other goods and services" (Environment Canada 2010, p. 1).

At the heart of sustainability as restoration is the assumption that the various demands of development can be balanced in an ethos of environmental conservation, what Allen Hammond (1998) described as "managing a planet and a global human civilization in ways that will sustain both" (p. 6). Implicit in this approach to action is an assumption that ecosystems damaged or lost through development *can* be restored and *should* be restored. Efforts at reforestation, for example, after clear-cutting, suggest an attempt to restore the ecosystem of a forest; in fact, this is explicit in the very term, "reforestation."

The idea that sustainable development is possible was advanced considerably with the publication of the United Nations World Commission on Environment and Development (1987), which defined sustainable development as, "development that meets the needs of the present without compromising the ability of future generations to meet their own needs" (Brundtland 1987, p. 8). Prior to this landmark publication, "sustainable" development focused on the *economic* sustainability of the development; that is, the financial viability of the resource extraction (Etymology Dictionary 2013).

As the twentieth century proceeded, it became increasingly clear that demands for products will continue to increase with increases in population, fueled by a "culture of globalization" based on the vision of "billions of people newly liberated to make their own choices in a market offering dignity and endless delights" (Derber 2002, p. 58). This growing demand requires continual development of a limited and dwindling supply of natural resources, a dilemma John Robinson (2004) identifies as the inherit "incommensurability" present in the term "sustainable development" (p. 370). Thus, while countries worldwide, including Canada, agreed in principle to the UN Report advocating development without compromising the future, actualizing this "sustainability" balancing act has proved elusive.

Consider, for example, the Tar Sands development. Nikiforuk (2010) reports that as early as 1977 university soil specialists warned companies developing the tar sands that restoration faces issues of soil salinity, oil contamination, erosion and a host of other consequences of the extraction that severely compromise any hopes of restoration. In fact, Nikiforuk found that investigations by the extraction companies themselves reached similar conclusions. Despite government regulations that the open-pit mines of the tar sands must reclaim the spent mines to equivalent to land capability, Alberta Energy reports that only 0.2 % of the land has been certified as "restored" by the Federal Government; the rest remains in use, waiting for "restoration" or certification as "restored" (Alberta Energy 2013b). As for the actual, practical, restoration of the consequences of tar sands mining, Nikiforuk (2010) concludes, "in truth, no one knows how to do it" (p. 106).

Areas strip-mined that often are considered "restored" are typically described as, "reforested." But "reforestation" is not the recreation of the original forest, for the trees planted to reclaim open mines are, in fact, destined for harvest. This is not, then, a restoration to the *original* ecosystem but *transformation* of the open-pit mine, whether for oil, metal or some other material, into continual business operation; in most cases, a tree plantation. While the owners of the Rio Tinto Alcan™ mine, for example, or companies involved in tar sands extraction may state a goal of sustainability, the very practice of extraction and development itself is *not* sustainable precisely because any hope of restoration to original state turns out to be an illusion. In practice, then, "sustainability" as restoration provides false assurances of repair, useful to advance and justify an agenda of resource extraction; but, overall, a use of the word "sustainability" that misleads the public to believe the effects of this extraction can be reversed. Simply put, the damage is irreversible and, thus, there is nothing "sustainable" in resource extraction, if by "sustainable" we mean restoration.

Yet political leaders cling to the hope and assurance in the phrase, "sustainable development." This became clear during a recent televised debate of candidates for leadership of a Federal political party. Candidates were asked to share their views on the environment and without exception each used the word, "sustainability" or "sustainable" to describe their approach to environmental responsibility. But the *meaning* of "sustainability" varied considerably with the candidate, from talking about the environment to extending the idea of sustainability to include federal social programmes (e.g., a "sustainable health care system") and accelerated capitalism in the form of job growth and increased GDP. In that debate, "sustainable" was a widely-used adverb intended, it seemed, to assure voters that the policies and approaches prompted and proposed will be somehow both environmentally and socially responsible, even though there is little evidence to support this possibility (and a great deal of evidence supporting the view that increased growth is simply *un*sustainable). And we might invest in the same hope when purchasing a glass jar, but the harsh reality is that when we buy this product—any product—we are participating in a discourse that is a *simulacra* of sustainability as restoration. We may be partially reducing the need for plastic, but we are still invested in globalized movement of goods and the resource extraction industries, ironically including oil extraction, that enables the manufacture of a glass jar.

To Hold onto, or to Let Go?

The word, "sustainability" or, "sustainable" comes from the Latin, *tenere,* which is the verb, "to hold"—as the French verb, *tenir*. To sustain is to hold onto what is, or to support what exists and in the form of justification can refer to a legal position, philosophical argument, ethical stance or action. Given the roots of this word, we might ask, "what are we holding onto, or justifying in a discourse of sustainability?" Certainly what remains in "sustainable development" is a belief in the value of development itself, in a *continual* development, indicated economically as a positive value for a nation's "gross domestic product" within a capitalist economic system. Suppose, however, we accepted and truly acted on a view of sustainability as "the ability of humans to continue to live within environmental constraints" (Johnston et al. 2010, p. 2); in other words, rather than attempt a balancing act where the environment typically suffers in the economic pressures of financial gain, we put the health of the environment *first*? This might require radical rethinking of our economic indicators, such as measuring a country's "sustainable health" as a *negative* or zero GDP and it might, even, mean not developing the resource at all. This, of course, means immediately fewer products for consumers and certainly less choice as well as a loss of those jobs associated with resource development.

To avoid this either/or scenario, consumers might consider purchasing items less disposable and more easily recycled when necessary. For example, what might be the consequences if products currently packaged in plastic, such as yoghurt, milk, or dish detergent were only available in glass? What if rather than recycling glass and then manufacturing more, our communities had depots where glass could be donated and collected? What if there was a sudden decline in the need for plastics, themselves a major by-product of oil extraction? Would mining decrease? Would consumers consume less as jars are used for other purposes? In fact, what we are beginning to imagine is the situation before plastics became ubiquitous, suggesting that an approach to sustainability as a return to a time when it was easier to live sustainably.

Sustainability as Returning

Romanticizing the Return?

The glass jar hearkens back to a time that was perhaps simpler . . . more stable and consistent; a time when a family would likely not have been scattered geographically and so could accumulate objects (furniture, heirlooms . . . glass jars) and pass them from one generation to the next. Could returning to such a way of life enable us to have a much smaller carbon footprint, and experience a rich sense of belonging? *Sustainability as returning* might be summed up by the following excerpt from the introduction of Helen and Scott Nearing's (1989) book, *The Good Life*:

> We have not solved the problem of living. Far from it. But our experience convinces us that no family group possessing a normal share of vigor [sic], energy, purpose, imagination and determination need continue to wear the yoke of a competitive, acquisitive, predatory culture. Unless vigilante mobs or the police interfere, the family can live with nature, make themselves a living that will preserve and enhance their efficiency, and give them leisure in which they can do their bit to make the world a better place. (p. 7)

There is beauty in the intentionality of the life the Nearings cultivated. They demonstrate a sense of responsibility for self and others, as well as a deep understanding of the impacts we all have on the world in which we live—and of the fact that we are an *integral part* of the world in which we live (Abram 1996). Sustainability as returning often conjures such images of harmony and coexistence. The Nearings' contribution to a more sustainable way of living continues to play an important role when it comes to resisting the single story of economic growth, individualism, and material accumulation. Motivated by ecological sustainability, physical health, and economic justice, they have demonstrated through their choices that more equitable and sustainable paths are not only possible, but quite gratifying.

Tracey Deutsch (2011) observes that, "pastoral nostalgia celebrates small farmers who have managed to retain or reinvent older forms of sustainable farming" (p. 167). She notes how food writers and those advocating slow and local foods regularly draw on the histories of women and their cooking, either implicitly or explicitly. Problems with the contemporary food system *and* solutions to those problems often point to women's food procurement and cooking (addressed further below). Images of grandmothers in the kitchen, the gratification of culinary activity, and the relational capacity of food production and preparation to weave generations together are all evoked when sustainability is conceptualized as returning.

Resisting the Return

But understanding these efforts as a 'return' might be a simplistic interpretation of what once was, including class and gender relations. Take, for instance, rural Newfoundland until the mid-twentieth century, which is a relatively recent example from which to draw: The landscape was harsh, the soil infertile, and goods were scarce. Communities were accessible only by boat, which meant health care and other services were simply not part of the reality for many of these communities. Work (for those who had it) was usually seasonal, which left most families reliant on loans from the local merchant for much of the year. This indebtedness left families extremely vulnerable as the margin of error was nearly non-existent: a poor crop, low fishing yields, or an illness could upset the delicate balance that kept a family afloat. Life required a great deal of participation from all family members, which meant children were often removed from school in order to contribute. Life was hard, options were limited, and hearts and bodies took a toll (Rompkey 2006).

In the 1950s, things began to change for rural Newfoundlanders: some communities were resettled and others became connected by roads. Funding opportunities for post-secondary education became available. Electricity, plumbing, refrigeration, telephones, automobiles. . .these and other modern conveniences undeniably contributed to certain improvements in the quality of life for many during those years. Options began to open up (particularly for girls and women) and lifespans lengthened. There was some resistance to these changes, to be sure, but for the most part the consensus was that things were looking up (Rompkey 2006).

Addressing the topic of food in particular, Deutsch (2011) casts light on the often-obscured gendered assumptions nested within notions of 'returning' in the sustainable food movement. With what might be seen as a misguided interpretation of the feminist slogan 'the personal is political,' the sustainable food movement seems to suggest the onus for change lies in the kitchen, and thus, in the hands of women. Deutsch cautions that

> [t]his emphasis on individual autonomy and taste runs through food writing. For instance, in light of what she considers to be the failure of feminism, [Barbara] Kingsolver does not call for structural change to food distribution, workplace policies, or even gender relations... Rather, she prescribes a change of attitude on the part of would-be cooks—identified in previous sentences as women. (p. 173)

Deutsch's (2011) historical analysis of gender dynamics in relation to food leads her to ask one important question: "what exactly is being sustained in sustainable food movements?" (p. 174). In response, she argues that if engaged with uncritically, the sustainable food movement is in fact quite actively serving to sustain an apolitical and neoliberal understanding of women, all those who engage in women's work, and markets in general. Its individualized rhetoric, which aims to be empowering, actually obscures the politics and policies that sustain a very conservative idea that "private enterprise, operating in an unregulated marketplace, is a more appropriate place to look for change than the state or publically controlled spaces" (p. 173).

The social imperative to replace our plastic containers with glass jars while overlooking the process by which we obtained them is a concrete example of how misguided this approach can be. Reflecting on the earlier discussions of sustainability about the production and distribution of glass, aluminium, and rubber, we need to understand the complicated political processes by which the glass jar ended up in our hands in the first place. Taking it at face value, we might pat ourselves on the backs for our ethical choices. Looking beneath the surface, however, we might see that sustainable paths also require that we critically engage on a systemic level, and recognize our parts in sustaining *un*sustainable patterns.

Complicating the Return

It is difficult to romanticize efforts to 'return' in conversation with people who endured hardships and now experience life as less harsh and more equitable due to modern conveniences. But it is also crucial not to romanticize modern developments and technologies, as the sustainability movement aptly reminds us. The same technological trends (towards growth and centralization) that led to infrastructure and health care in rural Newfoundland, for instance, also led to high-yielding off-shore dragnet fishing vessels, which contributed to the quick decline of cod stocks in the 1980s and 1990s (Rompkey 2006).

The ethics and practices inherent to sustainability as returning, however, need not be dismissed entirely with the acknowledgement that injustices existed when community life was (by necessity) more locally driven. Might we, instead, return to those stories of both sustenance *and* hardship in order to chart new and more sustainable paths?

Heike Mayer and Paul Knox (2010) observe that while social and political life in the eighteenth, nineteenth, and twentieth centuries was characterized by an 'either/or' principle (either nature *or* society, either economic growth *or* environmental stewardship, either us *or* them), we are now seeing glimpses of a 'both/and' principle at play (both nature *and* society, both economic prosperity *and* sustainability, both us *and* them). Noting a shift from what they refer to as the 'first modernity' to this, the 'second modernity,' these authors do not go so far as to suggest we are now reaching post-modernity. They simply observe that while the underlying motivations of competition and individualization of the first modernity continue to inform our policies and practices in many ways, globalization and related shifts have changed the context in which our economic and social activities take place. Rather than the nuclear family and the nation state, we are now operating on a much larger scale with less distinct boundaries among us. This means institutions we previously relied on are now being called into question, and collaboration across divides marks a major indicator of the 'both/and' principle that is currently at play.

Slow food, local food, back-to-the-land movements, community partnerships, and related activities are responses to these shifts. Embedded within these collective practices is an ethic of sustainability, but not only environmental sustainability: also the sustainability of small towns and groups of people (including but not limited to families), the sustainability of certain practices and knowledge, and the sustainability of alternative economies. In this sense, much of what might be seen as 'sustainability as returning' is also infused with values of social justice (Johnston et al. 2010; Mayer and Knox 2010). However, Deutsch's (2011) analysis of gender relations also serves as an important reminder to be vigilant in engaging with the political nature of these practices.

Mayer and Knox (2010) note that the shifts we are currently experiencing (in the 'second modernity') are associated with a set of new sensibilities. These sensibilities:

> [a]re based on sustainability, which functions as the categorical imperative and as an answer to overarching risks that are global in nature. In contrast to the first modernity, where zero-sum competition was rewarded, nowadays, collaboration aimed at sustainability and differentiation as well as economic growth is increasingly rewarded: a reflection of the shift to "both/and". Four principal sensibilities have emerged that are guiding collaborative efforts aimed at small-town sustainability. (p. 1549)

The four sensibilities they identify include: (1) slow and organic food, (2) environmentalism, (3) entrepreneurship, and (4) creativity. The implementation of these sensibilities is increasingly reliant on collaboration and the creation of networks and partnerships that range in scope. In this sense, activities that might on the surface appear to be a 'return' may, in fact, be coloured with different underlying sensibilities that can be seen as responses to current conditions, not a re-creation of those of the past.

Or at least, they *could* be.

Perhaps ethical social action need not be reduced to sustainability as returning to what once was, just as it cannot be simplified as the technological development of what might be. In his memoir, *Trauma Farm*, Brian Brett (2009) notes:

> Hunter-gatherer cultures weren't any more ecological than us. Their damage was usually only less because of smaller populations and either a cultural rejection of technology or a lack of it ... You can't practise 'traditional' whale hunting with high-powered rifles and harpoon cannons ... (p. 361)

With that observation in mind, sustainability as returning (in any pure sense) is both impossible at this juncture in history and missing the point. By striving for it, or simplifying notions of what we might return to, are we limiting possibilities for how we might creatively move forward *from here*? What other options exists?

How do we make our understandings of ethical social action more complex in relation to sustainability? How do we move beyond 'the glass jar' when asking ourselves what might be done differently?

Learning to Let Go of Sustainability

Sustainability is a thoroughly modern word, one that promotes dualistic thinking characteristic of modernity (Madison 1990). For example, one might argue that it is "more sustainable" to recycle paper instead of disposing of paper in the landfill. Situating as opposites one practice against another is characteristic of the current discourse-practice of sustainability; for example, it is "more sustainable" to use glass instead of plastic containers.

As we have seen, however, trying to live in ways less damaging to the conditions that support human life (and, of course, all life) is considerably more complex than either/or approaches to sustainability. Plastics are a good example of this complexity: "Plastic" is a generic term for a whole family of chemicals loosely related as products of a process of linking chemical units together called, "polymerization." An example is "ABS" or "acrylonitrile butadiene styrene"—a polymer made from chemicals fractionated from oil and coal. ABS has many slight variations and is used in a wide variety of applications, from LEGO bricks to housing for computers. Because this substance is stable over a wide range of temperatures and resistant to every chemical commonly found in homes (Plasticpedia 2013), one of the most common uses of ABS plastic is in the production of pipe for house drains, waste and vents. This system is easy to install, incredibly stable, shows no signs of erosion due to friction in decades of use and can be recycled. In plumbing, ABS replaces cast iron pipes, which eventually do rust and, of course, require mining extraction of iron ore. *Both* cast iron and ABS are thus products of extraction, and simply declaring all plastics as "unsustainable" misses the point: It's not one product versus another in some kind of generic categorization, but that the discourses-practices involved in *each* product must be considered carefully. This makes "living sustainably" considerably more complex than rhetoric of, "no plastics." In fact, it turns out that *some* plastics are incredibly useful and these may have less severe extraction and manufacturing footprints than other products, such as iron. The production of ABS plastics, for example, also suggests that oil and coal are *precious, non-renewable resources* that have important applications beyond the rather wasteful consumption of these resources as fuel.

Investigating a product for consumption reveals the complexity of our situation. As we have seen, sustainability as restoration is an illusion and a romantic past where we lived more sustainably is equally unavailable. Søren Kierkegaard (1983) demonstrated this inability to return to the past in his famous essay, *Repetition*. In Part One of his essay, he reflects on attempts to recollect the past, including a visit to his former lodging, but he finds it impossible to ever return to a previous state of being that one recollects; since even the slighted change reminds the visitor that the past, as one recollects it, always no longer exists. To think of the past from despair to hope, Kierkegaard realizes, means moving from recollection to what he calls, "repetition." Repetition and recollection are the same movement, he argues, but "in opposite directions, for what is recollected has been, is repeated backward, whereas genuine repetition is recollected forward" (p. 131).

In his essay on Kierkegaard's concept of repetition, John Caputo (1987) describes repetition as, "remembering forward"—a hermeneutics of the past that is in constant movement towards newness and insight. In contrast to the search for meaning in general hermeneutics, Caputo's radicalized hermeneutics as a "remembering forward" is more interested in "keeping the difficulty of life alive" (p. 3). This perspective challenges what is offered in the discourses of sustainability, such as claims that certain actions can "make a difference" or "save the planet," precisely because these acts fail to recognize the complexity and difficulty of living in ways less compromising to human existence and, thus, in result and effect, these acts are little more than superficial responses to the destructive trajectories of modern societies.

Letting Go of Sustainability

Sustainability is almost universally linked to "development" but the combination has a wide variety of meanings. Pearce et al.'s (1990) often cited *Sustainable Development* defines sustainable development as the minimum conditions of development where the "*natural capital stock* should not decrease over time" (p. 1; emphasis in text); by "stock" the authors refer to the "environmental and resource assets" of a country. The International Institute for Sustainable Development (2013), however, adds the dimension of social responsibility to development, including full access to education, religious freedom, and the absence of poverty as essential considerations of approaches to development. Such divergence of views of what constitutes "development" confirms Susan Owens' (2003) discovery that "there is no singular definition of sustainable development upon which all can agree" (p. 8).

Despite this lack of agreement and perhaps in the hopes of some shared meaning of the phrase, "sustainable development," UNESCO embarked on an ambitious project employing education as a vehicle for promoting sustainable development worldwide. The "UN Decade of Education for Sustainable Development" began in 2005 and runs until 2014. According to the UNESCO, the goal of this decade is to "to integrate the principles, values, and practices of sustainable development into all aspects of education and learning" (UNESCO 2005, p. 1). UNESCO also adopts a wide perspective on sustainable development, which includes education on environmental issues such as climate change and reduction in biodiversity but, to UNESCO, sustainable development also means reduction of poverty and the use of learning methods that "motivate and empower learners to change their behaviour and take action for sustainable development. Education for Sustainable Development consequently promotes competencies like critical thinking, imagining future scenarios and making decisions in a collaborative way" (UNESCO 2005, p. 1).

UNESCO provides examples of what might be called, "acts of sustainability" and "sustainable development" in education. These examples adopt a wide approach to sustainability that includes gender equity, promotion of health, peace and security, sustainable urbanization, indigenous knowledge and other related topics. A direct application of the UNESCO promotion is the Canadian non-profit *Learning for Sustainable Development* (LSD). Founded in 1991, LSD was created to "integrate sustainability education into Canada's education system. LSD's mission is to promote, through education, the knowledge, skills, perspectives, and practices essential to a sustainable future" (LSD 2013). LSD tries to accomplish this goal by offering educators lesson plans (for a nominal fee) and other resources in line with the areas for education outlined in the UNESCO Decade for sustainable development.

LSD's "resources for rethinking" include what they consider to be "exemplary classroom resources for sustainability" as well as "action projects" that "link education to action" (LSD 2013, p. 1). One of these programmes is, "EcoLeague," school-wide initiatives such as building a school vermiculture culture (to help composting), cleaning up of shorelines, greening school playgrounds using plants and the manufacture of reusable, cloth bags from old clothing to replace plastic shopping bags. The stated goal of these programmes is to help students to "save the planet" though, "community and school-based sustainability action projects."

Our hermeneutics of sustainability calls us to the original root of the word, *tenere*, or "to hold." This invites us to ask, "what is held onto in the action projects of LSD?" A closer look at the resources and calls-to-action suggest avenues for student actions within a global discourse of production and consumption that remains entirely unchallenged: Cleaning up a shoreline does not challenge directly the consumption of goods that end up floating in waterways and up on shores. Similarly, making shopping bags does not challenge the very act of shopping itself, an act that actually creates the need for changes in our actions. The resources of LSD are a start, perhaps, but when encouraging children to make reusable bags we hold onto the very fabric of a consumerist, shopping society that contributes to the conditions threatening the continuation of humans (and others). In this way, LSD and the myriad of "actions" available to educators via the Internet (including the examples provided by UNESCO) are in invitation to irony. A hermeneutics of these actions suggest that the actions promoted to educate for "sustainable development" do not fundamentally critique the very discourses that create the need for sustainability in the first place, and do not offer alternatives. Rather they enable us to 'hold' onto them, now feeling better about our own participation as we continue 'developing.'

All across the world, promoted in part by UNESCO, school children are thus engaged in "acts of sustainability." The danger in these actions is that with this engagement children may feel that they are doing enough to "save the planet." The reality is more complex and difficult: Children are, in fact, continuing to be induced into the cultural practices that overall contribute to the declining state of the planet and all these acts do is provide a false and misleading sense of security that the environmental problems facing humankind can be fixed with a garbage clean up here and there, or the planting of a few trees in schoolyard as parents drive up in their SUVs to take their children home. In this way, teaching for sustainable development is not only an illusion or neo-romanticism, but may be *dangerous* in the presumption that actions such as those by LSD and other agencies are making a positive difference in world conditions.

Remembering Forward as a Different Way of Being in the World

Although the pursuits of sustainable development that have been shared thus far vary greatly, what they seem to share is a common approach to sustainability that views change as a largely *technical-rational* process: define sustainability, identify a goal, and establish best practices aimed at achieving that goal. However, as David Blades (1997) observed when studying curriculum development, such approaches to change are in fact more likely to entrench existing practices than they are to change them. He says, "clearly change involves more than tinkering with an existing system, change is much, much more difficult: it is an effort to break from the systems in which we are trapped" (p. 95).

We can see evidence of this in the examples shared above, such as UNESCO's action projects and the purchase of mass-produced, globally distributed glass jars. If we teach our students to collect trash without intervening in the capitalist system that *relies* on the generation of waste, then we are in fact ensuring its perpetuation. We will effectively be educating an entire generation of citizens to participate in both ends of this process: the consumption and production that generates waste, and the clean-up process that allows us to believe we have repaired the damage it causes, allowing us to continue consuming at increasing rates. This does nothing to alter the course of action, as we can see by our nation's simultaneous participation in global conferences on climate change *and* increasing reliance on the oil and gas industry for economic development (see Canadian Association of Petroleum Producers 2012). In fact, it may be contributing to the seeming inevitability of the current course of events.

But it is only inevitable from *within* this system of thinking.

Blades (1997) suggests that in order to find a way out of this bind we might persistently ask different kinds of questions. He reminds us that, "Foucault deftly side-steps the hermeneutic question, "What does discourse *mean*?" by asking, "What does discourse *do*?", followed closely by the political question, "How else could it *be*?" (p. 107). In other words, our questions must shift from being technical in nature, to being *ontological*.

This means we are now free to let go of some of the things we've been holding onto (sustaining), many of which have been constraining this process. First, we can let go of the belief that we need to develop an accurate definition of 'sustainability' from which to work and a clear understanding of where we are heading as we pursue it. A technical approach to change would require that we seek consensus on 'right' next steps in order to proceed effectively towards them. The constraints of such an approach can readily be felt, and missteps can discredit the entire process.

An ontological approach, on the other hand, would allow for the fact that we have limited power to direct outcomes in this messy world in which we live. Instead of identifying and pursuing technical solutions to the problem of sustainability, we might *elicit* the just and equitable world we are seeking rather than attempt to 'produce' it. We might reflexively engage in each and every step—even as we take them—critically considering whether or not *how* we are engaging is congruent with *what* we are striving for.

Importantly, an ontological approach does not embrace a zero-sum conceptualization of power, but rather sees power as something that is constantly moving among us. This means we need not (indeed, cannot) achieve certainty about end results before proceeding. Caputo (2000), who suggests we might be better off considering what to do *in the meantime* than pursuing certain ends, urges us to recognize: "the secret is, there is no Secret" (p. 40). If we see power as dispersed rather than something anyone can ultimately possess, then our responsibility in the context of sustainability initiatives becomes considering how we might proceed *from here*—realizing this is a question that will have to be asked anew at each step along the way. As we critically reflect on our own pursuits and actions, we can more humbly embark on them, knowing that we can make changes when our new and ever-changing conditions call for them.

Such an ontological approach to change has been contrasted with a technical one by J. Michael Thoms (2007), who advocates for development of wise practices instead of best practices. The language of 'best practices' is probably familiar to most of us. We may work in contexts in which funding or resources are made available only if our work reflects what current research describes as 'best practices' within our fields. But what does this term mean? According to Thoms (2007), "a 'best practices' document takes into account all that is known about a subject, takes stock of lessons learned, and adds new knowledge drawn through the application of sound and effective research methods. The expectation is that 'best practices' in one situation can be replicated in a similar situation and have the same positive effects" (p. 8). Many sustainability initiatives that can be reduced to action plans and checklists can be described in this way.

Thoms (2007) offers a critique of the term 'best' when engaging with particularly vulnerable populations (his critique can be extended to the sustainability movement as well). His work has been primarily with two-spirit First Nations men. In his experience, he has found that the term 'best' can be seen as hierarchical. Such a hierarchy of knowledge and practices is dangerous because often what is touted by the evidence as 'best' is what has been trialled in large, well-funded academic contexts. Prioritizing this knowledge can marginalize Aboriginal and other valuable forms of knowledge.

The word 'best' also artificially decontextualizes our practices; this presents another danger. It assumes that what 'works' in one situation will work in another, even though there is a great deal of research that tells us our conditions continue to change, and what works in what time and place can be dangerous if taken prescriptively elsewhere. Thoms (2007) asserts that there are no perfect models that will fit all communities in all moments. He believes we might be better served by moving away from the language of 'best' when discerning how to engage with complex matters, and instead thinking in terms of *wisdom*, which he claims is non-hierarchical and contextually based.

'Wise practices' are Thoms' response to the critique of best practices. According to Cynthia Wesley-Esquimaux and Andrew Snowball (2010), wise practices are "a basis for creating a new dialogue within Aboriginal communities . . . and a way to foster culturally appropriate support and health care" (p. 121) more generally. Again, these ideas extend well into the context of discussions around sustainability, which—as has been articulated above—are intricately connected with matters of justice, ethics, and collective wellbeing.

When thinking of how to move forward with sustainability in mind, we might approach it *not* in terms of mastering best practices with any kind of a priori certainty. Instead, we might cultivate *'wise practices'*—which are those ways of engaging that enable us to be responsive to diverse conditions, needs, and strengths that are constantly shifting and changing. This uncertainty means developing practices that enable us to ethically and effectively respond *as* we encounter new situations, rather than thinking we can know a priori what sustainability will require of us by mastering existing, decontextualized knowledge.

Caputo's (1987) interpretation of Kiekegaard's "repetition" is a helpful way to think about how to act. Caputo reminds us that repetition is *movement*, a "remembering forward." This movement acknowledges that problems conceived as essentially a technical issue, such as working out steps to lower the production of greenhouse gases, will invariably lead to technical solutions. But as Heidegger argues, *thinking* in technical ways also leads to *living* and *being* technical, a condition he believes places humankind in danger (Heidegger 1977). Caputo (1987), agreeing with Heidegger's famous essay on technology, calls this shift as the way "technology comes to presence" (p. 232) as we seek to live in less destructive ways; a technical solution is what first comes to mind, which is, Heidegger claims, why we are in danger in the first place.

In this way, sustainability as restoring and sustainability as return represent *technical* approaches to living differently and thus continue the very thinking and being that endanger our existence. Caputo proposes "remembering forward" as an ethos of being but not, of course, a practice that one can spell out: that would be slippage back into the presence of technicality; neither, too, can we seek a new word or phrase to replace "sustainability"—in so doing we are simply changing garments but the body is still the same. Remembering forward requires *being* different in the world, it is an invitation to ontological change and thus is so much more difficult to live and to teach—it is the original difficulty of life Caputo seeks to restore.

As Caputo (1987) observes, we are "trying to restore the difficulty in life, not to make it impossible" (p. 209). Acknowledging the resident challenge of being different without specifying how to be different, we can find ways to engage and act that are less destructive to human existence and to the planet we share. To avoid slippage into technicality, we might initiate and support constant conversations of critical remembering the diverse understandings of what it meant by various communities to live well. This does not necessarily celebrate these lifestyles, but opens a conversation about such lifestyles that can be considered and examined. For example, our mothers used canning jars to preserve foods. Why did they do this? What was the value of this action? Why was this action largely forgotten with the next generation? Where did they obtain the glass jars? This conversation moves from technicality when we add to the conversation questions about the *ethics* of this past action: Was canning adding to the carbon demand, perhaps through the fuel needed to heat the boiling water used to heat the jars? Overall, would it be more effective just to purchase local foods available fresh seasonally—or would this practice require considerable driving to various locations, also adding to the carbon demand? To what extent were men involved in canning, if at all? "Remembering" is thus a large conversation that critically examines as many facets as possible about past actions.

The conversation eventually must consider, as well, our present situation to move forward. To avoid prescription, we first acknowledge the variability of each participant in conversations remembering forward. Perhaps embracing multiplicity and responsiveness when it comes to ethical social action can enable us to continue creatively engaging with our ever-changing planet and social realities, rather than foreclosing on fixed solutions prematurely. One example of this sort of moving conversation is the new *YouthXChange* (YXC) site supported by the United Nations UNESCO. This site is primarily for the youth of the world with Internet access, admittedly a select group but also a rapidly growing population though the use of inexpensive cell phones. According to UNESCO, the YXC site is designed to "help trainers and individuals understand and communicate on sustainable lifestyles" (p. 1). The website is currently offered in 18 languages and operates as a sort of global *agora* for youth to share, discuss and debate local social action towards lifestyles that are less damaging socially and environmentally. As with LSD, resources are available through YXC, but the primary focus of the site is a "participatory process based on interaction and cooperation between teachers and youth, discussion and learning from experiences" (p. 2). At the site, youth can share local situations and actions, opening up essentially to the world these actions for critical examination. The result is an on going, developing conversation that is constantly moving and shifting as it remembers forward.

An example is the current discussion on lifestyles. YXC acknowledges that, "our daily choices as consumers affect the lives of workers in distant places and the way people live" (p. 3). This has led to an involved discussion of the power of consumer choice and, particularly, options for doing things differently. The site emphasizes a pedagogical approach, inviting teachers to join youth in a discussion about:

> The products we buy; encourage curiosity about how and where goods are produced as well as what the working conditions are in the country of origin? How far are goods shipped to

reach the supermarket shelves? What is the environmental footprint of the production and transport of certain products?

But the conversation does not end in discussion; the site emphasizes changing our being in the world, a "transformative learning process" that "changes the way we interact with the world" through continual discussion and engagement that begins with critical remembering but move forward as action *informed* by critical conversations that cannot and should not be determined by any existing notions of "sustainability."

Coda: The Glass Jar

Perhaps buying a glass jar is an alternative to plastic containers; but, before making that purchase, we now have a lot of questions. We want to know why plastics came to be so ubiquitous. What prompted the switch from glass to plastic? How long do plastic containers last in landfills? What does it cost to recycle glass? What are other people doing about this in other regions of the world: We want to open up our conversation to global participation. But what is involved in doing *that?* Does this mean the mass acquisition of Internet-ready technology? What are the social consequences of this? We want to talk about buying a glass jar: About buying *anything* and even about participation in consumerist societies in the first place. What alternatives exist to continual consumerist actions?

Amidst this discussion, which must include children, perhaps we will buy some more glass jars after all—but, this time, we will look for any that are locally made and with a different, less damaging, system for securing and sealing the lid. Maybe we can borrow some jars someone is not using. Maybe we will collaborate with community members to create a glass depot for ease in sharing pre-used glass items. Maybe these used jars will not be appropriate and, in the end, we might need to buy some. If we find the product offering manufacturing with which we can live, we won't claim to be acting sustainably; instead, we do hope this choice, and all future choices we make in living will be *wise actions*, thought through and enacted through conversations of remembering forward that lead us to be less damaging in our actions to our continuation as a species and to the planet we call home.

References

Abram, D. (1996). *The spell of the sensuous*. New York: Random House, Inc.

Alberta Energy. (2013a). *About the tar sands*. http://oilsands.alberta.ca/about.html. Accessed 4 Feb 2013.

Alberta Energy. (2013b). *Reclamation*. http://oilsands.alberta.ca/reclamation.html. Accessed 4 Feb 2013.

Blades, D. (1997). *Procedures of power and curriculum change*. New York: Peter Lang.

Brett, B. (2009). *Trauma farm: A rebel history of rural life*. Vancouver: Greystone Books.

Brundtland, G. H. (1987). *Our common future*. Geneva: World Commission on Environment and Development.
Canadian Association of Petroleum Producers. (2012). *Canada's industry*. Retrieved February 15, 2013, from http://www.capp.ca/canadaIndustry/Pages/default.aspx
Caputo, J. (1987). *Radical hermeneutics*. Bloomington: Indiana University Press.
Caputo, J. (2000). *More radical hermeneutics*. Bloomington: Indiana University Press.
Derber, C. (2002). *People before profit*. New York: Picador.
Deutsch, T. (2011). Memorics of mothers in the kitchen: Local foods, history, and women's work. *Radical History Review, 110*(Spring), 167–177.
Environment Canada. (2010). *Planning for a sustainable future: A Federal Sustainable Development Strategy for Canada*. Sustainable Development Office Environment Canada. http://www.ec.gc.ca/ddsd/default.asp?lang=En&n=06E31414-1#s2. Accessed Oct 2010.
Government of Canada. (2008). *Federal Sustainable Development Act*. Full text available at: http://laws-lois.justice.gc.ca/eng/acts/F-8.6/page-2.html
Hammond, A. (1998). *Which world?* Washington, DC: Island Press.
Heidegger, M. (1977). *Basic writings* (D. Krell, Trans.). New York: Harper & Row.
International Institute for Sustainable Development. (2013). *About IISD*. http://www.iisd.org/about. Accessed Jan 2013.
Johnston, P., Everard, M., Santillo, D., & Robert, K.-H. (2010). Reclaiming the definition of sustainability. *Environmental Science and Pollution Research, 14*(1), 60–66.
Kierkegaard, S. (1983). *Fear and trembling/repetition* (H. V. Hong & E. H. Hong, Trans.). Princeton: Princeton University Press.
Learning for Sustainable Development. (2013). *EcoLeague*. http://r4r.ca/en/ecoleague/about-us. Accessed 1 Feb 2013.
Madison, G. B. (1990). *The hermeneutics of postmodernity*. Bloomington: Indiana University Press.
Mayer, H., & Knox, P. (2010). Small-town sustainability: Prospects in the second modernity. *European Planning Studies, 18*(10), 1545–1565.
Nearing, S., & Nearing, H. (1989). *The good life: Helen and Scott Nearing's sixty years of self-sufficient living*. Toronto: Random House, Inc.
Nikiforuk, A. (2010). *Tar sands: Dirty oil and the future of a continent*. Vancouver: Greystone Books.
On Line Etymology Dictionary. (2013). *Sustainability*. http://www.etymonline.com/index.php?allowed_in_frame=0&search=sustainable&searchmode=none. Accessed 7 Jan 2013.
Owens, S. (2003). Is there a meaningful definition of sustainability? *Plant Genetic Resources, 1*(1), 5–9.
Pearce, D. W., Barbier, E. B., & Markandy, A. (1990). *Sustainable development: Economics and environment in the third world*. London: Edward Elgar.
Plasticpedia. (2013). *Acrylonitrile butadiene styrene (ABS) and other specialist styrenics*. http://www.bpf.co.uk/Plastipedia/Polymers/ABS_and_Other_Specialist_Styrenics.aspx. Accessed 4 Feb 2013.
Rathje, W., & Murphy, C. (2001). *Rubbish! The archaeology of garbage*. New York: HarperCollins.
Rio Tinto Alcan. (2011). *Sustainable development report 2011*. Available from: www.riotintoalcaninbc.com
Robinson, J. (2004). Squaring the circle? Some thoughts on the idea of sustainable development. *Ecological Economics, 48*, 369–384.
Rompkey, R. (Ed.). (2006). *Literature and identity: Essays on Newfoundland and Labrador*. St. John's: DRC Publishing.
Thoms, J. M. (2007). *Leading an extraordinary life: Wise practices for an HIV prevention campaign with two-spirit men*. Toronto: Prepared for Two-Spirit People of the First Nations.
UNESCO. (2005). *UN decade of education for sustainable development (2005–2014)*. http://www.unesco.org/new/en/education/themes/leading-the-international-agenda/education-for-sustainable-development/. Accessed 1 Feb 2013.

Wesley-Esquimaux, C. C., & Snowball, A. (2010). Viewing violence, mental illness and addiction through a wise practices lens. *International Journal of Mental Health and Addiction, 8*, 390–407.
Yang, C. Z., Yaniger, S. I., Jordan, V. C., Klein, D. J., & Bittner, G. D. (2011). Most plastic products release estrogenic chemicals: A potential health problem that can be solved. *Environmental Health Perspectives, 119*(7), 989–996.

Part II
The Public Sphere

Steve Alsop and Larry Bencze

Preamble

In this section authors explore science and technology in public. Early discussions feature Street Medics and organic market farming culture as alternatives to neoliberalism and politics of competition and growth. This is followed by a post-constructivist analysis of a local environmental activist group and seventh-grade students outlining 'ethico-moral' dimensions of activism. While discussion of Occupy, as both a simultaneous virtual-physical phenomena, underscore the essential role of tech activists and open source technology in building a distributed global super-movement. An analysis of Canadian news media forms the focus of two chapters. These analyze trajectories in news print media production and how newspapers represent a prominent climate scientist. Together, they provide opportunities to reconsider media representations, identities, public expectations and scholarly conceptions of expertise. The section concludes with an analysis of transnational corporations as activists within an ongoing 'war against reality', and reflections on the discursive and action-orientated strategies of four kinds of social movement organisations associated with the Alberta tar sands industry (Aboriginal, Environmental, Religious and Labour organizations). As may be apparent from the above word cloud, our central themes of science, technology and activism are situated and analyzed within contexts of media production, social movements, and sites-of-struggle.

The Chapters

Chapter 12
Street Medicine as a Science Education for Activists

Matthew Weinstein

Abstract This chapter looks at the international street medic network, i.e., the network of activists who serve as emergency medics at protests, as a model for an educational movement that crosses scientific and activist knowledges. The chapter examines the medical practices of the street medics; reviews their origins in the United States Civil Rights Movement and briefly traces their history through the counter corporate globalization struggles of the 1990s and 2000s to the present. While medics are most well known for their work at protests treating tear gas and other medical emergencies, the chapter's main focus is on their work as educators. The medics engage in a wide variety of educational projects including trainings for new medics, workshops for activists, and general public health outreach through a variety of publications for their communities. The chapter also reviews the pedagogical practices that the medics use, including hands on experiences and, more importantly, narrative which serves a variety of functions in the education of medics. Finally, the chapter reflects on the insights that medics might bring to crafting a more socially just science education in the K-12 classroom. The chapter links the decaying social welfare systems experience by students under neoliberal governance to the conditions of protest to make a case for a science education guided by a street medic ethos.

Keywords Science education • Neoliberalism • Street medics • Activist education

This chapter explores the ways that science and science education has been organized in the service of resistance to corporate global agendas. My focus is on the United State's context, though the network I am describing is international with large nodes in Canada and Europe. This network consists of medics, i.e., people who deliver medical care, and includes MDs, nurses, alternative medical practitioners, and lay

M. Weinstein (✉)
Department of Education Program, University of Washington-Tacoma, Tacoma, WA, USA
e-mail: matthew@u.washington.edu

L. Bencze and S. Alsop (eds.), *Activist Science and Technology Education*, Cultural Studies of Science Education 9, DOI 10.1007/978-94-007-4360-1_12,

medics who organize to educate their community in medical self care and to actively provide medical service in the chaos of mass protests against neoliberal restructuring.

My interest in the medics began as a way of empirically exploring frustrations I have had with the discourse of social justice within the field of science education. These frustrations included inattention to the corporatization of institutional schooling (Apple 2001, 2010) and the limits of social justice in science education's focus on projects largely isolated from the more radical challenges to global economic and cultural restructuring. Social Justice in the "Rethinking Schools" mode has focused on curricular levers to permit teachers to air progressive points of view, e.g., for emphasizing environmental work. But such efforts are perpetually at risk of being side lined as schools become (again, I am focused on the U.S. context to which I can bear witness) increasingly driven by scripted curricula, judged solely on test measures, and put into perpetual crisis as faculty and students are put into permanent migration as institutions fail to meet Annual Yearly Progress in high stakes tests (failure to meet specific standards forces schools to close under the No Child Left Behind law). As such, at best social justice is shoe horned in as a curricular topic.

To better understand this shoe horning of social-justice-into-projects I wanted to examine science education – I take science, following Bruno Latour (1987), to include those related and co-dependent networks of medicine, engineering, etc. – that was developed for social justice, not social justice developed for education. In other words, what does science and engineering look like when it is produced to serve struggles for social justice and not merely applied to it post-hoc. Since this was education for a social movement, it would by necessity happen outside of formal schooling, so that its time, place, and content could be tailored to the needs of activists. This lead to an examination of social movement literature to identify moments of intersection between technoscience and movements for radical social justice and social transformation (Weinstein 2010). While several projects of interest emerged from this search, my focus over the last several years has been on the Street Medic network, i.e., the medics organized, among other things, to provide support for anti-neoliberal globalization protesters at the meetings of the G8, G20, World Trade Organization, International Monetary Fund, and related organizations (here in the U.S., that includes both the Democratic and Republican Parties).

Know Your Street Medics

So I want to provide a short, breezy history of the street medics so that ultimately I can turn to the question of how their work could help science teachers' meet their students needs better. Street medics emerged initially in the 1960s as a protester support auxiliary of the Medical Committee on Human Rights (MCHR). The MCHR had originally been organized to fight for the racial integration of the American Medical Association, which had been a whites only organization. In 1965 it organized the Medical Presence Project to provide (1) medical support for the march from Selma lead by Dr. Martin Luther King and (2) legitimacy to the Civil Rights Movement by having doctors (i.e., middle class professionals) present at the protests (Dittmer 2009a; McCay 2007). Over the 1960s, as the focus of protests broadened and the New Left developed, the MCHR became increasingly divided between those working on policy and legal issues, on the one hand, and those involved in street demonstrations, opening free clinics, and otherwise providing services directly to the mix of social movements known affectionately in the 1960s as The Movement, on the other. These latter medics are generally called street or action medics. The MCHR was also divided between those who came from traditional (or as medics would say "allopathic") medical backgrounds and those trained in other medical systems, e.g., Chinese medicine or herbalism (Dittmer 2009a). In New York City, one group of action medics formed a street clinic called the Broome St. Collective. It included "Doc" Ron Rosen (Doc), a doctor of Chinese medicine who over his career served as medical support in Selma, at the Democratic Convention in Chicago, 1968 (which was the site of huge protests), and the Siege of Wounded Knee in 1973 by the American Indian Movement. Doc, after the MCHR terminated in the 1980s, went on to form the Colorado Street Medics and helped revive the street medic tradition in the early part of the twenty-first century.

The MCHR disbanded in the early 1980s. However, a second wave of street medicine began as a response to a re-emergence of police use of tear gas, pepper spray and other weapons starting most visibly at the 1999 World Trade Organization (WTO) Ministerial Meeting in Seattle. Doc and other medics present at Seattle felt a pressing need for training and organization in the face of the more violent encounters that protesters were likely to face at globalization protests; their concerns were validated in the bloody encounters with police at the Free Trade Area of the America (FTAA) Meetings in Quebec City and Miami, FL. Doc's trainings provided the second wave with both a continuity to the first wave of street medicine and explicit norms and ethics for street medic practice, ethics such as a radical reading of "do no harm" (see discussion of "do no harm" at Street Medic Wikia 2007) and an acceptance of alternative as well as allopathic medical traditions.

Before exploring street medicine both as a practice and as an educational movement, the very dire circumstances in which it is carried out must be made clear.

States of Emergency: Where Only Street Medics Dare to Tread

Street medics are called on to operate when normal legal rules are suspended and the military or the government declares a state of emergency. These suspensions of the normal rule of law happen frequently around the large globalization demonstrations. They also happen in natural disasters, and street medics are eager to serve in both circumstances. In addition to the globalization protests (most famously in

Seattle in 1999, but they are active at meetings of the G8, G20, WTO, and NATO), they were the first to set up clinics after hurricanes Katrina and Ike. They participated in medical brigades in Haiti after its terrible earthquake in 2010.

While the street medics' special skills of improvising treatments in difficult terrain serve them broadly, they are primarily organized to support protesters in these increasingly frequent states of emergency declared at counter-globalization protests. While the suspension of law has been temporary at the local level, the state of emergency has been a more general problem in the last 30 years as around the globe governments have moved to free market policies. The backers of these polices are often suspicious of democracy and popular need/expression. As David Harvey (2005) notes:

> Neoliberal (aka free market) theorists are, however, profoundly suspicious of democracy. Governance by majority is seen as a potential threat to individual rights and constitutional liberties... Neoliberals therefore tend to favour governance by experts and elites. A strong preference exists for government by executive order and by judicial decision rather than democratic and parliamentary decision-making. (p. 66)

But even the experts are problematic here. Consider global warming or the health effects of tobacco, scientific and academic; authority is gladly undermined when seen as inconvenient to the most powerful (this is explored extensively in Mirowski 2011). Public accountability of the most elite is seen as something to be avoided. The inconvenience that the most powerful feel regarding democracy, legal and ethical accountability, and public governance more generally, has lead to an increasing use of states of exception or states of emergency. The United States has been in a continuous state of exception legally since September 12, 2001, after the attack on the Trade Towers of New York City, thrice extended by President Obama, which has lead to an increasing legal gray area where there is little to no accountability, e.g., in warrantless wiretaps and what are known as National Security Letters – which force internet service providers to spy on their customers – of which about 50,000 per year have been issued between 2003 and 2006, not to mention the now leaked news of massive data collection of email and telephone information.

Of course, this shift to the use of suspension of law to accomplish more aggressive profiteering (Naomi Klein shows how in the case of globalization, this profiteering happens at both the local level and internationally (2004, 2007)) affects the poor much more brutally than those better off. Henry Giroux (2009), drawing on philosophical writings by Italian political theorist Giorgio Agamben (2005), has argued that in this "New Gilded Age" the poor are basically thought of as "disposable" populations, people who can not contribute to the consumer economy and therefore should be treated as without rights. This argument emerges from his analysis of what happened to the poorer quarters of New Orleans after Hurricane Katrina where the devastation was the worst. Entire sections of the city were basically left to drown or flee.

The antiglobalization movements are, at their heart, resisting the broad set of policies of this new transnational global order focused on free markets and de-democratization. The street medics enter the picture because, as in their work in the American South in the Civil Rights struggle, they are trained in how to operate both legally and medically in states of emergency or, as Agamben calls them, states of exception, meaning exceptions to the rule of law.

Street Medicine and the State of Exception

Street medics are creatures of the state of exception. Most have no medical license, they are just highly trained citizens. Normally they would not be allowed to treat, but in emergencies special laws – Good Samaritan Laws – pertain. In addition, as street medics tell it, normally licensed doctors cannot provide service within states of emergency without the permission of the governing authority at risk of loss of license. This barring as told by medics is encoded in the Geneva Convention. My own untrained reading of the UN Convention for the Amelioration of the Condition of the Wounded and Sick in Armed Forces in the Field finds that there certainly are provisions that demand the mutual consent of both sides before medics can serve (Chapter IV, Article 7); however, it's unclear if the "zone of no protest" (the phrase used in Seattle, 1999) in which activists protest qualifies as armed conflict, or what laws remain to protect them. The bigger issue is that in crossing over into the state of emergency, medical licenses are de jura suspended. As noted, this is not a new problem for medics. When Northern doctors operated in Southern states in the Civil Rights struggle of the 1960s, they also were working without license, since medical licenses are granted by states (Dittmer 2009b). The doctors in these circumstances were also reduced to operating under "Good Samaritan Laws."

The Good Samaritan Laws do not provide carte blanche. There are rules for who can be a Good Samaritan, and these rules shape the script that street medics must use in the field. For instance, the provider of help must operate with consent and within his or her competence. These constraints provide a critical part of the training of street medics: learning allowable ways of gaining consent and knowing the limits of ones ability to diagnose and treat. It should be also clear that Good Samaritan Laws are limited; recently, the Supreme Court of California allowed civil suites to continue against volunteer aid providers (aka good Samaritans) (Williams 2008).

One effect of operating only as good Samaritans is that people with a wide variety of backgrounds are able to serve as medics. While doctors and nurses certainly do participate, most second wave street medics have shorter, often more specific, trainings for the conditions in which they will act. These can vary from several week Wilderness First Responder (WFR) Trainings, which are understood as providing the most important skills street medics need – wilderness medicine like action medicine works outside the "golden hour," i.e., the time after injury when treatment in a hospital is most effective – to 20 h street medic trainings, usually offered by medic collectives twice a year or prior to major protest actions. Medics often come with backgrounds both in traditional and non-traditional medicine. Others, assisting medics, may be untrained, but take responsibility for carrying and obtaining clean water, bandages, etc.

Street medicine is a kind of scientific practice for activists; it is developed and tested within the chaos of the eruptive violence of protests. The street medics' purpose is to protect first protesters and then all others, but also to re-enable protesters, to have them treated and re-confronting police and the National Guard. In ethnographic interviews with a street medic collective, I call – a pseudonym to protect their identities – the Seaview Street Medic Collective (SSMC) this was expressed in the joyous pride they had in seeing on television protesters with multiple streaks from successive exposures to pepper and teargas:

> Carin: we treated a lot pepper spray to some and could patch up...
>
> Bonnie: But if you look at like videos from the last Seaview protest you see the people who have the whites around their eyes and running down. You see multiple lines they're different colors. That's because they were eye washed multiple times from being pepper sprayed multiple times. And it's amazing when I watched the news after that how many people I saw. I cured that guy! I cured that guy! (laughs) (9/18/2008)

Medics often emphasize that they are not protesters and that their work is non-political. Often they seek to be unaligned in the often-fractious debates between radical groups doing the work of organizing protests. They also know that they may need to provide care to police, counter protesters, etc. Here, however, in Bonnie's joy, a real politics is made visible. Medics sustain protesters and, hence, protests. Police are aware of this, and they have taken to targeting (literally) medics. Medics are seen as group leaders (and while they may deny playing leadership roles within coalitions, they are treated with enormous respect; one amateur video about medics is called "All Hail the Street Medics") and thus are at higher risk of arrest and injury in protests. To protect themselves medics often "run unmarked" meaning serve as medics without crosses or caducei or other symbols to identify them.

So part of street medicine is a practice honed by the needs to resist small and large-scale violence – as well as the chaos of emergencies that are political, natural and both simultaneously. However, much of street medicine happens away from the riotous atmosphere of protests and hurricanes. That part is primarily educational, some of it also involves scientific research on treatments. It is these aspects of street medicine to which I now turn.

Street Medicine Is Education

In terms of time and energy spent, street medicine is as much an educational activity as a medical practice within demonstrations. Street medics engage in at least three types of explicit educational practices: "trainings" that produce new street medics (or strengthen the skills of existing medics), "workshops" that educate activists and members of alternative communities about medical self care around issues pressing for those communities (these workshops range from proper dress for and diet prior to demonstrations, bicycle safety, and medicine for traveling – a significant part of their communities are nomadic), and, finally, publications, i.e., outreach through flyers and websites on topics ranging from free health resources to issues of rape and consent in college communities. The workshops and flyers point to a broader interest in medicine on the part of street medics rather than a narrow concern with demonstrations. As one founding member of the SSMC explained:

> Amy: You'll hear more about this later but the flyers that I was showing Bonnie and Carin earlier that there's an herbal clinic starting up, a free herbal clinic downtown and there's also [group name] and they work with needles, needle exchange, and people on the streets and that kind of stuff and, you know, they're like we really want to do stuff with you; like come and do workshops. We want to set up regular workshops. Like we have so many plans and so many requests for the educational aspect of our job. You know we're almost – *we're not only a resource to come out and keep people safe physically in a protest or riotous situation.* (9/18/2008, emphasis mine)

Carin, who worked on issues of Latin American liberation politics, went beyond this and emphasized that her work as a medic was an extension of Freirian (1968) popular education, that education was central to her medical work.

Trainings for new medics typically draw between 10 and 20 people. Much of the training builds on that of mainstream EMT and wilderness first responder skills. The education consists of short lectures on protocols: patterned interactions that medics use to treat specific injuries. These protocols are boiled down to acronyms to help medics remember the procedures, acronyms such as ABC (airways, breathing, circulation) and LOC (levels of consciousness). This reduction to standard scripts was important, since, as was emphasized repeatedly in my own training, in the chaos of demonstrations medics become paralyzed, or as one of my teachers explained, "Eyes turn to light houses, hands turn to flippers. Go back to really dumbed-down [models]. Sing the ABCs (one of the acronyms for a key protocol)" (10/26/2008). In other words, in the swirl of risk, violence, and disorder of demonstrations, medics easily can become too focused and literally lose fine motor muscular control. Only by working with easy to access, algorithmic procedures could medics function in a field of pain, violence, and panic.

In addition to lectures, trainings consist of medics practicing key skills such as approach, consent, bandaging, splinting, and documenting. While most of the medicine we learned was "Western," some techniques were from Chinese medicine including a treatment for hypothermia and asthma. At the training's end medics worked in teams to practice treatment in complex scenarios that tried to simulate the informational and physical noise of demonstrations. These short practice sessions and simulations were as close to student centered learning as the trainings ever got. The pedagogy was closer to a master-apprentice relationship, in which we, the students, tried to consume as much information as the battle-hardened teachers were willing to provide. Central to their teachings were stories, accounts of specific encounters with police, counter-protesters, the military, and protesters themselves. These stories at once credentialed our teachers, but also became the material for understanding the function and specificity of the practices we were learning.

Contemporary science education has placed an emphasis on inquiry as a method and content for science education. In our trainings the emphasis was on protocols and scripts, for the reasons just explored. The fact is, the medics have used inquiry, most explicitly in clinical trials conducted to find treatments for tear gas and pepper spray (chemical weapons). One collective, the Black Cross, did blind trials to assess the effectiveness of different treatments for both skin and eye exposure to chemical weapons (Black Cross Health Collective 2003). These results are widely known among street medics. Elsewhere I have discussed these trials and problematized their applicability to action-medicine (Weinstein 2011). The fact is, the trials and the inquiry upon which they depend and the work of action medicine exist in almost parallel universes: in the first there is flexible time, reflection, and, critically, social stability; in the second, there is chaos, danger, and a thousand pressing demands, including some with life and health consequences depending upon medics' responses. It is not that these two universes do not inform each other; they do, though not as directly as medics or science educators may like (many things that work in the clinical space fail in the field). This model of two universes: one under martial law, the other under civil law provokes the question: in which do students and schools operate? Which version of science and time is appropriate for students?

The Street and the School

At my medic training, I was approached by several of the teachers who all said some variant of "this is what high school science should be." Urban public schools, like many public institutions, and many poor communities, have been challenged to the extreme in the current free market embrace, e.g., through regimes of testing, No Child Left Behind forced school closures, budget starvation, and enforced pedagogies that resemble factory work more than teaching. Science education has tended towards embracing these "reforms" (though the focus on inquiry actually has helped deflect calls for behaviorist, scripted direct instruction regimes that have hegemony in math and language arts).

My growing understanding of the medics' urging of street medicine-as-science education is that they were not simply saying that students should be street medics. Also, it was not just a call for practicality, but also a radical pushing of a science education that fosters resistance to the dire legal, political, and material world of current moment, meaning a world in which law and social stability are under attack. School science, I believe, needs to focus on a community driven agenda of reclaiming knowledge in the wake of the abandonment by much of the state, or better said: left to corporate plundering and waste. Science educators are in a good position to transform schools into sites where communities can collectively develop resources to address questions of food, shelter, and medicine. If there is one domain where I see this kind of work flourishing in schools it is in the areas of food politics. Many schools in my area have used their lands as the sites of community gardens. These gardens connect local diets, cultures, and technoscientific practice in ways that directly benefit the members of the community. Schools have also become sites where corporate food is contested: vending machines fought over and lunch menus resisted. At the moment this is for me the most tangible example of mixing of science, health, multiculturalism, and community funds of knowledge. It can serve as suggestive models perhaps for science students as community health workers, as community ecologists, as community storm/earthquake mitigation experts, as community planners with knowledge of energy and transportation flows that materially sustain communities.

In all of these domains, however, teachers will have to adopt the street medics' eclecticism. Western medicine, diet, and construction practices alone will not do. To be effective, not as measured in test scores, but in the health of urban communities (I cannot speak to the circumstances of rural communities, which face their own neoliberal fracturing – and fracking) teachers must be in open hearted dialogue with the neighborhoods they are serving, communities which have their own cosmologies (plural).

Of course teachers are very much under pressure to produce standardized, consumer students. I do not want my call for teachers to formulate a pedagogy of resistance and reclamation to be taken as naïve. I know, in the era of hyper-standardization, there are fewer places and times for resisting; more of the curriculum is under high stakes monitoring. I think that there are still opportunities. For novice teachers these may be at the end of the year when the state has measured every one; or in classes not covered by the test; or in after school clubs (community gardens rather than robots, which seem the popular STEM after school thing at the moment). The emerging standards in the U.S. (the so-called Next Generation Science Standards), in fact have spaces for exploring ecology, connections between science and society, and the development of instrumental (useful) knowledges, however pushed to the margins they are.

The street medics seek to create a technical craft wholly organized to promote social resistance to corporate power – much of that power embodied by the state. Science educators, being employees of the state, cannot simply discard their given curriculum and merely function in resistance to the powers-that-be, but they can act in guerilla ways to find spaces and times within curricula to develop practices that are grounded in the intersection of community desire and technical know-how. That I believe is what we in formal science education can take from the medics' brave work.

References

Agamben, G. (2005). *State of exception*. Chicago: University of Chicago Press.

Apple, M. W. (2001). *Educating the "right" way: Markets, standards, God, and inequality*. New York: RoutledgeFalmer.

Apple, M. W. (2010). *Global crises, social justice, and education*. New York: Routledge.

Black Cross Health Collective. (2003, November 24). *Brave? Crazy? Black Cross' pepper spray trials*. http://blackcrosscollective.org/OCTrials. Accessed 29 Jan 2009.

Dittmer, J. (2009a). *The good doctors: The medical committee for human rights and the struggle for social justice in health care*. New York: Bloomsbury Press.

Dittmer, J. (2009b, February 9). [Personal conversation].
Freire, P. (1968). *Pedagogy of the oppressed* (M. B. Ramos, Trans.). New York: Seabury Press.
Giroux, H. A. (2009). *Youth in a suspect society: Democracy or disposability?* (1st ed.). New York: Palgrave Macmillan.
Harvey, D. (2005). *A brief history of neoliberalism*. Oxford: Oxford University Press.
Klein, N. (2004, May 2). *The rise of disaster capitalism*. http://www.thenation.com/doc/20050502/klein. Accessed 27 July 2008.
Klein, N. (2007). *The shock doctrine: The rise of disaster capitalism* (1st ed.). New York: Metropolitan Books/Henry Holt.
Latour, B. (1987). *Science in action*. Cambridge: Harvard University Press.
McCay, M. A. (2007). *Radical health activism: The Boston chapter of the Medical Committee for Human Rights, 1964–1981*. MH Masters, Tufts, Boston.
Mirowski, P. (2011). *Science-mart: Privatizing American science*. Cambridge, MA: Harvard University Press.
Street Medic Wikia. (2007). *Street medic history and philosophy*. http://medic.wikia.com/wiki/Street_medic_history_and_philosophyw. Accessed 13 Sept 2010.
Weinstein, M. (2010). A science literacy of love and rage: Identifying science inscription in lives of resistance. *Canadian Journal of Science, Mathematics, and Technology Education, 10*(3), 267–277.
Weinstein, M. (2011). *Invited address: States of emergency, zones of civil disturbance, and sciences for "badly" behaving subjects*. Paper presented at the CSSE Springer Forum, Orlando, FL.
Williams, C. J. (2008, December 19). *California Supreme Court allows good Samaritans to be sued for nonmedical care*. http://www.latimes.com/news/local/la-me-good-samaritan19-2008dec19,0,4033454.story. Accessed 29 Jan 2009.

Chapter 13
Why Science Education Mediates the Way We Eat

Michael P. Mueller

Abstract Neoliberalism is at the heart of the North American food system. This understanding drives food choices that are concomitantly embedded within education and not generally mediated by school. Neoliberalism's metaphors of competition, individualism, and the economic mindset drive conventional farming practices, which rely heavily on synthetic herbicides and pesticides and inexpensive labor. These metaphors influence how people in North America eat, how food is produced, and how much is wasted. Around the world, many farmers use culturally and environmentally responsible methods of cultivating their produce. These trends are breaking "new ground" in North America with local, organic, responsibly nurtured, fresh, farmers' market vegetables and fruits. This chapter explores organic market farming culture as a 'pocket of resistance,' a place/context for dynamic polysemic knowledge that evolves in social concert with change/adaptability, positionality/relationality, and ecological condition. The curricular trajectory of science education in the farmer's market is a site for children to investigate whether their cultural traditions and skills serve to protect them from hyperconsumerism or overreliance on the dominant types of produce. What we eat affects how we understand and the way we behave in relation to it. Our eating has the potential to transform school science. This transformation can be an activist force in our society.

Keywords Citizen science • Ecojustice • Farmers' market • Organic farming • Neoliberalism • Place-based education • Relational epistemology

---

M.P. Mueller (✉)
College of Education, University of Alaska Anchorage,
3211 Providence Drive, Anchorage, AK 99508-4614, USA
e-mail: ak.mikem@gmail.com

L. Bencze and S. Alsop (eds.), *Activist Science and Technology Education*,
Cultural Studies of Science Education 9, DOI 10.1007/978-94-007-4360-1_13,

Introduction

> How much would you pay for your food? Would you reduce the size of your portions to eat more expensive food? Would you slow down your eating to mindfully think about the quality of your food and its impact on society?

Neoliberalism survives on the indulgence of society's understanding. This understanding involves the taken-for-granted metaphors of competition, individualism and an economic mindset. These metaphors are not challenged by the ways most people in North America eat, how food is produced and how much is wasted. We implicitly contribute to the status quo when we place oversized garbage bags in bins on the side of the road or in the dumpster for others to take away to places we cannot or do not want to see. We contribute to the status quo when we do not analyze the corporate stickers on fruits that describe how far our food comes from or how we get 'freshly picked' or 'natural' foods during the winter. We contribute to neoliberalism by not thinking mindfully about the ways that the food we eat harms people and ecosystems elsewhere.

It is clear that food choices are driven by a deeply embedded matrix of neoliberal cultural assumptions, which are concomitantly embedded within education and not generally mediated by school. Given the potential problems associated with commercial food systems and the ways food is connected to every aspect of science, it is problematic when we deemphasize the study of massive farm tracks or agricultural wastes from fertilizers, pesticides, soils and animal manure that travel in streams across major continents. Consider how prolific herbicides such as Syngenta's *Atrazine* can now be traced in most of the waterways in the United States and which have been linked with cancer (Mueller 2009a). What are the economic, health and social costs of participating in the high stakes policies of conventional farming? Do science educators stop short with the *economic* costs of participating in conventional food systems without delving into the real hardships of farmers, the shortcuts that many farmers have to take to minimize their costs associated with food or animal meat production, and the hidden costs for workers? Consider the many neighborhoods surrounding farmlands where toxic chemicals are driven by wind through windows of homes. Are these things talked about in schools, and if they are, where are they discussed, in what classes, and by which groups of children? Withstanding the ways that the market and industry influence food choices, many financially secure people have choices about how they live and eat, but they do not think twice about how these choices affect economically vulnerable people's access to responsibly grown food. "It follows that those who have more options for where they spend their money cannot escape some of the responsibility for other significant expenses associated with conventional food even they do nothing to bring them about" (Mueller 2009b, p. 1004). In short, organically (or responsibly nurtured) grown foods reduce air and water pollution, pesticide residual, and the loss of biodiversity. In general, organic farms use less energy for a given yield and the soil stores more carbon, which may offset emissions that contribute to climate change. There are also higher costs associated with the scientific research inculcated with genetically modified organisms (not used on organic farms), which are sold to conventional farms.

Many farmers around the world already use culturally and environmentally responsible methods of cultivating their produce. In Malawi, for example, these practices are incorporated into the science curriculum (Glasson et al. 2006). However these practices are still not widespread in the curriculum of schools despite the need. In North America, this results from the predominately middle class neoliberal cultural attitudes, values and assumptions that are largely taken for granted in science education. This chapter considers this connection between neoliberalism, food and science education. With an increasing human population worldwide, there are fewer agricultural and natural resources, and yet more choices with trade-offs, such as loss of biodiversity associated urban development and sprawl. Degraded ecosystems worldwide challenge science teachers to begin thinking about ways in which today's students face perils that their teachers did not confront. How do teachers collectively and science educators, more specifically, share some responsibility for cultivating, participating, and advocating for change within local communities? These actions might begin by addressing with student-generated research and solving problems, such as degraded stream systems or protecting indigenous species, or deciphering which species should receive community resources—perhaps already stretched thin.

Too much emphasis in school has been on 'consuming things.' We teach kids at a very early age how to consume things. In science education (National Research Council 1996), the notion of consuming things is reinforced through the *consumption of science as scientific literacy* ideology. Consider how high we hold in esteem the professional knowledge and technology of experts, which is presented to children through the process of schooling. The media concomitantly perpetuates an inherent faith in science and technology to solve the world's problems. Over time, children become over-reliant on science and technology to resolve all of their problems and accept the notion that there are experts who can and will deal with problems before they become too large. The neoliberal approach of allowing others-out-of-reach to deal with social problems does little to catalyze responsibility. It certainly does not create the best atmosphere for engaging youth in activism. While I don't want to be misunderstood as saying that this is the only approach to teaching science, it is the dominant approach that is undergirded by neoliberalism (Bencze and Carter 2011). Although the dominant talk about education and scientific literacy is situated with the means to make decisions using science, it is rarely, if ever, focused on agency and activism. More often this talk inadvertently perpetuates consumption as a *globalized* or standardized form of science learning. In other words, within the ideology of neoliberalism, 'consumption' consumes science education today!

Perhaps this neoliberalized story of science education is so deeply enmeshed within the fabric of society that it survives on the indulgence of society's understanding—that is, namely, *hegemony*. My chapter will explore this hegemony of neoliberalism by discussing organic farming and food in relation to it. My philosophy draws on over 2 years of my direct participation in a local organic market farming culture in Southern Appalachia, USA, which involves hundreds of hours working to raise animals (some for meat), growing plants, and providing vegetables, fruits and flowers for an urban farmers' market in a major metropolitan city, Atlanta. My eclectic philosophical method is mostly aligned within the enthnomethodological and phenomenological traditions in ecojustice studies and science education (Mueller 2009b). To make sense of this organic market farming culture and my experiences, I took hundreds of photographs and used artwork, farming and seed journals, and my notes from participant observations. I argue for science education that mediates the way we eat. After all, what we eat affects how we understand and the way we behave in relation to it. Our eating has the potential to transform school science. This transformation can be an activist force for society's understanding.

Neoliberalism in Science Education, *In Brief*

I begin with a short description of neoliberalism in science education. Larry Bencze and Lyn Carter (2011) note how science education, especially in North American, emphasizes epistemology congruent with individualism, competition, economic superiority and consumerism. This *commodified* notion of teachers, students, grades, and knowledge "stands in stark contrast to notions of *social epistemology*, which posit that knowledge is historically and socially constructed and, in a sense, belongs to everyone past, present, and future" (p. 654). Bencze and Carter explain that neoliberalism commoditizes everything (desires, success, democracy) in terms of capitalism and economy. The commodities of professional products from science and technology are deliberately emphasized in school science—not by some accident—but by the advertent selection of particular social and political agendas. This means that the current curricular trajectory restricts students' abilities to critique, suggest revisions, and take action against business practices, goods, services, and media aligned with the 'science-as-a-god,' and I add, *our deepest cultural assumptions that support this.* Teachers and students 'police' themselves by adhering to strict regulation of best practices, testing culture, criteria and school governance. Further, teachers' subjugate their curriculum to the neoliberal and neoconservative elite, who act as governors and gatekeepers. Consider Jack Johnson's (2003) lyrics:

> "It wasn't me", says the boy with the gun
> "Sure I pulled the trigger but it needed to be done
> Cause life's been killing me ever since it begun
> You cant blame me cause I'm too young"
>
> "You can't blame me sure the killer was my son
> But I didn't teach him to pull the trigger of the gun
> It's the killing on this TV screen
> You cant blame me its those images he seen"
>
> Well "You can't blame me", says the media man
> Well "I wasn't the one who came up with the plan
> I just point my camera at what the people want to see
> Man it's a two way mirror and you cant blame me"

"You can't blame me", says the singer of the song
Or the maker of the movie which he based his life on
"It's only entertainment and as anyone can see
The smoke machines and makeup and you cant fool me"

It was you it was me it was every man
We've all got the blood on our hands
We only receive what we demand (*excerpt from Cookie Jar,* Johnson 2003).

Let's face it—we can't be saints. Most people oscillate between different positions based on what is known about an issue. Philosophically speaking, we may hold some positions longer than others, but our worldviews are not crystalized to the opinions and social influences of others. Consider the science teacher educator who uses as a rationale for educating teachers the way they do the high stakes pressures of testing and conformity to national standards: 'We must prepare teachers to survive in a world of standardized tests!' In this case, new teachers may be inadvertently prepared to exclude their students from tapping into diverse perspectives or building on the social strengths of the community. Bencze and Carter (2011) suggest that these things occur to the detriment of the local community, planet, and future sustainability.

How do youth even begin to resolve local problems when they do not even know what is wrong? Today's children have very little sense of the status or condition of their local cultures, commons, or habitats (Mueller 2009b). Recall that most children can identify more corporate labels than trees and plants in the local ecosystems (Louv 2008). Where are children taught to investigate whether their cultural traditions are degraded?—Think gardening, canning, preserving food, seed saving, and eating a meal face-to-face (the old way!). What about their community?—think talking with neighbors, creating a walking-safe path for kids to get to school without having to ride a car or bus, or bartering for materials or vegetables. What about their environment?—degraded habitats, species' movements in response to climate change, or youth who have never touched an insect or seen a bee. In fact, the grocery store is the closest that some children will ever get to pollinators.

My point here is that there is a lot to be gained when we begin to assess the hegemony of science education and critique how it is almost exclusively focused on students' future contributions to the economy. Withstanding some of the ways that science educators are challenging the neoliberal front, namely, by teaching with socioscientific issues, for ecojustice, or for sociopolitical action (Hodson 2011), the vast majority of science educators still have not considered what it means to teach in ways that contribute more fully to the decision-making processes implicated by science teaching standards. Wrestling with the larger ideology is one way to approach this problem, creating small pockets of resistance is another possibility. Below, I analyze one of these small pockets of resistance that I believe has the potential to transform science education worldwide.

One simple activity—how we eat—can change everything.

A Pocket of Resistance—Local and Organic Food

My interest in local and organic food grew out of my interactions with a vibrant food movement in a southern Appalachian town where I was teaching prospective science teachers. As part of a methods course, I decided to incorporate *farming as a way of life* that I wanted my mostly urban students to begin to appreciate and experience in an attempt to get them to think more fully about incorporating local farming into their teaching of biology, chemistry, earth science and physics at the secondary levels. We visited a local *community supported agricultural* project or CSA and helped the farmer plant garlic and harvest eggplant each fall. In exchange, the farmer provided his epistemological perspectives on the idea of local food and the science embedded in the farming.

This is where I learned with my students, again and again, that the term 'local' and 'fresh' have something to do with a *relationship*. It became apparent over several years that this relationship was between a farmer and his/her customers. (My students were given produce for their hard work each time they visited the farm, which helped to solidify the idea that eating locally is part of a larger relationship between persons who eat fresh food and their farmer.)

Five years of cultivating teachers' understandings of this local relationship led to a specialized course focused on ecojustice, in which graduate students would further investigate the 'relationship idea.' As part of this course, we invited guests who were part of the burgeoning local food movement in and around Atlanta: Slow Food Atlanta, Farm255 Restaurant, PLACES, etc. These people, representing these organizations, shared with us their ecojustice story, which often involved bringing people together around food. Students in the class developed food research and activist projects. It was through this work that I discovered many people in the community who were resisting the poisons of corporate farming through 'guerilla gardening,' school and neighborhood gardens, grocery cooperatives, charitable groups, non-profits, and clubs.

In the mean time, I was contemplating a move to a local farm with my family. After trying to cultivate small gardens in woodsy areas or in one case—on the land surrounding our home where home development ceased because of the economy—we decided that more land was needed (at least 2–5 acres) to raise fresh food. So, after several moves, we finally found a suitable place to farm. I decided to align my interests and begin a 2-year research project to understand local and organic food as well as conventional farming. I was also interested in learning Appalachian knowledge and skills, which occurred after developing trusting relationships with my neighbors—only accessible by living the farmers' market lifestyle. While hesitant at first, my family quickly became accustomed to this lifestyle and deeply embedded in their love for the farming culture. This immersion would allow me to experience science education as it is situated within an organic market culture in ways that I would never have anticipated or imagined before the project.

The next part of this chapter is written in present tense to capture the essence of epistemological growth in my farm journey.

Organic Farmers' Market Culture

Going into this project, I already understood and had begun to research further dangers of conventionally farmed food and devastating impacts of herbicides and pesticides such as Atrazine (Mueller 2009a). I understand the impacts of farming with harmful chemicals on the health and welfare of farm-field laborers. I understand that poultry farming is paramount to the Southeast USA and also that raising chickens in crowded chicken houses is inhumane. There are many times that I become faced with these ecologically destructive, socially unjust, and inhumane conditions nearby as a reminder that growing plants and raising animals in the most "natural" (without synthetic chemicals etc.) way possible is the most ethical way to engage this work.

As a result, we begin this project by adopting many of the adages of the organic farming movement without going through the expensive and lengthy process of getting certified 'Organic.' However, we do decide to go through the process of getting 'Certified Naturally Grown' (or CNG), required by most of the major organic farmers' markets to sell produce locally. Part of the process is having your soil tested by agricultural extension and completing an on-site inspection by a knowledgeable person. The soils report and inspection presents promising conditions for organic farming. We cultivate the fields and plant a cover crop of Crimson clover and rye for the winter months. We also repair an abandoned greenhouse on the property to begin our seedlings.

©2011 Michael Mueller, *Cover Crop of Crimson Clove and Rye: A Place to Relax and Play*

A Responsibly Nurtured Organic Place

While place-based education is an emerging trend in science education (Tippins et al. 2010), it is often critiqued for being unclear and relatively undefined in science education. When you live and work on a farm, the idea of *place* becomes much clearer. The responsibility that we have to the places where we live becomes evident. Small things influence your place. This farm is a place to relax, enjoy nature, and play. Crimson clover and rye cover the ground that we will begin to plant throughout the spring with seedlings from our greenhouse. The 'cover crop' protects the soil from eroding away during the winter, and adds and conserves nutrients such as nitrogen through its roots. The clover has deep roots and breaks up the hard red Georgia clay. Throughout the spring, we till the loamy soil and make beds for the new seedlings from the greenhouse. We water these beds by hand, which provides for time to think, experience the surroundings, and feel the cool weather of spring. Eventually it will get so hot that we will have to get up at the wink of dawn and work hard for 4–5 h before eating breakfast and working on other indoor tasks.

During the winter months, there are a lot of things that we have to prepare for the farmers' market, including repairing old baskets, crafting signs, and so forth. We begin scavenging the local antique stores for old farm baskets, unique glass jars and other artifacts that have been historically used to display produce at farmers' markets. We also take great care of our chickens and pigs, which eat our food waste and create manure that we compost for fertilizer. Our chickens roam freely on the farm and despite that we lose some of them to predators (foxes, opossum, coyotes, etc.), they seem happy when they can move freely around the fields grubbing for insects. We learn the hard way that pigs are great escape artists, but easily convinced to go back into their pen after they have their turn exploring the farm (we are lucky they never eat our vegetables!) We raise chickens for meat and eggs, and I learn to harvest them by hand, but this is never easy. In my mind, I wrestle with how to discuss 'harvest' with my children, but it becomes clear that they should understand where their meat comes from and they realize the significance of caring for animals for meat and to not waste it. We take many of the 'rejects' from the chicken farmer down the road, which are the grey and black chickens that would be normally culled as chicks. Industrial meat companies prefer white-feathered chickens and these chickens have been genetically modified to grow to full size in 7–9 weeks. My children adopt one of the chickens as their 'friend,' but we learn the difficult lesson that chickens genetically modified to grow with large breasts cannot support themselves into adulthood and their legs become dilapidated. In this part of the country, we are regularly faced with seeing chicken trucks loaded with hundreds of chickens in clear view cages on their way to the nearby meat processing plants. Because we know about the lives that these chickens live in the chicken house and the way they are confined to live in darkness and in their urine and feces, and the way they are grabbed by the wings or legs by factory workers, we cringe often at the sight or look away from the trucks. We have even witnessed chickens falling out of a moving truck, not to mention the ways they are treated when they get to a factory.

The animals with which we share this farm are as much a part of the memory-in-the-making as the vegetables, fruits, and flowers that I discuss below. They remind us to be constantly mindful that our meat comes from a relationship between humans and animals, and to never waste our food. Consider how much food is wasted in neoliberal-influenced-middle-class-homes, restaurants, or grocery stores, and then think about why it is possible to throw away food. Why is it possible for a chicken's life to be lived for nothing, only to end up in the garbage because of portion, package or plate size? These questions become vitally important to the person who has a relationship with animals or who understands the needs of animals. These relationships constitute food knowledge, but are lacking for many children in science education. Without the contextualization of human-animal relationships, it is difficult to imagine how meat would be considered inseparable from the place where it is raised. For many people, meat comes from the store. This is the first place they encounter it and where they get their education (Rowe 2011).

©2011 Michael Mueller, *Grey and Black-feathered Cornish Rock Chickens*

Local Knowledge

A small farm takes a lot of creative vision and the integration of plants in a way that will provide the most produce for the least amount of space. We can calculate how much food will be grown based on seed companies and conversations with local farmers, but we do not have the life-long experiences that many farmers have.

There are several farm supply stores nearby and these places prove to be a good source of information about what to plant, when, etc. Most of these stores are owned locally by elder gentlemen farmers, and in many cases, they share the responsibilities of the store with their children who have grown up in the business of farmer culture. There is much to gain by 'hanging around' the farm store: stories, sage advice and perspectives on raising animals and planting seeds. We also read books on organic gardening and companion planting, which include traditional knowledge about what to plant together. We draw many maps and design plans, chart planting dates, and use intuition to solve problems associated with late frost, insects, and other complexities associated with growing healthy local food.

Interdisciplinary Knowledge

The place where we live and work has much to do with the ways in which the knowledge, experience, narratives, traditions, and other plans come together in a synthesis of farming activity.

©2012 Michael Mueller, *Shards of Local History Tilled Up in the Soils*

The disciplines traditionally separated in schools and in classrooms cannot be separated in the farmer's field. They merge through the constant interplay or reciprocation of education and lived activity.

Reading the Farmers' Almanac, traditional anecdotal stories of companion planting, scientific reports, mathematic charts, and the lay of the landscape over time create the rationale and intuition for knowing when to evaluate and take particular actions over others.

Learning the cultural history, planting skills, Native knowledge, and playing in the present and future context of the farmer's field takes time, appreciation, and respect (and valuation) of knowledge. Working together as a family, cooking and creating meals with the vegetables and fruits of our labor, and preserving vegetables, is a bonding experience but also supports intergenerational knowledge. The human geography, physical geography, geology, climate and weather, sunlight, shade, runoff, dry and wet areas of the field are considered together with human exertion, fatigue, and excitement. Clothing is as significant as insects and these things are all placed in the context of the science education embodied within the farm. Consider the following journal entry (*personal journal 3/5/12*):

> Planted 100 tomatoes. Discussed how to identify one from the other by leaf, shape, coloration at [different places on the] stem, etc. Sunscreen always an issue, hats, shirts, increasing intensity of sun. Fire ants are a regular thing, easier for the adults to deal with, but our kids are new to them. [Our daughter] ended up in a [fire ant] mound. 20–30 stings later!!!

©2012 Michael Mueller, *The Effects of Shade on Part of the Row of Sunflowers*

Change and Adaptability

Another journal entry demonstrates how experience on the land leads to new explorations and knowledge about particular insects:

> Watering takes forever! One hose and water can. Find caterpillars on broccoli (*personal journal 4/6/12*).

The learning process occurs again and again on the farm as we encounter new problems and learn how to deal with them. The slow process of watering by hand allows for interactions that would easily be missed if we had a mechanical watering system—missed interactions with birds, rodents, cats, chickens, hawks, etc. More importantly, the slowing down of many of the processes on the farm allows for interactions with insects, disease, and so forth. The insect encountered on the broccoli turns out to be a cabbage moth—interestingly, we have hundreds of little white "butterflies" flying around the broccoli and their caterpillars are picked by hand. Chickens walking around the garden also help with insect control, especially with the caterpillars and other insects that rapidly destroy our vegetables. We learn that everything has its window to flourish but that all things will eventually succumb to local change. This increasing understanding of change and adaptability helps us to plant particular vegetables in succession knowing how long they will survive (or how long we will survive pulling insects off or treating diseases) before they become overwhelmed by change. Because we take risks with frost and early planting, we are able to beat the insects to productive vegetables and fruits in most cases. We learn that early planting is possible with the warming soils:

> Well, we didn't make it . . . but only a few, maybe 3–4 plants suffered from cold damage. Tonight we have another cool evening; we think maybe the warmer soils helped keep our plants warm and hay mulch; but two nights in a row could be different. My heal split (the skin) on the back—it's painful. No more bare feet for while to heal my sole (*personal journal 4/12/12*).

Change is more than what happens to a place, it is an embodied, lived experience that creates knowledge and reciprocal values. We change a place and it changes us in the process, it is an interaction.

In school science, we teach and learn about environmental change, but we may not *emphasize* the way it influences and is influenced by culture. While we explore it in a past or present context, we do not explore how it is lived and becomes part of our future lives.

Embodied Change and Positionality

Change occurs in the farmer's body. My body aches and I cannot stand up straight. I learn to crouch low to weed and work with soil. My feet need as much care as my brain and my body tells me about the conditions—the physics, geology, and biology—of the ground. I am also misled by my bias about particular 'pests,' and at times, reminisce about the misguided ways that I've changed the place.

> Today I killed a 5' Rat snake. I felt bad about doing it, but it was so close to our house, probably lives under the house killing rats, mice and so forth. Maybe I should have just scared it away. Rat snakes can be pretty aggressive (*personal journal 5/18/12*)

We strive to understand this place in terms of life and death, which is also a vital aspect of change on the farm. Plants spring to life and wrestle with each other to establish their place on the farm. The neighbor's dog kills our chickens and at one point I am faced with the hard decision of whether the dog or the chickens will live. We don't want anything to die; we love this place and all that belongs here but there are inherent tensions between the farm's inhabitants. The farm cats kill just about anything that moves—rabbits, rodents, birds, and drop them off on the farmhouse porch:

> Its amazing how our cats hunt around here. When they find a rat, mouse, bird, chipmunk, etc. and kill it, we either find it on the porch or somewhere in our field or garden. Its one way to see these animals up close and examine their characteristics or teach my kids about them. This time of year we can figure out how long animals have been dead by looking at the blowfly (a green fly) and its larva (*personal journal 5/16/12*).

Environmental Condition

The farm is a place where we are able to photograph and document the life of insects, animals, and plants and understand their interaction.

©2012 Michael Mueller, *Citizen Science is a Daily Part of Life on the Farm*

This environmental monitoring or citizen science (Mueller et al. 2012) provides valuable information about the changing seasons, weather, and periods of time within seasons. We learn, for example, that flowers are better for the farmers' market if they are harvested before the bloom fully unfurls. With organic produce, fruits and flowers, everything is about timing and within a few hours, things can change significantly. The variety of citizen science investigations that go on continuously on a farm are astounding. These experiments are bound by the questions derived through the contextualization of problems, issues and aspirations.

> Morning is a great time to photograph and explore pollinators. While the hummingbirds are already "humming," many bees and wasps have camped out over night in flowers remaining motionless until the sun warms them. This morning I found Mason bees, honeybees, wasps and many other solitary bees including an Eastern Carpenter (*personal journal 6/29/12*).

©2012 Michael Mueller, *Monitoring a Swallowtail Caterpillar on a Swath of Dill – They Eventually Eat It. There is a Fine Balance between Insect Control and Pollination – We Can Do This Because Dill is Planted in Different Places*

The weather, temperature, cloud cover and time of day play a large role in these daily investigations. We become very keen to the changes in the weather such as rain because it affects our work:

> Rain sweet rain! We see rain clouds creeping up in the distance. They're moving fast! "Let's get more glads [gladiolus bulbs] in!" So we push hard to get a whole row of glads. The rain is coming faster now. We have a great routine, Heather plants and I make a hole. We are about 2/3rd done and rain pours down, the sky gets dark—it's 5:15 PM. Heather uses a bucket at first to hold the bulbs but they go in the ground faster from her shirt! That's what a shirt is for on the farm. We run inside just as the rain has drenched our clothes. Heather says, "were we just out in the thunderstorm with a steal tool planting glads?" (*personal journal 3/3/12*)

While we laugh about it after the fact, the reality is that the soil, seeds, weather and people interacting in this place exert a science experience whether one realizes they are engaging in science learning or not. The ground is hard clay, so we plant in the rain. We plant as many gladiolus corms as possible because of the ease in which they go into the ground. They get watered at the same time. The rain feels good and is a bonding experience for two people who enjoy the experience of being in one place at one time.

©2012 Michael Mueller, *Our Children Regularly Walk Around with or Eat Vegetables—This is Our Son's Zucchini (That We Cannot Sell at the Market!)*

Probably the most profound characteristic of being part of a farm is that it inspires a "love-hate" relationship (or 'both/and' way of thinking-in-relation, see Thayer-Bacon 2003). There are many hardships faced by organic farmers—Fire Ant stings and scars, Red Wasp stings and allergic reaction, and tearing a fingernail of the finger. These things bring stories and memories-in-the-making to the significance of farming. There is a constant question of whether 'this is all worth it?' As the body begins to heal, there is a continual renewal of analysis around our decisions and actions. It is also renewed continuously in the process of participating in a farmers' market where other people's choices and aspirations play a large role in the ways we justify some actions against others. Even the soil, weather, vegetables, fruits, flowers, and animals 'scream' at us so 'loudly' we cannot deemphasize or ignore them.

> This weekend we did very well with flowers and tomatoes. Can definitely see how different flower creating styles affect the buying decisions of our patrons. The weather has been slowly getting hotter each week and there are some vegetables and especially flowers that just can't take the heat. Glads looked really worn this week because of the heat. Sunflowers were not as full stemmed because of a different location in the garden. Peppers are growing in now, tomatoes needed stakes, and we are removing squash. People seem to be sick of squash now (*personal journal 6/23/12*).

A Market Creates a Relationship with Food

People's tastes shift throughout the farmers' market season. Their aspirations change with the variability of diverse weekly produce. We never miss a farmers' market, no matter how discouraged or sickened by weekly events, and neither do most of the regular vendors and patrons. It is a weekly gathering filled with the joy, love, and passion of people who are committed to organic farming. The market is a lively downtown metropolitan experience. For most patrons, the farmers' market is the closest they will ever get to a relationship with a farmer who grows their food or the soil and land that nurtures it. We bring photo albums and share stories with our regular customers who believe in the importance of local food. The farmers' market starts early in the morning and is filled with activity—exercise programs in the park, cooking demonstrations, weekly educational classes, cultural events, and lively music. People come to the farmers' market to renew their lives. Bartering, trust, and moral reciprocity are essential to this market experience.

> Wow, the temperature reached 106 degrees Fahrenheit at the farmers' market yesterday. After a busy morning, the market became quiet—the "dead" of summer and the heat was hard to bear. We often barter for our weekly supplies of vegetables we need for the family, by either trading with other farmers etc. or buying produce at reduced rates. Some customers come by the stand and want our produce but don't have enough cash to pay. Often we ask them to come back the following week and pay, many who do. This practice is a good way to encourage folks to come back each week and it's the sort of relationship that few places offer in the corporate world. Where could a customer walk away with produce or other items and not pay—or be trusted to pay later? Often a long week of work on the farm and frustration with 3 weeks of dry [weather] in Georgia, the farmers' market almost instantly revitalizes my spirit and desire to create produce so that I can participate in the experience the next week. I loved the "surf" "sublime" music this week at the market! (*personal journal 6/30/12*)

©2012 Michael Mueller, *We Know About and Care for Our Market Patrons*

The farmers' market is very diverse. We talk with people from all walks of life and learn about their stories, interests, aspirations, and personal circumstances. Many people have children, dogs, or other pets—one boy and his mother bring a different insect, lizard, or other creature they have caught in the park to share with us. We get to know people with diverse identities—student, pregnant, gay, mother, immigrant, and so forth—they all *share* affection with us. They participate in a social imagination surrounding a world filled with the air of romanticism for responsibly nurtured organic foods. Because of this acquired affection we have a unique opportunity to educate the people we come to care for about what organic food looks like (e.g., insect damage), seasonality, and the culture itself.

> The farmers' market really serves as a gauge about people's attitudes. It helps regenerate people each week. One customer [says that she]. . ..hasn't missed a farmers' market date since moving to Atlanta. We see a lot of the same people and the farmers' market becomes a place to socialize and get to know people as much as anything. These people return week after week to immerse themselves in a context because it is special to them (*personal journal 7/21/12*).

A Critique of Neoliberalism

The farmers' market is a place to reinvigorate the mind-and-body-in-relation-to-others. This idea of relationality comes from Barbara Thayer-Bacon (2000), who uses the quilting bee to illuminate a similar kind of world where people's mind-bodies are constructed in the *social epistemology* also emphasized by Bencze and Carter (2011). The ideal philosophy of responsibly nurtured organic food is inherently social—it's a belonging metaphor of shared existence. Through my participation in the organic farmers' market culture, it becomes evident that it is a shared consciousness of context, local knowledge, interdisciplinary knowledge and experience, awareness of positionality and relationality, and environmental understanding. In contrast to the root metaphor of neoliberalism that shapes the ways most people participate in purchasing and mindlessly eating their food, this social epistemology embraces people's desires and democratic social imagination that transcend economic capitalism.

This praxis of individuals-in-relation becomes a source of resistance against neoliberalism whether fully recognized or not (Weinstein 2013). The practices associated with organic farming are 'policed' by and large—in the same ways that neoliberal factions watch over and ensure the practices of Bigbox superstores. Consider the absence of 'markings' that would normally indicate insect damage, normal scars, fruit and vegetable size, or packaging (and the ways that meat, fruits, vegetables etc. are preserved). Corporate personnel, supermarket owners and managers, and customers regulate the generally accepted (read neoliberal) produce practices that ensure prices are subsidized and so forth. Although there are small factions of people who gravitate to the organic sections of the Bigbox, or read the back of the packaging to look for ingredients such as corn syrup or to count calories, the vast majority of consumers do not consider how their education has contributed to or mediates the ways that they eat.

At our organic farmers' market the market manager hired by the city, farmer vendors, and patrons also play a large role in 'policing' the practice by attending the market weekly, knowing what can be legitimately grown during particular seasons and not others, and by asking questions and learning about what comprises an organic farming experience (what organic looks like and so on). As part of this project, I learn about farmers who are also violating the trust of patrons and discover that the market manager also knows about these breaches of trust. For example, one farmer sells produce that is purchased from another farmer not using organic practices, and another farmer purchases food from another region of the country. These farmers are quickly exposed by consumers and asked to never return. Breach of trust is a serious violation.

Farmers work hard to make a living in the organic market culture. They work very hard to connect with people who value highly the ways they protect farm workers, animals or ecosystems from being degraded—in many ways they revitalize these systems. Organic food is expensive if the comparison is superficially related with the Bigbox industry. Prices fluctuate at the farmers' market depending on availability, supply and demand for specific produce. Prices differ between farmers despite that we try to do a quick survey at the beginning of every market to get a sense of the 'going rate.' The point is that patrons have choices and some people do go from farmer to farmer looking for the 'best deal.' One woman asks us "where she would get some *normal* tomatoes!?!" She is surprised by the price of heirloom tomatoes, which go for about $1.00 per pound. Most people, however, understand that they are buying more than meats, fruits, and vegetables—they are supporting a cultural livelihood and contributing to the improved welfares of farm workers, animals, and physical environments. They are supporting the livelihood of the market as a significant pedagogy (i.e., the market as a place, tool and site of learning).

Consider how children are led around the farmers' market with specialty cooks to learn how to select and purchase greens, or the special diet needs of individuals who have been instructed by their doctors to purchase organic vegetables, or tourists craving culture. People who say that they have a limited income visit the farmers' market as much as people who say that they do not worry about it. When asked about whether price matters, of course social and ecojustice issues come up, but more often patrons discuss the ways that purchasing fresh organic food from a farmers' market can be justified. They discuss smaller plate or portion size, and slowing down to enjoy their food with family and friends. We sold pansies and other edible flowers and herbs, which always were hot items!

Contrast these understandings with the hegemony of neoliberalism, which emphasizes individualism, competition, and the economy as the highest good. On the whole, people involved in farmers' markets from the city manager of the market to farmers and patrons understand that they are intimately involved in collectivism—the sharing of lives and experiences—or equity and fairness for others. They develop the capacity to resist neoliberalism because of a lived curricular trajectory implicit in the science education of the farmers' market that capitalizes on their ability to critique, suggest revisions, and take action against business practices, goods, services, and media that go contrary to their acquired affection for responsibly nurtured organic food. Through this science education, they develop the underlying cultural assumptions and understand the root metaphors of sharing, trust, moral reciprocation and so on, which mediate the ways they eat. This social epistemology creates the conditions for resolving local problems and issues that might come up even in a farmers' market such as fraudulent practices. A critical mass of resistance protects the livelihood because of the constant monitoring and in-depth immersion of the culture itself. This science education is highly contextualized. Interestingly, children who regularly visit the farmers' market are learning and teaching others around them about the status or condition of their local cultures, communities, commons or habitat. In many ways, these students' science education mediates the ways that they eat and vice versa.

Implications for School Science

The curricular trajectory of science education in the farmers' market incorporates an in-depth understanding of place/context, dynamic polysemic knowledge that evolves in social concert with change/adaptability, positionality/relationality, and eco-knowledge. This context for science education is a site for neoliberal resistance as children investigate whether their cultural traditions, narratives, events, and skills serve to protect them from hyperconsumerism or overreliance on the Bigbox supermarket to meet all of their needs. They learn to garden, can, cook, preserve food, save seeds, and eat face-to-face with loved ones. They have not lost the art of talking with their neighbors (the old way!), they care about the community and the ways that people and animals are treated. They mindfully conserve physical environments—their *habitus* the commonplaces.

The conditions for science education are often discussed in terms of a science classroom. This emphasis on science classrooms creates a significant problem for the anthropology and phenomena of science education communities that serve to mediate knowledge and social action from within sites of resistance. In many ways, it is society's understanding that inadvertently perpetuates neoliberal forms of science education and the scholars who support this ideal. What is the right environment for learning? What environment motivates students to engage in particular conceptual or relation understandings versus others? What habitus creates social action? Rarely, however, are the conditions of science education discussed in the same way that farmers explore the rain, wind, sun and shade. Science education that mediates the ways we eat is rich with future sociopolitical action and ecojustice, withstanding its contributions to the status quo, or almost exclusive focus on economic capitalism (for more, see Derek Hodson's (2011) text, *Looking to the Future*).

The farmers' market is but another metaphor for the many ways that people participate in activities outside of the neoliberal push. These activities have the ability to mediate science education by creating a powerful inertia that activates people in social change and more importantly, science education for ecological nurturance. What we eat certainly affects how we understand and the way we behave in relation. Our eating has the potential to transform school. This transformation is an activist force for society's understanding, and when realized more fully it offers the imagination necessary to catch a glimpse of the world that would provide for future peoples.

References

Bencze, L., & Carter, L. (2011). Globalizing students acting for the common good. *Journal of Research in Science Teaching, 48*, 648–669.

Glasson, G. E., Frykholm, J. A., Mhango, N. A., & Phiri, A. D. (2006). Understanding the Earth systems of Malawi: Ecological sustainability, culture, and place-based education. *Science Education, 90*, 660–680.

Hodson, D. (2011). *Looking to the future: Building a curriculum for social activism.* Rotterdam: Sense.

Johnson, J. (2003). *On and on.* Universal City: UMvd Labels.

Louv, R. (2008). *Last child in the woods.* New York: Algonquin.

Mueller, M. P. (2009a). On ecological reflection: The tensions of cultivating ecoJustice and youth environmentalism. *Cultural Studies of Science Education, 4*, 999–1012.

Mueller, M. P. (2009b). Educational reflections on the "ecological crisis": Ecojustice, environmentalism, and sustainability. *Science & Education, 18*(8), 1031–1055.

Mueller, M. P., Tippins, D. J., & Bryan, L. A. (2012). The future of citizen science. *Democracy & Education, 20*, 1–12.

National Research Council. (1996). *National science education standards*. Washington, DC: National Academies Press.

Rowe, B. (2011). Understanding animals-becoming-meat: A disturbing education. *Critical Education, 2*, 1–25.

Thayer-Bacon, B. (2000). *Transforming critical thinking: Thinking constructively*. New York: Teacher's College Press.

Thayer-Bacon, B. (2003). *Relational (e)pistemologies*. New York: Peter Lang.

Tippins, D. J., Mueller, M. P., van Eijck, M., & Adams, J. D. (Eds.). (2010). *Cultural studies and environmentalism: The confluence of EcoJustice, place-based (science) education, and indigenous knowledge systems*. Dordrecht: Springer.

Weinstein, M. (2013). Sciences for the red zones of neoliberalism. *Cultural Studies of Science Education* (Online First). doi:10.1007/s11422-013-9490-y.

Chapter 14
From-Within-the-Event: A Post-constructivist Perspective on Activism, Ethics, and Science Education

Wolff-Michael Roth

Abstract The purpose of this chapter is to sketch out a post-constructivist perspective on activism, science education as/for socio-political action, and the associated ethico-moral dimensions. I begin by providing glimpses at one local environmental activist group, which had taken the environmental health of the main watershed in which its municipality is located as its object, and at seventh-grade students who, following a call by the activists, contributed to realizing the common goals that these articulated. I then respond to the rhetorical question whether community-based activism is something to feel morally good about before articulating theoretical perspectives on activism, the eventness of events that orients us to continual becoming, and on ethics from classical and from-within-the-event perspectives on activism. I conclude that the post-constructivist perspective emphasizes the ethico-moral dimensions of activism, which does not inherently do good, but whose actions are subject to the same kind of precautionary principles that activists often jut into the faces of (sorcerer-apprentice) scientists.

Keywords Community-based activism • Participation • Life-as-event • Eventness • Responsibility • Precautionary principle

In Western culture, it is pervasive to think about learning specifically and human activities more generally in terms of agency and, therefore, in terms of cause–effect relationships. For example, an environmental activist group may be said to have the intent of changing some aspect in/of their community, such as making a local creek the viable habitat that it has not been for decades because of chemical and biological pollution. This intent will be said to be at the origin of the group's actions, which, for example, might be a proposal for changing the official

W.-M. Roth (✉)
Faculty of Education, University of Victoria, MacLaurin Building A567, Victoria, BC, V8P 5C2, Canada
e-mail: mroth@uvic.ca

L. Bencze and S. Alsop (eds.), *Activist Science and Technology Education*, Cultural Studies of Science Education 9, DOI 10.1007/978-94-007-4360-1_14,

community plan by inserting additions that represent its vision. Thus, in one municipality, an environmental activist group petitioned the town council and official community planning team: "A statement/policy is needed that shifts maintenance of storm drain ditches toward managing all watercourses for improved diversity, complexity and stream function. (11.1.5)" (Lee and Roth 2001, p. 337). When such a change is made, it subsequently is attributed to the actions and intentions of the activists. They might feel vindicated and appreciate the fact that they have done something for the common good by bringing about a "necessary" change of water management policy.

Participating in community-based activism has increasingly received attention by educators interested in taking science education outside of the classroom; thereby not only making it more relevant to the students but also allowing the latter to learn while contributing to the common good (e.g., Roth and Barton 2004). There is evidence that students who participate in activism positively evaluate and assess the changes they have brought about in the real world (Posch 1993; Roth and Lee 2004). What the students do, for example, in a green energy program, not only is attributed to their agency but also is said to develop their agency (e.g., Barton and Tan 2010). This research also suggests that students develop their identities and, in the course, learn some science, too. Moreover, in the not-too-distant past, I had thought that in and through their environmental actions, which implemented their intentions to contribute to the common good, students had exhibited responsibility and, therefore, enacted high ethico-moral values. They had, I thought, articulated an ethico-moral perspective on collective, societal life by "*aiming at true life with and for others in just institutions*" (Ricœur 1990, p. 211, original emphasis, my translation).

There is something fundamentally wrong, however, with this picture: It overemphasizes the role of human beings in a changing world by making them the source of changes in the world and in/of themselves (Roth 2013b). The problem in such thinking comes from the misconceived relation between actions, the causes, and what the actions bring about, the effects, can only be constructed after the fact (Nietzsche 1954). In fact, the subject, to be properly understood, needs to be theorized as the effect of events that always exceed the powers of the individual (Romano 1998); thus, "the subject, produced as a residue at the side of the machine . . . is not itself at the center, which is occupied by the machine, but on the periphery, without fixed identity" (Deleuze and Guattari 1972/1973, p. 27). A simple thought experiment about the science curriculum brings out why there is something wrong with the figure of agency and where it is wrong. As science teachers, we prepare lessons in more or less elaborate ways – beginning teachers have to write out their lesson plans to be approved by their supervisors or mentors. Science educators have come to know this as the "planned curriculum," against which the "enacted curriculum" is assessed (e.g., Kurz et al. 2009). That is, despite the fundamental gap that has been said to exist between plans, on the one hand, and situated actions, on the other hand, even when only one agent is involved (e.g., Suchman 1987), university researchers and supervisors have negatively assessed practising and beginning science teachers for differences between the two forms of curriculum.

This unbridgeable gap comes about because we can grasp an unfolding event as a whole – as an inner- worldly fact the nature of which we can assess – only when eventing is already completed. As far as living events are concerned, their com/-prehension always is delayed; the delay between living and comprehended experience is constitutive of understanding (Romano 1998). The curriculum has in fact an *emergent* quality about it (Roth 2003), which means that *new, unforeseen,* and *unanticipated* phenomena enter the experiences of participants: participants are *affected* by the emergent event as much as they affect it (Roth 2013a). When we consider activism generally and science education as/for socio-political action (Roth and Désautels 2002) more specifically *from-within-the-event, then we have to fundamentally* change how we theorize a science education curriculum that takes an activist stance and the pertinent ethico-moral issues. This perspective also radically changes the way in which (radical, social) constructivist positions theorize the agential subject of activity: it is also *subject* and *subjected to* the activity, that is, it is *patient* and *advenant,* to whom events happen in unforeseen and unforeseeable ways. The subject is, as Deleuze and Guattari suggest, a residual of the machine of life and is located on its periphery; and for Nietzsche the subject is the result of a total abstraction. I therefore denote the emerging theory that looks at activism *from-within-the-event* by the adjective *post-constructivist.*

The purpose of this chapter is to sketch out a post-constructivist perspective on activism, science education as/for socio-political action, and the associated ethico-moral dimensions. I begin by providing glimpses at one local environmental activist group, which had taken the environmental health of the main watershed in which its municipality is located as its object, and at seventh-grade students who, following a call by the activists, contributed to realizing the common goals that these articulated. I then respond to the rhetorical question whether community-based activism is something to feel morally good about before articulating theoretical perspectives on activism, the eventness of events that orients us to continual becoming, and on ethics from classical and from-within-the-event perspectives on activism. I conclude that the post-constructivist perspective emphasizes the ethico-moral dimensions of activism, which does not inherently do good, but whose actions are subject to the same kind of precautionary principles that activists often jut into the faces of (sorcerer-apprentice) scientists. In re/writing the issues of activism and ethics that I had produced earlier (e.g., Roth 2008a, 2010), this text also is testimony of the unforeseeable changes in a scholar's theoretical understanding that arises from participating in scholarship-as-event, that is, seeing scholarship from-within-the-event.

Environmental Activism at Work

Environmental activism has become an important social movement that emerged from a variety of movements concerned with the preservation of wildlife, the constitution of national parks, concerns with the impact of (nuclear) technologies, and other issues threatening the well-being of the natural and societal worlds. Although many citizens become conscious of activism when there are large or spectacular protests (e.g., ramming of whaling boats by the Sea Sheppard Conservation Society), there are many activist groups that are interested in local issues, such as cleaning up and preserving the watersheds they inhabit or in revitalizing some inlet previously destroyed by industrial activities by planting eel grass beds that produce habitat and promises a return of species. Such activities have become sites for science education such that in some jurisdictions, environmentalism and "green goals" are "seeping into the state's curriculum" (McMahan 2008). These forms of science learning are not limited to school students but extend to all community members who feel called upon to participate (e.g., Boyer and Roth 2006). After sketching the work of one environmentalist group and a curriculum aligned with its goals, I ask the question about the ethico-moral dimensions of the ways in which nature comes to be represented in activism.

Transforming the Practices in One Municipality

In the municipality of Central Saanich, British Columbia, where I have lived for 17 years, water has been an issue for a long time. The climate is characterized by long favoured hot dry summers and wet winters, with concomitant shortages and excesses of water available to residential areas and farms. Historically, the watershed consisted of bogs and meandering watercourses, teaming with wildlife that provided food to its traditional inhabitants, the W̱SÁNEĆ (Saanich) First Nations, which are part of the Coast Salish peoples. However, after their arrival in the mid-1800s, the white settlers turned the creeks into ditches to drain away the water, turning the bogs into farmland. However, the water that used to fill the local aquifers during the rainy winter months now are no longer filled because the water runs off too quickly to be absorbed; and increasing urbanization has led to a greater amount of impervious areas from which rainwater is directly taken into the drainage system and into the ocean. Moreover, heavy irrigation schedules during the summer put further pressure on the water resources. Along with the rainwater, storm drains and ditches channeled pollutants of suburbia, lawn chemicals, and car leakage into the watershed and ocean; and the drains of the machine shops and biotechnology labs in an industrial park emptied into Hagan Creek (the W̱SÁNEĆ call it ḰENES, pronounced /qʷənəs/ [kwanus]) further taxing the environmental health of the area. Over the years, Hagan Creek–ḰENNES[1] had been deepened and straightened, and much of the covering vegetation has been removed, thereby increasing erosion and pollution from the surrounding farmers' fields. As a result,

[1] Although a website of the SENĆOŦEN aboriginal language spells the name ḰENES, the activists and the academic unit doing related research refer to it as ḰENNES. I use one or the other spelling appropriate to the context in which the name is used.

erosion and silt load had increased in the wet winter months and were responsible for low water levels and high water temperatures during the dry summer months. As a consequence, water flows had become extremely ranging, from more than 8,000 l/s in the winter to less than 10 l/s during the summer months. Every year, moderate to extreme water advisories limit the amount of water available to residents. Those with individual wells have found their water biologically and chemically contaminated and sometimes have to get their water from gas stations about 5 km away.

The Hagan Creek–ḰENNES Project arose from the concerns about water quality of three watershed residents: a farmer, a professor of environmental law and policy at the local university, and a research scientist working at a nearby lab. This group wrote a proposal, and obtained funding, for restoring Hagan Creek–ḰENNES. The Hagan Creek–ḰENNES Project – which, since my original research, has become part of a regional association of creek preservation – was an environmental group that had as its mission the change of attitudes and practices regarding water and the watershed, but without engaging in confrontation as can often be observed with other local but especially national and international groups (e.g., Greenpeace or Sea Sheppard Conservation Society). A coordinator and a 5- to 7-member steering committee headed the Hagan Creek–ḰENNES Project. It enlisted the support of many other people (e.g., hired high school and university students to collect data as summer jobs), institutions within the region, and, importantly, classes of seventh-grade students from the local middle school. (I was the mediator, being acquainted with the coordinator and simultaneously working with middle school teachers towards improving science education in their school.) The members of the Hagan Creek–ḰENNES Project said at the time that they worked in and against an adverse political climate, where the interests of farmers, industry, and other landowners are often opposed to those underlying the Hagan Creek–ḰENNES Project. But as my research showed, rather than seeking confrontation, the environmentalists intended to build and maintain good relationships with the greatest number of stakeholders possible, which they considered paramount for successful *event*uation of the intended changes. They networked with many people and groups to increase the likelihood of success and, intending to build continuity of their ideas into the community that would exist long after they had gone, formed a training and support group for watershed stewards.

Each year, the Hagan Creek–ḰENNES Project organized an open-house event in which their work and those of their allies was featured. For example, the water technician hired with the above-mentioned funding for the Hagan Creek–ḰENNES Project, using one farm along the lower reaches of Hagan Creek– ḰENNES as her operational base, displayed a pen-chart recording of the water levels that she recorded in the course of a year (Fig. 14.1a). Project members also built a model of the watershed to teach young and old alike about water flow patterns in the watershed, the role of impervious surfaces (e.g., roofs, roads, and driveways), and the effects heavy rainfall has on the siltation of the creek (Fig. 14.1b). That is, the environmentalists not only sought to create groups and change policy but also informed the general public about their achievements and ongoing projects.

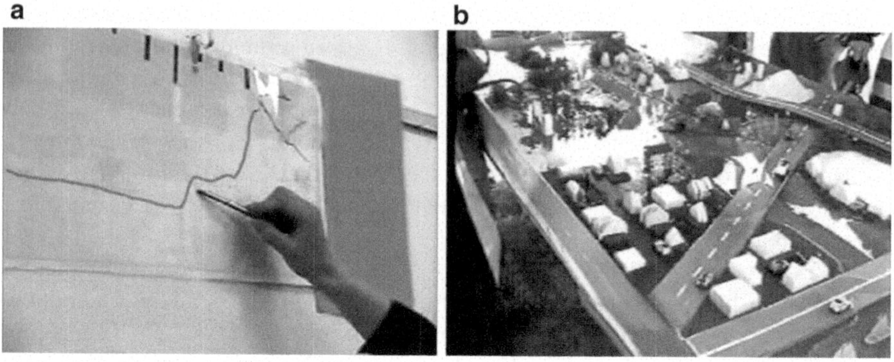

Fig. 14.1 As part of their fight for environmental and human health in Central Saanich, local activists represent Hagan Creek–ḰENNES, here at an open-house event, in the form of a graph depicting its depth over the course of a year (**a**) and a cardboard model (**b**) (©Wolff-Michael Roth, used with permission)

The Hagan Creek–ḰENNES Project also was successful in involving some First Nations elders or artists, such as the one who carved a logo of the creek, which, from his native perspective, was not Hagan Creek known to the white people but was ḰENES, the place of whales, which used to come to the mouth of the creek to feed. His representation differed in distinct ways from those that the white people used:

> Visitors who talked to the carver likely entered into discussion about the animals that once lived in the valley, the landscape features pre-settlement, the place the creek had in their spiritual lives, and the effect of treaties on his Nation's lifestyle. His representation of [Hagan] Creek, a logo commissioned by the Project, represented the animal life in artistic form and included geographic features that were of significance to his people. The carver frequently pointed to the representation and talked about the ways, for instance, in which his people used the small island just off the mouth of the creek. He provided visitors with opportunities to understand the creek, elsewhere in the Open House represented by dissolved-oxygen levels and coliform counts, in terms of a its connections to a different culture and way of life. In his person, there existed a resource to link science to history and First-Nations culture. (Lee and Roth 2003, p. 412)

Science Education as/for Participation in the Community

We developed an activism-centered curriculum that took students into the community in part driven by the recommendation that science education should involve students in experiencing the results of their action in the community (Hodson 1999), in part driven by a recommendation to *deinstitutionalize* science education (Roth and McGinn 1997). The curriculum was to enable students to contribute in significant ways to society by making available the results of their work to politicians and media, and thereby to influence environmental policy and decision-making

(McGinn and Roth 1999). That is, community-based activism has been proposed as an important way of rethinking science education under the slogan of "science education as/for participation in the community" (Roth and Lee 2004). I had coordinated my school-based intervention that was to be a test case of the at-the-time relatively new idea concerning activism with the coordinator of the Hagan Creek–ḰENNES Project.[2] She came to the school, talked to the students about the need of having the community get involved; and the students also read an article from the local newspaper that featured the coordinator, and which ended in her pitch for community involvement. In the following, I focus on the work of one student and its after-the-fact attributable impact it has had in and on the community. (Details of the curriculum can be found in Roth and Barton 2004.)

Graham, an eighth-grade student at the middle school where my intervention took place, became interested in what we were doing and decided to do a science project in the context of the seventh-grade curriculum and its activist goals. He was interested in measuring biological contaminant levels in different parts of the creek. Initially, he used a school kit for assessing coliform bacteria amounts; but, when this turned out not to be a very reliable method, he was able to negotiate access to a university laboratory with the result that he generated reproducible coliform bacteria counts from his samples.

Graham presented results of his analyses during one of the open-house events that the Hagan Creek–ḰENNES Project had organized in the community using a poster format common in scientific conferences and science fairs (Fig. 14.2). In his poster, he published the results, including the name of one farmer on whose property large amounts of fecal coliform entered the creek – as evidenced by the negligible count upstream from this farm and more than twice the limit for safe bathing just a little downstream from the farm. Upon finding out about this publication, the farmer no longer allowed students to return to his property for conducting further research. Graham hypothesized that horses, which were able to step into and drink from the creek, were contributing to the coliform levels that he had determined.

During an interview that two graduate students conducted with him about 1 year later, he rendered account of his work concerning the coliform levels:

> We did one [sample] along Malcolm Road, which, I guess, park area, which has been graveled down the sides. Pretty nice and I got a sample there and a sample- that's along one arm. I took a sample from the other arm and this was right before the confluence of the two arms and then one at Gordon Godfrey's farm, which is right after the confluence. And we found, I guess there wasn't extremely high levels at all [on the upper reaches]. They meet up with each other at Gordon Godfrey's. Uh, but then we found there were really high levels at Gordon Godfrey's but there was some I guess, the two arms, I guess it's a narrow pathway, they meet just after it. You see one arm and it goes under the road, I guess in a big pipe, and the other one just been manicured to make it look all good and parky. So, I didn't notice this at first but later, when we came back, there was a pipe running, I guess, about three meters ahead of where we took our sample, and out of it there was, I guess, brown stuff and I followed the pipe back to the field that was right across from the, uh, right across there was a

[2] I came to understand only much later that what would unfold was not caused by my actions and what would come out of all of this exceeded the powers of my research group that thought about activism as a context for science education.

Fig. 14.2 At an open-house event organized by the environmental activists, Graham, gazing at the results from his study posted as part of his display, talks about the high coliform levels he found just downstream from one farm, whereas he had not identified coliform bacteria in the reaches of both creeks just above the farm (©Wolff- Michael Roth, used with permission)

> horse area and the pipe was coming in from that direction and so I'm assuming there was something gross. Anyway it all leads up to there were really, really high levels leading up to Gordon Godfrey's which is after the conjunction so something between the three meters back from that pipe and the um, Malcolm Road, the two samples something was getting in the water which it have higher levels of coliform – fecal levels of coliform. (Graham)

In this narrative account of his experience of doing the research, Graham describes how he had sampled two creeks upstream from Godfrey's farm, and then had sampled Hagan Creek after the confluence downstream from the farm. On the upper reaches, he did not find "extreme levels" of coliform but there were "really high levels" on the lower reach. He also stated having noted a "big pipe," where water exited that he characterized by the expression "brown stuff." He was measuring fecal coliform levels just 3 m downstream from the pipe. As he had followed the pipe upstream, he discovered "a horse area," which he hypothesized to be a potential source of the contaminants.

The Morality of Community-Based Activism: Is It Something to Feel Good About?

Activists often understand themselves as "good citizens," where forms of morality come to be integral parts of the discourse that weaves together science and the environment (Lee and Roth 2003; McGinn and Roth 1999). We also had linked student participation in activism to the development of responsible citizenship (Roth and Désautels 2004). In my research group, we had felt satisfied about what a tiny environmentalist group was able to achieve in terms of changing practices consistent with our own beliefs about the value of improving environmental health through activism. However, we were taken aback and began to reflect about the morality of activism when an aboriginal friend – who had lived on the reserve bordering the creek and whose brother, a well-known native artist, had a study and gallery there – reacted to our account of what the Hagan Creek–ḰENNES Project was able to achieve despite its limited resources.

> The activists are doing the same thing that the farmers did when they first cleared the forests, drained the swamps and channelised the stream. They are perpetuating the dynamics of colonialisation. They haven't taken the time to educate themselves through dialogue with the Coast Salish people who've lived there for hundreds of years and who probably have stories about the birth of the creek. They've spent a summer measuring it with their meters and yardsticks and now they've got their machines in there, changing it. They haven't taken time to build relationships with the people who first inhabited the land. I do not understand how this can be called a democratic process. (Lee and Roth 2001, p. 349)

In the article from which we quoted our friend, we concluded that the activists had not adapted their practices to make them consistent with the values and beliefs of the First Nation that still inhabits the lands. From our aboriginal friend's perspective, activism is, in a necessarily different form, a repetition and return of the same sort of unfolding event, the "dynamics of colonialization," a continuity of an unfolding diasporic experience and the real or symbolic violence that comes with it (Roth 2008b). As another study among aboriginals involved in environmentalism showed, this is often experienced as but a perpetuation of colonialization in one of its ever-renewed forms (van Eijck and Roth 2009). In the actions of the activists, and inherently in our own actions that enabled students to work at the elbow of the activists, as in the ways in which they described the creek and what they were doing, there was a form of racism. We concluded that this aboriginal woman taught us to "understand that our claims and hopes for a science in the service of all is not an end game that will have its conclusion in a Utopian society in harmony with all its peoples and non-peoples" (Lee and Roth 2001, p. 349). Elsewhere, we concluded: "Simply returning the creek to the state in which it had been some 100 years before does not address a more fundamental issue concerning the relationship between people and their lifeworld" (Roth and Lee 2002, p. 50). The farmer Gordon Godfrey, too, had been affected, as per his response of discontinuing the access to his farm; and not only Graham but also all other students participating in our project were affected by this discontinuation, an affectation that is possible only because of the students' and our response-ability, that is, ability to respond.

I note above that the local aboriginal people call the place ḰENES, place of the whale. Before colonialization, it not only was a place where the whale came to feed or where the W̱SÁNEĆ fished 20-in cutthroat trout that no longer exist today, hunted fowl, and collected shellfish (at the mouth of the creek where the whales) but also had been a place for spiritual cleansing. There are considerable differences between the ways in which the indigenous peoples of the area and Western science think about and see the land, differences that are embodied in different chronotopes in the telling of places. In science, as in the Western culture that has given rise to it more generally, places tend to be accounted for by their geographical locations and names that only denote the place but do not matter to our understanding thereof (van Eijck and Roth 2010). For example, we wrote about a place not far from ḰENNES, which is also integral to Saanich Inlet, and which is known to most (white) inhabitants of the area called "Tod Inlet." It is but a name for a place where boaters anchor off to spend a night, a place where some boaters have discharged their toilet effluents, a practice that stopped only when activists made available a boat, Pumpty-Dumpty, which they used to pump the contents of the septic tanks from the boats and discharge them in environmentally friendly ways (and places) (Roth 2010). For the indigenous peoples of the area – the W̱JOȽEȽP (Tsartlip) First Nation that is part of the W̱SÁNEĆ First Nation – the place is SṈITȻEȽ (pronounced "sneakwith") a name not only integrally associated with the collection of food and water but also a place of absolution and prayer, with renewal, and a place where young warriors practised their skills, in other words, a sacred place.

Neither the Hagan Creek–ḰENNES Project and our students, who concretely realized the goals of the Project, nor we, in our elated accounts of the ethico-moral ideals that activism enacts, intended to offend the aboriginal peoples; but our actions, perhaps naïvely oriented to contribute to a greater good, had been experienced as a form of colonialization, and, therefore, dealt the First Peoples of this land but another blow in a series of blows – an eternal return of differences and their relations (Deleuze 1968) – that they had to endure since the first white folks came and took their land. As pointed out in the introduction to this chapter, our activist intentions and actions had brought about effects made public in the response of an aboriginal person that were in excess of our intentions. But, although unintended, we are responsible for the effects of our actions which, because inherently unforeseeable, given the openness of the horizon that characterizes "life-as-event" (Bakhtin 1993, p. 10) seen by participant- witnesses from-within-the-event. This, then, calls for a new understanding of activism, which has to confront the fact that its own actions are to be responsible in the face of an inherent irresponsibility, deriving from the fact that activists, as much as those who are affected by their action, are advenants, those to whom unforeseen effects of actions *advene* in unanticipated and unexpected, sudden ways.

From Activism to the Eventness of Events

Activism

> The philosophers merely interpreted the world, in various ways; the point is to change it. (Marx and Engels 1958, p. 535)

In a recent paper, I propose activism as an analytic category – as distinct from using 'activism' to name something individuals do when they are strongly advocating and fighting for this or that issue (e.g., women's rights, gay and lesbian rights, or AIDS and environmental activists) (Roth 2010). As a category of a new learning theory, activism is the smallest unit that allows us to make sense of unfolding, culturally and historically situated, and contingent processes. Seen with a theoretical gaze after the fact, the constituent moments of this category include *subject, object/motive, means of production, division of labour, community,* and *rules.* Importantly, the object/motive is a composite constituted by the material (conditions) at hand and the ultimate goal of the productive activity. Thus, for example, the Hagan Creek–ḰENNES group had as its goal the transformation of the material condition of the creek and the watershed it drains, which in many parts were no more than ditches, into a "healthy creek." This object/motive inherently organizes emotion and motivation, thereby integrating two affective moments that other learning theories need to import as external factors that mediate learning. This is so because what we do is shaped by two forms affect. On the one hand, our bodily states influence how well we do what we do and even what we might decide to do; on the other hand, what we do is oriented towards some goal, and reaching the goal comes with positive feedback on our affective states, whereas failure tends to constitute negative feedback (Roth 2007a; Roth and Radford 2011). This is especially the case for actions that contribute to the common good.

The important aspect of the activism category is that goal-directed actions and contextually determined (mental, physical) operations that constitute the former cannot be understood independent of the societally motivated activism. Thus, even Graham's dipping of a test tube into the creek, which older theories consider being a (tacit) skill, can be understood only through the lens of activism and the actions that realize it. By participating in activism, individuals' operations change – e.g., become more fluent, increase in terms of competence, are more adapted to bring about change – and, with it, the individuals transform. These transformations are both material (embodied skills) and ideal (the practical understanding that the participants develop). Participating in the *field* of activism shapes the way in which new and old activists are dispositioned to see and act toward the environment, that is, their *habitus* changes. There is a mutually shaping relation between the material and societal-cultural dimensions of a field and the structured structuring disposition (habitus) that develop in and through participation (Bourdieu 1980). We comprehend the world because the world comprehends (includes) us (Bourdieu 1997).

Most scholarship in the social sciences is concerned with agency, mobilization of the human power (capacity) to act upon its intentions – as evident from the primacy of actions in constructivism, sociological agency|structure theories, and even common usages of cultural-historical activity theory. Agency is only one part of the equation for understanding the human life form: In contrast to animals, humans not only live under and are determined by (are subject and subjected to conditions) but also have the capacity to change the conditions under which they live and to which they are subjected and subject (Marx and Engels 1958). The category of activism does include this passive dimension, whereby we are changed physically and ideationally, generally without noticing it at the instant, by engaging in and developing forms of human practice.

From-Within-the-Event or the Eventness of Events

In their rethinking of human nature generally and ethical questions more specifically, philosophers with very different theoretical, historical, and cultural backgrounds – e.g., M. M. Bakhtin (1993), G. Deleuze (1968) or C. Romano (1998) – orient us not merely towards events as completed and grasped phenomena in the world but more specifically to the eventual nature of events – their eventness.[3] That is, they orient us to a continual becoming, which, when understood as becoming, no longer allows us to identify states (e.g., knowledge, intention, cause, effect). The fundamental reason for approaching events through a perspective from-within-the-event is to come to grips with the fact that we cannot *grasp* events from within because they are not yet completed; and because events are not completed, because we do not know (i.e., comprehend) what will become, we do not know what effects there will be and, therefore, what causes to attribute to them. As participants, who see activism from-within-the-event, we are but witnesses confronted with and affected by events in ways that always are in excess of any foreseeable effect. Members of the activist group and I initially were shocked (i.e., affected) by what our actions had done, which we neither anticipated nor foresaw. After the fact, we could understand, because we determined what our actions had caused from the effect that they had brought about – perpetuated colonialism. That is, what we grasped when everything was said and done was different than what we witnessed while we participated as activists and their ethnographers – we felt good about doing something for the environment and getting kids to contribute to society rather than copying notes from a chalkboard or writing stuff on pieces of paper that would sooner-or-later end up in the garbage can. That is, our theoretical grasp of the events – participating in activism and bringing about a activist science curriculum – was delayed, and, in fact, only could arise because of this delay.

Educators, in fact, made a step toward recognizing the eventness of events when they pointed to the unbridgeable gap between the planned and enacted (lived) curriculum (e.g., Roth 2007b). They have not taken the next step required, however, in thinking curriculum from-within-the-event, which, because its future course and states are unknown, makes events seen through the lens of the unfolding event, a very particular phenomenon with yet-to-be-fathomed implications. Thus, at the instant when a science teacher tells her student that her way of figuring out the valence of chemical elements is not generalizable and insists on her way of figuring out that number she cannot know that the situation will develop into a major argument (Roth and Tobin 2010); the science teacher who tells the heretofore

[3] In his original Russian, Bakhtin (1993) uses the term *sobitijnost' sobitija* (событийность события), which the translators of his book rendered as "eventfulness of the event"), but which other translators render as "eventness of the event." In his French language, Romano (1998) uses the term *événementialité*, a neologism based on the adjective and noun *événementiel* ("of events") that was constructed to emphasize the event us something unfolding and inherently graspable in its extent. The English translation of the book renders the French term by means of the adjective "evential."

best student in his class that he received 60 % on his last test cannot foresee that a conflict will arise that almost leads to the expulsion of the student, who only remains at the school because of the dialogue that the research team present allows to unfold between the two protagonists (Roth et al. 2004). Implicit in this approach is that we cannot assign causes to events until we know the effects, which explodes the entire metaphysics underlying the common ways of understanding actions and ethics (Nietzsche 1954). We do not know what activism brings about and why (e.g., repeating colonialism in a new form); and we do not know what a living curriculum involving children and students in activism will do to the material world (e.g., when they plant trees), society (e.g., when they report findings in the way Graham did), or to themselves (e.g., who they come as ethico-moral beings). That is, we can no longer give primacy to intention as causal origin of practical actions and its effects. Rather, in and through the response of the material and societal worlds, the actors come to be confronted with dimensions of the unfolding event to which their actions have contributed to constituting. From-within world-as-event (Romano 1998) and life-as-event (Bakhtin 1993), the subject is on the margin, itself a product rather than a cause of the event eventually understood as inner-worldly fact. That is, what is traditionally the agent (i.e., subject of action) comes to be subjected and subject to the unforeseeable effects of his/her action so that even the agent *becomes* in ways and outcomes unseen and unforeseen: through our participation in environmentalism we had become colonizers rather than doing something good for the community, including its First Peoples. Seen from-within-the-event, these effects, exhibited in the response of the material world or other human beings, necessarily are in excess of the agent's intentions. This also requires us to rethink responsibility.

Activism and Ethics ...

... *Classically Understood*

The Kantian imperative, viewed from the dialectical relation of individual and collective, leads us to understand that events have effects not only on the natural environment, but also on the agent, who is transformed in and by acting, but also on the collective, which is transformed whenever individuals are changed. This fact:

> [l]eads us to an ethical relation – what I do affects us all, what any other individual does affects me, too. Thus, any action of a child that improves environmental health not only changes the child but also the community; and any action of a polluter – such as the farms that my children have discovered to increase fecal coliform counts in another creek to unsustainable levels – affect them and the community as a whole. (Roth 2010, p. 286)

Even dialectical approaches that place a primacy on practical actions do not lead us out of the quagmire of agency. This is quite evident when Ricœur (1990) states that "practical wisdom consists in inventing conduct that will best satisfy the exception required by solicitude, by betraying the rule to the smallest extent possible" (p. 312, my translation). Here, too, the agent '*invents* conduct,' which means, finds out or produces, by mental activity, something like a plan for that is to guide subsequent conduct; and it is this possibility that allows courts of law to impute causes to agents.

A different conceptualization of ethics characterizes a feminine perspective – i.e., caring for, which is the approach typified in the mother's ("the one-caring") care for the child ("the cared-for") (Noddings 1984/2003, p. 4). Here, primacy is placed on the *relation* as the fundamental ontological category where "both parties contribute to the relation; my caring must be somehow completed in the other if the relation is to be described as caring" (p. 4). In this approach, caring is fundamental to the human condition "toward which we long and strive, and it is our longing for caring . . . that provides the motivation for us to be moral" (p. 5). Thus, "we want to be *moral* in order to remain in the caring relation and to enhance the ideal of ourselves as one-caring" (p. 5). In these quotes, it is quite evident that even the radically different approach to ethics through the category of caring does not deviate from the traditional approach of giving primacy to intentions and motivations that precede the actions. These are considered moral because "we *want* to be moral," in order (a) "to remain in the caring relation" (intention) and (b) "to enhance the ideal of ourselves." Here, moral beings are theorized in terms of what they want to do, which is to be moral beings; and they do so for specific intentions, that is, to remain in the caring relation and to enhance what we think about ourselves (ideal). It is "this ethical ideal . . . that guides us as we strive to meet the other morally" (p. 5).

The problem of the classical perspective is that plans (intentions) are thought as the causes that precede and bring about actions. In considering plans, agents also consider or ought to consider all possible implications that arise from what they intend to do. This approach is embodied in current legal practice, where agents are held accountable for what they have done and the implications that arise from the actions, even if those were inherently unseen and therefore unforeseen and unforeseeable – unless the court recognizes some mitigating circumstance such as mental incapacity to understand the consequences of actions. However, the legal dimensions of answerability – according to which actions can be imputed to the agent who should have known better – are but a subset of a more encompassing, and ethico-moral answerability (Kant 1956). Being ethical means acting in a manner that could govern the behaviour of all human beings even in the absence of specific regulatory laws.

. . . *From-Within-the-Event*

In the subsection on the eventness of events, I note that the agential subject can know what s/he has done only in and through the response of the Other (world, human being). Because of the *ability* to respond, their *response-ability*, other agents and the world become accountable for something that they themselves could not foresee. In acting, the subject not only affects "once-occurrent being"

(Bakhtin 1993) but also exposes him-/herself to the unforeseeable effect. This leads us to a form of responsibility that is more radical than anything we traditionally associate with this term. This is so because the actor is in a position of having to be responsible in a situation that is marked by *ir*responsibility, something s/he cannot foresee and therefore be made responsible for; and it attributes responsibility to those who are exposed to and affected by the original actions, because, in their response, they affect the agent (Levinas 1978/2004).

When we consider activism from-with-the-event – i.e., from the perspective of becoming and the eventness of events – our perspective on ethics is forced to change. We no longer are afforded to think of ethics in some abstract or even practical sense divorced from our participation in "the once-occurrent real Being of an event" (Bakhtin 1993, p. 18), a participation that makes us as much advenants, to whom events happen, as it makes us agents that transform the world. That is, as activists, we are not only irresponsible agents of change but also responsible respondents, that is, subjects subject and subjected to the changes in the societal and material world to which we respond because we have the ability to respond, *response-ability*. Even more fundamentally, as apparent from the category of activism, we are changed in acting even without having to intend it, which exhibits a much more fundamental concept of responsibility, which is also the origin of a different nature of the Self: a consequence rather than the source of responsibility (Romano 1998). The Self, from-within-the-event is itself an unfinished process – a "being-as-event" (Bakhtin 1993) – that "presupposes my answerable participation" so that it is "only from *within my participation* that Being can be understood as an event" (p. 18, emphasis added). But this Self, as seen *from-within-the-event*, cannot be seen when we look at the event as a grasped phenomenon, as an inner-worldly fact. This dislocates the constructivist agential self – source (cause) of changes to itself, its cognition, and the world – to a Self that is the result of the ability to respond; this also dislocates ethics, now understood as an integral dimension and manifestation of relation. The constructivist agent is in a position to weigh the consequences of his/her action before acting; the post-constructivist advenant is subject to effects of actions that are always and already in excess over its intentions, so that the advenant is not in the position to deliberate good and bad. That is, my participation in activism is answerable, is subject to an answer on the part of the Other, and I am answerable for it, even though I cannot anticipate the answer, as seen in the reaction of my aboriginal friend to the activists' and my work. I have recently developed the implications of such a radical, post- constructivist approach to ethics in the curriculum (Roth 2013b).

Conclusion

Activists often use the term "precautionary principle" as a discursive resource in the attempt to bridle all-too-eager scientists who endeavour populating the world with genetically modified organisms, gigantic technological objects, tools, and devices, "ultimate" drugs, and monster creatures that grow twice as fast to twice their normal size (e.g., genetically modified trout and salmon). This principle states that when a policy or action runs some risk causing harm to the material or societal world, those acting or making policy need to proof that what they intend is *not* harmful. The ethico-moral authority appears to lie with activists, who, in advocating the principle, emphasize that it is better to "err on the safe side." In this chapter I outline a post-constructivist theory of activism and ethics, which suggest that activists, too, are subject to the same precautionary principle that they hold up to and wave before the eyes of scientists. As participant in "life-as-event," that is, in life that has an inherently open horizon with respect to what happens even seconds hence, activists are answerable for their non/action as much as scientists are. To be useful, the precautionary principle needs to be applicable to the precautionary principle as a policy for action: precaution itself is understood as giving rise to potential harm (e.g., if a potentially harmful drug were withheld even though it ultimately might turn out that it could have saved lives). The precautionary principle, to make any sense as a plausible policy for action in a democracy, is subject to the same precautionary principle, that is, to the same kind of responsible irresponsibility as any other form of policy or action. The precautionary principle, seen from-within-the-event of activism or science education as/for socio-political action, always has to come with an open horizon, for when it no longer changes, it inherently is dead and no longer participates in life-as-event.

For science educators, this post-constructivist approach to activism seen from-a-within-the-event perspective has considerable implications. We may no longer simplistically feel good about ourselves when we enable students to participate in activism and science education as/for socio-political action. Science educators, science teachers, and their students need to understand that even those actions that to the best of knowledge serve the common good and therefore represent the general – in other words, are consistent with Kant's categorical imperative – have an open horizon seen from within-life-as-event so that they may understood as turning out to be detrimental (e.g., contribute to the eternal return of colonialism). They therefore need to be conscious of the fact that whatever they do – participate or not participate in activism – seen from within-life-as-event, has consequences they have to answer to and for even though there is nothing from within their horizon that would allow anticipating any negative effect whatsoever.

References

Bakhtin, M. M. (1993). *Toward a philosophy of the act* (V. Liapunov, Trans., V. Liapunov & M. Holquist, Eds.). Austin: University of Texas Press.

Barton, A. C., & Tan, E. (2010). We be burnin: Agency, identity and learning in a green energy program. *Journal of the Learning Sciences, 19*, 187–229.

Bourdieu, P. (1980). *Le sens pratique* [Logic of practice]. Paris: Les Éditions de Minuit.

Bourdieu, P. (1997). *Méditations pascaliennes* [Pascalien meditations]. Paris: Éditions du Seuil.

Boyer, L., & Roth, W.-M. (2006). Learning and teaching as emergent features of informal settings: An ethnographic study in an environmental action group. *Science Education, 90*, 1028–1049.
Deleuze, G. (1968). *Différence et répétition* [Difference and repetition]. Paris: Presses Universitaires de France.
Deleuze, G., & Guattari, F. (1972/1973). *L'anti-œdipe: Capitalisme et schizophrénie* [Anti-Oedipus: Capitalism and schizophrenia]. Paris: Les Éditions de Minuit.
Hodson, D. (1999). Going beyond cultural pluralism: Science education for sociopolitical action. *Science Education, 83*, 775–796.
Kant, I. (1956). *Werke Band IV* [Works Vol. 4]. Wiesbaden: Insel Verlag.
Kurz, A., Elliott, S. N., Whby, J. H., & Smithson, J. L. (2009). Alignment of the intended, planned, and enacted curriculum in general and special education and its relation to student achievement. *Journal of Special Education, 44*, 131–145.
Lee, S., & Roth, W.-M. (2001). How ditch and drain become a healthy creek: Representations, translations and agency during the re/design of a watershed. *Social Studies of Science, 31*, 315–356.
Lee, S., & Roth, W.-M. (2003). Science and the "good citizen": Community-based scientific literacy. *Science, Technology & Human Values, 28*, 403–424.
Levinas, E. (1978). *Autrement qu'être ou au-delà de l'essence* [Otherwise than being or beyond essence]. The Hague: Martinus Nijhoff.
Marx, K., & Engels, F. (1958). *Werke Band 3* [Works Vol. 3]. Berlin: Dietz.
McGinn, M. K., & Roth, W.-M. (1999). Towards a new science education: Implications of recent research in science and technology studies. *Educational Researcher, 28*(3), 14–24.
McMahan, K. (2008, August 12). Environmentalism in North Carolina K–12 education: Science or activism. *Carolina Journal Online*. http://www.carolinajournal.com/articles/display_story.html?id=4917. Accessed 14 Feb 2011.
Nietzsche, F. (1954). *Werke Band 3* [Works Vol. 3]. Munich: Hanser.
Noddings, N. (2003). *Caring: A feminine approach to ethics and moral education*. Berkeley: University of California Press. (First published in 1984)
Posch, P. (1993). Research issues in environmental education. *Studies in Science Education, 21*, 21–48.
Ricœur, P. (1990). *Soi-même comme un autre* [Oneself as another]. Paris: Éditions du Seuil.
Romano, C. (1998). *L'événement et le monde* [Event and world]. Paris: Presses Universitaires de France.
Roth, W.-M. (2003). Scientific literacy as an emergent feature of human practice. *Journal of Curriculum Studies, 35*, 9–24.
Roth, W.-M. (2007a). Emotion at work: A contribution to third-generation cultural historical activity theory. *Mind, Culture, and Activity, 14*, 40–63.
Roth, W.-M. (2007b). Toward a dialectical notion and praxis of scientific literacy. *Journal of Curriculum Studies, 39*, 377–398.
Roth, W.-M. (2008a). Agency and passivity: Prolegomenon to scientific literacy as ethico-moral praxis. In A. Rodriguez (Ed.), *The multiple faces of agency: Innovative strategies for effecting change in urban school contexts* (pp. 135–155). Rotterdam: Sense Publishers.
Roth, W.-M. (2008b). Bricolage, métissage, hybridity, heterogeneity, diaspora: Concepts for thinking science education in the 21st century. *Cultural Studies of Science Education, 3*, 891–916.
Roth, W.-M. (2010). Activism: A category for theorizing learning. *Canadian Journal of Science, Mathematics, and Technology Education, 10*, 278–291.
Roth, W.-M. (2013a). To event: Towards a post-constructivist approach to theorizing and researching curriculum as event*-in-the-making. *Curriculum Inquiry, 43*, 388–417.
Roth, W.-M. (2013b). Toward a post-constructivist ethics in/of teaching and learning. *Pedagogies: An International Journal, 8*, 103–125.
Roth, W.-M., & Barton, A. C. (2004). *Rethinking scientific literacy*. New York: Routledge.

Roth, W.-M., & Désautels, J. (Eds.). (2002). *Science education for/as socio-political action*. New York: Peter Lang.
Roth, W.-M., & Désautels, J. (2004). Educating for citizenship: Reappraising the role of science education. *Canadian Journal of Science, Mathematics, and Technology Education, 4*, 149–168.
Roth, W.-M., & Lee, S. (2002). Scientific literacy as collective practice. *Public Understanding of Science, 11*, 33–56.
Roth, W.-M., & Lee, S. (2004). Science education as/for participation in the community. *Science Education, 88*, 263–291.
Roth, W.-M., & McGinn, M. K. (1997). Deinstitutionalizing school science: Implications of a strong view of situated cognition. *Research in Science Education, 27*, 497–513.
Roth, W.-M., & Radford, L. (2011). *A cultural-historical perspective on mathematics teaching and learning*. Rotterdam: Sense Publishers.
Roth, W.-M., & Tobin, K. (2010). Solidarity and conflict: Aligned and misaligned prosody as a transactional resource in intra- and intercultural communication involving power differences. *Cultural Studies of Science Education, 5*, 805–847.
Roth, W.-M., Tobin, K., Elmesky, R., Carambo, C., McKnight, Y., & Beers, J. (2004). Re/making identities in the praxis of urban schooling: A cultural historical perspective. *Mind, Culture, & Activity, 11*, 48–69.
Suchman, L. A. (1987). *Plans and situated actions: The problem of human-machine communication*. Cambridge: Cambridge University Press.
van Eijck, M., & Roth, W.-M. (2009). Authentic science experiences as a vehicle to change students' orientation towards science and scientific career choices: Learning from the path followed by Brad. *Cultural Studies of Science Education, 4*, 611–638.
van Eijck, M., & Roth, W.-M. (2010). Toward a chronotopic theory of "place" in place-based education. *Cultural Studies of Science Education, 5*, 869–898.

Chapter 15
#OccupyTech

Kate Milberry

Abstract OccupyWallStreet grew rapidly from an internet meme and tent city in the heart of America's financial sector to a global revolt against neoliberal capitalism. Focusing on economic inequality and corruption in the banking industry, Occupy drew attention to the plight of the majority of people suffering under neoliberal globalization. Its slogan, "We are the 99 %," references the growing concentration of income and wealth among the top one percent of income earners in the United States. Within weeks, the protest had sclf-replicated, with occupations cropping up in 900 cities around the world. Evidently, it spoke a common language of hope, rage and refusal that had been unleashed by the Arab Spring almost a year earlier. The internet was crucial to the birth and proliferation of the Occupy Movement, enabling protestors to overcome the initial media embargo against OccupyWallStreet and began airing their concerns via social media. The #Occupy hashtags were powerful signifiers that enabled the ideas, sentiments and spirit of the protest to diffuse, evolving from a movement tactic into a global phenomenon. This chapter traces Occupy's roots in the recent history of internetworked social movements and examining its dual nature as a simultaneously virtual-physical phenomenon. It considers the essential role of tech activists in building the technical infrastructure of Occupy, using free and open source technology (FOSS) as well as corporate social media to bridge the online-offline divide. Finally, this chapter discusses Occupy as a distributed platform upon which a global super-movement is currently being built.

Keywords Internet • Social movements • Tech activism • Occupy • Philosophy of technology • Social media • Free software

---

K. Milberry (✉)
Faculty of Extension, University of Alberta, 10230 Jasper Ave NW,
Edmonton, AB T5J 4P6, Canada
e-mail: kate.milberry@gmail.com

L. Bencze and S. Alsop (eds.), *Activist Science and Technology Education*,
Cultural Studies of Science Education 9, DOI 10.1007/978-94-007-4360-1_15,

From an unlikely encampment of hippies, hipsters and homeless to a global revolt against neoliberal capitalism, OccupyWallStreet took many by surprise. Focusing on economic inequality and corruption in the banking industry, Occupy drew attention to the plight of the majority of people suffering under the neoliberal global order. Its slogan, "We are the 99 %," refers to the growing concentration of income and wealth among the top one percent of income earners in the United States. From the outset, Occupy bemused and befuddled the corporate mainstream media. Pundits refused to cover the initial encampment at Zucotti Park in New York City, instead spilling ink and filling the airwaves with a debate over the rationale for this non-coverage (Randall 2011). Meanwhile the protest had self-replicated, with occupations cropping up in 900 cities around the world (Adam 2011). Evidently it spoke a common language of hope, rage and refusal that had been unleashed by the Arab Spring almost a year earlier. Occupiers and supporters noticed the media embargo, and began airing their concerns via social media, most notably the microblogging website Twitter. Thus was born the #occupy hashtag and meme, powerful signifiers that enabled the ideas, sentiments and spirit of OccupyWallStreet to diffuse, evolving from a movement tactic into the global phenomenon it is today. This chapter surveys the birth of internetworked social movements seeking to create civilizational alternatives, tracing Occupy's roots and examining its dual nature as a simultaneously virtual-physical phenomenon. It considers the essential role of tech activists in building the technical infrastructure of Occupy, using free and open source technology (FOSS) as well as corporate social media to bridge the online-offline divide. Finally, it discusses Occupy as a distributed platform upon which a global super-movement is currently being built. Occupy's story of resistance and revolutionary imagining is one that can inform science and technology education as teachers prepare their students to critically engage with-and challenge-the world around them. In particular, the lessons of innovative use of internet technology drawn from #OWS can inspire students to OccupyEducation, in the classroom as well as the streets.

The Rise of Internetworked Social Movements

It is almost cliché to speak of Twitter™ revolutions and Facebook™-fuelled uprisings. Nevertheless, internetworked social movements have a history dating back to the Global Justice Movement, and its 'coming out' party: 1999s massive street protests against the World Trade Organization in Seattle, WA. This moment signalled the rise of the 'newest social movements,' organized around 'non-universalizing, non-hierarchical, non-coercive relationships based on mutual aid and shared ethical commitments.' Activists and affinity groups within these movements engage in *prefigurative* politics, modelling the change they sought to bring about in the broader society (Day 2005, p. 9). The newest social movements are further characterized by their general opposition to capitalism as the dominant mode of socio-economic organization, as well as their use of the internet as space and tool for communication, organization and mobilization. Structurally, the newest social movements tend to resemble the internet: nodal, networked, decentralized and leaderless, which has made them resilient in the face of state repression.

The Global Justice Movement was not the first globally networked social movement, having roots in the earlier transnational fight against the Multilateral Agreement on Investment. Nor was it the first movement that coalesced around technology: the Free Software Movement that emerged in the mid-1980s focused on the inherent freedoms embedded in free and open source software (FOSS). In the mid-1990s, the Zapatistas were the first social movement to rally global support using computer networking. Internetworked social movements differ in that they organize through and on the internet, using technology that mirrors many of their social justice objectives. Tech activists have been central to these movements, facilitating the novel combination of interactive digital technology and social justice activism, and bridging the divide between geek and activist communities online, and between users and designers of internet technology. Self-identified as geeks, tech activists adopted the philosophy of the Free Software Movement, with its ethos of freedom and culture of collaboration, even as they (re)constructed the internet, using FOSS to build the digital infrastructure of the newest social movements. There is an explicit understanding among tech activists that the both technology they use, and the open source process by which it is constructed is 'deeply political,' prefiguring the progressive social change they want to see in society (Henshaw-Plath 2001).

Prefigurative Politics and Technology

The politicization of technology is a hallmark of the global justice movement. It is evident in the way tech activists design the democratic goals of the movement into the very technology used to pursue those goals. This is an example of 'prefigurative politics,' a defining characteristic of the global justice movement. John D.H. Downing (2001) defines "prefigurative politics" as "the attempt to practice socialist principles in the present, not merely to imagine them for the future" (p. 71). It draws upon anarchist philosophy, with its foundational idea that the means and methods employed for achieving a goal must be consistent with the goal itself. Within the Global Justice Movement, as with Occupy, internal process takes precedence over external demands; thus both movements emphasize the enactment of values such as gender and racial equality, environmental sustainability and economic justice, which also represent the movement's future goals.

Extending the idea of prefigurative politics to technology, the transformative potential of free and open source software is evident, insofar as it is used to build an alternative technical infrastructure and inculcate new social relations. FOSS is prefigurative in that it fosters democratic practice and actualizes many of the goals and values global justice activists strive for, including freedom, participatory democracy, autonomy, self-organization, decentralization, collaboration, and mutual aid, all of which contribute to the greater objective of freedom. In this way, FOSS both anticipates and embodies an alternative to the present mode of social organization. "Are we enabling of movement work or are we itself movement work?" asks one tech activist. "I do think we are in some ways acting in the way we want organizations to be structured at the same time as building infrastructure" (interview, Tufted Puffin 2008). Many movement projects coalesce around FOSS, such as Indymedia, the online media making network built on FOSS, radical tech collectives like Riseup and Resist!, and a variety of communication solutions that emerged from Occupy, discussed below.

Hacking the Technical Code

The radical potential for social change lies in the technical code of free and open source software, which is tech activists' software of choice. FOSS both reflects and advances the social justice goals of the newest social movements, contrasting the social and technical requirements of capitalism—profit, competition, control, exclusivity, inequality and individualism—that are condensed in the technical code of capitalism (Feenberg 2002). The technical code of capitalism relies on a technological rationality that aligns the construction and interpretation of modern technology with a system of domination. Thus "capitalist hegemony is an effect of its code" (p. 76). In their design and creation of FOSS applications, tech activists are reconstructing the internet in the image of the better world they seek. They do so by altering the technical code of the internet. Whereas earlier constructivist notions, like momentum (Hughes 1994) and path dependency account for certain technological trajectories, the technical code refers to the values and concerns that prevail in the design process, and concretize in the technology itself. "Technical codes define the object in strictly technical terms in accordance with the social meaning it has acquired. These codes are usually invisible because, like culture itself, they appear self-evident" (Feenberg 1999, p. 88). In capitalism, the technical code translates dominant social interests into technical terms, invisibly sedimenting "values and interests in rules and procedures, devices and artefacts that routinize the pursuit of power and advantage by a dominant hegemony" (Feenberg 1991, p. 14).

The technical code reveals an opening, however, engendering resistance in those shut out of the design process whose needs are not met by the current technical regime. In response to the internet as an increasingly surveilled, censored, controlled and proprietary space, FOSS developers, in particular tech activists, have begun to hack the technical code of the internet, reclaiming the network as a free and open space for democratic communication and action, a plane of equality and non-discrimination, and a tool for liberation. Tech activists therefore comprise a "recursive public"—an independent collective that challenges power through the production of "actually existing alternatives," contributing to a new "moral and technical order" (Kelty 2008, pp. 3, 301). What distinguishes recursive publics is "their focus on the radical technological modifiability of their own terms of existence" (p. 301). The free software movement is an exemplar of a recursive public. Tech activism as it arose in the global justice movement and migrated to the occupy movement, offers another.

The Repertoire of Electronic Contention

Like the social justice movements it supports, tech activism ebbs and flows according to a complex configuration of social, political and economic opportunities and constraints. Tech activists once again reformulated as a recursive public, supporting Occupy from its inception as a movement tactic through its evolution into a global movement. While they built upon the existing 'repertoire of electronic contention,' they nevertheless added to the virtual toolkit used to support social justice work both online and off. A repertoire of contention comprises the ways in which people work together to advance shared interests (Tilly 1995, p. 26). This includes well known social movement tactics such as sit-ins, boycotts and protests. With the rise of internetworked social movements, however, the concept has been used to understand how these tactics have been adapted in cyberspace (Meikle 2002). Cyberactivism, variously called hacktivism, electronic civil disobedience or cyberjamming, has a repertoire of *electronic* contention, including cyberpetitions, virtual protests, sit-ins and blockades, email bombs, web hacks, parody sites, computer viruses and the distributed denial of service (DDOS) attacks made infamous by the hacker group Anonymous (Costanza-Chock 2004). The tech activist repertoire of electronic contention includes building and maintaining websites, wikis, mailing lists, servers and mirrors. It also involves tech training, and the establishment of temporary media centres, hacklabs and squats (Obscura 2005). Uniquely, it involves the design, development and customization of free and open source software to meet the special needs of social justice activists as well as the construction of online spaces that embody and advance movement goals.

Occupy, like the global justice movement before it, would not have been possible without the internet. The very idea for an occupation of the American financial district was hatched in a series of email exchanges between the founder and editor of *Adbusters*, a Canadian magazine whose goal is to "advance the new social activist movement of the information age" (About Adbusters). In June of 2011, they sent an email to subscribers stating: "America needs its own Tahrir" (Schwartz 2011) and registered the domain occupywallstreet.org. The magazine created the iconic image of the protest—a ballerina poised on the famous "Charging Bull" sculpture near Wall Street in New York City—which would go viral along with the Occupy meme. The poster read: "What is our one demand? Occupy Wall Street. Bring tent." On September 17, about 5,000 people from all walks of life gathered in Zuccotti Park, which was quickly renamed Liberty Plaza, and struck camp. On the heels of the Arab Spring was born the American Fall.

The electronic repertoire of contention deployed for Occupy consisted of both proprietary and FOSS web tools. Occupy organizers relied heavily upon the commercial web, including corporate social media services like Facebook™, monopoly suites such as Google™, and microblogging sites like Twitter™, as well as proprietary (if free) email, list servs and blogs. "In the months leading up to the first occupation, and in the year afterward, Occupy established an online presence unmatched in the history of social action, leveraging multiple online spaces to stage protests and to generate a distinctive counter-public and alternative polity" (Massey and Snyder 2012, np). Facebook™ was useful for spreading information and generating interest in OWS, providing an agile and accessible platform on which to organize. By connecting potential supporters and distributing information, Occupy-related Facebook™ pages helped facilitated both the creation of local occupations and the organization of protests and marches (Caren and Gaby 2011). It was Twitter™, however, that broke the corporate mainstream media embargo. The hashtag #OccupyWallStreet, the first of more than 100,000 Occupy-related hashtags (Dugan 2011), allowed for "the spontaneous assembly of strangers on Twitter™ and other internet platforms" (Massey and Snyder 2012). Many of these were city-specific, such as #occupyboston and #occupyvancouver, as nodes linked in to the emerging online network of occupations. The top ranked Occupy hashtags were #occupywallstreet, #ows #occupywallst and #occupy, reflecting the meme-like quality that characterized the movement early on and appeared to give it staying power.

There are negative consequences to using commercial web tools, especially for activists, who are often subject to state surveillance as part of the predictive policing used to contain large scale contemporary protest (Milberry and Clement, forthcoming). It is well known that law enforcement uses internet to monitor activists (Rawlinson 2012). Google's™ *Transparency Report* shows that "government surveillance of online lives is rising sharply," with the US topping the list of states around the world requesting user data (*BBC News* 2012, np). Social media have been a boon to police, allowing them to easily monitor protest activities, including the dates and times of meetings, fundraisers and rallies, and gather information on protest activities in real time. According to the US Justice Department when trying to get evidence from social networking sites, law enforcement found that Facebook™ was "often cooperative with emergency requests" while Twitter™ had more privacy protective policies (Hoffman 2010, np). In a recent case, a New York court ordered Twitter™ to hand over the tweets of an Occupy protester charged with disorderly conduct during a mass arrest on the Brooklyn Bridge (Ax 2012). It was a cautionary tale—and not the first—that activists must be careful of the trail they leave in cyberspace, as it one day might be used as evidence against them.

OccupyTech

Tech activists are keenly aware of the political nature of technology, and the potentially serious implications of using the corporate web: "In the Facebook era, we seem to be faced with the difficult conundrum of activists using tools fundamentally opposed to their goals...The best thing we could do as part of the Occupy Movement is make people aware that technological choices reflect political choices as well" (Ross 2011). Since Occupy's inception, tech activists have gathered online and face-to-face to create movement-specific solutions. Websites, although not the latest or sexiest 'killer app' are the glue of the internet and have served an important anchoring function for Occupy. While most of the 1,500 occupations rely on Facebook for their online presence, the larger ones have their own websites. These are built by tech working groups using open source content management systems (Wordpress) and web hosting services (Github). Aside from the local occupations, there are also hundreds of occupy-related websites, devoted to a range of themes (e.g. OccupyResearch), practices (e.g. OccupyDesign), and social ills (e.g. OccupyStudentDebt).

A few websites in particular were key to Occupy's early success, helping the fledgling movement coalesce online. Launched and operated by an organizer in Philadelphia, occupywallst.org describes itself as the "unofficial *de facto* online resource for the growing occupation movement happening on Wall Street and around the world." Like many Occupy tech projects, it is run by an "affinity group committed to doing technical support work for resistance movements" (occupywallst.org). It acts as an online gathering space and archive, with a newsfeed providing movement updates and an "info tent" featuring links to key occupations, allies, events, campaigns and founding documents. The site promotes interactivity through its internet relay chat (IRC) channel and discussion forum. The LiveStream page links to Occupy-related footage streamed in real time from around the world, while the HowTo page is the playbook for setting up new encampments. The Maps page features the custom built OccupyMap, discussed in more detail below.

New York City General Assembly (nycga.net) is the official website of the working groups that comprise Occupy Wall Street—the first, and arguably the biggest and best organized node in the Occupy movement. It is run by tech activists in TechOps—the Technology Operations Group that supports the online communication and organization needs of #OWS and the NYCGA. "We seek to provide online tools that promote participation among occupiers and beyond by extending communication streams and promoting the exchange of information" (NYC General Assembly 2011). TechOps' choice to use free and open source software in building and maintaining the technical infrastructure of #OWS is political. "We need to own the means of production, which is a very socialist, Marxist idea. But that idea is actually taking place and has been taking place over the past 20 years on the internet, in the technology world," explains tech activist Drew Hornbeim (Occupy Brookly TV 2012). The essential freedoms embedded within free and open source software prefigure social change when applied to social relations. Says Hornbeim of FOSS:

> It is free—doesn't cost any money, it's gratis. It's libre, as in freedom—you can do whatever you want with [it]. And it's open, meaning that anyone can look at the components that make it up, can audit the source code, can see what's going on, how the product works, what it does. With those principals, taken outside the internet—that's the road to solving the many problems that we face today (ibid).

The general assembly (GA) is a defining characteristic of the Occupy movement. It is a directly democratic practice that has roots in anarchist practice. The GA was a unique experiment in mass participatory decision making, operating on a consensus basis, with assemblies sometimes reaching into the thousands. "Unlike at rallies, where protesters convene to listen to speeches, we directly participate in relating the needs of our movement ourselves. Direct democracy is integral to the process of articulating what we the 99 % are asking for, what we want as a people" (NYCGA FAQ n.d.). A daily ritual of the encampments, GAs were run by a rotating facilitation team, hearing proposals from working groups and prioritizing traditionally under-represented groups.

Another key website is occupy.net, a hub of tech activism and a focal point for the development of movement-specific contributions to the repertoire of electronic contention. "Occupy.net provides people with software tools that align with the values of the #occupy movement. All of the tools offered here are free/libre/open source: part of the global information commons, maintained by communities, not corporations." One of these projects is the OccupyDirectory, a public listing of all known physical occupation sites, culled from publicly available sources. Tech activists within Occupy "identified a need to standardize the data collected, liberate it from Google, and assist data submission and editing with form validation, editorial workflows, and community participation." (Occupy Directory *About*). More than a year after the birth of Occupy Wall Street, there were 1,495 occupations listed. The data is freely available, and the directory intended to be a service layer upon which anyone can build other web applications.

The Federated General Assembly is an ongoing project of occupy.net. Its goal is to digitize and automate the face-to-face process of the general assembly, creating a distributed social platform that promotes trust and embodies the values and meaning of the movement. Like the global justice movement before it, Occupy is unique in the history of modern social movements in that it has refused to issue demands, beyond the very general demand of an end to corporate exploitation of people and the planet. What participants have focused on instead is the process by which they will create a better world—hence the centrality of the GA. With the inevitable eviction of 'bricks and mortar' occupations, however, came the redoubled importance of the internet for evolving movement.

Achieving good process is hard, even under the best of circumstances. And Occupy is hardly the best of circumstances: *many* millions of voices representing various forms of involvement, experience, identity, and (of course) desperation, all spread across a huge geographic area. The movement is absolutely committed to creating substantial, far-reaching change, but is equally committed to maintaining the grassroots process. Satisfying both goals at once is a daunting challenge. Scaling participatory process to thousands in an individual occupation is already a herculean task; scaling it to the millions who would participate in the movement as a whole requires a major leap in organizing and communication techniques (Federated General Assembly n.d.).

With the FGA project, tech activists envision a new social web, a holistic and integrative platform that combines "community organizing techniques and ideas, lessons and patterns from social networks, web standards and best practices, all together with the very real ecosystem of Occupy itself: occupations and their working groups, the values and principles, and all the coordination and communication challenges" (Federated General Assembly n.d.). Based on the FOSS content management system Drupal, the FGA seeks to reflect the lived experience of Occupy and to nurture meaningful communication in a globally distributed movement (FGA Introduction). Still under development, the FGA is intended to be a next generation General Assembly site capable of being customized and deployed by local Occupy groups. Building upon the networked model originating with Indymedia, each installation will become a node in the digital confederation of Occupy, enabling the sharing of content, strategies and ideas, and anchoring locals while interconnecting them in the global "network of trust."

Disruptive Technologies: Building Tools for Revolution

In addition to creating websites and other web projects to support Occupy's process-based social justice work, tech activists contribute to the repertoire of electronic contention by developing movement-specific web and mobile applications. These are disruptive technologies that both incorporate and actualize their social justice values and goals. OccupyMap is an activist mapping tool developed by TechOps using Ushahidi, a FOSS project originally built for use in crisis and disaster response efforts. Combining social justice activism, citizen journalism and geospatial information, OccupyMap enables activists to publish reports on Occupy actions and events worldwide, acting as a source of up-to-the-minute news while creating an archive of the movement's history (map.occupy.net). People can contribute via Twitter by tweeting their location with links to news stories, videos or photos to @Occupy_Map, or by submitting their content through the web. During the height of the protests,

> [t]hese crowdmaps visualized Occupy participants and camps as discrete elements that aggregated to form a global phenomenon. They associated people, texts, images and videos with particular places, constructing hypergeographies of action and potential. Animated timeline features encouraged users to visualize themselves and local events as part of a process of "#globalchange." (Massey and Snyder 2012)

OccupyMap represents the evolution of online movement journalism that began with Indymedia and its open publishing software, enabling eyewitness accounts from inside protests to be uploaded to the internet and published without editorial gatekeeping. Like Indymedia before it, anyone can post, making OccupyMap "a place to post first-hand accounts of events related to the global Occupy Movement."

Another Occupy-specific technology is Kune, a distributed social network built on FOSS and designed around group collaboration. Kune, which means *together* in Esperanto, is a response to the problem of using free commercial internet services and web applications, or having to pay a technical expert in order to get an internet experience free from state surveillance and corporate data mining. It attempts to "extrapolate the philosophy of collaborative development in free software projects to other initiatives...and to promote social movements. Thus, its internal logic is the seeking and cooperative building of openly participative solutions, collaborating in a horizontal way" (Kune n.d.). In other words, Kune is an attempt at creating a "free/open distributed ecosystem of resources, methods, designs, tools and knowledge that can enable a free/open society" (Kune 2012). In this way, tech activists create prefigurative technology that is consonant with the movements in which they participate.

Communications

The internet was critical to Occupy's success from its inception, not only as a tool for organizing and communicating internally, but also as a broadcast medium for public outreach. As mentioned, Twitter™ was important in the early days of the encampment due to a corporate mainstream media blackout. When news of protesters occupying the heart of America's financial district failed to make headlines, people began tweeting about it. Key hashtags included #ows, #occupywallst and #occupywallstreet though thousands of Occupy-related hashtags would emerge. "In the months leading up to the first occupation, and in the year afterward, Occupy established an online presence unmatched in the history of social action, leveraging multiple online spaces to stage protests and to generate a distinctive counter-public and alternative polity" (Massey and Snyder 2012). Only after Occupy became a social media sensation did the corporate news media pick up on it, after which it was a major news item. As well as broadcasting information to people outside the physical encampments, Twitter™ was a source of on-the-ground communication, enabling protesters to communicate breaking decisions, police movements and calls for support. Although it is a commercial enterprise, Twitter™ is modelled on TXTmob, a text messaging system designed by tech activists to allow rapid, anonymous communication during protests (Henshaw-Plath 2008). Unlike TXTmob, however, Twitter communications are not anonymous; in fact, Twitter has become a valuable surveillance tool for law enforcement. Although Twitter™ tries to protect its users' privacy, in one high profile case a New York criminal court ordered the company to surrender the tweets and data of an arrested Occupy activist (Williams 2012).

In addition to broadcast-style communications, the need for internal communication within the movement also arose. Inter-Occupy, a web-based conference call system, evolved to foster communication among Working Groups and local General Assemblies across the movement. Interoccupy.net grew from an unwieldy conference call with more than 100 participants representing 40 occupations in October 2011 to a sophisticated website coordinating multiple weekly calls organized around themes. Infused with the spirit of the movement, Inter-Occupy uses direct democratic and horizontal decision-making processes to facilitate the calls (About InterOccupy). Thus is an attempt to reproduce "the mechanisms of communication and organization which are practiced in the movement at a local level, including straw polls, twinkles, direct responses, and points of process"

(*InterOccupy FAQ*). The result is the technical translation of embodied movement values, principles and process to the virtual realm. Layering atop local Occupy networks, interoccupy.net continues to foster movement-wide communication, even after the coordinated mass evictions in November 2011, when police dismantled camps and evicted occupiers. Since then, interoccupy.net has fostered the growth of an interconnected yet distributed global movement, helping to organize other direct actions including the West Coast Port Shut Down and the May Day "general strike" (Donovan 2012). More than a year later, interoccupy.net's calendar is dotted with upcoming calls and events and its newswire is populated with recent Occupy-related news stories, suggesting the movement is very much alive.

From Technology to Technique: Prefiguring Change

The examples above demonstrate how tech activists 'occupied' technology, adapting and creating technology that at once reflected and advanced their vision of progressive social change. The internet provided the technical platform on which the Occupy movement was built and was essential for its growth as a global phenomenon. However, while free from the brutality and control that characterized the state's response to the physical protest, the internet as a space and tool for organizing came with its own problems. It enabled state surveillance, as discussed above; it also limited participation to those with online access. Importantly, however, some of the affordances of the FOSS tools adapted and created to help foster the movement online translated into offline practice. Open source values such as freedom, autonomy, decentralization, collaboration, cooperation, voluntarism and mutual aid were essential in the building and maintaining of the physical encampments. After the striking of tents, occupations rapidly cohered into functioning communities complete with kitchens, libraries, medical care, and skills sharing run by autonomous yet coordinated working group. Self-organized by local activists, individual occupations operated on a voluntaristic economy, each developing to different levels of scale and sophistication. In the spirit of open source they all adopted and modified the New York model, which was itself inspired by the knowledge, resources and ideas garnered from the uprisings in Spain, and the Arab Spring. Independent yet connected, encampments shared a mode of governance based on consensus and direct democracy, with the General Assembly as the central decision-making body.

Because the model of Occupy, including its directly democratic mode of governance, was nodal and replicable, this enabled the original Wall St. protest to reproduce with no formal membership, leadership or institutions. In this way, Occupy is more of a platform or interface, allowing other people and groups to build upon the existing infrastructure of the movement. Like an API (Application Programming Interface) that enables web developers to create an open architecture for sharing content and data amongst communities and applications, #OWS generated a template upon which the entire movement could build, developing, remixing and sharing content—including democratic practices. Occupy's function as platform rather than merely a protest site was key to the survival of the movement after the mass evictions, fostering its maturation into a lasting, if diffused, movement.

The numerous Occupy offshoots continuing today use the scaffolding of the movement to push it forward. OccupySandy was a particularly effective—and highly visible—remix that revived the skills and processes developed during the original occupation (Feuer 2012). Organized by #OWS veterans in New York, OccupySandy mobilized to help victims of Hurricane Sandy, which hit the Northeastern US about a year after Occupy's birth. OccupySandy was largely coordinated by Inter-Occupy and #OWS TechOps. Once again, social media were heavily relied upon, with extensive use made of Facebook and Twitter to post updates on what supplies and assistance were needed and in which affected areas. OccupySandy set up distribution sites at two Brooklyn churches, which also served as temporary shelters serving daily meals and providing clothes and blanket to hurricane victims. In the style of Occupy, communications, medical and reconstruction working groups formed. A "laterally organized rapid-response team" that some said rivalled established charities and FEMA in effectiveness, OccupySandy represents the ongoing evolution of the movement, the continued enactment of social and political beliefs (Feuer 2012).

Strike Debt is a coalition of Occupy groups, including OccupyStudentDebt, fomenting popular resistance to all forms of debt imposed on people by the banking system. "We are building a movement to challenge this system while creating alternatives and supporting each other. We want an economy where our debts are to our friends, families, and communities—and not to the 1 %" (strikedebt.org). Rolling Jubilee is one of Strike Debt's key campaigns. Described as a "bailout of the people by the people," it transforms Occupy's criticism of the injustice and corruption of the American financial system into the practice of debt forgiveness. Using money from donations, Rolling Jubilee purchases consumer debt for pennies on the dollar and then cancels it. "Debt resistance is just the beginning. Join us as we imagine and create a new world based on the common good, not Wall Street profits" (rollingjubilee.org). In contrast to the highly controversial bank bailouts—one of the original triggers for Occupy Wall St.—the people's bailout is an example of prefigurative politics that actualizes movement values in practice.

Conclusion

Despite the mass evictions of local encampments, the Occupy movement endures. Since its birth in 2011, it has morphed from a physical protest destination around which the original memes took shape, into a diffused movement that can reconverge—either online or in the streets—at any time. The movement remains a distributed phenomenon, connected in time and space via the internet: this flexibility has been its greatest strength. Tech activists and tech savvy Occupiers have tapped the internet for all that it has to offer, contributing to the electronic repertoire of contention that propels contemporary social justice activism generally and internetworked social movements in particular. Open source web tools and corporate social media have both been essential to Occupy—as much in its infancy as today, in its more mature incarnations. The open source nature of Occupy enabled it to be disseminated and recombined, creating a platform on which new and fertile resistances can be built using shared practices of direct democracy.

A simultaneously a virtual and physical phenomenon, Occupy laid claim to many contested terrains: the internet, the financial district, the corporate mainstream media—even a hurricane. Most importantly, Occupy captured the global imagination, building upon the hope inspired by the Arab Spring and the Spanish uprising, and galvanizing later resistances, including Quebec's Erable Printemps and Idle No More, a protest movement of Canada's aboriginal people. All of these are internetworked social movements building upon a shared model of direct democracy that infuses their technology and social practices. All share a commitment to ending the exploitation of human beings that the natural environment that supports all life. Through the internet, a technical platform that has enabled a social revolution in communication, they have found solidarity, connection and common cause, converging into a global super-movement, saying in distributed unity: 'No more.' The lesson for science educators is this: the line between the classroom and the "real world" is increasingly blurry. Socio-political realities impose upon contemporary education, regardless of any efforts to sanitize curricula. Teaching students to integrate formal learning with lived existence is essential if the social ills identified by Occupy are to be addressed. The internet and social media not only offer a bridge between these two worlds, they can help create a new, better world.

References

About Adbusters. (n.d.). Retrieved from http://www.adbusters.org/about/adbusters

About Interoccupy. (n.d.). Retrieved from http://interoccupy.net/about/mission/

Adam, K. (2011). Occupy Wall Street protests go global. *The Washington Post*. Retrieved from http://www.washingtonpost.com/world/europe/occupy-wall-street-protests-go-global/2011/10/15/gIQAp7kimL_story.html

Ax, J. (2012). Twitter, privacy advocates eye Occupy case after guilty plea. *Reuters*. Retrieved from http://ca.news.yahoo.com/twitter-privacy-advocates-eye-occupy-case-guilty-plea-192903560.html

BBCNews. (2012). *Spike in government surveillance of Google*. Retrieved from http://www.bbc.co.uk/news/technology-20319505

Caren, N., & Gaby, S. (2011). *Occupy online: Facebook and the spread of Occupy Wall Street*.

Costanza-Chock, S. (2004). The whole world is watching: Online surveillance of social movement organizations. In P. Thomas & Z. Nain (Eds.), *Who owns the media? Global trends and local resistance*. London: WACC & Southbound.

Day, R. J. (2005). *Gramsci is dead: Anarchist currents in the newest social movements*. Toronto: Between the Lines.

Donovan, J. (2012). How Occupy birthed a rhizome. *Waging Nonviolence*. Retrieved from http://wagingnonviolence.org/2012/09/how-occupy-birthed-a-rhizome/

Downing, J. D. H. (2001). *Radical media: Rebellious communication and social movements*. Thousand Oaks: Sage.

Dugan, L. (2011). #OccupyWallSt, #OWS or #Occupy? Over 100 K different hashtags used in Occupy Protests. *All Twitter*. Retrieved from http://www.mediabistro.com/alltwitter/occupywallst-ows-or-occupy-over-100k-different-hashtags-used-in-occupy-protests_b15109

Federated General Assembly. (n.d.). *Introduction*. Retrieved from http://projects.occupy.net/projects/fga/wiki/Introduction

Feenberg, A. (1991). *Critical theory of technology*. New York: Oxford University Press.

Feenberg, A. (1999). *Questioning technology*. London/New York: Routledge.

Feenberg, A. (2002). *Transforming technology*. Oxford/New York: Oxford University Press.

Feuer, A. (2012). OccupySandy: A movement moves to relief. *The New York Times*. Retrieved from https://www.nytimes.com/2012/11/11/nyregion/where-fema-fell-short-occupy-sandy-was-there.html?pagewanted=all&_r=0

Henshaw-Plath, E. (2001). *IMC-Tech summary for November 16th 2001*. Independent Media Center. Retrieved June 22, 2009, from http://archives.lists.indymedia.org/imc-summaries/2001-November/000028.html

Henshaw-Plath, E. (2008). *TXTMob gets subpoenaed: Data retention in the surveillance era*. Retrieved from http://anarchogeek.com/?s=TXtmob

Hoffman, M. (2010). EFF posts documents detailing law enforcement collection of data from social media sites. *Electronic Frontier Foundation*. Retrieved from https://www.eff.org/deeplinks/2010/03/eff-posts-documents-detailing-law-enforcement

Hughes, T. P. (1994). Technological momentum. In M. R. Smith & L. Marx (Eds.), *Does technology drive history? The dilemma of technological determinism*. Cambridge, MA: The MIT Press.

Kelty, C. (2008). *Two bits: The cultural significance of free software*. Durham/London: Duke University Press.

Kune. (n.d.). *About*. Retrieved from http://kune.ourproject.org/about/

Kune. (2012). *New release of collaborative distributed social network Kune: "Ostrom"*. Retrieved from http://kune.ourproject.org/2012/10/new-release-of-collaborative-distributed-social-network-kune-ostrom/

Massey, J., & Snyder, B. (2012). Occupying wall street: Places and spaces of political action. *Places: Forum of Design for the Public Realm*. Retrieved from http://places.designobserver.com/feature/occupy-wall-street-places-and-spaces-of-political-action/35938/

Meikle, G. (2002). *Future active: Media activism and the internet*. New York: Routledge.

Milberry, K., & Clement, A. (2014). Policing as spectacle and the politics of surveillance at the Toronto G20. In M. Beare, N. Des Rosiers, & A. Deshman (Eds.), *Putting the state on trial: The policing of protest during the G20 summit*. Vancouver: UBC Press.

NYC General Assembly. (2011). *GA guide*. Retrieved from http://cryptome.org/0005/occupy-guide.pdf

NYC General Assembly. (n.d.). *FAQ*. Retrieved from http://www.nycga.net/resources/faq/

Obscura, V. (2005). *From free software to street activism and vice versa: An introduction*. Retrieved September 28, 2009, from http://garlicviolence.org/txt/drkvg-fs2sa.html

Occupy Brooklyn TV. (2012). *Season 1, Episode 7*. Retrieved from https://www.youtube.com/watch?v=27-SWAbWloA

Randall, E. (2011). http://www.theatlanticwire.com/national/2011/09/media-non-coverage-occupy-wall-street-gets-lots-media-coverage/43013/

Rawlinson, K. (2012). Activists warned to watch what they say as social media monitoring becomes 'next big thing in law enforcement'. *The Independent*. Retrieved from http://www.independent.co.uk/news/uk/crime/activists-warned-to-watch-what-they-say-as-social-media-monitoring-becomes-next-big-thing-in-law-enforcement-8191977.html

Ross. (2011). What are the political ideas we, as techies, contribute to the movement? *Occupy Technology*. Retrieved from http://occupytechnology.org/discussions

Schwartz, M. (2011). Pre-occupied: The origins and future of Occupy Wall Street. *The New Yorker*. Retrieved from http://www.newyorker.com/reporting/2011/11/28/111128fa_fact_schwartz?currentPage=all

Tilly, C. (1995). Contentious repertoires in Great Britain, 1958–1834. In M. Traugott (Ed.), *Repertoires and cycles of collective action*. Durham: Duke University Press.

Williams, M. (2012). Twitter complies with prosecutors to surrender Occupy activist's tweets. *The Guardian*. http://www.guardian.co.uk/technology/2012/sep/14/twitter-complies-occupy-activist-tweets

Chapter 16
Trajectories of Socioscientific Issues in News Media: Looking into the Future

G. Michael Bowen

Abstract Socioscientific issues, such as global warming, constitute some of the most significant issues we face in our culture, yet they often appear to be poorly understood by the general public. Why might this be? The public knowledge of socioscientific issues in large part derives from the coverage of these topics in the news media – newspapers, radio and video broadcast (either in the traditional formats or on-line). This chapter, using various case studies, discusses common practices engaged in by journalists (and news media companies) – practices which are essentially within the very DNA of journalistic practice – that contribute to problems with accurate representation of socioscientific issues in the news media and the difficulties the general public has in understanding those issues when they use the news media as a source of information. In discussing the case studies questions arise such as if news media should be trusted to convey those topics, if news media reports should be used to teach subjects such as global warming to students and if so how, or even if students should be learning about those issues in science classrooms at all or if some other venue is more appropriate.

Keywords Journalism • News media • Accuracy • Socioscientific issues • Science education • Journalism education

---

This chapter is an update and considerable reorganization of "Insights on the media's practices and representations of (global warming) science: Confusing the public, educating school children?" in 2011, Volume 3(1), pp. 52–79 of the *Journal for Activist Science and Technology Education*.

G.M. Bowen (✉)
Faculty of Education, Mount Saint Vincent University, Halifax, NS, Canada
e-mail: gmbowen@yahoo.com

L. Bencze and S. Alsop (eds.), *Activist Science and Technology Education*, Cultural Studies of Science Education 9, DOI 10.1007/978-94-007-4360-1_16,

Introduction

> We consider a socioscientific issue to be one which has a basis in science and has a potentially large impact on society. (Ratcliffe and Grace 2003, p. 1)

Socioscientific issues – science issues with cultural, political, personal and moral dimensions – are omnipresent in our news media. Socioscientific issues include topics such as global warming, new drug development and testing, pollution issues, genetic engineering of animals and foods, drilling for oil in nature preserves and on and on.

Most people learn little about socioscientific issues in schools (in schools they are often referred to as STSE or "science, technology, society, and environmental" issues), particularly in the integrated fashion that actually makes them socioscientific issues. Apart from that, the majority of voters (looking at voting trends across age groups and the size of various age cohorts) were out of school for some years before current SSI issues – with the personal, political and social dimensions – came to be thought of as important. Thus, the general public's understandings of socioscientific issues emerge almost entirely from news media coverage. Why might this be an issue?

The "Black Box" of News Media

To most people the practices of science are a "black box" into which they have few insights, and they are generally aware of this. But what most don't realize is that the news media and its practices represent just a different type of "black box" with similar unknown practices; the public often generally accepts the practices news media engage in, particularly with science reporting, essentially without question. I believe that this is a major problem, one that undermines almost any attempt to sway public opinion on socioscientific issues through any form of activism and in this chapter I will present a number of reasons why I think this is the case. I deconstruct media reporting practices on socioscientific issues by drawing on my personal experience as a student in a journalism school as well as on numerous examples drawn from published and broadcast science news.

How I first became interested in socioscientific issues such as global warming science and my route to becoming a student in a journalism program are germane to the conclusions I have reached.

I can clearly remember the first time I read something that made me think about the issue of global warming. In 1984 I had completed a BSc in marine biology and was starting work on an MSc in behavioural toxicology. The sub-disciplines of biology I was most interested in (both then and now) were ecology and animal behaviour. At that time I subscribed to various nature and science magazines, read newspapers voraciously, and read journal articles both as part of my graduate work and my broader interests in science. I can recollect first reading about the issue of global warming in the later 1980s in an article in the magazine called *Seasons* which was published by the Federation of Ontario Naturalists. The article mentioned that increasing winter temperatures meant that there was probably not going to be a commercially viable downhill ski industry in Southern Ontario by the mid 1990s. Although I wasn't much of a skier, this article caused a profound impact on my thinking and a focus about the global climate. Since then I've attended to the research on global warming reading both secondary and primary source materials.

In the earlier, halcyon times of the 1980s global warming was not really on anyone's agenda. For instance, in a book focused on global environmental issues entitled "The Earth Report: The Essential Guide to Global Ecological Issues" (Goldsmith and Hildyard 1988) there were chapters dealing with the politics of food aid, nuclear energy after Chernobyl, acid rain and forest decline, and the availability of potable water. One other chapter, written by James Lovelock, dealt with "Man and Gaia" but devoted only a single page to the issue of the greenhouse effect. Keeping up on the writing about global warming in those times was not a particularly big endeavour and there was little or no public awareness of it being an issue.

Following my MSc I became a middle and high school science teacher and then, after completing a PhD, a professor of education. My interests moved from teaching kids about ecology to understanding issues involved in teaching teachers to even broader interests in the public understanding of science and what factors intrude on that understanding. My personal interest in global warming science and related issues led me to becoming interested in understanding what role the media plays in how the public understands socioscientific issues.

For this chapter I'm first going to conduct a brief review of the literature on news media and socioscientific issues (with somewhat of a focus on global warming, because of the preponderance of sources that have been produced on that topic recently). In the next two sections will be a discussion of the sociocultural embedded practices which influence the production of news stories. These will involve an examination of practices taught (or not) in a journalism school as well as narratives and data analysis detailing the practices enacted by working journalists. In the concluding section some of the overall implications of these issues for both the public and science classrooms will then be discussed.

The Media and Presentations of Science

There is a substantial body of literature critiquing the media and how it constructs "news" for the public as well as a specific body of literature critiquing how the media portrays socioscientific issues, particularly global warming science. For this review I will briefly summarize four main issues regarding the media and its portrayal of socioscientific issues: influence of the media, manipulations of the media, commercial interests of the media, and the lack of knowledge/incompetence in journalism. I readily accept the argument that this parsing of topics is personally derived and that others are indeed possible.

Influence of Media

For much of the public the news media (which these days would include the internet) is a significant source of information about science (Boyes and Stanisstreet 1992; Dispensa and Brulle 2003; Lewenstein 2001; Schibeci 1990) and plays a significant role in shaping the discourse on climate change (Boykoff 2008). This is problematic however as the shortcomings of news media reports about global warming contributes to public misunderstanding about global warming science (Boykoff and Mansfield 2008; Dispensa and Brulle 2003) and other socioscientific issues (Oreskes and Conway 2010). This affects more than adults as high school students often have a poor understanding of climate change science and have acquired most of their knowledge about global warming (Adams 1999; Gowda et al. 1997) and other socioscientific issues (Reis and Galvao 2004; Bencze et al. 2009) from the media.

In classrooms teachers use newspapers as part of their approach to teaching science (Jarman and McClune 2002; Kachan et al. 2006) and some researchers suggest that media reports of science should be used to augment science textbooks, particularly to help students develop "critical reading skills needed to interpret argumentative text" (p. 432) exposing them to science meta-language (Penney et al. 2003) reflecting an overall trend in the increase of use of media in classrooms (Yore et al. 2003). Further, given the various initiatives to provide students' active and ongoing internet access in classrooms (such as providing them netbooks or tablets), access to news media in classrooms is likely to increase. However, issues with how science is presented in the media, both in tone and accuracy, make this problematic.

Manipulations of the Media

George Monbiot (2006), in a book discussing global warming and its "deniers" in some detail, describes how a significant number of organizations which contribute to the spreading of disinformation about global warming, including to reporters and others writing in newspapers, received their funding from Exxon (p. 27). Presenting numerous examples, Monbiot concludes that "a total lack of scientific knowledge is no barrier to publication" (p. 23) such that when checked with authorities one published claim was responded to with the statement "This is complete bullshit" (p. 24). In the chapter "The Denial Industry" Monbiot details a campaign – essentially orchestrated by "professional deniers" – designed to influence public opinion by using websites, "fake" citizens groups, and media manipulation with the result that public pressure to deal with climate change has been considerably reduced. Monbiot argues that "[B]y dominating the media debate on climate change during seven or eight critical years...by constantly seeding doubt about the science...they have justified the money their sponsors spend on them many times over. I think it is fair to say that the professional denial industry has delayed effective global action on climate change by several years" (p. 39).

In their book detailing how a "handful of scientists obscured the truth on issues from tobacco smoke to global warming," Naomi Oreskes and Erik Conway (2010) argue that the mass media became "complicit" in undermining global warming science (p. 214) to such an extent that "reporting on climate in the United States became biased *toward* the skeptics and deniers". They argue that the presentation of "balance" in the majority of media articles (drawn from 1988 to 2002; reporting on a study by Boykoff and Boykoff 2004), despite consensus amongst scientists involved in climate science research, made it "easy for our government to do nothing about global warming" (p. 215). In part this occurred because providing that balance lends credibility to the minority side (Dearing 1995). The bias present in American news is even more significant when political leaders such as our Prime Minister state that they don't watch Canadian news – "I tend to watch mainly American news because I don't like to watch Canadian news" (MacCharles 2009) – so any improved coverage of socioscientific issues in Canada would likely have little influence on those involved in Canadian politics anyway. Alison Anderson (2009), concluded that the negative influence of public relations activities "has played a highly significant role in the climate change debate and claims-makers are employing increasingly sophisticated strategies to target the media" (p. 171).

Commercial Interests of the Media

The phrase "commercial interests" is used in the context of the media finding ways by which maximizing profit influences decisions made about how to conduct the business and practices of journalism. The influence of these "commercial interests" can be profound – "commercial interests of the mass media can be incompatible with the social responsibility one should expect from journalists" (Roll-Hansen 1994). These commercial interests drive reader/viewership and undoubtedly explains why there is a "focus on drama, aberration, and controversy in much reporting about science and technology [reflecting] the quest of journalists to make their articles more entertaining" (Nelkin 1987, p. 119; also McBean and Hengeveld 2000; Curtis 2007). It is not difficult to see how this plays out in news articles about global warming science, especially the tendency towards showing "balance" by journalists introducing a denier perspective into a majority of news articles (88 % of newspaper articles between 1988 and 2002 had a "denier" perspective included; Boykoff and Boykoff 2004; also see McBean and Hengeveld 2000; Curtis 2007) because by offering both perspectives the news media source is effectively broadening the potential range of consumers of that media (particularly given the current tendency for people to consume news media that parrots their own perspectives rather than challenging them with alternative perspectives (see Hindman (2009) for a discussion of this). The commercial interest of media might also result in the downplaying of global warming issues, or perhaps that introduction of "balance", because of a fear of the loss of advertising from industries who contribute to global warming (Gelbspan 2005; Monbiot 2007). One could also argue that radio and television documentaries on global warming are designed to dramatically entertain as much as they are to inform (McBean and Hengeveld 2000).[1] Although the literature I've cited describes the treatment by the media of global warming, such practices would also hold true for smoking, drug testing, nuclear energy, and other socioscientific issues as well.

The Competency and Practices of Journalists

There is a detailed literature discussing whether most journalists who report on science issues are equal to the task with respect to their own understanding of the topic. It should be noted that science and scientists being misreported and misrepresented by journalists is not a recent phenomenon; complaints of misrepresentations by journalists date back more than a century (e.g., Hyslop 1899, p. 696).

Misreporting of science is serious because it often does not occur for reasons of style but rather because of insufficiency in the backgrounds of journalists with regards to both the science itself as well as their understanding of science as a social practice. Recently, the results of the longitudinal "Interphone" study from the World Health Organization, which examined cell phone usage and cancer risk, was reported by some media as inconclusive, and by other media as finding that there is a link (Tyson 2010); clearly it cannot be both, and my reading of the articles suggests that it is a lack of understanding of both the science content and how it is reported that led to that problem.

There are other instances where lack of journalist knowledge on environmental topics has led to inaccurate reporting (Roll-Hansen 1994) with one research report suggesting that one story in six contained significant misreporting (Bell 1994), although one could argue that the inclusion of "denier" perspectives in 88 % of newspaper reports (Boykoff and Boykoff 2004) means that the "significant misreporting" rate is much higher. In Canada the "vast majority" of journalists covering science have no science training, and the structure of Canadian daily newspapers is "not supportive of the style of reportage required for quality science writing, nor of the development of such writers" (Saari et al. 1998, p. 61). One might well suggest that there is a serious shortage of science-dedicated reporters (McBean and Hengeveld 2000), and that this leads to inadvertent misreporting on science issues, such as failing to distinguish between scientific debate about small details and that about larger issues and by so doing presents the appearance of controversy where there really isn't any (ibid).

---

[1] In my journalism program I attended a talk by one of Canada's most prominent science journalists at which he indicated that he saw his role as a science journalist as being mainly that of entertainment, with informing the public about science being somewhat incidental rather than of necessity.

In Canada the media also tends to focus on the aspects of socioscientific issues where there is controversy and disagreement *no matter how minor* rather than the much greater areas in which there is agreement, leading to a "public perception of scientific uncertainty that significantly exceeds that perceived within the scientific community itself" (McBean and Hengeveld 2000, p. 11). The news media's tendency to write articles on climate change science during unusually warm periods (Shanahan and Good 2000) also likely misdirects the public on the consequences of global warming. Finally, in socioscientific issues such as global warming, how the press deals with issues such as scientific uncertainty, and represents them to the public, may very well lead to inaction by the public (Zehr 2000).

So, clearly the news media, whether through ineffective practices or manipulation by other agents, has a considerable effect on how the public perceives socioscientific issues and, consequentially, on what public policy develops. The question then arises, why do these issues occur? What is it about the culture of journalism, and how that culture develops, that leads to stories with these sorts of issues? In the following two sections, "Prelude to a News Story" and "Production of a News Story" I will explore parts of the culture and practices of journalism that lead to a presentation of socioscientific issues to the public that differs quite substantially from how the scientists who produced the research would understand the issues.

Prelude to a News Story

Every profession has a culture – implicit practices, particularly social ones, which are engaged with transparently without any real awareness of their existence. They are often accepted without question, without any interrogation regarding whether they are the best way of doing things. In that regard, journalism is no different than other professions. It has aspects of its cultural practices which are accepted as appropriate. This section discusses several of those accepted cultural practices in journalism, but problematizes those practices asking, implicitly, whether their influence perhaps biases the stories that emerge from the journalistic process.

The J-School Experience: Insights into Media Practices Reporting Science

During my sabbatical year I was interested in exploring my familial roots (my father, who passed away when I was 13, was a journalist; my mother studied radio broadcasting) so I spent time participating as a student in journalism school for the first half of the year-long program.[2] During those months I participated in courses teaching the

[2] Two years later I finished the program, graduating in 2013.

"basics" of print and broadcast journalism and then participated in a 6-week "radio workshop" which was responsible (with respect to content, production and reading) for providing the daily evening news for a local community radio station.

As a professor in education I involve my students in an integration of the concepts (derived from research) and the related practices/skills associated with effective teaching. Given the apparent similarities that journalism has to education (i.e., performing the role of conveying information to others in a manner they can understand it), I expected somewhat similar approaches in a journalism program. In fact, given the rather extensive critique of the practice of journalism present in both the research literature and trade books, I particularly expected a rather thorough discussion of those critiques in a journalism program so that practices criticized in current journalism were either not replicated. . .or at least were engaged with in full knowledge of what was being done. Much to my surprise this was not to be the case – in fact there were only a few circumstances where research on journalism was referred to either explicitly or implicitly – in general the journalism classes and workshop focused almost entirely on procedural methods with almost no conceptual perspectives whatsoever. There was also almost a complete lack of meta-thinking/talk or reflection about journalism or the role it serves in society in either the "basics" part of the program or in the workshop. In several instances, discussions with many of my faculty instructors made it clear that they themselves were unaware of any specific critiques of journalistic practices. This likely explains why we were often encouraged to engage in the very actions/practices that the literature identified as problematic.

As an example, the instructor of a radio workshop frequently said that a story had to "focus on one person, doing something, for a reason." This tremendously affected the types of stories that were covered because, as he forcefully pointed out early on, "process stories are boring." In other words, stories about individuals and what they were doing were encouraged and stories about ideas or events were discouraged (unless seen through the eyes of an involved individual). Problematically, science stories are often about ideas, ideas that formed from the practices of collections of scientists, so this meant that unless an individual scientist could be interviewed about a particular story, then the story itself wasn't approved. . .and generally a strong human-interest perspective was necessary in any story. So, for instance, a proposed story about the Atlantic launch of the Science Media Centre of Canada (an organization formed to improve the reporting of science stories in Canadian media) and the panel presentation/discussion with science journalists, with interviews proposed with several people (a science journalist, a science blogger, the SMCC director, etc.) was supplanted by a single interview with one of the science journalists participating in the panel discussion that coming evening which then, unsurprisingly, mostly focused on that journalist (as opposed to the SMCC launch). Science stories that were permitted were most often not about broad issues, but about the actions or findings of a single individual – in complete contrast to how science claims arise from the collective actions of (often many) individuals operating within a community of practitioners. This focus on "individual" scientists undermines and devalues the collective work that scientists do as a community and leads to an inappropriate understanding of the practice of science in the public (Charney 2003).

Modifiers and "Verbs of Saying"

I was also surprised by what topics were not covered in the journalism program... the very program designed to guide new journalists in what their practices were to be and how they were to think about those practices. For instance, courses did not really involve discussions about the nuances of using language or how to effectively use background sounds (and associated fade-in and fade-out and volume issues) in a radio news story to improve story impact. As a non-English major (having studied science) it was the lack of discussion around the subtleties in the use of language I found most surprising. For example, consider the nuances of using language to convey what others are saying (see Table 16.1).

In each bulleted point in Table 16.1 the third word, known as a "verb of saying",[3] was different. Reading each of those statements one by one the importance of which word is used in that third position becomes apparent, because each influences the

Table 16.1 Changing the "verbs of saying"

And she warns that if B.C. and Canada adopt a cap-and-trade system, domestic businesses will be out of step with the direction of U.S. regulation.[a]
And she feels that if B.C. and Canada adopt a cap-and-trade system, domestic businesses will be out of step with the direction of U.S. regulation.
And she says that if B.C. and Canada adopt a cap-and-trade system, domestic businesses will be out of step with the direction of U.S. regulation.
And she insists that if B.C. and Canada adopt a cap-and-trade system, domestic businesses will be out of step with the direction of U.S. regulation.
And she believes that if B.C. and Canada adopt a cap-and-trade system, domestic businesses will be out of step with the direction of U.S. regulation.
And she states that if B.C. and Canada adopt a cap-and-trade system, domestic businesses will be out of step with the direction of U.S. regulation.
And she argues that if B.C. and Canada adopt a cap-and-trade system, domestic businesses will be out of step with the direction of U.S. regulation.
And she claims that if B.C. and Canada adopt a cap-and-trade system, domestic businesses will be out of step with the direction of U.S. regulation.
And she suggests that if B.C. and Canada adopt a cap-and-trade system, domestic businesses will be out of step with the direction of U.S. regulation.
And she predicts that if B.C. and Canada adopt a cap-and-trade system, domestic businesses will be out of step with the direction of U.S. regulation.

[a]Original statement from the article "From carbon steam to cash flow: Companies implementing new technologies could see a profit in a cap-and-trade system" by Patrick Brethour posted in the Globe and Mail online edition 7:00 AM EDT ON 29/03/07

[3] My thanks to Robert J. P. Lyon, former English Department Head of Fergus High School, for introducing me to this term and discussing the issue with me. This discussion helped clarify my thinking on the significance of this issue.

interpretation of the subsequent statement somewhat and thus conveys a slightly different meaning. Consider, for instance, the difference when using "warns" as compared to "insists" – the first sounds helpful, the second somewhat desperate. Given the importance of choosing the appropriate term, one might expect that a journalism program would discuss that the accuracy of quotes and statements lies not just in the words being used, but also in the tone being set by the speaker and, just as importantly, how the journalist chooses to represent that tone. However no such discussion occurred either in the "basics" courses or in the radio workshop itself.

The following case example of problems with the use of such modifiers by journalists comes from a broadcast news item. It illustrates the way in which those "verbs of saying" can have a significant impact on how a statement by an interviewee is portrayed and subsequently interpreted. This excerpt comes from an interview broadcast on CBC Radio (December 21, 2012, The Current) on the topic "Should polar bears be banned from international trade?" The following comments by the host, 10:32 into the show, was a summary of the earlier description provided by Terry Audla, president of Canada's national Inuit association (Inuit Tapiriit Kanatami) about his recent visit to a UN body in Brussels. Audla had described presenting his case that the polar bear should not be added to the Convention on International Trade in Endangered Species in order to solve a problem with polar bear populations in other countries because the numbers and trends of the population of polar bears in Canada did not warrant such an addition. The host then made the following summary of Terry Audla's statements:

Host: So, you feel that the bear doesn't meet the criteria of endangered, and you also feel that the populations are being managed properly and that there is some financial benefit to be gained by the Inuit. Plus, I guess the preservation of their traditional way of life. You traveled to Europe. You were just in Brussels to make this case, how were you received? [underlining by author]

Audla, seemingly completely aware of how his data-derived arguments were being misrepresented by the host's summary, in the radio broadcast tried to break in a couple of times after the host made the statements with the word "feel" in them but he was ignored by the host who kept talking. Audla then waited until the host finished asking his question and responded with the following statement which corrected the statement by the host:

Terry Audla: [11:41] Well, I'm received quite well. They appreciated that they could put a face to an issue. And it's not that I feel that the criteria are not being met or that I feel that the polar bear population in Canada are healthy; I *know*, it's actual fact. When you consider that just five years ago the population estimate in Canada was at fourteen thousand, now the estimates today are at sixteen thousand – that's an increase in the number of bears. [*italics* indicates tonal emphasis by the speaker; underlining of significant "verbs of saying" by the author]

As can be seen from this exchange, what seemed problematic as far as Terry Audla was concerned was that statements he had made, which were derived from data and thereby represented scientific claims, were being reduced in significance of meaning by the host who was summarizing them as deriving from an affective domain – that Audla "felt" these things – which misrepresented Audla's population count based argument for not including the polar bear on the CITES list. Unsurprisingly, it appeared that Audla expected the science that he was describing should be accorded the same significance as the arguments provided by others and he made it clear that he did not accept the diminution of his arguments through the choice of verb made by the host of the show.

In the journalism program I participated in only a few of my 50 classmates had academic backgrounds within which such subtleties of language would have had the opportunity to arise as part of the subject matter. Based on my observations in the program, such a discussion was clearly warranted for the journalism program as I observed numerous instances where I felt that a stronger "verb of saying" was chosen than was appropriate; often in the aid of contributing to making a stronger statement than I felt the speaker in the recording intended. Do I think there was anything necessarily capricious in this? No, in fact I don't even think the journalism students consciously noticed what they were doing. But generating more drama (see Nelkin 1987; Curtis 2007), as media these days seems prone to do to generate a greater emotional reaction in the reader/listener, can distort the very issues that journalists are supposed to be objectively reporting on and the practice is misleading – there is no objectivity if emotional "impact" for dramatic purposes takes precedence over accuracy in conveying the meaning of the text. Both the content of what was spoken (known as the "referential function": Jakobson 1960) and the emotional state of the speaker (known as the "emotive function": Ibid.) are important components of effectively conveying the meaning of a statement. In fact, I would argue that conveying both (the textual meaning and the emotive meaning) accurately is requisite for the recipient to understand the meaning in the manner intended by the individual being quoted. Consequently, expecting that journalists have a conscious awareness of this issue and that journalism students should have their work examined to ensure they are accurately conveying the meaning and emotive tone of someone being quoted would not seem to be an unreasonable expectation. But such was not done.

Being the "Instant Expert"

In the radio workshop although we received feedback of some sort on every broadcast (not, note, every story either collectively or individually) from the supervisor/instructor each week we also received feedback from someone else, either a working reporter or producer, who listened to our show and critiqued each piece in a full-class discussion. One week one of these reviewers had just that week gone back to being a beat reporter after being a producer and host of his own show (because of organizational changes at his broadcaster). To our workshop he described his assignment as a reporter for later that day (90 min or so after providing feedback to us); he was to attend a government press conference and then write a story on it for later that afternoon about a topic about which he knew little. In discussing this assignment he made the following comment:

> I'm going to have to go and learn about rural economic development. I don't know anything about rural economic development, but I'm supposed to go and listen to the Premier speak on it. (recorded in class notes)

He also made other comments about this topic including how unprepared he felt to do this story and the responsibility he had for getting it "right." This theme of needing to be an "instant expert" within very constrained timelines was a constant refrain of the instructors throughout the "basics" courses taken in journalism school. Clearly, based on the comments of the above journalist and the constant emphasis on this in the j-school it seems that an acceptable part of journalism culture is to consider oneself someone who is capable of essentially being an "instant expert."

In the radio workshop it was obvious that this "instant expertise" often played out in the most superficial manner you can imagine, and as a consequence science issues (not to mention others) were not necessarily represented accurately. In one case I happened to be overhearing the recorded script and interviews for a story that had been "pitched" a couple of days earlier – the reporter's involvement in other activities meant that he had longer than usual to prepare the story (the only instance I'm aware of this happening) and so had more time than usual to conduct background research. That pitch derived from a sign that the reporter had seen in a window about a product that a company wanted to test on people who had already been treated for cancer. For the story the reporter interviewed a senior company official about the "trial" of the product (Which had advertised for curiously low numbers compared to trials one normally read about in the media) and then interviewed a local medical researcher about what conditions were necessary to effectively test a medical product.

However, while listening to bits of the interviews and recorded script as the story was being constructed I began to wonder about the story, because the tone of the piece seemed to suggest that the product being tested was something which could contribute to treating cancer. Curious, I asked the journalist if the product he was discussing was curative (i.e., treatment) or palliative (i.e., something intended to improve the quality of one's life as one recovered from cancer treatment), and then had to explain the distinction between those terms. What became obvious, 90 min or so before the broadcast deadline was that the reporter didn't actually know whether the product was curative or palliative and hadn't considered that issue. A follow-up phone call to the company president seeking clarification resulted in his understanding that the company's product was palliative resulting in his changing his story and script. My sense was that this was happening not because of the pressures of time for producing a piece (because the reporter had several days preparation time) but rather because of the journalist culture that *accepts* that being an instant expert, without necessarily doing any in-depth reading on the topic, is sufficient for being a reporter on a topic.

This belief in the acceptability of being an instant expert contributes, I believe, to many of the errors about science which appear in the North American media and exposes an implicit arrogance that what they can do under pressure and with short timelines is sufficient for informing the public. Further, one could also speculate that the lack of expertise (and the lack of recognition that lacking that expertise is a big problem) also contributes to poor questioning when conducting interviews for stories. I will note that there was little evidence that pre-interviews were required in the workshop (we were told they were, but most workshop participants never conducted pre-interviews before their pitches and comments were never made about that) nor was there ever any apparent mechanism to ensure that that stories were being checked factually, that copy stories accurately reflected original source material, that anyone had gathered sufficient background to write about the story authoritatively, or even that interviews were accurately represented in the brief segments that were chosen for the story. In general, all of this suggests that the "instant expert" who is implicitly trusted is an accepted part of journalism culture as conveyed by the journalism school – which also means that the ethical standards of the journalists, even in journalism school, are assumed to be sufficient[4] and are taken at face value.

The Reducibility of Complex Relationships

This lack of in-depth understanding of a topic is exacerbated by another "belief" that journalists appear to have: that any topic, no matter how complex, no matter how abstract, can be reduced to simple, concrete relationships that the general public can understand. This belief was voiced several times throughout the journalism "basics" courses, and even underlay some of the comments made by science journalists in the Science Media Center panel discussion (see below). Although I won't go into great depths here critiquing that belief, the idea that complex systems can be fundamentally and effectively understood through reduction to a few simple ideas or relationships misrepresents the very complexity of our world, and can lead

[4] I could, but will not, relate many instances which lead me to believe that such an assumption about the ethical standards being sufficient is completely unwarranted.

to great misunderstandings. In climate change comment threads it is quite common to see comments such as "How can there be global warming, it's only October and I'm seeing snow" which is of course, not at all inconsistent with the more irregular patterns in weather that global warming will result in. That there are a range of mixed outcomes in multivariate climate change models that portray complex systems seems to be seen as evidence by many that global warming is not happening, and the portrayal of simplistic systems in the media (including, for instance, the removal of qualifiers such as 'surface' and 'satellite' temperatures), one could argue, have contributed to this perspective. A final aspect of this that I find interesting is that this seems to be a perspective held more by North American news media than in other cultures.

Production of a News Story

In this section of the chapter I am (loosely) following the trajectory of how a socioscience-related news story is produced from the inception of the idea to the production and broadcast of the news item.

Demonstrating the Need for Science Journalist Experts in News Media

None of the issues I've described above or that in the literature about journalism are necessarily helped by trends in who becomes a journalist and what types of jobs they do when they become a journalist. In the recent class of 50 journalism students in the 1-year program I attended there were 3 with a background in science (notably all in biology) and 1 with a background in health science (i.e., nursing qualifications). This is, I'm told, not unusual and is consistent with most science reporters in Canada (for print or broadcast) having little or no background in science. At the Atlantic Canada launch for the Science Media Centre of Canada, an audience member asked the discussion panel (which had four current science journalists on it) if someone who was interested in becoming a science journalist should study science and then become a journalist, or whether they should become a journalist and then focus on learning to write about science. Their collective response essentially indicated that they didn't see that it mattered whether one had a science background and then became a journalist, or whether one's science literacy was developed on the job.

Although I'm not necessarily convinced of the equality of the routes discussed above, as I believe that many of the nuances of language and practice in science are less likely to emerge in the latter route, inadvertently one of the panelists in response to another question at this event highlighted an issue with regards to science journalism that effectively negates my concerns about the equality of those routes to becoming a science journalist because it demonstrated that even a knowledgeable science reporter has their stories ultimately chosen by an "unknown" other who would be even less likely to have a science background.

The following segment is a presentation made by a current national science reporter at the Atlantic SMCC launch which was broadcast on local radio. The broadcast of those comments was preceded by the radio announcer who described why there was a need for the Science Media Centre of Canada:

> The need to understand the science behind everything from invasive species to the latest treatment for MS has never been greater. Yet, the ability of the media to tell complicated science stories is being compromised by the demands of the 24 h news cycle, shrinking numbers of newsroom staff, and the fact that few media outlets have the luxury of allowing reporters to specialize in science reporting. (Originally broadcast 16 November 2010; transcribed from a streaming audio record at http://www.cbc.ca/maritimenoon/2010/11/telling-science-stories-phone-in-pain-management.html on February 14, 2011)

In the introduction the host highlighted that science stories are becoming more and more important, and that it is getting harder and harder to report on them effectively, concluding that the Science Media Centre of Canada will therefore fill an important role (similar to comments also made by the executive director of the SMCC at the live event; so it is possible that the news reader introducing the recording of the event was reading a script provided by the SMCC).

Following the news reader's introduction he played a recording of the panel of science journalists who were speaking to a live audience at the launch. The second panelist (identified below as "Panelist A") spoke about the difficulties associated with being a medical reporter. The narrative is interesting because it highlights that even the voice of the "expert" on reporting science can be over-ruled by an "unknown" other (in this case probably the show producer).

> Panelist A: But first I want to tell you a little story. And that is something that happened at the end of last week. Two fairly big medical stories were breaking late in the afternoon one day. One was to do with a new drug called {drug x} which is something that everybody has been waiting for. It's an oral, a pill you can take as a blood thinner. As opposed to having to take warfarin that you need to have your blood tested. It's a big deal. So with only a few hours I was trying to figure out should we get something up on this, how could we do that. At the same time I got word about a study that showed the first evidence of ability to detect lung cancer at an early stage. And of course lung cancer is the deadliest cancer because it's undetectable when it's early and treatable. So, two stories, up the pipe they go to Toronto, "What do you like?" Well, I was disappointed. Toronto wasn't as excited about the lung cancer story as I was so they were just going to do a little copy. So then we turn our attention to the {drug x} story,[5] which you know is an interesting story which will have impact for a

[5] It is worth noting that 3 months later the Canadian Cardiovascular Society released new guidelines for treatment of atrial fibrillation, which affects 250,000 Canadians, and in those recommended that "drug x" be prescribed instead of warfarin to reduce stroke risk because it has fewer [known] side effects (Picard 2011).

> lot of people. So then you've got to find your experts and say "What do you think about this? Is this, it feels big, is this really big?" and you know I talk to the author of the original research, talk to other experts that I know, some of whom may be here tonight. Uh, and, you know, the consensus was this is huge. People have been waiting for this. So this was more or less the story I did. And I have to say I think this was the first time in ten years of doing this that I did a "Ooh, there's a new drug on the market story". And I felt a little bit uncomfortable about that. And sure enough, the next day, I get an email from somebody who says, "We actually have some concerns about this research." And I thought "I knew there was going to be somebody out there." I just couldn't find them when I had three hours to mount a story. So I feel as though I didn't really get all the angles covered on that. And that is the worst feeling. You know, I hyped a drug. Uh, that now, somebody says "That could potentially harm someone. There's a signal that maybe there's an increased risk of myocardial infarction." That's a bad feeling as a health reporter. So, we do struggle to get it right, and there are a number of challenges that can impact our ability to do that.... My story is a minute and a half long. Well, what can I tell you about a really complex story in a minute and a half?"...It's a tough job to do even when you do it every day. But there are 100's of general assignment reporters out there who are asked to do this at the drop of the hat. "Here's this great breaking medical news story. Go get it." [They] don't know where to begin, don't know the right questions to ask, don't understand what the potential biases of the research are. All of those things. So I guess that's where I see the Science Media Centre is a very valuable organization for people who are dropped into what could be complex kinds of stories to tell.

Panelist B then asked "one of the things I'd like to know is, what is driving those decisions?"

> Panelist A: Uh-hmm. Ya ya. Um, in terms of story selection, it would make my job a lot easier. Sometimes it's who's on the desk, sometimes it's that there may be some relation to a previous story that was of interest, or maybe they don't want it because there was a previous story somehow tangentially related "Oh we've heard that before." So you can't really predict. And sometimes it's maybe my pitching. Maybe I'd didn't pitch it right. Maybe I didn't say the right things that made somebody say "Oh, that'd be good."
>
> Panelist B: And it would depend who you were pitching it to.
>
> Panelist A: Absolutely, yeah.
>
> Panelist B: So see there's a lot of behind the scenes stuff that makes it kind of tricky.

This anecdote was particularly interesting coming from such a senior journalist voiced in a public forum as it describes several issues that are common in science journalism. However, this sort of story is not uncommon; other journalists have related similar stories to me about being pressured by those above them (such as editors or producers) to produce stories, or slants in stories, that will lead to the most drama in the piece and, supposedly, most interest amongst the consumers irrespective of whether what they were asked to focus on represented the most accurate version of the whole story or not.

The statements made by Panelist A are interesting for several reasons. The first reason is the way in which the concept of "big" in reference to the story, as the reference is presented as if the concept of what was an important story was separate from the interests and focus of the journalist-panelist and whatever resources they had to learn about stories. This de-personalized presentation of why it was chosen from any number of potential stories available on that day would thus decrease any critique a listener might have as to why it was chosen, as the statement was made unambiguously. Consider, for instance, if the journalist had said "There were two stories that I thought were really big that I learned about late in the afternoon one day." In that instance an observer might both question why that journalist thought they were big and how the journalist came to learn about them. By using the statement that was used any impact of public-relations mechanisms in how that information comes to the attention of the journalist is considerably downplayed and, I would argue, falls out of consideration by the listening public, as does any bias of the journalist in their interests or background and how something might be considered a "big" story.

The second feature of the panelist's statement that is worth discussing is the process by which decisions are made about "stories" which may not reflect the more involved understanding of the person who will be responsible for the story.[6] In this current example two stories were pitched by the science specialist reporter to a seemingly unknown decision apparatus "up the pipe" in Toronto and the consequence was that one story was picked and the other (despite the interest of the reporter) was not – for reasons unknown to the journalist. It is unlikely, given the rarity of journalists or producers with science backgrounds,[7] that whoever made the decision had a science or medical background so the reporter's background and interest was essentially meaningless in this context. In general then what this means is that the availability of science specialist reporters is more-or-less meaningless if their expertise is ignored when choosing what stories are significant.

Apart from these issues, one also must wonder what would have happened if those two stories, rather than appearing on the same day, had appeared on sequential days. Would they have both run, one on each day? Science stories are not broadcast every day on newscasts, so was it just bad timing in the release of the journal articles? And even given that, if both are "big" stories (as the journalist stated) then what would the problem be with running both stories, but on different days? There would appear to be no reason not to run the stories on separate days, as a science story being "news" one day when the research paper (or press release) about it is released doesn't diminish its value on a subsequent day. Both "big" stories being released on a single day is serendipitous, and inconsequential with regards to their relative worth. . .yet, in the world of journalism both stories being released on the same day *does* influence the opportunity for the public to hear about both of them. In this particular instance I think of no reason that one or both stories could not run on different days. This "first past the post" and "fast out the door"

[6] Such a condition also exists in the "Copy Story Modification" section of this chapter where the author's more thorough understanding of the topics (through having read both Guardian newspaper stories in detail and other background knowledge) were over-ruled by the host/producer who had read neither.

[7] A science journalist who hosts a science show on a speciality network related to me that only 1 of his 14 producers (for different segments of the show) had a background in science. Given this, it is highly unlikely that a news show producer "up the pipe" for general news broadcasts has a science background.

practice that permeates journalism seems completely counter-productive for doing science stories well, if they're done at all. In this context it is amusing to consider how sports reports would be different, and games played differently, if the score from only a single game for each sport was deemed worthy to report on each day.

The time factor also plays out in yet another way. The journalist in this excerpt describes how the short timeline she had (which is imposed as a cultural practice in journalism, not because science news that is a day old becomes meaningless) meant that she "missed" a critique of the product and thus she might have hyped a drug that may have actual problems. One must ask why the story wasn't given 27 h to come together rather than 3 h so that the "big" news was done thoroughly and well by specialist reporters, not to mention the non-specialists she refers to ("general assignment reporters" who would have many fewer resources to draw on to create the story). At the end of the day one can see the argument that plane crashes, sports scores, weather and other such events must be broadcast with some sense of immediacy, but failing an impending and unexpected collision with a meteor it is difficult to understand the sense of immediacy that newscasts wrap around reporting on the majority of science issues.

Connections Between the SMCC Anecdote and the Radio Workshop

In the broadcast work of the radio workshop similar patterns of behaviour were evident in creating reports as those which were described by the senior science journalist on the SMCC panel. Although this suggests that the environment of the workshop paralleled that of a commercial broadcast station, this is not necessarily the best way to enculturate newcomers to a profession (see Lave and Wenger 1991), particularly in a situation where changes from the current culture of practice are perhaps necessary. The issue of "what is important" for broadcast (see Gans 1979) was highlighted through other events in the radio workshop. For instance, if too many stories were prepared for broadcast some were dropped, yet I am aware of no instances where they were introduced in a later broadcast even when promoted by the journalist who conducted the interviews and did the story. There is, for many stories, no reason for a later day broadcast not to occur that I can think of, and therefore such practices would appear to be more of habit than necessity. Unless a piece is "timely" (i.e., something happening today) I cannot think of a reason, particularly in a journalism school but also in real-life, that prepared science pieces should not be broadcast on a different day, given the time to be prepared thoroughly, and given the length to be effectively descriptive. In the radio workshop every story that was successfully pitched was given a length and a "type", but that did not mean that they necessarily got broadcast (and if they didn't get broadcast you also didn't receive academic credit for them as only broadcast pieces were assessed. . .so woe to those whose news items the host/producer located near the end of the broadcast). Additionally, pieces were sometimes given much less time than necessary to do a proper job of representing the topic so that all accepted stories could be fit in. This created both a rush-to-finish environment and contributed to errors (both editorially and factually) in the stories. When stories were successfully pitched it might make sense to have one or two fewer pieces each day (where some were scheduled for a subsequent day) that would then also provide sufficient time in the broadcast for the pieces on that day. However, in the journalism school students were not taught to consider what length they would need to do a good job on the piece, but instead settled for whatever they were given (which conversations with journalists suggest is usually the practice in real-world journalism too). Given these practices and others, I would argue that if the role of media is to inform the public then the practices being taught in the journalism school run against effectively engaging with this task and it is therefore not surprising for working journalists to just accept the times and topics afforded to them by producers who make decisions knowing few of the details about how "big" the story is. Consistent with this, as described by Panelist A, despite thinking that the cancer detection story was more significant, she did not describe going and arguing the decision with whoever made the decision.

Once a story is chosen then the journalist starts drafting the script (which may also involve conducting interviews and incorporating them into the script). However, the journalist does not control the final product that turns into the news piece as producers/editors can request edits so that it fits their needs (with respect to length, lead in, language and so forth). In the following section I relate how a copy story for radio based on two newspaper articles broadcast was prepared for broadcast.

Copy Story Modification

This section describes the manner by which a copy story – a short news item – was created for radio broadcast. The copy story was written following a news broadcast the previous day which had included a "live" phone interview with someone who was on location for an upcoming international climate change conference.

It is common practice for "copy stories" to be written by someone delegated to the task. Nowadays that person or persons usually uses the internet (accessing newspapers) or "wire" stories (from various news agencies; e.g., Reuters, Canadian Press, etc.) to draft stories to be read by the news reader for the radio news broadcast. These copy stories are often around 20 s long, although they may be somewhat longer depending on station policy, the topic, or the wishes of the host (newsreader) and producer.[8]

[8] Note that this means that the limitations of the production of the print news story subsequently acts as a mediator for what is possible for the content of the radio newscast story.

The copy story being described here was drawn from two separate articles in the *Guardian* Newspaper in England. The first article was "2010 on course to be joint hottest year since 1850" (Reuters 2010a). The second article was "UN: Greenhouse gases at highest level since pre-industrial times" (Reuters 2010b). In choosing to write about this topic the Copy Story Writer was following up both on a previous day's story as well as his and the Host/Producer's interests (not to mention the interests of others in the newsroom).

The Copy Story Writer's responsibility is to find the "nugget" that is central to the story, and summarize it so that it can be read out-loud by the news reader or host. However, between the copy story writer and broadcast lays the show producer (responsible for the content and management of the entire show) and the news reader (who reads the stories on-air, announces the time, etc). In some cases, or for some shows, one person may serve both roles. Each of them read each copy story to make sure that the words flow in a way that works for the reading style of the host and that the content "fits" with other items chosen for the news broadcast. The producer is responsible for the show length and tone (knowing all of the pieces that are going to be incorporated in the show), and so may have changes or suggestions that involve both content, length and/or "fit" with other news pieces.

This copy story was slightly longer than the norm because it incorporated two related news media stories. The Copy Story Writer drafted the story on the topic of climate change and then presented a hardcopy to the Host/Producer who edited it and returned it. The Copy Story Writer then revised the story and again presented it to the Host/Producer. This iteration happened three times for this copy story.

In editing the first version of the story (Table 16.2, Column 1) the Host/Producer attended to the top half of the story (this is a consequence of other demands being made simultaneously) and, following an interruption, the Copy Story Writer took the edits on the top half and revised the text according to the provided feedback.

The text revisions on the initial draft are significant for two reasons. Firstly, in the second paragraph the replacement of the provided text with "it was the warmest October on record worldwide" significantly changes the meaning of the text. The original story (Table 16.2, Column 1) was explicit in mentioning "surface" temperature, and that was included by the Copy Story Writer because it has a specific significance in climate change science. In the segment below the Copy Story Writer discusses changes proposed by the Host/Producer for the climate change copy story:

> Host/Producer: {reading paragraph, saying some of the text out loud, and crossing text out while reading it} "October's surface temperatures", no no.
>
> Copy Story Writer: {interrupting} I was going to explain why it used "surface."
>
> Host/Producer: It's fine, I don't care.
>
> Copy Story Writer: This stuff is the stuff that's measured on the surface, as opposed to measured from satellites. That's the difference. That's why it's surface temperature; it's measured by stations around the globe that have been recording temperature for a long period of time.

Table 16.2 Editorial revisions of copy story by host/producer

Version 1	Version 2	Version 3	Broadcast version
With no snow yet in XXX, you probably won't be surprised at this news.	With no snow ~~yet~~ ON THE GROUND in XXX, you probably won't be surprised at this news.	With no snow on the ground in XXX, you probably won't be surprised at this news.	With no snow on the ground in XXX, you probably won't be surprised at this news.
According to *NASA* scientists ~~October's surface temperatures set a new record high world-wide.~~ IT WAS THE WARMEST OCTOBER ON RECORD WORLDWIDE.	According to *NASA* scientists it was the warmest October on record [half-pause] world-wide.	According to *NASA* scientists it was the warmest October on record [half-pause] world-wide.	*NASA* scientists say it was the warmest October on record [half-pause] world-wide.
This means that 2010 is ~~still tied~~ NECK AND NECK with 1998 ~~for having the highest average temperature for this point in a year~~ AS THE HOTTEST YEAR.	This means that 2010 is neck and neck with 1998 for having the highest average ~~yearly~~ temperature.	This means that 2010 is neck and neck with 1998 for ~~having~~ the highest average temperature.	This means that 2010 is neck and neck with 1998 for the highest average temperature.
~~In a related story~~...	ACCORDING TO ~~T~~the UN weather agency ~~also announced today that~~ greenhouse gases have reached ~~the highest recorded levels THEIR HIGHEST LEVELS~~[a] THEIR PEAK SINCE THEY STARTED RECORDING THEM.	According to the UN weather agency greenhouse gases have reached their highest recorded levels ~~since the industrial age started.~~	According to the UN weather agency greenhouse gases have now reached their highest recorded levels.
The ~~United Nations~~ UN'S weather agency ~~—the World Meteorological Organization—~~ also announced today that greenhouse gases have reached the highest recorded levels.	These reports were both released just before the 2010 UN Climate Change Conference[,] WHICH starts next week in Cancun, Mexico.	Both of these reports were released just before the 2010 UN Climate Change Conference which starts next week in Cancun, Mexico.	Both of these reports were released just before the 2010 UN Climate Change Conference which starts next week in Cancun, Mexico.

(continued)

Table 16.2 (continued)

Version 1	Version 2	Version 3	Broadcast version
These reports were both released just before the 2010 United Nations Climate Change Conference starts next week in Cancun, Mexico.	Last year's CONFERENCE IN Copenhagen ~~climate conference did not result in an agreement between countries~~ WAS A BUST on how to deal with climate change.	Last year's conference in Copenhagen was a bust.	Last year's conference at Copenhagen was a bust.
Last year's Copenhagen climate conference did not result in an agreement between countries on how to deal with climate change.	COUNTRIES COULDN'T REACH AN AGREEMENT. ~~And expectations for success at the Cancun conference are mixed~~.	~~It did not result in an agreement between countries~~ COUNTRIES COULD NOT AGREE on how to deal with climate change.	Countries could not agree on how to deal with climate change.
And expectations for success at the Cancun conference are mixed.	HOPEFULLY CANCUN CAN DO BETTER.	LET'S HOPE ~~Hopefully the Cancun conference will~~ THAT THIS YEAR THEY CAN do better.	Let's hope that this year they can do better.

Legend: Strikethrough indicates removed, Upper Case is text added by the Host/Producer.
[a]Reporter intervened and said they'd been at higher levels in the past, so host/producer inserted following text.

> Host/Producer: …we don't need to go into all of the background. If someone wants the background they can go into it. We're just trying to tell the story. We're not trying to tell the story about the difference between surface temperature and satellite temperature. (From notes made in workshop notebook)

In that first editing of the copy story the Host/Producer also significantly edited the second paragraph crossing out "highest average temperature for this point in a year" and changing it to "as the hottest year".

Again, the Copy Story Writer explained to the Host/Producer why that specific language had been used (it was only the first 10 months of the year that were the "highest average temperature for this point in the year" compared to the first 10 months of records from previous years). And again the Host/Producer indicated that the change to "as the hottest year" was sufficient for the purposes of the news broadcast. The Copy Story Writer took the written edits, edited the file, and returned it to the Host/Producer.

Version 2 of the story (Table 16.2; Column 2) was presented to the Host/Producer and a colleague by the Copy Story Writer and further significant edits were made to the second half of the copy story.

The first paragraph was changed from "yet" to "on the ground" because the Host/Producer pointed out that a few weeks before there had been snowfall that had stayed on the ground for an evening, and thus "on the ground" was a more accurate description.

The first significant edit changed the reference to greenhouse gases' "highest recorded levels" to "their peak since they started recording them" (this was after the Copy Story Writer pointed out that the initial edit to just "highest levels" was misleading).

The second last paragraph removed the detailed description of "climate change conference did not result in an agreement between countries" and changed it to "was a bust" and "countries couldn't reach an agreement."

The final paragraph ("And expectations for success at the Cancun conference are mixed") was changed to "Hopefully Cancun can do better."

These edits were incorporated into Version 3 (Table 16.2; Column 3) and then submitted to the Host/Producer for feedback. Three significant changes were made:

- In the fourth paragraph the reference to "the industrial age" was removed.
- In the second last paragraph the generic reference to the conference not resulting in an agreement was removed and replaced by a specific reference to "Countries" not being able to agree.
- In the final paragraph the referent was changed from the conference to "Countries" doing better through the addition of "they".

The final copy (Table 16.2; Column 4) was what was read for the news broadcast.

Analysis of Copy Story Modifications

The changes made to this copy story as it progressed from the first submitted draft to the final broadcast are an interesting insight into news media representations of global warming science and how it is portrayed to the public. In this particular case the modifications were made by participants in a journalism school and both individuals had a background in science and a stated interest in science journalism, suggesting that when writing their own future stories they would make the same sort of decisions for the same reasons as in this case.

In the initial draft (Version 1) of the story there were eight paragraphs. In the first edit the fourth paragraph was eliminated, and in the second edit the original seventh paragraph was split into two statements resulting in the final copy having eight paragraphs. The original draft was 124 words (including the bridging words that tied together the two related news items) and the final broadcast version was 106 words long. However, making it somewhat shorter changed some of the accuracy of the content of the science and also, through removing the bridging phrase "In a related story. . ..", made it more of a longer single story than two related stories (copy stories are normally 18–20s; this one was almost double that). Overall five of the eight final paragraphs contained information about global warming that could be considered either inaccurate or misleading. The inaccuracies and their implications warrant further analysis.

The changing of the phrase "surface temperature" to "temperature" means that the listener no longer knows where the temperature was measured. In global warming research the phrase "Surface Temperature" has specific meaning, and NASA's use of the term has even greater specific meaning for how they collected and analyzed the temperatures. Removing the specific reference to "surface" temperatures means that the listener may assume that the temperature was collected and determined in other ways. Clearly, based on the described discussion about the change wanted by the Host/Producer, the Copy Story Writer was aware of the distinction and wanted attention paid to that detail, but the Host/Producer thought that the difference was irrelevant. In fact, in examining the records of global warming "deniers" collected for an earlier paper (Bowen and Rodger 2008) one can see that it is this exact form of change in specifics that provides a foundation for "deniers" to criticize not just the work of the journalist but also allows them to raise doubt and uncertainty in the mind of the media consumer about the credibility of the original research itself. This highlights the importance for those producing news items to not just understand the specifics of the language used by the scientists, but also to understand how the media's practices end up influencing the public.

The third paragraph was altered so that the understanding that the temperature was warmest comparing the first 10 months of the years for which temperature had been recorded was now lost. Instead, the listener would likely understand that the average temperature of the first 10 months of 2010 was warmer than the 12 months of any previously measured year – not necessarily surprising might think a North American as the colder months of November and December in 2010 were not included in that average. Again, it is this very sort of poorly phrased statement that allows "deniers" to influence public understanding of what the journalists have reported.

The fourth and fifth paragraphs in the final version did not contain any misleading or inaccurate information.

The sixth paragraph in the final version, "Last year's conference at Copenhagen was a bust" is both a value judgment (particularly in comparison to the original paragraph which was accurately descriptive) and to understand it in isolation the listener would have to know what the outcome of that conference was. However, the following (seventh) paragraph somewhat provides that context.

The seventh paragraph in the final version, as edited by the Host/Producer, stated that "Countries could not agree on how to deal with climate change." Unlike the original statement, this phrasing implies that there was broad disagreement between countries, however factually only a few countries did not agree with the drafted agreement and as such it was not adopted because it needed full consensus. Additionally, it further implies that the "country" could not agree, whereas in some cases it was particular negotiators or government officials that did not, not necessarily the overall electorate. For instance, one might argue (based on polls at the time) that a majority of Canadians were interested in dealing with climate change, but that their government at the time did not. However that is not the type of understanding that would emerge from that final version of the copy story. Thus, there is a subtle but important difference between saying that "Countries could not agree..." and the original "...did not result in an agreement..."

The final paragraph in the final version again reflected a value statement (i.e., "hopeful") that might implicitly derive from either the broadcaster and/or the host (and notably might alienate audience members who do not support government making changes to address global warming; which the original statement would not have done) or might be taken to reflect the perspectives of either NASA or the UN weather agency. Although thinking that it reflects the host or broadcaster may not necessarily be inaccurate or problematic, to ascribe "hopefulness" to an outcome to either the UN or NASA scientists could lead to the perception that they have a bias in their science (as has occurred in many other such instances).

To summarize, clearly the changes that were made from Version 1 to the Broadcast Version could very well lead to different understandings in the public than would occur from the original press articles about either the global warming science itself or policy issues around those. More problematically, several of these inaccuracies or misrepresentations (particularly the third and fourth paragraph, and possibly the eighth) were such that if they appeared in print media they are the type that provide a foundation, based on comments collected for research for the Bowen and Rodgers (2008) paper, for deniers to seed suspicion about the motives of the scientists and the accuracy of their science. This further suggests that journalists reporting on socioscientific issues, such as global warming, need to have both a firm grasp of the science and an understanding of how the work of journalists is being used by others, either directly or indirectly, to influence public debates.

The Media Making Sense of Science: A Case Study of Two Publications

Various issues about the social and material practices in journalism and how those might influence the production of "news" stories have been identified throughout this chapter. In this final section we will examine how two different print media outlets portray a short data-based research study on a socioscientific topic (i.e., testing in schools).[9]

Although the purpose of this article is to discuss the news media representations of broad socioscientific studies to the general public, studies of this sort are often both quite long and quite complex making it difficult to summarize issues of news

[9] Socioscientific research studies on topics such as global warming are often quite complex with large data sets. For the sake of this chapter I've chosen a socioscientific study (one related to education) which has a data set which is manageable for this discussion. Issues present in the news media reporting of a small study such as the one discussed here are as likely (or, I would suggest, more likely) to occur when discussing larger and more complex research reports on issues such as global warming.

Science DOI: 10.1126/science.1199327

REPORT

Retrieval Practice Produces More Learning than Elaborative Studying with Concept Mapping

Jeffrey D. Karpicke* and Janell R. Blunt

± Author Affiliations

*To whom correspondence should be addressed. E-mail: karpicke@purdue.edu

ABSTRACT

Educators rely heavily on learning activities that encourage elaborative studying, while activities that require students to practice retrieving and reconstructing knowledge are used less frequently. Here, we show that practicing retrieval produces greater gains in meaningful learning than elaborative studying with concept mapping. The advantage of retrieval practice generalized across texts identical to those commonly found in science education. The advantage of retrieval practice was observed with test questions that assessed comprehension and required students to make inferences. The advantage of retrieval practice occurred even when the criterial test involved creating concept maps. Our findings support the theory that retrieval practice enhances learning by retrieval-specific mechanisms rather than by elaborative study processes. Retrieval practice is an effective tool to promote conceptual learning about science.

Received for publication 20 October 2010.
Accepted for publication 10 January 2011.

Fig. 16.1 News media article

media representations of that type of research. To address this problem, I've chosen to discuss the media representation of a shorter research article (4 pages long) without the co-varying interval-ratio variables or mathematical models such as is often found in socioscientific studies (i.e., climate change research) – in the chosen study there are four discrete treatments and a limited number of measured outcomes.

This analysis will involve an examination of two "news media" articles[10] written about a research article published in the journal *Science*. Above (Fig. 16.1) is a screen capture of the journal website depicting the article and summary information around it (captured 19 February 2011; located at http://www.sciencemag.org/content/early/2011/01/19/science.1199327.abstract).

Although the print version of the journal publication (Karpicke and Blunt 2011) is dated February 11, 2011, Volume 331 (pp. 772–775), according to the publisher's website the article was available online prior to that on January 20, 2011. A news article about the research in the journal article appeared in the *New York Times*

[10] My thanks to Don Duggan-Haas for the conversation from which this topic arose. I had posted the "Wired" article on my Facebook wall and he copied it to his wall with the comment that it "has a different spin than an earlier article" he had read. Then a friend of his posted the *New York Times* article on the discussion thread; that article was the article which Don had read. I read both of the popular news media articles, then the original research article and supplementary materials. This analysis and discussion derives from those readings.

(online edition) on January 20, 2011 and in *Wired* Magazine (online edition) on January 21, 2011. This apparent "rush" to publication, also described and discussed in an earlier section, is seemingly typical of news media reporting on science research – It is after all, as one instructor in the journalism program kept repeating, "The news, not the olds."

However, as argued earlier, one does wonder exactly why there is the rush to publish articles such as this in the popular press because the research itself is hardly time-sensitive. One might, however, infer that the publisher or the author of the original article was "pushing" the research into the media because the graphs used in the *Wired* article (Keim 2011) were marked with the text "Embargoed until 2:00 PM US ET Thursday, 20 January 2011" suggesting that the news media had received advanced copies of the article. An advanced receipt of the journal article may also explain the length of the *New York Times* article, as the more time a writer has to produce a story the longer it is (Long 1995).

This type of promotion of science articles used to be unusual, but scientists are now being encouraged to promote their science into the public sphere by people such as Nancy Baron (2010) who runs workshops and gives talks to scientists about promoting their research into the public news media. However, what this means for the public is that it is not the reporter determining what is relevant or significant research (a role the news media used to serve), that is now being determined by an outside agency or individual who provides "significant" research to the journalist. So, now a news media consumer must question the motives of those involved in determining what research ends up being promoted in the popular press. There is, for instance, a lot to gain for an assistant professor applying for tenure if their research is promoted in the popular press; any review committee would look on that favourably. But, does that warrant the research being written about for the general public? And why this piece and not others? In many cases one should probably also ask whose agendas are being forwarded by those research articles being written about.

The *New York Times* article (Belluck 2011) was titled "To Really Learn, Quit Studying and Take a Test". It was 1,133 words long (using the MS Word "word count" feature) and in addition to interviewing the lead author – Karpicke – also involved interviews with five other cognitive science researchers commenting on the findings of the research study. Despite frequent use of the word "test" throughout the article, the activity engaged in by the students does not reflect a "test" as most would think of it. Firstly, students didn't receive any mark, nor did the assessment by the researchers "count" for anything; for the students it was definitely a "low stakes" assessment. Secondly, there were no questions asked in that particular treatment – students spent 5 min reading a passage of text, and then had 10 min to write down what they could recall from it. They then repeated that sequence once more. In that context I would argue that calling that activity a "test" would be inappropriate and that then suggesting that "To Really Learn, Quit Studying and Take a Test" is a relevant summary of the research is a misleading representation of what was reported in the research. And although I'm not suggesting that the basic premise of the research would not provide insights useful to the researcher's broader agenda, as described it has little application to a classroom (despite content in the article both directly and indirectly suggesting it does) as it tests four completely discrete conditions which do not, in my view, represent realistic studying conditions. For instance, I've never heard a teacher or a researcher suggest that using a concept map as the only form of studying was sufficient. Or that just reading text was sufficient. None of this type of critique is apparent in the *NYT* article, nor is the way in which the material was tested a week later critically examined at all (in the first "experiment" 84 % of the calculated score on the final test derived from questions which were low-order memorization questions). I'd also suggest that the reporter did not adequately interrogate the reported results. For instance, suggesting that "students who read a passage, then took a test asking them to recall what they had read, retained about 50 % more of the information a week later than students who used two other methods" ignores that the improvement is only around 30 % for the higher-order inferential questions (only two inferential questions were asked; worth 2 marks each) and that this was masked through the researcher collapsing the question types together. Overall, my reading of the *New York Times* article suggests that what is being promoted by the author is a pro-testing agenda, which was not really the purpose of the *Science* research article and so the research itself is being misrepresented.

The *Wired* article (Keim 2011) was titled "To learn best, write an essay" and is a much shorter article (221 words) relying entirely on material provided in the *Science* journal article. Yet, despite the length in some ways the *Wired* article more accurately represented the findings of the research stating, in its lede, "Trying to remember what you've just studied, then writing it down, may be a surprisingly good way to learn." Arguably this is a good one-sentence summation of the research article. However, the *Wired* article ignored presenting any statistical analysis of some of the data resorting to a comparison of the means: "Students who originally wrote essays performed best. Next came the crammers, then the concept mappers." This avoidance of statistical comparisons and focus on the comparison of means could lead to misunderstandings in the reader. For instance, the quoted sentence (i.e., "Students who originally...") suggests to the reader that the differences between treatments were unambiguous. Yet, despite the inference to the contrary in the statement, there was no statistically significant difference between the "crammers" and the "concept mappers" and both were significantly different (i.e., lower) than the "essay writers". In addition, the writer does not distinguish between the two different studies reported in the journal article, and by so doing conflates the findings of the two different studies. However, unlike the *NYT* article, the *Wired* article does include the graphs depicting the findings of the first study, so it is possible for the reader to understand the degree of some the differences and draw some of their own conclusions.

Although both news article were consistent with normative media practices of reporting only on a single research article, I would argue that neither news media article did a thorough job of either representing or critiquing the findings reported in that journal article. This is problematic because in the first case the general public doesn't have access to the journal article, and in the second case doesn't have the background to allow them to effectively understand the research article in the context of other research. Thus, in neither case would the public be able to accurately understand the findings of the article and what the implications of that research were. If issues like this arise reporting on such a small and simple study, where the news media distorts the original findings and don't place them in a critical framework, just what can we expect of them in representing more complex research, such as that of global warming, to the public? More importantly, just why do these issues arise? I suspect that they arise because of the issues discussed previously, including journalists' belief that they can be experts on any issue in short order, that they understand 'what the public wants', that there is an immediacy to writing articles for news release (either audio, video, or print), and that any complex issue can be reduced to a simplified understanding. Collectively these lead to less than ideal presentations of the findings of socioscientific issues, or even, apparently, more simple studies than those.

Conclusions and Implications About Science in the News Media

I'm going to discuss the implications of the various cases and topics discussed throughout this chapter under three sub-headings: science in the media; what j-school suggests about future journalism, and; what the implications are for using media in the classroom with students to help them learn about socioscientific issues such as global warming.

Implications for Understanding Science from the News Media

There are any numbers of issues that arise from the previously discussed cases, and not one of them suggests that the public should take at face value any of the science research presented in the media. First there is the question of how it was chosen, or even if it was (given the current influence of press releases and public relations activities; see Baron 2010), by the journalist. Their personal biases and interests remain unexplored despite the influence it might very well have on what science is portrayed in the media with much of the science in the news media being either medical or environmental (Einsiedel 1992) apparently reflecting either the interests of the journalists or the editor/producer (who might be operating on some sense of "what the public wants" that from personal experience I suspect is a "seat of the pants" assessment for most of them).

Additionally, the qualifications of journalists to write about certain topics is almost never critiqued, the assumption that they can be "instant experts" means that someone with a background in English can one day be an education reporter for several years and then be assigned to be a science reporter at the whim of an employer without any apparent expertise in either area. To pretend that the details can be understood with minor preparation by working on the stories themselves reflects either an ignorance of what it takes to become an expert or arrogance about their own abilities to become one. Either way, the assumption that such is possible is unwarranted. In concert with this is the naïve belief, perhaps coupled with arrogance, that complex relationships are always reducible (in a few column inches or a few minutes of broadcast) by *them* to simple relationships which nonetheless can accurately and effectively inform the public about the issues. This latter belief alone illustrates why it has been so difficult for the media to inform the public about socioscientific issues in the last few decades as the issues themselves have become more and more complex. Despite the fact that research suggests that if journalists had a better understanding of science and the work that scientists do it may result in a reduction in the negative perceptions of science and scientists contributed to by agents such as the media (Reis and Galvao 2004), there is little evidence that news media organizations are interested in developing this (Saari et al. 1998) or, seemingly, that journalism schools are either (at least the one I attended).

Apart from all of this, the apparent demands of commercial media, which reassigns (or constantly assigns) journalists at a whim whilst providing them little time or resources to develop insights, suggests that the news media organizations accept doing an inadequate job of conveying information accurately to the public as part of their business; apparently accuracy and depth can be sacrificed on the pyre for the sake of commercial interests. This means that however dissatisfied one might be with the state of science journalism, there's little chance that it's going to get better, at least in the traditional news media.

As briefly mentioned earlier, this has some considerable implications for the public with respect to socioscientific issues and public opinion. In my earlier paper (Bowen and Rodger 2008) although I noted that "deniers" often picked up on and exploited language used by journalists (as opposed to that in the original science) as a way of arguing against global warming (including against policies that may address carbon production) the many cases described above provide fresh insight into that. Overall, I do not think I am making an egregious claim when I argue that in many cases it is the loose language used by journalists, for whatever media expediency they may believe in or are accommodating, that provides a foundation for those who wish to argue against the science of global warming or other socioscientific issues. Consequentially, I agree with Oreskes and Conway (2010) that journalists are complicit in the negative public opinion that exists in North America regarding dealing with socioscientific issues such as global warming. This, one would think, would seemingly have consequences for the preparation of new journalists.

So how does "media coverage" of socioscientific issues need to change? Firstly, inaccuracies (either factual or in nuances of language) in reported stories often provides an access point for "deniers" to argue against global warming science (or policy) and either the reporter or the news media organization needs to do better due diligence in the use of language and the misrepresentation of science in news items. Secondly, I wish to propose that there has been a shift in what would be considered the "unit" of a news media item. Prior to the internet and these interactive technologies, the readership of an article only read and engaged with the content of the article in the media. The "unit" of media was merely that of the journalist's article itself and the reading of it by each individual person. Now, however, I would argue that the "unit" of media is constituted by the reader and his or her examination of the interaction between the article and the appended discussion forum…that all three components constitute the final understanding of the issue that is gleaned and, consequently, the combination of them now constitute the "news" unit. This means then that a "news item" is no longer static in the way it used to be but is now a dynamic entity that is influenced by who posts in the discussion forums and which posts the reader ends up reading. I believe this has implications for journalists, because I now believe that journalists are not only responsible for "getting the story" right in the first place, they also have a responsibility to address how their story is engaged with by others. If they have taken journalistic short cuts to portray information (such as some of those described in earlier sections), and if these are then used by others to re-frame the meaning of their story, then the journalist (or the media outlet) has a responsibility to redress this in the discussion forum through directly engaging the readers and their representation of the news article. I would argue that one cannot both claim a special role in society to inform the public, as journalists and the news media have, and then turn around and ignore that their work is now contributing to *mis*-informing the public through being redefined by others in texts appended to their articles. Readers now form their perspectives on the topic through reading both the article and the comments, so if the intent of the article is to lead to a particular understanding in the average reader (who apparently reads the forums but does not post in them), then the responsibility of the media to be 'accurate" in what they report now also includes redressing misrepresentations of the story content in the discussion forums.

Looking at the Preparation of Journalists

In my experience faculties of education take a determined look at what goes on in schools, and then ask what they should be doing with their own Bachelor of Education and their graduate students so that the problems that exist in current schools do not continue. I will readily admit that despite their best efforts it may be difficult to advance change in schools through programs in faculties of education, including for reasons that are embedded within faculties of education themselves. However in some way or another faculties of education take on that task as part of their role in preparing education professionals. After progressing through a journalism program I cannot say that I think that the program I attended, or other programs similar to it (and others were described to me both by professors and other students as being similar[11]), make any serious attempt to try and address issues in current journalism practices and the effectiveness (or lack thereof) by which the media informs the public about socioscientific issues, or any other issues for that matter. For the most part, I wasn't convinced that many of my instructors were even aware of the academic critiques of the practices of journalists and news media companies in any detail whatsoever, and consequently made no efforts to address those; in many instances I was actually taught, forced in fact, to engage in practices which were the exact ones identified in the research literature as problematic. Thus, it's not just that we didn't learn how to address problematic issues in journalistic practice in our j-school program, it's that we had those very same problematic issues entrenched in our journalistic practices through what we were taught. This suggests that journalism programs such as that which I attended are actually part of the problem; which is not the role that university programs are supposed to serve in our society.

Apart from the actual content issues, the instructional methodologies were also problematic. For instance, few of the practices suggested by Lave and Wenger (1991) for effectively enculturating newcomers to a profession were present in many aspects of the program I attended (there was one exception, and that workshop was an exemplary case). From an assessment perspective there was no evidence that work was being checked for accuracy of either content or in reflecting the content or emotive nature of the interviews conducted by program participants. In the end, the problems I've described in the j-school program in this paper paint a bleak picture for the future of science journalism, because if journalism schools are not going to prepare future journalists to address problems with media reporting on science issues, then who will? Saari et al. (1998) make the argument that effective science writing is unlikely to emerge from media organizations themselves. What this suggests, again, is that the problems that exist in science journalism now are unlikely to be improved upon in the future. This leaves one asking where does this leave schools with regards to using news media in the classroom? Where does it leave the public in understanding science as it is portrayed in the news media? As I see it, neither question has a satisfactory answer.

[11] That's not to say that there are not J-schools that take a more conceptual approach to teaching journalism so that its students are more reflective about their practices. For instance, Columbia Journalism Professor Judith Matloff (2011), commenting on her academic program, said "...generally you have to remember the curriculum of the school is a much more theoretical one and it's more craft-oriented, safety training is not something which is usually incorporated in an academic program" suggesting that they deal with more conceptual and theoretical issues in the Columbia School of Journalism program than did the program I was in.

Implications of Using News Media in the Classroom for Studying Socioscientific Issues

Although many suggest that students should learn more actively about issues such as global warming by using news media in the classroom, particularly print media, teachers themselves probably do not have a background sufficiently strong to help students address incorrect statements in the media about global warming science (see Fortner 2001; Khalid 2001; Norris et al. 2003). Given the arguments above, this is problematic, and for reasons that go beyond media representation issues.

There is another way to look at this. Perhaps we are asking too much to ask students to engage critically in ideas, and particularly evidence, about global warming or other socioscientific issues at all. Troy Sadler et al. (2004) report that "just under one-half of students sampled were not able to accurately identify and describe data" (p. 402) and there is little reason to suspect that they have any experience with anything other than single dependent variable relationships from their experiences in schools (since that is the type of study that student teachers design when asked to design an investigation). Not only are there potential issues with student's data literacy, there is also little evidence that teachers will have the requisite skills to address those data literacy issues (see Bowen and Bencze 2008; Bowen et al. 2010, 2013; Miller and Wynne 1988) suggesting that teachers are perhaps ill-prepared to scaffold students on the complexities and nuances of science, not to mention the multivariate complexities of global warming science or other socioscientific issues.

Yet, clearly this is a defeatist attitude. Rather than viewing global warming science from a data literacy perspective, perhaps getting students to engage with global warming science through critical media literacy activities might be productive. For instance, using the Google news aggregator students could find multiple articles on the same topic and compare them for differences and then critically compare those to the original news release or publication (such as was done above with the *New York Times* and *Wired* article analysis). Using such an approach would actually help develop critical media literacy as well as data literacy because students would have to examine original source documents and define terms so that they could best understand the media representations.

However, it is still easy to see where problems could arise with this approach. The media, and critics of global warming science, rapidly grabbed onto a statement in emails which were (illegitimately) downloaded from the Climate Research Unit's email server in which Phil Jones, a climate scientist, wrote, "I've just completed Mike's Nature trick of adding in the real temps to each series for the last 20 years" with many taking the word "trick" to mean that something inappropriate was being done with the data and therefore that the claims emerging from the data were incorrect resulting in a "scandal" referred to as "Climategate."[12]

[12] Formal investigations by multiple authorities (both governmental and academic) concluded there was nothing in the emails regarding inappropriate analysis or the conclusions that were drawn about global warming from those analyses. Wikipedia has a full discussion for those interested in this issue.

Reading the offending email, and with my science background in analysis involving multivariate statistics, I found nothing offensive about the statement as it colloquially suggested there was some sort of transformation or use of a covariate – a completely normal practice in science statistical analysis. Indeed, my assumption was consistent with the argument of others with a science background (see http://www.skepticalscience.com/What-do-the-hacked-CRU-emails-tell-us.html; see the interview at http://voices.washingtonpost.com/capitalweathergang/2009/12/gerald_north_interview.html). However, that simple statement about a "trick" was used repeatedly in the media and blogosphere to suggest that there were the problems with claims emerging from the CRU about global warming. Given that my background in data analysis, as historical now as it may be, is uncommon for a teacher to have, one is left wondering just what a teacher would do with students who brought that quote of using a "trick" forward, how a teacher would explain it. Even doing media comparisons, or comparisons to source information, in this instance could leave a teacher in considerable quandary with a high likelihood of not being able to adequately explain or defend the context of how the word "trick" was used.

So where does that leave us? I think we need to re-think the wisdom of having socioscientific issues being dealt with in science classrooms. School science curricula, in both the information portrayed in graphs and activities in which students are engaged, deals almost exclusively with single dependent-variable relationships. Nothing suggests to me that most teachers or students have a background sufficient to effectively engage with discussions around global warming science that incorporated the broad public perspectives that the topic has, nor that students or teachers could engage with available datasets in any effective manner. So, perhaps phenomena such as global warming need to be dealt with as a sociological construct by those who are more used to dealing with those; such as social studies teachers. Maybe social studies classrooms, not science classrooms, is where it would be useful to use news media to discuss the issue of global warming. I'm no longer as convinced, particularly given the media commentary of the last few years, that teaching about complex socioscientific topics should be done in science classrooms at all given the complexity of resources (such as misrepresentative statements in news media) that exist for students to access. The manner in which the public and the media engage socioscientific issues in contrast to the science (see Beacco et al. 2002) leaves me more and more convinced that sociological not scientific perspectives provide a better foundation for students to engage with those issues.

References

Adams, S. (1999). Critiquing claims about global warming from the World Wide Web: A comparison of high school students and specialists. *Bulletin of Science, Technology & Society, 19*(6), 539–543.

Anderson, A. (2009). Media, politics and climate change: Towards a new research agenda. *Sociology Compass, 3*(2), 156–182.

Baron, N. (2010). *A guide to making your science matter: Escape from the ivory tower*. Washington, DC: Island Press.

Beacco, J.-C., Claudel, C., Doury, M., Petit, G., & Reboul-Toure, S. (2002). Science in media and social discourse; new channels of communication, new linguistic forms. *Discourse Studies, 4*, 277–295.

Bell, A. (1994). Media (mis)communication on the science of climate change. *Public Understanding of Science, 3*, 259–275.

Belluck, P. (2011). To really learn quit studying and take a test. *New York Times* [Online edition]. http://www.nytimes.com/2011/01/21/science/21memory.html?_r=1&src=me&ref=science. Accessed 10 Feb 2011.

Bencze, J. L., Alsop, S. J., & Bowen, G. M. (2009). Student-teachers' inquiry-based actions to address socioscientific issues. *Journal for Activist Science and Technology Education, 1*(2), 78–112.

Bowen, G. M., & Bencze, J. L. (2008). Engaging preservice secondary science teachers with inquiry activities: Insights into difficulties promoting inquiry in high school classrooms. In: W.-M. Roth & K. Tobin (Vol. Eds.), *The world of science education: Handbook of research in North America* (pp. 587–609). Rotterdam: Sense Publishers.

Bowen, G. M., & Rodger, V. (2008). Debating global warming in media discussion forums: Strategies enacted by "persistent deniers" and implications for schooling. *Canadian Journal of Environmental Education, 13*(1), 89–104.

Bowen, G. M., Bartley, A., MacDonald, L., & Fraser, D. (2010, March). *Improving your students' graphing and graph interpretation practices*. A workshop presented at the annual conference of the National Science Teachers Association, Philadelphia, PA.

Bowen, G. M., Bartley, A., MacDonald, L., & Sherman, A. (2013, manuscript). Experiences with activities developing pre-service science teacher data literacy (To be submitted).

Boyes, E., & Stanisstreet, M. (1992). Students' perceptions of global warming. *International Journal of Environmental Studies, 42*, 287–300.

Boykoff, M. (2008). Media and scientific communication: A case of climate change. *Geological Society of London, Special Publication, 305*, 11–18.

Boykoff, M., & Boykoff, J. (2004). Balance as bias: Global warming and the US prestige press. *Global Environmental Change, 14*, 125–136.

Boykoff, M. T., & Mansfield, M. (2008). 'Ye Olde Hot Aire': Reporting on human contributions to climate change in the UK tabloid press. *Environmental Research Letters*. http://www.iop.org/EJ/article/1748-9326/3/2/024002/erl8_2_024002.html. Accessed 5 May 2008.

Charney, D. (2003). Lone geniuses in popular science. *Written Communication, 20*, 215–241.

Curtis, D. (2007). *It's not news, it's fark: How mass media tries to pass off crap as news*. New York: Gotham Books.

Dearing, J. W. (1995). Newspaper coverage of maverick science: Creating controversy through balancing. *Public Understanding of Science, 4*, 341–361.

Dispensa, J. M., & Brulle, R. J. (2003). Media's social construction of environmental issues: Focus on global warming—A comparative study. *International Journal of Sociology and Social Policy, 23*(10), 74–105.

Einsiedel, E. F. (1992). Framing science and technology in the Canadian press. *Public Understanding of Science, 1*, 89–101.

Fortner, R. W. (2001). Climate change in school: Where does it fit and how ready are we? *Canadian Journal of Environmental Education, 6*, 18–31.

Gans, H. J. (1979). *Deciding what's news: A study of CBS Evening News, NBC Nightly News, Newsweek and Time*. Evanston: Northwestern University Press.

Gelbspan, R. (2005). Snowed. *Mother Jones* [Online publication]. http://motherjones.com/politics/2005/05/snowed. Accessed 21 Feb 2011.

Goldsmith, E., & Hildyard, N. (Eds.). (1988). *The Earth report: The essential guide to global ecological issues*. Los Angeles: Price Stern Sloan Inc.

Gowda, M. V. R., Fox, J. C., & Magelky, R. D. (1997). Students' understanding of climate change: Insights for scientists and educators. *Bulletin of the American Meteorological Society, 78*(1), 2232–2240.

Hindman, D. B. (2009). Mass media flow and differential distribution of politically disputed beliefs: The belief gap hypothesis. *Journalism & Mass Communication Quarterly, 86*, 790–808.

Hyslop, J. H. (1899). Newspaper science. *Science, New Series, 10*, 695–696.

Jakobson, R. (1960). Closing statements: Linguistics and poetics. In T. A. Sebeok (Ed.), *Style in language* (pp. 350–377). Cambridge, MA: MIT Press.

Jarman, R., & McClune, G. (2002). A survey of the use of newspapers in science instruction by secondary teachers in Northern Ireland. *International Journal of Science Education, 24*(10), 997–1020.

Kachan, M. R., Guilbert, S. M., & Bisanz, G. L. (2006). Do teachers ask students to read news in secondary science?: Evidence from the Canadian context. *Science Education, 90*, 496–521.

Karpicke, J. D., & Blunt, J. R. (2011). Retrieval practice produces more learning than elaborative studying with concept mapping. *Science, 331*, 772–775.

Keim, B. (2011). To learn best, write an essay. *Wired Magazine* [Online edition]. http://www.wired.com/wiredscience/2011/01/learning-methods/. Accessed 10 Feb 2011.

Khalid, T. (2001). Pre-service teachers' misconceptions regarding three environmental issues. *Canadian Journal of Environmental Education, 6*, 102–120.

Lave, J., & Wenger, E. (1991). *Situated learning: Legitimate peripheral participation*. Cambridge: Cambridge University Press.

Lewenstein, B. V. (2001). Who produces science information for the public? In J. H. Falk (Ed.), *Free-choice science education: How we learn science outside of school* (pp. 21–43). New York: Teachers College Press.

Long, M. (1995). Scientific explanation in US newspaper science stories. *Public Understanding of Science, 4*, 119–130.

MacCharles, T. (2009, October 23). If the news is Canadian, Harper's not watching. *Toronto Star*. http://www.thestar.com/news/canada/2009/10/23/if_the_news_is_canadian_harpers_not_watching.html. Accessed 21 Feb 2011.

Matloff, J. (2011, February 23). Discussing taboos surrounding sexual assault on female reporters in conflict zones. *"Q" on CBC*. http://www.cbc.ca/q/episodes/. Accessed 25 Feb 2011.

McBean, G. A., & Hengeveld, H. G. (2000). Communicating the science of climate change: A mutual challenge for scientists and educators. *Canadian Journal of Environmental Education, 5*, 9–25.

Miller, R., & Wynne, B. (1988). Public understanding of science: From contents to process. *International Journal of Science Education, 10*(4), 388–398.

Monbiot, G. (2006). *Heat: How to stop the planet from burning*. Toronto: Doubleday Canada.

Monbiot, G. (2007). The editorials urge us to cut emissions, but the ads tell a very different story. *The Guardian* [Online edition]. http://www.guardian.co.uk/commentisfree/2007/aug/14/comment.media. Downloaded 21 Feb 2011.

Nelkin, D. (1987). *Selling science: How the press covers science and technology*. New York: W. H. Freeman.

Norris, S. P., Phillips, L. M., & Korpan, C. A. (2003). University students' interpretation of media reports of science and its relationship to background knowledge, interest, and reading difficulty. *Public Understanding of Science, 12*, 123–145.

Oreskes, N., & Conway, E. M. (2010). *Merchants of doubt: How a handful of scientists obscured the truth on issues from tobacco smoke to global warming*. New York: Bloomsbury Press.

Penney, K., Norris, S. P., Phillips, L. M., & Clark, G. (2003). The anatomy of junior high school science textbooks: An analysis of textual characteristics and a comparison to media reports of science. *Canadian Journal of Science, Mathematics, and Technology Education, 3*(4), 415–436.

Picard, A. (2011, February 18). New guidelines to treat heart condition stress quick, aggressive approach. *The Globe and Mail*, Globe Life section, p. L5.

Ratcliffe, M., & Grace, M. (2003). *Science education for citizenship: Teaching socioscientific issues*. Berkshire: McGraw-Hill.

Reis, P., & Galvao, D. (2004). Socioscientific controversies and students' conceptions about scientists. *International Journal of Science Education, 26*(13), 1621–1633.

Reuters. (2010a, November 25). *2010 on course to be joint hottest year since 1850*. http://www.guardian.co.uk/environment/2010/nov/25/2010-joint-hottest-year-global-warming. Accessed 19 Feb 2011.

Reuters. (2010b, November 24). *UN: Greenhouse gases at highest level since pre-industrial times*. http://www.guardian.co.uk/environment/2010/nov/24/un-greenhouse-gases. Accessed 19 Feb 2011.

Roll-Hansen, N. (1994). Science, politics, and the mass media: On biased communication of environmental issues. *Science, Technology & Human Values, 19*, 324–341.

Saari, M.-A., Gibson, C., & Osler, A. (1998). Endangered species: Science writers in the Canadian daily press. *Public Understanding of Science, 7*, 61–81.

Sadler, T., Chambers, W. F., & Zeidler, D. L. (2004). Student conceptualizations of the nature of science in response to a socioscientific issue. *International Journal of Science Education, 26*(4), 387–409.

Schibeci, R. A. (1990). Public knowledge and perceptions of science and technology. *Bulletin of Science and Technology Society, 10*, 86–92.

Shanahan, J., & Good, J. (2000). Heat and hot air: Influence of local temperature on journalists' coverage of global warming. *Public Understanding of Science, 9*, 285–295.

Tyson, B. (2010). World Health Organization report says 30 minutes cell phone use daily may increase risk of brain cancer. *Helium: Medical Concerns and Issues*. http://www.helium.com/items/1835360-world-health-organisation-report-30-minutes-cell-phone-use-daily-may-increase-risk-of-brain-cancer. Accessed 24 Feb 2010.

Yore, L., Bisanz, G., & Hand, B. (2003). Examining the literacy component of science literacy: 25 years of language arts and science research. *International Journal of Science Education, 25*(6), 689–725.

Zehr, S. C. (2000). Public representations of scientific uncertainty about global climate change. *Public Understanding of Science, 9*, 85–103.

Chapter 17
The Perils, Politics, and Promises of Activist Science

Bernhard Isopp

Abstract This paper considers "activist scientists": those who become socially or politically active and transgress traditional scientific cultural norms of impartiality and neutrality. Such overt political positions are often connected to instances in which scientists bypass usual lines of scientific communication and popularization, and take research findings or expert opinions directly to the public. This paper examines the case of Andrew Weaver, a prominent Canadian climate scientist who has become an active proponent of climate change action, as well as a vociferous critic of the perceived inadequacy of government policy. His activism garnered him a significant amount of unflattering attention which ostensibly related to the appropriate scope of scientists' activities. Historical reflections on the relationship between ecology and the environmental movement suggests that such activism is typically tied to "crisis situations," which often lead to major boundary reworkings regarding the proper role of science. Such boundary reworkings present an opportunity to consider the ways that scientists imagine their own identities and how these compare to public expectations of scientists, as well as challenge certain STS conceptions of expertise.

Keywords Activist science • Climate change • Science popularization • Public engagement • Public understanding of science

Introduction

This chapter explores the intersections of science communication, public scientists, and activism. In particular, it looks at the activities of climate scientists who have, for various reasons, decided to take a pronounced public role in the promotion of

---

B. Isopp (✉)
Graduate Program in Science & Technology Studies, York University,
218 Bethune College, 4700 Keele Street, Toronto, ON M3J 1P3, Canada
e-mail: bisopp@yorku.ca

L. Bencze and S. Alsop (eds.), *Activist Science and Technology Education*,
Cultural Studies of Science Education 9, DOI 10.1007/978-94-007-4360-1_17,

climate change activism especially through media like newspapers, television, and websites and blogs. Such cases open up a set of generative and provocative questions. An immediately apparent point of inquiry is the recognition that activist scientists transgress traditional norms of scientific behaviour, most notably impartiality and political neutrality. Given the penchant in Science and Technology Studies (STS) for questions of boundary making, this is an analytically rich starting point.[1] What contingent sociological, political, and cultural factors account for these deviations from accepted standards of scientific practice? How exactly are the transgressed boundaries conceptualized by scientists? What is it about activism that appears to conflict with scientific norms? How do activist scientists attempt to cope with these tensions? What kinds of boundary reimaginings occur? What risks do activist scientists face in taking on public, activist roles?

As other contributors to this volume will likely note, the role of activism in science pedagogies has received scant attention. There has similarly been a dearth of research on the relationship between activism and science. Those studies that do exist typically explore the interactions between citizen activist groups and scientists, and these case studies tend to focus on the conflicts between activist or public interest groups and scientific experts (Kroll-Smith and Floyd 1997). There are studies that consider instances of cooperation, but are typically framed in terms of lay-expert divides, or uneasy alliances, or the enlisting of scientific expertise for activist purposes (e.g., Delgado 2010; McCormick 2007). Rarely are scientists themselves considered as activists (Frickel 2004).

Overall, this chapter explores these issues from a position of reconstructivist STS. This denotes a critical perspective that is sympathetic to, or aligned with, activist positions. At its most pronounced, reconstructive STS is explicitly normative. It does not, in the name of methodological impartiality or scholarly objectivity, shy away from making value judgments about scientific and technological practices, controversies, and assemblages. Indeed, it actively seeks to reimagine and reconstruct ethical technoscience. More modestly, it is reflexive about the normative potential of STS analyses – how it might be enrolled by various agents – and its own socially constructed nature. As such, this chapter is generally aligned with climate change activism, and may offer some modest lessons to practicing scientists who have also enrolled themselves in this cause.

While this analysis does not situate itself explicitly in practices and theories of activist STME, especially not in a formal way, there are insights for those thinking about such things. Firstly, while formal educational structures are integral to understanding the role and possibilities of activism in science education, it is important to note that when the public understanding of a particular scientific (or socio-scientific) issue becomes a concern, the majority of this public falls outside the reach of formal educational structures. Hence the need to look at broader attempts to educate or "raise

[1] Here I don't mean to favour any particular disciplinary formation; by "STS" I am referring to all work that examines science and technology from diverse historical, sociological, anthropological, cultural, etc. perspectives.

awareness" among the public (see Bowen, Chap. 16). Secondly, STME and STS have mutually benefitted from numerous interactions in recent years. In particular, the teaching of STM has been increasingly grounded in socio-scientific issues, and so STME has increasingly shared case studies, resources, and goals with STS. Furthermore, effective activism takes place at multiple sites and from multiple perspectives, and this chapter aims to contribute to such cross-pollinations. This chapter considers two brief case studies of science activism and advocacy that lend themselves to analysis from multiple perspectives. These cases ultimately complicate STS notions of symmetry and impartiality, revealing that the diverse ways in which science is political demand different ethical considerations, and arguably establish the need for clear demarcations between science and politics.

The Perils of Activist Science

Consider the case of Andrew Weaver, a Canadian climate modeller working in the School of Earth and Ocean Sciences at the University of Victoria. Weaver is an established climate scientist, with over 130 peer-reviewed publications and numerous research awards, and author of two popular books on the science and politics of climate change. He was also a lead author on various chapters in the second, third, and fourth Intergovernmental Panel on Climate Change (IPCC) Assessment Reports, and will serve as a lead author on the upcoming Fifth Assessment report.

Starting in the mid-1990s Weaver was occasionally recruited by news organizations to be a source in stories about climate change, mainly because of his role on the IPCC. He soon ended up on newswire source lists, and became a go-to source for stories about climate change in Canadian media. For various reasons, Weaver eventually took a more proactive role in public engagement. In 1996 he co-penned with his colleague Ken Denman a brief letter to the editor of the *Globe and Mail* in response to disparaging remarks made by then business columnist Terence Corcoran about the review process of the 1995 IPCC assessment report, for which Weaver was a co-author (Denman and Weaver 1996). Weaver and Denman rejected Corcoran's claim that the IPCC had expunged the concerns and doubts of sceptical scientists in order to make the case for anthropogenic global warming. Instead, they argued that the IPCC report had been inherently conservative in its proclamations, precisely because they had overplayed sceptical concerns in order to facilitate approval from reluctant member countries. Since then Weaver has become more vocal in his attempts to bring attention to the issue of anthropogenic climate change and increasingly critical of what he sees as inadequate political action. In his book, *Keeping Our Cool*, Weaver professed his belief that "global warming is the single biggest issue facing humanity today" and was highly critical of the Canadian government's "obstructionist" positions and policies on climate change (Weaver 2008, pp. 28, 274). Due to such public engagement activities, 13 years later Weaver would again find himself in a dispute with Corcoran, now a columnist for the *National Post*.

The *National Post* is typically considered to be a "conservative" or "right-leaning" newspaper (Uzelman et al. 2005; DiFrancesco and Young 2011). Editorially they have taken a pronounced sceptical view of climate change, frequently running editorials which challenge scientific theories of anthropogenic climate change, denigrate climate scientists and institutions like the IPCC, and offer ample space to scientists offering alternative theories of global warming. Ironically, one of Weaver's first more pronounced forays into public engagement was a opinion piece published in the *National Post* regarding an earlier piece by skeptical scientist Fred Singer (Weaver 1999). The first bit of negative attention regarding Andrew Weaver in the *National Post* was a few brief paragraphs in a vociferous and scathing attack by Corcoran of an article that ran in the *Globe and Mail* written by Charles Montgomery, which critiqued the activities of Canadian climate skeptics (Corcoran 2006a; Montgomery 2006). In the editorial, without any clear segue, Corcoran suggests that Montgomery's "lapse on facts" had been influenced by Weaver's questionable (as alleged by Corcoran) criticism of the prominence afforded to climate sceptics in Canada, as well as perceived Government inaction. Corcoran goes on to state that Weaver is amongst the most politically-driven players in climate change debate, loudly implying that his scientific views have been compromised by partisanship.

Weaver responded with a letter asking for corrections to be made, mostly regarding factual errors regarding his place of employment (Weaver 2006). The *National Post* issued an editorial statement acknowledging error and offered corrections to information regarding Charles Montgomery, but not to Weaver. Corcoran himself penned a mock apology that conjectured that no errors had been made (Corcoran 2006b). When he called Weaver a "civil servant," for example, Corcoran claimed that he did not mean it in the conventional sense as somebody who works for a government agency, and suggested that any discerning reader would be able to pick up on that fact that Weaver was a metaphorical servant of the state because he has received funding from the government at various points in his career. Corcoran also repeated a claim that had appeared in an earlier piece in the *Post* that Weaver had called a paper allegedly debunking the so-called Hockey-Stick Graph "pure and unadulterated rubbish." Weaver pointed out that the quotation was erroneous and had already been the subject of any earlier editorial correction and retraction, but the *Post* decided not to re-issue a correction in this instance.

After 3 years of relative calm, Peter Foster (2009), another *National Post* columnist, wrote an attack piece about Weaver that labeled him "Canada's warmist spinner-in-chief" and said he had become part of the "left coast Suzuki-PR-industrial complex" (for non-Canadian readers, this reference is to David Suzuki, Canada's most well-known environmental activist, and most derided by the likes of the *National Post*). The focus of the article was a recent set of break-ins at the University of Victoria, which Weaver speculated were targeting climate scientists in an attempt to discredit or intimidate them (the break-ins occurred shortly after stolen e-mails between members of the University of East Anglia's Climate Research Unit and other climate scientists were released to the public). Foster claimed that Weaver blamed the fossil fuel industry for the break-ins, and called on him to produce evidence for the allegations, though Weaver had never made any such accusation. Foster then sarcastically tried to draw a parallel between

Weaver's flimsy evidence for his allegations and his apparent without-basis belief in anthropogenic climate change. The following day, Corcoran wrote another piece about Weaver, repeating the false claim that he was "publicly blaming the oil industry for the break-in at his office," and ridiculed Weaver after it was revealed that there had in fact been multiple break-ins in various buildings at the University of Victoria campus, suggesting that Weaver's office was not intentionally targeted (Corcoran 2009).

The following month the *National Post* wrote a front-page story incorrectly suggesting that Weaver was leaving the IPCC and wrongly claimed he was calling for "the replacement of IPCC leadership" and "institutional reform." It went on to again falsely state that Weaver had concocted a "cockamamie" story that the fossil fuel industry was responsible for the break-ins to his office. The piece accused the IPCC of fraud, manipulation, and distortion, and implied Weaver's agreement with this accusation was the cause of his alleged departure from the IPCC (Corcoran 2010). Finally, a few days later the *Post* ran another editorial claiming Weaver's "accusations" about the break-ins into his office were meant to distract from the attention being given to "climategate." It further went on to claim that Weaver's career was dependent on "global warming panic," and implied that Weaver helped manufacture such panic for financial gain. Overall, the attacks in the *National Post* can be interpreted as an attempt to convey Weaver as incompetent, dishonest, fraudulent, and manipulative. In response to these attacks, Weaver contacted the *National Post* asking them to retract various false statements. The *Post* refused, and Weaver ended up suing them for libel. The case is still pending (Littlemore 2010).

Almost a year later in an editorial written for the *Canada Free Press*, an online conservative tabloid, Timothy Ball, an emeritus geography professor from the University of Winnipeg, repeated the false claims first printed in the *Post* about Weaver leaving and criticising the IPCC.[2] Ball also implied that Weaver had bribed grad students with research funding in order to secure personal financial benefits from further government funding, and refused to debate in public out of fear that it would expose his incompetence. Overall, Ball conveyed Weaver as intellectually deceitful, lacking in expertise, and corrupt. He flatly conjectured that Weaver "knew very little about climate." After being contacted by Weaver, the *Canada Free Press* promptly removed Ball's piece from its website, and issued a public retraction and apology. Ball offered no personal apology or retraction, and was subsequently also sued for libel by Weaver. The case is also still pending.[3]

[2] The original piece has since been removed by the *Canada Free Press*, but their subsequent retraction and apology (which is also unable to be found on the *Canada Free Press* website) can be found here: "Andrew Weaver Wins One Against Canada Free Press, No News on National Post Libel Case," *Carbon Fixated* (blog), January 21st, 2011, http://carbonfixated.com/andrew-weaver-wins-one-against-canada-free-press-no-news-on-national-post-libel-case. Timothy Ball was also the focus of the above mentioned Charles Montgomery *Globe and Mail* article about Canadian climate change skeptics.

[3] Many of the details of the original piece can be found in Weaver's Statement of Claim regarding the lawsuit (Littlemore 2011)

Weaver receives so much attention from sources like the *National Post* because he has been a vocal critic of the perceived ineffectiveness of government policy, and has engaged in various advocacy and public-engagement activities. It might be worth noting that there is a classic asymmetry in the *National Post*'s treatment of climate change. Evidence of Weaver's deviance is not merely that he has engaged in political activities thereby transgressing scientific standards of neutrality and impartiality, since the *Post*'s editors do not consistently criticise the political activities of those scientists whose views they champion. The evidence of Weaver's transgression is simply that he believes that human-caused climate change is real. In the eyes of the *Post*'s editorial team, this belief is the contemptible political act that betrays the ideals of "objective" science. According to the *Post*'s view, the fact that is wholly taken for granted is that anthropogenic climate change is not real, and thus anyone denying this fact must have been compromised by political ideology, financial pursuits, or incompetence. The point will not be lost on STS scholars that others make similar (though arguably better substantiated) arguments about the beliefs about climate change sceptics and deniers (Oreskes and Conway 2010).

Overall, these particular pieces, and the general editorial position on climate change of the *National Post*, offer little in the way of serious and substantial critiques surrounding the many concerns surrounding climate change or any other politically-complex socio-scientific issues. The criticisms of Weaver are mostly simple ad-hominem attacks, aimed to undermine his credibility and thereby indirectly bolster doubts about theories of anthropogenic climate change. Above all else, Weaver's being singled out is chiefly a function of his effectiveness; the public was listening to what he had to say. This is the risk that attends any activist. The more effective one is at promoting their cause, the more they will receive unwanted attention, regardless of how reasonable, well-intentioned, or fair their position.

Activist scientists are particularly vulnerable to dismissive critiques, since the ethos of activism is assumed to be antithetical to the ethos of science. And to the small extent there are any substantive arguments being offered by the *Post*, they revolve around (in this case, incoherently applied) rules of scientific propriety. In short, the ostensible issue at hand is a classic one: to what extent do political or financial interests influence, or corrupt, science? However, given that many STS analysts will object to the framing of this question, since, arguably all science is political, the more interesting sociological inquiry would be to understand the ways that concerns about political interference and scientific propriety shape scientific practice. And for activist scientists a more productive question is: how can scientists best navigate public controversies?

Historical Lessons from Ecology

Various researchers have remarked that in the decades following Second World War, science came to occupy, in various ways, a much more public and political role. While the reasons for this shift are complex (and the extent of this shift is also debatable), commentators often point to two prominent factors: First, the increased scrutiny and demands for accountability of scientists working in military capacities, which stemmed from reflections on the development of nuclear weaponry; second, a related development, the emergence of the environmental movement (Kasperson et al. 1980, pp. 11–23). Books like Rachel Carson's *Silent Spring*, Barry Commoner's *Nature, Man, and Technology*, and Paul Ehrlich's *The Population Bomb* were symbolic of the new public and political roles demanded of and taken on by scientists (it is less than a coincidence that STS, with its critiques of traditional conceptions of the politically neutral nature of scientific knowledge, began to emerge alongside these developments). Dorothy Nelkin explored these changes in an early science studies paper considering the ways in which public and political demands for ecological knowledge affected the professional activities of American ecologists (Nelkin 1977). While many ecologists and ecologically-minded scientists (typically, various kinds of biologists) welcomed or embraced their newfound public and political import, or were even actively engaged in cultivating it, others found the attention disconcerting.

Operating under an assumption that politics was an "alien element, essentially destructive of scientific endeavour," or at the very least a potentially dangerous element, ecologists attempted to more strictly define what constituted ethical ecological practice, especially insofar as public or governmental consultations were involved (Haberer 1969, cited in Nelkin 1977, p. 81). Certainly, a large part of these moves were largely for professional protection. Consulting firms and think-tanks proliferated, some based on suspect-credentials, to capitalize on the public and governmental demand (and research grants and consulting fees) for expert ecological knowledge. In 1974, there were 1,130 private consulting firms actively working on environmental issues (Nelkin 1977, p. 81). Ecologists lacked any sort of governing body that offered professional certification, and many feared that the emergence of these so-called "instant experts" would jeopardize the legitimacy of professional and academic ecologists, or perhaps even pose as competitors in the market for expert knowledge.

But behind these issues of professionalization lay more fundamental concerns about the proper role of scientists, about the inherent conflict between politics and science, and the constitution of scientific knowledge and practice. Ecologists were being recruited by various groups, from environmental activist organizations like the then-fledgling Greenpeace, to various federal American agencies operating under the recently established National Environmental Policy Act, to a multitude of industrial corporations looking for scientific assessments of their environmental impacts.

Most ecologists accepted that ecology was an inevitably politically-entrenched science. It was integral to directing environmental policy, something many ecologists felt was of paramount importance. Ecology did not seem to have the luxury of feigning impartiality or neutrality. It was inescapably normative, insofar as it constituted an integral component of considerations of how to best organize the relationship between humans and their environments. This is why Paul Sears called ecology a "subversive science." It had the power to challenge the "assumptions and practices of modern societies, whatever their doctrinal commitments" (Sears 1964, p. 11). However, in order to meaningfully, sincerely, and honestly make these challenges, it needed to maintain certain standards of scientific knowledge. And so came the boundary making. An internal-external divide was erected, meant to protect the integrity and credibility of ecologists. This divide not only followed classical boundaries, protecting questions and principles supposedly internal to scientific practice, such as hypotheses, methodologies, and theoretical frameworks, from external influences like vested ideological or industrial interests, but also concerned the proper locale of ecologists' activities. Many ecologists were wary of working directly for corporations, since there were few guarantees about how their work would be used or presented, but sometimes felt that such arrangements might put them in a better position to influence change and have their voices heard. The dilemma was whether to operate "inside the system in hopes of preventing destructive decisions, or to remain outside, relatively powerless but at least maintaining integrity" (Nelkin 1977, p. 83).

Additionally, there were deeper concerns about the overall effect of policy-driven research on the character of ecological scholarship. First, there was a concern that ecological research was being conducted outside of the established parameters of scientific research, especially in consultation arrangements. Studies would be published by private consulting firms, or by industry-contracted scientists, that did not go through peer-review, but were nonetheless being considered in policy discussions. Furthermore, even the research that did meet peer-review criteria, was largely conducted ad-hoc for the purposes of short-term exigencies, like measuring pollutants. The result was that basic research was being marginalized, and little work was being done to establish fundamental ecological theories with predictive value, which was widely seen as necessary for establishing the long-term viability and credibility of ecology.

It was not just threats from the most egregious misappropriators of scientific credibility, or those employing the authority of ecology for disreputable or misleading ends, that were the cause of consternation among scientists and heightened boundary-work. While many ecologists were deeply sympathetic to, or even active in, projects for environmental protection and sustainability, they had at times uneasy relationships with citizen environmental groups, and worried about the effects this would have on their credibility as scientists. Just as the likes of Rachel Carson in the midst of the environmental movement were called "eco-doomsters," today scientists like Weaver are pejoratively dismissed as "alarmists." Carson's approach and positions were vehemently attacked, even at times by ecologists sympathetic to the environmental movement. Her scientific credibility was frequently the focus of such critiques. Biologist Ira Baldwin, though similarly concerned about the health risks associated with pesticides, took to the pages of *Science* to critique the "dramatically written emotional appeal" of *Silent Spring*, and urged concerned citizens instead to consult the "careful study" and "sound judgement based on facts" to be found in the National Academy of Science reports on pesticides (Hecht 2011, p. 292). Weaver has taken such threats to credibility to heart, professing at the onset of *Keeping Our Cool*, that he would not "sensationalize [climate] science with outlandish claims of apocalyptic proportions" (Weaver 2008, p. 27).

However, despite these risks, the historical case of ecology demonstrates that the contextual demands of activism and advocacy prove inescapable, even for the most ostensibly apolitical scientists. These demands emerge from the exigencies of societal concerns in which certain kinds of expert knowledge are more implicated than others. Just as the public role demanded of ecologists was seen as unavoidable, climate scientists find themselves deeply implicated in public debates, whether they would pursue them or not.

A View from the Front Lines

As a part of a larger project exploring the relationship between scientists' interaction with the media and their research practices, I interviewed a climate scientist who had found himself occasionally called upon as an expert authority for media stories, despite making no concerted effort to pursue such relationships. His thoughts offer valuable insights into the dynamics of science popularization and public engagement, boundary work, and the conflicting ethical or normative demands encountered by scientists.

As various critics have noted, "[activist groups] who need scientific expertise do not necessarily share scientific values" (Nelkin 1977, p. 83; Latour 2004; Giddens 2011). Indeed, various activist or advocacy groups do not necessarily even share the same social values as activist scientists. For those holding or producing expert knowledge in high public demand, the threat of misappropriation is a persistent concern. Like the ecologists who were wary of the surge of non-peer-reviewed research that was able to proliferate because of increased public and political demand, my interviewee also worried about the effect that climate change debates were having on established means of making a scientific argument:

> If you want to make a scientific argument, the way to do it is to publish a paper in a peer reviewed scientific journal. That's the way that science works in every single field, the peer review system, you just don't take people seriously unless they're willing to submit to peer review. For most science, peer review works pretty well – it may take a few years, and things will get published that are wrong and have to get corrected. If you want to engage in that debate you have to step into the arena and argue in that way. You just can't sit by the sidelines and throw stones. You know, scientists find it incredible that anybody would be taken seriously that wouldn't publish serious articles, but of course, the press doesn't make a distinction.

Even more of a concern was that activism or advocacy presented a threat to images of credibility. The possibility of disparaging personal critiques, or, more importantly, the danger of causing reputational damage to the community of climate scientists were causes for serious apprehension in deciding whether or not to interact with the media or pursue public engagement activities. He believed that such activities were often cited as evidence in "coverage of climate science from the skeptics, or from the *National Post*" to present an "image of climate scientists as this club that just wants to fund themselves [or are] going for the limelight."

However, my interviewee felt that the credibility of science was inherently self-regulating. Scientists who spent inordinate amounts of time in popularization activities, or those that inflated their credentials to capitalize on policy interests in climate were "never taken as seriously by the rest of the community, anyway. Typically, [their work] tends to be kind of flashy, or the phrase we tend to use for that kind of science is 'quick and dirty.'"

Despite these concerns, like the ecologists in the midst of the environmental movement, my interviewee felt that there was a common sentiment among the community of climate scientists that they had an obligation to speak up if they saw their knowledge was of pressing public interest. The question was not whether scientists should engage in advocacy, but how, and "how much":

> I don't think you can tell scientists to stop being activists; that would be crazy. Scientists are also human beings and citizens and if they feel that people aren't taking the science seriously enough, they're going to say something. [...] Some believe that there's a crisis – that we're just heading towards a cliff – and they feel that they just have to speak out. So, I think there's the well-intentioned advocate who crosses the line, because they just feel that they have to."

For my interviewee, while "crossing the line into advocacy" was perfectly admissible, it was imperative that the line remain as clear and distinct as possible. Science involves "conducting sober analyses" and "proving things to a very high level of confidence," but one has to "be able to somehow separate the advocacy role from the sober reports which do not get into advocacy."

Our conversation did not turn to exploring what my interviewee saw as the fundamental basis for establishing the essential divisions demarcating sober scientific analysis from advocacy. Much work in STS and the philosophy of science has been done to problematize attempts to establish this division in some internal logic of scientific knowledge (e.g., Barnes 1974; Shapin 1992). The majority of discussion about internalism-externalism in early STS focussed on the analytical legitimacy of this categorical division, that is, how useful is it in explaining historical and sociological developments of scientific knowledge. Much of this work was philosophically tinged, implicitly (or at times explicitly) weighing in on normative questions surrounding rules of scientific discovery. Despite claims to impartiality and analytical distance from the categories of sociological actors, this supposedly descriptive work often contained implied prescriptive critiques of the conceptual ordering devices of scientists. That is to say, my interviewee's desire for clear demarcations between science and advocacy might be implicitly rejected as conveying a philosophically dubious notion of science. However, again in the name of impartiality, STS scholars would often absolve themselves from the reconstructive task of imagining productive and useful boundary-making. In recent years there have been various encouraging exceptions (Lahsen 2005).

For what it's worth, it is doubtful that my interviewee supposed that the demarcation between science and activism followed any neat rubric, as he admitted that establishing this distinction was a constant challenge: "You have to be able to *somehow* separate the advocacy role from the sober reports which do not get into advocacy [...] I think the only answer I would give is that *somehow* people have to be able to wear two different hats and maybe separate those hats." What is clearer is that for my interviewee, boundary-making served an important pragmatic function.

The "internal" integrity of scientific research is part and parcel of broader concerns about expert credibility. Furthermore, the integrity of the "sober analyses" and the credibility of scientists are crucial for achieving the aims of advocacy and activism. Indeed, the effectiveness of the latter is seen as wholly dependent on maintaining a clear and distinct boundary from the former. Thus, for my interviewee the ethical responsibility incumbent on scientists does not necessarily open up the possibility of more porous boundaries between science and activism, nor indeed challenge those boundaries. Instead, demarcations are of paramount importance in maintaining scientific credibility, upon which depends effective advocacy and activism.

All Science Is Political, but Politics Are Complicated

It is in the separation between questions of sober scientific analysis and activism that lies my interviewee's hope for "science-based advocacy." Criticisms from STS hold such separations to be social constructions, not inherent to the nature of either science or politics. In practice, policy rarely flows directly from "sober scientific analyses." There are innumerable border crossings, and indeed, acts of construction are intractably political. But, for my interviewee this is precisely why political acts and social constructions are so integral to maintaining the credibility that ultimately allows for meaningful advocacy.

Thomas Gieryn argues that such boundary work is largely rhetorical, and he chiefly places the rhetorical force of science in the purview of scientists. The image of science cultivated by scientists functions at least in part to establish credibility of certain kinds of expert knowledge. A simplistic conclusion of these arguments is that science isn't "really" value-free, or objective, or politically neutral, but these are rather things that scientists say about science so as to bolster the authority of their knowledge claims. But, of course, such rhetoric is only effective if the public values such conceptions of science.

The intensely discussed "climategate" is revealing here. Though typically intemperate, the *National Post* captured much of the public sentiment concerning this incident, claiming it revealed the scientists involved to be "intensely preoccupied with politics" and "perpetrators of fraud." Most STS scholars found the episode rather unremarkable, since decades of historical and sociological studies have made STS scholars acutely familiar with the conclusion borne out by the "climategate" episode, namely that "scientists are not infallible and that they can be idiosyncratic and petty" (*Toronto Star*, August 31, 2010; see also Ryghaug and Skjolsvold 2010). Moreover, based on these constructivist conceptions of science, most STS scholars do not expect scientists or science to be politically neutral. But publics often do, or at least they expect scientists to aspire to certain ideals. "Climategate" would not have been so intensely discussed otherwise (see Boykoff 2011).[4]

Some will tie public expectations for ideals like political neutrality in science into broader rhetorical strategies. They are simply a result of the self-reinforcing authority of science. It is rhetorical through and through. But, again, there are largely unexplored normative consequences to these kinds of conceptions. Most importantly, to challenge conceptions of political neutrality as socially constructed is to imply that scientists are wrong to think like this. The public is also wrong to think like this. In some sense they have been "fooled," by the rhetorical success of scientific experts and others who have a vested interest in maintaining the cultural authority of science. This poses challenging questions for reconstructive or activist STS, most notably, how to reconcile critical STS perspectives with democratic ideals. Engaged STS has traditionally encouraged democratic public engagement in cases where publics have been marginalized, or these publics' conceptions of science are congruous with STS conceptions of science (Wynne 1992). But what about cases when the public's demands or conceptions of scinece are in conflict with lessons from STS? How can STS engage with publics it "disagrees" with?

Furthermore, while Gieryn and others are arguably correct about the rhetorical nature of scientific credibility, there is a risky tendency of thinking this means that it is "merely" rhetorical. This downplays the degree to which the "communication end" of science is based on the "production end." Rhetorical credibility is predicated on practices, institutions, epistemologies, and ethics. And rhetoric can have varying degrees of substance. Similarly, the realization that divisions between politics and science are not clearly delineated by some universal logic of scientific knowledge should not lead to the conclusion that such constructed boundaries are ineffective or unnecessary. Constructions and rhetoric are tied to not-inconsequential actions, and can facilitate or hinder not-inconsequential activisms.

The statement that "all science is political" is a common, but arguably tautologous, refrain. What constitutes something being "political" is so varied, it does not say very much in particular. Originally, the recognition that science is inescapably political was seen as a dangerous or subversive notion as it conflicted with longstanding ideals of political neutrality or impartiality. It helped expose rhetorical appeals to authority. And it can still lend itself to reconstructive STS to be enlisted for activist projects. But the symmetrical sociological analysis eventually masked the normative work being done, as well as hindered reconstructive possibilities. Symmetry may have inadvertently introduced a kind of moral equivalence. Not all science is political in the same ways. Just as Patrick Hamlett (2003, pp. 112–130) notes that the jump is very small from recognising that all science is socially constructed to imagining ways to reconstruct it, the recognition that all science is

---

[4] In the period from 2000 to 2010, "climategate" was most likely discussed more than any other issue surrounding climate change.

political is a liberating one. Acknowledging the inescapable political nature of scientific knowledge, the question then becomes, in what ways *should* science be political?

Science is mobilized in the name of particular interests, or in public debates, or for specific projects, all of which involve political boundary work and social constructions, and all of which are contingent and situated. Some mobilizations are desirable, others damaging. Some are likely, others latent, but none are inevitable. GMO research is not destined to be co-opted by dominating corporations, nuclear research is not destined to become a focal point of foreign policy and international relations, and climate change action is not destined to be stalled by *certain kinds* of politics. Thus, concerns about *certain kinds* of politicization of science are not unreasonable or naive. Notwithstanding intractable philosophical issues, and not forgetting the complex nuances of science revealed by constructivist analyses, science is at least in small part about producing useful and robust accounts of the world. The possibility that science can be co-opted for nefarious or selfish purposes is a real and persistent concern. Pharmaceutical companies downplay the risks of drugs, oil companies cover-up the extent of ecological disasters, and governments justify inaction on pressing environmental issues with the authority of scientists willing to be enlisted for these ends. The frequent inability of status-quo social, cultural, political, and economic systems to effect these ethical considerations is what compels science activism. This is what makes some constructions, some boundaries, some politics, more desirable than others.

Activism and Power

Edward Woodhouse and colleagues (2010) define activism as inherently grassroots, or at least, belonging to the domain of the relatively disenfranchised: "By the term 'activism,' we refer to a range of methods used by groups with relatively little institutional power attempting to influence opinion, policy or practice toward democratic and other normative ends" (pp. 297–319). The often not-explicitly-stated goal of many STS analyses of science has been to undermine the cultural authority of science, since this power is largely unjustified in that it is unanswerable to publics in democratic ways (Jasanoff 1990). How then, are we to understand activist scientists?

Massimiano Bucchi argues that deviations from traditional lines of science communication occur typically in crisis situations that demand unorthodox activities or the transgression of established boundaries. As my interviewee conveyed, this is precisely the chief motivation for scientists engaging in advocacy: they believe we are "heading off a cliff," and while perhaps not ideal, there is little alternative than to engage in advocacy. But the feeling of crisis not only stems from the severity and pressing nature of the problem of climate change, but also from the perception that official lines of action are failing. Contrary to the authoritative image of science presented in so much STS work, climate scientists find themselves relatively powerless in the face of competing interests. Weaver and other climate scientists, while comparatively powerful and authoritative in relation to the others compelled to take action on climate change, cannot, even with their influence, sufficiently compel governing bodies to take climate change seriously. Generalizations about the authority of science are thus not particularly helpful here. Some science and scientists have immense, and arguably unjustified, social, political, and economic authority. Others are relatively powerless. Contra Gieryn, the authority is not chiefly rhetorical; it is dependent on various social, political, and economic assemblages. Scientific knowledge can be marginalised if it poses a threat to powerful interests. Is climate change a case in which scientists have too little power, not too much?

However, power differentials cannot in themselves be a standard which determines what interests to rally behind, though questions of power have proved to be a pervasive and seductive starting point for discussions of the relationship between ethics and expertise. What if things were actually how climate sceptics imagined? That sceptical voices were being silenced in the name of powerful environmental activists, with vested economic and social interests that were served by the authoritative claims of climate scientists? Here we perhaps run up against the limits of symmetry since climate sceptics argue their case along similar lines as so many STS critiques of the excessive and co-opted cultural authority of science. According to sceptics the activities of powerful institutions like the IPCC are socially unjust: they promise to stall economic development, cost jobs, unfairly re-distribute wealth, and ultimately hinder human well-being and prosperity. Of course, this is not to say that sociological analysis is not immensely valuable. Symmetrical methodologies need not substantiate symmetrical conclusions. This is the whole point of recognizing that while both climate activists and climate sceptics are engaged in politics, some politics better promote "more democratic, environmentally sustainable, socially just, or otherwise preferable civilizations" (Woodhouse et al. 2010, p. 298).

References

Barnes, B. (1974). *Scientific knowledge and sociological theory*. London: Routledge & K. Paul.

Boykoff, M. (2011). *Who speaks for the climate? Making sense of media reporting on climate change*. New York: Cambridge University Press.

Corcoran, T. (2006a, August 23). Hockey sticks and hatchets. *National Post, p. FP18.*

Corcoran, T. (2006b, August 31). The hockey stick: They shoot, don't score. *National Post*, p. FP17.

Corcoran, T. (2009, December 10). Weaver's Web II, *National Post, p. FP11.*

Corcoran, T. (2010, January 27). Climate agency going up in flames. *National Post, p. A1.*

Delgado, A. (2010). Activist trust: The diffusion of green expertise in a Brazilian landscape. *Public Understanding of Science, 19*(5), 562–577.

Denman, K., & Weaver, A. (1996, July 22). Global warming. *Globe and Mail*. Letter, p. B2.

DiFrancesco, D. A., & Young, N. (2011). Seeing climate change: The visual construction of global warming in Canadian national print media. *Cultural Geographies, 18*(4), 517–536.

Foster, P. (2009, December 9). Weaver's Web. *National Post, p. FP13.*

Frickel, S. (2004). Scientist activism in environmental justice conflicts: An argument for synergy. *Society and Natural Resources, 17*, 359–366.
Giddens, A. (2011). *The politics of climate change*. Malden: Polity.
Haberer, J. (1969). *Politics and the community of science*. New York: Van Nostrand Reinhold.
Hamlett, P. (2003). Technology theory and deliberative democracy. *Science, Technology & Human Values, 28*(1), 112–130.
Hecht, D. (2011). Constructing a scientist: Expert authority and public images of Rachel Carson. *Historical Studies in the Natural Sciences, 41*(4), 277–302.
Jasanoff, S. (1990). *The fifth branch: Science advisers as policymakers*. Cambridge, MA: Harvard University Press.
Kasperson, R., et al. (1980). Public opposition to nuclear energy: Retrospect and prospect. *Science, Technology, and Human Values, 5*(31), 11–23.
Kroll-Smith, S., & Floyd, H. H. (1997). *Bodies in protest: Environmental illness and the struggle over medical knowledge*. New York: New York University Press.
Lahsen, M. (2005). Technocracy, democracy, and U.S. climate politics: The need for demarcations. *Science, Technology, & Human Values, 30*(1), 137–169.
Latour, B. (2004). *Politics of nature: How to bring the sciences into democracy*. Cambridge, MA: Harvard University Press.
Littlemore, R. (2010). Climate scientist sues national post. *DeSmogBlog*. Accessed at http://www.desmogblog.com/climate-scientist-sues-national-post
Littlemore, R. (2011, February 4). Andrew Weaver sues Tim Ball for libel. *DeSmogBlog*. Accessed at http://www.desmogblog.com/weaver-sues-tim-ball-libel
McCormick, S. (2007). Democratizing science movements: A new framework for mobilization and contestation. *Social Studies of Science, 37*(4), 609–623.
Montgomery, C. (2006, August 12). Meet Mr. Cool: Nurturing doubt about climate change is big business. Globe and Mail, p. F4.
Nelkin, D. (1977). Scientists and professional responsibility: The experience of American ecologists. *Social Studies of Science, 7*, 75–95.
Oreskes, N., & Conway, E. (2010). *Merchants of doubt: How a handful of scientists obscured the truth on issues from tobacco smoke to global warming*. New York: Bloomsbury Press.
Ryghaug, M., & Skjolsvold, T. M. (2010). The global warming of climate science: Climategate and the construction of scientific facts. *International Studies in the Philosophy of Science, 24*(3), 287–307.
Sears, P. (1964). Ecology: A subversive subject. *BioScience, 14*(7), 11–13.
Shapin, S. (1992). Discipline and bounding: The history and sociology of science as seen through the internalism-externalism debate. *History of Science, 30*, 333–369.
Toronto Star. (2010, August 31). Climate debate: Time to move on. Editorial, p. A14.
Uzelman, S., Hacket, R., & Stewart, J. (2005). Covering democracy's forum: Canadian press treatment of public and private broadcasting. *Critical Studies in Media Communication, 22*(2), 156–169.
Weaver, A. (1999, September 2). Clash over climate change: Singer article clouds the picture. *National Post, p. CO7*.
Weaver, A. (2006, August 31). My information was not false. *National Post*. Letter, p. FP17.
Weaver, A. (2008). *Keeping our cool: Canada in a warming world*. Toronto: Viking.
Woodhouse, E., et al. (2010). Science studies and activism: Possibilities and problems for reconstructivist agendas. *Social Studies of Science, 32*(2), 297–319.
Wynne, B. (1992). Misunderstood misunderstanding: Social identities and public uptake of science. *Public Understanding of Science, 1*, 281–304.

Chapter 18
Passive No More

Leo Elshof

Abstract This chapter will examine from a critical science and technology education perspective how corporate activism in collusion with neoliberal thinking has driven a 'war against reality' (Grant J, Denying science. Prometheus Books, New York, 2011) when it comes to working on the toughest collective action problem humanity has yet faced, climate change. Although I will examine the issue primarily from a Canadian context, similar situations exist in nearly all developed countries. In posing a number of critical questions for educators concerned with activism it will explore the role of media, public relations, agency and citizenship, transnational corporations and activism in education. Towards this end, this paper will focus on disclosing power as a precursor to activism in the public interest. Specifically it will question the wisdom of advancing an anti-scientific agenda in Canada concerning climate change, and what lessons young people might learn by this examination.

Keywords Critical education • Corporate activism • Propaganda • Astroturf • Public good

Introduction

Our ability as citizens to collectively address major issues of vital public concern like climate change is increasingly drawn into question when industry propaganda abetted by neoliberal political ideology attempts to frame and set the boundaries of public discourse around this issue. Educators need to help young people understand how openness and transparency in both science and in politics is essential to the development of balanced and effective public policy. 'Good science' alone in the

L. Elshof (✉)
School of Education, Acadia University, 15 University Avenue,
Wolfville, NS B4P 2R6, Canada
e-mail: leo.elshof@acadiau.ca

L. Bencze and S. Alsop (eds.), *Activist Science and Technology Education*,
Cultural Studies of Science Education 9, DOI 10.1007/978-94-007-4360-1_18,

case of climate change is clearly insufficient as short-term profits and short-term economic thinking dominate public policy.

From the Canadian government's perspective it's not 'if' we should continue to expand the Alberta tar sands, but rather how fast can existing environmental laws, review procedures and safeguards be gutted and re-written to permit the maximum exploitation as soon as possible. The Canadian Association of Oil Producers (CAPP) having worked with other business and industry groups through lobbyists for years to stall and eventually kill the Kyoto Accord in Canada, now take out full page ads in Canada's newspapers extolling the virtues of the tar sands for Canada's business community, "The oil sands are tremendous for Ontario…It's great to be part of a winning team" one businessman gushes, fitting the public relations theme of "Energy at work for all Canadians" (CAPP 2012a, b).

It is crucial that young people become aware of the various forms of corporate activism that work to disempower citizens, to obscure and hide activities ranging from influence peddling to environmental destruction from public accountability and scrutiny in the interests of concentrating wealth and power. When the interests of the oil and gas industry are framed by the Canadian government as 'Canadian interests', hegemony is at work, as Orlowski (2011) defines it:

> Hegemony refers to the ideal representation of the interests of the privileged groups as universal interests, which are then accepted by the masses as the natural political, and social order. (p. 2)

When it comes to actually listening to what the scientific community has to say concerning the problems of the day as Henderson (2012, p. 7) argues: "Precisely what politicians think is less important than how they think". When it comes to politicians actually employing the best scientific advice available to inform the climate legislation policies they develop, the last 20 years have been an abysmal failure.

In terms of activism, the current Federal government is one of the most activist in a generation, quickly moving to polarize Canadians who care about the environment. Activism is often thought of as something protestors engage in, the right-wing media throws the word 'activist' about as a pejorative label associated with left of center causes. In this paper we will explore neoliberal and corporate activism that seeks to protect the status quo and minimize change to the existing economic and political hierarchy.

Awakening Dissonance

The question of whether to promote *activism* in science and technological education is important. On a basic level if as teachers we're not preparing students to be 'activists' with respect to helping them activate and actualize their knowledge of science and technology within their communities and in the wider world, then what exactly are we preparing them for? On a more fundamental level, in order to become 'active' young people need teachers who will help them understand how the real world, not an idealized or theoretical world, actually operates. Informed activism in science and technological education begins with revealing the world in all its political, cultural and ideological diversity, all posing constraints on the practice and application of science and technological development to today's problems.

There are ample opportunities for passivity with respect to teaching and learning processes. Poor pedagogical practices can make the most interesting of subjects boring and listless, sterile, devoid of meaning and connection to real life activities and experiences for youth. Poor curricula can promote passivity, leaving young people struggling to understand how any of the material they are learning connects to the world they live in. Passive curricula promote a monoculture of understanding, a way of understanding the world that is often very closely aligned with the dominant worldview and the 'received wisdom' of neoliberal capitalism. Passive curricula doesn't encourage young people to ask disruptive 'jugular' questions related to power, and influence, it fails to ask the difficult questions of neoliberal progress, namely whose world benefits and whose world loses? Passive curricula can reinforce feelings of powerlessness in young people leaving the impression that they have little to meaningfully contribute in terms of analytical provocative insight or new innovative ideas to apply to the planet's innumerable problems. Passive curricula supports the status quo situating young people at the nexus of consumption, consuming prepackaged notions of what a meaningful life consists of, consuming and internalizing models for behaviour, careers and the economic means to attain it.

Any discussion involving students in topics related to social activism often elicits howls of indignant complaint from the political right, accompanied by charges that teachers are involved in the indoctrination of young people and are thus abusing their authority. Canada's right-leaning national newsmagazine, *Macleans* recently ran a cover story whose headline screamed, "Why are schools brainwashing our children? Protesting oil pipelines, celebrating polygamy: is the new 'social justice' agenda in class pushing politics at the expense of learning? (Reynolds 2012, p. 1)." The authoritarian thinking which informs this perspective implies that learning should be 'fact-based', decontextualized, unconnected to real-world issues of power, privilege and justice. However critical education should be involved in assisting young people to see the traps of indoctrination, groupthink and the development of ideological and disciplinary blind spots.

The litany of global ecosystem damage continues unabated, even accelerating in spite of our knowledge. Despite this, the stories of the scientific research that painstakingly documents our collective stupidity are ephemeral in our media saturated world. The 'bad news' stories may break through to our consciousness for a few moments before we're on to the next media spectacle that we're told should command our undivided attention. Passive science and technological education may leave students unaware that the rejection of science, the rejection of critical rational thinking has become commonplace in the highest political offices in North America.

The rejection of science and evidence is of course not limited to the political right, left of centre opponents of genetically modified foods, vaccines, nuclear energy and modern pharmaceuticals are also open to the charge of being 'anti-scientific' (Berezow and Campbell 2012). What young people may not be aware of is how many corporations employ large teams of public relations experts, and lobbyists of all sorts to advance their narrow interests, to shape public opinion, to diminish the notion of 'the public good' in order to maximize their profits.

There are many diverse paths to activism, all involve an awakening of dissonance, what they have in common according to Ricketts 'is the moment when a person stops and thinks, this is not right and I'm going to do something about it' (Ricketts 2012, p. 249). Breaking the 'habit of passivity' involves overcoming self-limiting beliefs that preclude action, some of these self-limits may include feelings of inferiority because we may not be 'experts' or we 'don't know enough to take a stand' (Ricketts 2012). Education has an enormous role to play in helping young people think deeply and critically about contentious issues involving science and technology, society and environment, the aim being that when they come to a decision concerning their role as a citizen-activist, their stance is informed by science, ethics and justice.

Activism for Inaction, a Complacency That Will Kill and Destroy

Concerned citizens might ask why despite repeated public polls indicating that North American citizens want governments to seriously tackle climate change, is so little being done? It would be a mistake to conclude that governments that are doing little or nothing to confront the problem are politically inactive. Over the past 6 years Canada's government has been very active in marginalizing climate science and promoting oil and gas interests above all other interests. Canadians are currently saddled with one of the most autocratic and secretive governments in their history (Nadeau 2010). Prime Minister Stephen Harper has the dubious distinction of leading the first Canadian government in history to abrogate an international environmental treaty by withdrawing from the Kyoto Accord. It is the first Canadian government to put the full diplomatic weight of the country at the service of promoting the energy industry and to publically state their goal of turning Canada into a carbon energy "superpower". Canada's Green party leader Elizabeth May gives some perspective:

> In the past, we've always had a reputation for environmental achievements that exceeded what we actually achieved, but we were at least compliant with global treaties until recently. . .The withdrawal from Kyoto is a devastating blow and a blot on our reputation in our role, as well as being a significant threat to our kids future. (Cited in DeSouza 2012, p. 7)

Many Canadians are concerned about what the Alberta tar sands, the largest industrial development on the planet, are doing to Canada's carbon footprint and to Canada's international reputation concerning environmental stewardship, environmental justice and responsibility. In this case 'many' Canadians does not include Canada's federal government, its actions clearly indicate that climate change and the environmental fallout of the tar sands, are minor concerns that interfere with it's perceived mandate to serve as an unapologetic booster of anything related to coal, oil and gas development. It has done nothing to educate Canadians about the perils of uncontrolled greenhouse gas emissions, in fact it cancelled the only small scale education project set up by a previous government that attempted to engage Canadians to reduce their enormous carbon footprint (one of the planet's largest) by an insignificant one ton of emissions. The federal government has done nothing to educate Canadians about sustainable consumption or its impact on our carbon footprint, despite the fact that Canadians have the eighth largest ecological footprint on the planet (WWF 2012).

Cognitive dissonance runs deep in political circles, despite the fact that the U.S. was experiencing it's hottest year in recorded history in 2012 (Morello 2013), climate change as a topic of discussion failed to break through into the televised presidential debates. The Republican party full of oil and gas industry lobbyists even denied that climate change was a real problem, preferring to attack wind and solar energy programs. The previous Canadian election also failed to elevate climate change onto the national political agenda, largely because in the previous election the highly successful politics of character assassination coupled with ridiculing the carbon tax waged by now Prime Minister Harper leader of Canada's Conservative Party against the Liberal's leader Stephan Dion, a proponent of a carbon tax, ensured that the issue remained on the sidelines. Again the reason for this lies in activism, the enormous political lobbying power of the oil, gas and coal industries that spends tens of millions lobbying politicians to do their bidding.

Growing Up on an Overheated Planet

In 2011 global carbon dioxide emissions rose another 3.2 % to 31.6 billion tonnes the highest ever, with China and its expanding use of coal leading the increase with a 9.3 % increase in GHG emissions (Rose 2012). New science indicates that thawing permafrost which underlies one quarter of the Arctic may release billions of tonnes of methane, an even more potent greenhouse gas than carbon dioxide (UNEP 2012).

Coal is still the dirtiest and worst fuel in terms of its impact on global warming, even so, coal's share of the global energy mix is rising. The International Energy Agency (IEA) estimates that by 2017 coal will surpass oil as the world's top energy source. Incredibly, by 2017 the world is expected to burn 1.2 billion more tonnes of coal than today, this is more than the current annual consumption in the U.S. and Russia combined (IEA 2012). A third of all the carbon dioxide emitted by humans is absorbed by the world's oceans and humans are making them more acidic than they have been for tens of millions of years. The Great Barrier Reef has lost half it's corals since 1985, partly as a result of climate change (De'ath et al. 2012). Oceanic phytoplankton which provides half of the oxygen we enjoy, has declined 40 % over the last century and about 6 % in the last three decades, (Boyce et al. 2010). Arctic sea ice coverage has reached a new record low (Fischetti 2012), the Western Antarctic region is one of the fastest warming regions on the planet, new research indicates that it is warming at three times the global average (Bromwich et al. 2012). The Western Antarctic ice sheet, one of the planet's largest is rapidly melting and may be vulnerable to long-term collapse with huge consequences for global sea level rise (Gailus 2012). The Amazon rainforest is showing the first signs of large-scale degradation as climate change is drying parts of it out (Buis 2013).

The environmental and social changes ahead being forecast by the planet's most eminent scientific and economic organizations forecast are by all accounts sobering. The conservative OECD in it's "Environmental Outlook to 2050" (OECD 2012) focuses on four major areas of global concern, climate change, biodiversity, water and the health impacts of environmental pollution.

> The costs and consequences of inaction are colossal, both in economic and human terms. These projections highlight the urgent need for new thinking. Failing that, the erosion of our natural environmental capital will increase the risk of irreversible changes that could jeopardise two centuries of rising living standards (OECD 2012, p. 4)

The projections are stark:

> [u]nless the global energy mix changes, fossil fuels will supply about 85 % of energy demand in 2050, implying a 50 % increase in greenhouse gas (GHG) emissions and worsening urban air pollution… Global water demand is projected to increase by 55 % to 2050. Competition for water would intensify, resulting in up to 2.3 billion more people living in severely water-stressed river basins. By 2050, global terrestrial biodiversity is projected to decline by a further 10 %. (OECD 2012, p. 4)

The OECD concludes: 'The reality is that, if we fail to transform our policies and behaviour, the picture is rather grim' (OECD 2012, p. 3).

While individual efforts to reduce material consumption and consumption of fossil fuels is laudatory and necessary, these efforts will fall far short of what is necessary to avert widespread ecological collapse in the absence of government intervention and economic restructuring around carbon taxes of some sort. Allowing fossil fuel interests to maintain their economic stranglehold on the status quo and thus effectively veto progressive climate policy is collectively ecocidal.

The Neoliberal Activist Attack on Climate Science and the Environment

Canada is a global laggard when it comes to meaningful policies with respect to climate change, it ranks 58th, dead last of OECD countries in terms of climate performance (Burck et al. 2013). Achieving this dubious distinction it must be pointed out, requires genuine committed activism. Over the past 20 years scores of corporate activists working with pliant politicians have managed to derail any meaningful commitment to climate change legislation in Canada. In the U.S. an even more intractable situation exists with nearly unlimited corporate funds flowing into political coffers.

Over simplifying complex ideas and framing public policy discussions with citizens into simplistic binaries is not only insulting but leads to poor public policy. This is especially true when politicians pit oil industry jobs against climate change policies. Despite projections that climate change could cost Canada between $21 billion and $43 billion each year by 2050 if the government fails to come up with an effective domestic climate change plan (National Round Table on the Environment and the Economy (NRTEE) 2012), the Canadian government ignored the findings and decided to kill the proverbial messenger. It eliminated the NRTEE, a high-level organization bringing business, environmental and community leaders together to tackle economic-environmental issues of national importance. The roundtable was dissolved by the government after producing a series of non-partisan reports which highlighted the utter ineffectiveness of Canada's current climate change initiatives. Foreign Affairs Minister John Baird had this to say about the NRTEE:

> Why should taxpayers have to pay for more than 10 reports promoting a carbon tax, something that the people of Canada have repeatedly rejected?... It should agree with Canadians. It should agree with the government. No discussion of a carbon tax that would kill and hurt Canadian families (Scoffield and Ditchburn 2012).

While an obvious example of political hyperbole and a reflection of the antagonistic relationship between evidence driven public policy and the current government, detached from both science and reason, a carbon tax would of course not 'kill and hurt Canadian families', in fact the government of British Columbia has initiated an effective carbon tax that is not killing families. The notion that a non partisan group would be expected to craft recommendations that simply reflect the government's own ideological position, reveals the nature of the quality of thinking underlying government climate policy in Canada today. Similarly Canada's Minister of Natural Resources Joe Oliver began a PR offensive on the cusp of the public hearings concerning the 'Northern gateway' pipeline to transport tar sand's bitumen to the Pacific ocean for export to Asia. In an open letter Oliver charged that:

> Unfortunately, there are environmental and other radical groups that would seek to block this opportunity to diversify our trade...Their goal is to stop any major project no matter what the cost to Canadian families in lost jobs and economic growth. No forestry. No mining. No oil. No gas. No more hydro-electric dams. (Payton 2012)

According to Oliver these "radical groups":

> [t]hreaten to hijack our regulatory system to achieve their radical ideological agenda," stack the hearings with people to delay or kill "good projects," attract "jet-setting" celebrities and use funding from "foreign special interest groups." (Payton 2012)

Demonization of ordinary citizens and NGO groups who care about the environment as 'radicals' and 'hijackers' of regulatory processes by government ministers is disconcerting. It is however a tactic consistent with the ruthless and authoritarian public relations campaigns waged by corporations in defense of their behaviour. The rationale for this PR offensive was to position environmental NGO's and Aboriginal groups as 'suspects' to the Canadian public, working against Canada's interests. It was also used to provide cover for the gutting and rewriting of environmental laws to suit the interests of the petroleum lobby. Concerned about the loss of habitat protection in the new Fisheries Act, 625 scientists including many of Canada's most renowned ecologists and aquatic scientists urged the government in an open letter that proposed changes would be 'a most unwise action, which would jeopardize many important fish stocks and the lakes, estuaries and rivers that support them': they were summarily ignored (Perkel 2012, p. 5).

The conservative government also shut down the 'Polar Environment Atmospheric Research Lab', or PEARL, the most northerly civilian scientific research lab in the world and a facility used by scores of international and Canadian scientists. It's closure has severely hampered research in arctic contaminants, climate change and ozone depletion in northern Canada, it has also damaged Canada's reputation for supporting science (McDonald 2012; Nature Editorial 2012). Harper also closed the 'World Ozone and Ultraviolet Data Center', a scientific group operated by Environment Canada that has measured ozone and UV radiation since the mid-1950s (Bagley 2012a).

Governments are elected in part to make economic budgetary decisions based on what they perceive public priorities to be. As Canadian atmospheric scientist Thomas Duck explains:

> PEARL costs $1.5 million per year to operate, and EC's (Environment Canada's) ozone program likely costs about $1 million per year. Compare these costs with those of the government's flagship programs: The budget for purchasing and servicing one F-35 fighter jet ($246 million) would power PEARL half way into the next century...The budget for the War of 1812 celebrations ($28 million) could have supported either program for decades. (Dechene 2012)

By critically exploring the nexus of economic and political ideology and support for science, students can gain important real-world insight into how the enterprise of science and politics are intertwined at the national level. Important parallels can be drawn between the political interference with the arms-length funding of science in the G.W. Bush and the Harper regimes (Union of Concerned Scientists 2007; Donaghy et al. 2007).

In an unprecedented move in July 2012 over 2000 scientists marched in Ottawa to protest government cuts to scientific research, and the move toward funding more applied research at the expense of basic research. The scientists held a mock funeral for "the death of evidence" in Canadian public policy (Nature Editorial 2012).

Eighteen leading American climate scientists have written to the U.S. President arguing that the proposed Keystone XL pipeline to take Alberta tar sand's bitumen to the U.S. Gulf coast is counter to 'both national and planetary interests' (Hansen 2013), because it will facilitate an enormous expansion of the tars sands development. The Canadian Prime Minister quipped that U.S. approval of the Keystone XL pipeline was a "no brainer" (Harper 2011, p. 8). The depth of oil industry jingoism in Canada was revealed when the opposition environment critic questioned the wisdom of building the Keystone pipeline in Washington, wherein she was summarily accused of taking a "treacherous course" by the Minister of the Environment (Harper 2011).

Students can ask about the nature of the science informing current public policy with respect to the tar sands development, namely whose interests are being served in the short term and whose interests will suffer in the long term?

Disclosing Power, the Courtiers to Climate Disaster

One of the most important aspects of activism lies in its potential to reveal how power is hidden and how it acts upon science and technological policy. Despite the ubiquity of unequal power relationships within neoliberal economics, there is an absence of an analysis of this power in economic discourse as Haring and Douglas explain:

> Power. It is ubiquitous, yet mainstream economics—despite having been made into a cold war weapon by the US—is highly limited and one-sided in how it models power relations. Monopolists and unions are always bad. Consumption and competition are always good. Taxes are always bad. More money is always good. Government is held to be coercive, so it is generally bad. Markets are held to mean freedom, so they are generally good. (Haring and Douglas 2012, p. viii)

Introducing young people to the ubiquity of unequal power within neoliberal economic relationships reveals whose decisions 'count'. When the short-term interests of neoliberal economics and multinational corporations conflict with long-term intergenerational interests, what should governments do? How does intergenerational interest manifest itself in decisions concerning science and technology?

The 'Inquisition' against climate science (Powell 2011) has been financed to the tune of hundreds of millions of dollars by giant corporations like Exxon Mobil and Chevron, by petro billionaires like the Koch brothers, who fund a variety of industry 'think tanks' like the Fraser Institute in Canada and the American Enterprise Institute among others. According to the Union of Concerned Scientists:

> Industry players have been known to intimidate or openly attack scientific researchers, to skew analyses of the costs and benefits of proposed regulations, or to undermine the regulatory process itself. (The Union of Concerned Scientists 2012, p. 3)

In learning about the nature of organizations that work against the public understanding of science, young people can begin to understand that the existence of a 'public interest' is a notion that many powerful people do not share.

Today we have technologies that can measure carbon dioxide concentrations across the planet, measuring its inexorable rise year by year. As spectators we can watch the unprecedented climate change driven rapid destruction of glaciers in high definition video, one researcher witnessing the largest iceberg calving ever filmed, a massive 7.4 km^3 of ice falling off the Ilulissat glacier in Greenland says 'it's like watching Manhattan breaking apart in front of your eyes' (Orlowski 2013). We passively watch these events unfold as politicians feed us bromides about how economically prohibitive it will be to transform our economies into low carbon ones, or how our 'lifestyles' would be inconvenienced by attempting to slow the enormous rise of greenhouse gas pollution. National politicians of all political stripes in North America have served up 'sustainability' platitudes while the fossil fuel industry has spent hundreds of millions to convince these same politicians and the public that inaction is the most prudent policy going forward.

That the paid apologists for Big Carbon (coal, gas, oil) industries should actively promote the status quo and a do-nothing climate policy is hardly surprising. There is a long history from tobacco manufacturers to major pharmaceutical companies of advertisers, company 'scientists' and public relations spokespeople being well paid to deflect scientific truth whenever it appears to conflict with maximizing corporate profits (Oreskes and Conway 2010). The difference between the tobacco is good-for-you scientists of yesteryear and the climate change denialists and apologists today is that the stakes are so much higher for both the whole of humanity and the non-human world as well. It is crucial that young people learn how science can be used to advance ideological agendas of all sorts. Without adequate public oversight powerful interest groups can hold enormous sway over how a participatory democracy functions or more importantly act in dysfunctional ways to obscure the truth in order to advance corporate interests. As McChesney has observed:

> In many respects we now live in a society that is only formally democratic, as the great mass of citizens have minimal say on the major public issues of the day, and such issues are scarcely debated at all in any meaningful sense in the electoral arena. (McChesney 2000, p. 260)

We are being lulled into a state of climate policy inactivity by the same forces that attempt to muzzle democracy and to promote plutocracy. Citizens have been fed a number of narratives; "it's too expensive to do anything meaningful about reducing GHG emissions"; "'it doesn't matter what they do because the root cause is 'natural'"; "our contribution is too small to make a discernible difference", and on and on. The overall objective is to mute the message that immediate and effective action is both necessary and possible.

Reinventing Participatory Democracy

Increasingly citizens are experiencing a form of authoritarian executive democracy (Giroux 2013) wherein people have the opportunity to vote in a federal election every 4 or 5 years and are then provided little or no meaningful opportunity to participate in the decision making processes of that government. Outside of activism there are very few opportunities to influence the decision making processes of authoritarian governments. Governments of all stripes routinely enact major policy decisions without anything but token input from citizens. Students should be encouraged to ask critical questions concerning how 'the public interest' is framed by politicians and corporate spokespeople, they certainly need to debate the wisdom of neoliberal economic policies that eliminate scientific organizations which provide vital scientific evidence that should inform public policy.

There is an urgent need for young people to reinvent participatory democratic government, to reassert 'the public good' that private interests have fought so hard to eliminate from public discourse. This is especially true when it comes to dealing with issues like climate change and other global environmental problems. Rothkopf (2012) warns of the trouble that governments around the world are in today, they are he argues 'calcified... still very much in the national, centralized, hierarchic forms that first took shape centuries ago' (Rothkopf 2012, p. 362). Rothkopf argues that governments today are:

> [t]rapped by their inability to adapt and by the unceasing and frequently successful efforts of private interests to contain or alter those dimensions of public power that might restrict their growth and freedoms (Rothkopf 2012, p. 362).

Neoliberal politicians like Canada's Harper, former U.S. President George W. Bush, and former Australian Prime Minister John Howard, have all capitalized on the disenfranchisement of young people with the political process, all worked to weaken global transnational institutions that reflected inter and intragenerational interests like the U.N. and the authority of it's agencies that provide global environmental leadership like the 'Intergovernmental Panel on Climate Change' (IPCC), as well as the 'United Nations Environmental Programme' (UNEP). They ignored and in Harper's case of abrogating the Kyoto Accord, international environmental treaties in order to provide even more generous unfettered operating environment to their favoured Big-carbon industries. As Rothkopf explains:

> Global private actors have evolved so that, while lacking many of the legally enshrined powers of governments, they nonetheless can rival, challenge, defeat, or sidestep public power or, alternatively they can simply manipulate it to serve their needs. Those actors, great corporations and financial institutions, operate far more nimbly on the global stage than governments still trapped within their own borders (Rothkopf 2012, p. 362).

Canada, one of the original signatories to the Kyoto Accord now takes an ethically and morally indefensible international position in support of the oil and gas industry, at the expense of poorer nations struggling to cope with the multiple impacts of climate change. Science and technological education for activism engages young people in exploring the nature of how science and technology shape and inform or subvert effective public policy. Activist education would help them examine the interests at work in determining in terms of public policy what science is deemed important, and what science is disregarded or marginalized. An education for activism would examine the political barriers to more sustainable forms of technological development, and investigate the powerful interests behind the reentrenchment of the status quo.

Activist STSE education would help students deconstruct simplistic political hyperbole to ask questions concerning the role of science and vested interests in balanced decision making. For example, a carbon tax is widely viewed by many of the world's leading scientists and economists as the most effective and fair regulatory approach that governments could take to reduce greenhouse gas pollution

(Hsu 2011). However to a staunch opponent of meaningful national and international climate change legislation like Canadian Prime Minister Stephen Harper, the answer is simple:

> Canadians and people across the globe know we have a government smart enough to reject dumb ideas like a $20-billion carbon tax. (Scoffield and Ditchburn 2012)

Developing a Skeptical Way of Reading the Media

It is difficult to overemphasize the importance of critical thinking in terms of deconstructing modern media or as Kovach and Rosenstiel (2010) refer to it as 'skeptical knowing' and verifying. They suggest a process that involves several systematic questions:

1. What kind of content am I encountering?
2. Is the information complete; and if not, what is missing?
3. Who or what are the sources, and why should I believe them?
4. What evidence is presented, and how was it tested or vetted?
5. What might be an alternative explanation or understanding?
6. Am I learning what I need to? (Kovach and Rosenstiel 2010, p. 75)

Kovach and Rosenstiel (2010) argue that a shift in responsibility has occurred from the journalist being gatekeeper of verifiable information to the consumer or citizen becoming her own gatekeeper, underscoring the necessity of acquiring a skeptical way of knowing. According to Kovach and Rosenstiel the first step of asking 'what kind of content am I encountering?' entails identifying which type of journalism is under scrutiny. The four main types in common use include the:

Journalism of Verification, a traditional model that puts the highest value on accuracy and context.

Journalism of Assertion, a newer model that puts the highest value on immediacy and volume and in so doing tends to become a passive conduit of information.

Journalism of Affirmation, a new political media that builds loyalty less on accuracy, completeness, or verification than on affirming the beliefs of its audiences, and so tends to cherry-pick information that serves that purpose.

Interest-group Journalism, which includes targeted Web sites or pieces of work, often investigative, that are usually funded by special interests rather than media institutions and designed to look like news. (Kovach and Rosenstiel 2010, p. 75)

The atomization of so-called news channels into propaganda vehicles for narrow political ideologies is well underway in North America. The ability to fragment and target political ideologies has taken on a sinister form as separate realities can be constructed with information filtered and crafted ever more precisely to cater to pre-existing beliefs, values and worldviews. Examples of the 'Journalism of Affirmation' includes 'Fox News' (Brock et al. 2012; Feldman et al. 2011; Stromberg 2010) in the U.S. and the 'National Post' newspaper and 'Sun Media' television in Canada. All regularly feature and promote the views of climate change deniers and oil industry insiders. All have been involved in attacking the scientific consensus position regarding climate change as well as the scientists involved in climate change research.

The public relations push by the oil industry is quite extensive in Canada. The video series 'Energy in Action' is produced by the Canadian Association of Petroleum Producers (CAPP) to promote "an energy literacy program to teach students in Grades Four and Five about our oil and natural gas resources, and the importance of environmental stewardship" it naturally features smiling faces and lots of green, and of course no images of the tar sands pollution realities or the impacts it has on the boreal forest and its wildlife.

> Energy in action bring communities and industry together to teach to learn, to grow and build something important that will last a long time. Energy in action is community engagement in action building understanding, growing roots in the community, reinforcing reputations, ensuring our social license to operate. Skilled educators and a curriculum linked to energy realities, opens eyes and opens minds. (CAPP 2012a, b, p. A3).

The corporate public relations campaigns aimed at selling Canadians on the benefits of the tar sands have been unrelenting and have intensified dramatically under the Harper government. The advertising campaign bears a striking resemblance to the equally disingenuous "clean coal" media campaign in the U.S., largely funded by libertarian climate denial front groups.

It is estimated that there are 1,500 communications staffers working in Canadian ministers' offices and departments, including 87 in the Prime Minister's Office and the Privy Council Office (Ryckewaert 2011). According to one national affairs columnist:

> The government of Stephen Harper has gradually increased the level of political control over public information to an extent that is unprecedented in Canada or similar countries, to the point that we are starting to think it is normal. (Maher 2011, p. 9)

One of Canada's most senior political correspondents Craig Oliver, a reporter with 50 years of experience with ten prime ministers said:

> In the last 30 years, the size of "media control" operations has greatly expanded... There's a whole infrastructure at every level of every department, of people whose job it is to manipulate and massage media. Highly paid people...hundreds of people. Their only job every day is try to manipulate a message. (Ryckewaert 2011)

The federal government working with industry and the government of Alberta developed an *'Oil Sands Advocacy Strategy'* to push carbon intensive tar sands oil into whatever markets they could, it included:

> Coordinated strategy sessions with some of the largest oil companies in the world, high level meetings with political leaders, hundreds of lobby meetings with decision makers, and a public relations campaign that includes bill boards in Times Square and full page advertisements in major newspapers. (Climate Action Network Canada 2012, p. 3)

The most troubling aspects of this tar sands propaganda is their focus on undermining democratic decision making in other countries like the clean energy and greenhouse gas mitigation policies in the U.S. and the E.U.:

> Intensive lobbying, underhanded pressure, and public relations campaigns have been targeted at a number of policies including California's low carbon fuel standard, section 526 of the United States Energy Independence and Security Act, as well as the European Union's Fuel Quality Directive. (Climate Action Network Canada 2012, p. 3)

Among the many half-truths and deceptive information practices employed by the tar sands PR campaign, Canadian oil producers went as far as to produce ads comparing tar sands tailings to yogurt, before complaints to the Advertising Standards council of Canada caused them to withdraw the ad (Polczer 2010).

The public relations business is a $10 billion global industry, growing at 8 % per year (Holmes 2012). Eight of the world's largest PR firms are headquartered in the U.S. where in the wake of the Supreme Court's 'Citizen United' decision, the manipulation of democracy by its wealthiest citizens is virtually unhindered.

> Public relations was created to thwart and subvert democratic decision making. It was a means for 'taking the risk' out of democracy. The risk was to the vested interests of those who owned and controlled society before the introduction of voting rights for all adults. Modern PR was founded for this purpose and continues to be at the cutting edge of campaigns to ensure that liberal democratic societies do not respond to the will of the people and that vested interests prevail. (Dinan and Miller 2007, p. 11)

This fundamental insight helps explain why despite consistent polling over the last 10 years indicating that citizens want their elected officials to craft effective climate change policies to reduce and slow the impacts of climate change, nothing effective has been forthcoming. Efforts to reduce greenhouse gas emissions, even the weak and imperfect Kyoto Accord have been vehemently and effectively opposed by many of the largest business and industry lobbies on the planet, as well as by neoliberal politicians across the developed world. Vested interests, namely the 'Big carbon' industries and their courtiers continue to advance business as usual as the planet teeters ever closer toward a climate catastrophe. Dinan and Miller level a number of incisive charges against the public relations industry, they include:

1. It is overwhelmingly carried out for vested powerful interests, mainly corporations.
2. It is not open and transparent about its means or even about its clients and interests it is working for.
3. It characteristically involves deception and manipulation.
4. It does not engage in democratic debate, but rather seeks to subvert it in the interests of its clients.
5. Corporate social responsibility (CSR) and other 'ethical' activities are all subordinated to corporate strategy.
6. PR has played a crucial role at the cutting edge of corporate power in the neoliberal revolution (Dinan and Miller 2007, p. 12)

While not all PR workers are implicated in these charges, the same cannot be said for the main groups involved in spinning the media narratives on behalf of big oil and the 'ethical' tar sands.

> Dependency on the tar sands as Canada's prime economic driver has occurred simultaneously, and in some ways has come as a result of, corporate involvement in how public policy is designed and implemented by the Canadian government. (Cayley-Daoust, and Girard 2012, p. 3).

Students can examine the ideological stance of Canada's government from a number of ethical standpoints including 'global equal per capita emission entitlements' and the rights to 'subsistence emissions' versus 'luxury' emissions (Gardiner 2011). On both counts Canadians have one of the largest per capita carbon emissions footprints on the planet, a significant portion of their footprint resulting from luxury goods produced by developing countries.

Young people should be encouraged to 'follow the money' in terms of fossil fuel funding of the climate disinformation industry. Resources like "Dirty Energy Money" (Oil Change International 2013) and 'ExxonSecrets' (Greenpeace 2012) allow users to track the money going to politicians from the big carbon lobby and their astroturf front groups. 'Dirty Energy Money' tracks campaign contributions and politicians voting record on legislation related to fossil fuel industries. Sites such as these bypass corporate media which quite deliberately ignores the connection between political influence, corporate public relations and public subsidies and tax give-aways. Mapping out the network of relationships between fossil fuel 'power holders' (Ricketts 2012) can reveal where activism is probably best focused in order to change the dynamics of oil money influence and public policy.

Whose Long-Term Interests Are Being Served?

Young people are very capable of exploring the questions: Why is their elected government ignoring the best scientific advice available when it comes to climate change? Is the basis for governmental inaction rooted in scientific uncertainty or ideological intransigence? The rejection of science and the construction of an ideological, and in this case a neoliberal PR smokescreen, has become commonplace in the North American media marketplace (Brock et al. 2012; Grant 2011; Otto 2011). What is ultimately sacrificed is the broader notion of the public good, intergenerational justice, the development of balanced responsible public policy and effective environmental regulation.

Citizens have every right to be concerned about exactly whose interests their elected representatives are serving, for example when the annual Canadian federal-provincial government conference of energy ministers was held in Alberta in 2011 it was 'sponsored' by national and transnational oil companies (Corbella 2011). When public officials are accessible by powerful corporate groups in ways that smaller non-governmental citizen-led groups are not, critical questions related to democratic fairness and a 'democracy deficit' (Aman 2004) should ensue. Similar oil and coal friendly circumstances occurred during the Bush Cheney regime, where behind closed doors public energy policy was crafted under the heavy influence of the transnational oil, coal and gas industries, with only token representation from environmental interest groups (Abramowitz and Mufson 2007; Millbank 2005).

When politicians and the corporate elite complain about the cost of environmental regulation and environmental protection, young people should be encouraged to learn about how plutocrats can avoid paying taxes that could be used to resolve some of the planet's most pressing environmental and social issues. A common refrain of North America's corporate elite over the past 20 years is that effective climate change legislation or even weak measures like the Kyoto Accord, are too expensive to implement, prohibitive in cost and in the words of the leader of the Canadian Chamber of Commerce would "destroy the Canadian economy" (Toronto Star Editorial 2002).

Shell plans on expanding its tar sands operations by a third by the end of this decade (Welsch 2012), it's poised to become the first major company to drill for oil in the Arctic despite a track record of chronic widespread environmental destruction in Nigeria that the U.N. says has not been cleaned up (United Nations Environmental Programme 2011; Vidal 2011). One of the most insidious is the packaging of thinly disguised corporate propaganda under the guise of informal science education. For example, in 'Energy 101' Shell tells us that 'Life takes energy. In the end, we want to leave more than we take, building a positive legacy for all Canadians to enjoy for generations to come' (Shell Canada 2012, p. 1).

In their 'Fueling Change' campaign Shell urges us to 'Help us change the world one click at a time' (Shell Canada 2012). Yes, a multi-billion dollar corporation which spends tens of millions on lobbyists, public relations and advertising wants citizens to help it 'change the world'. There are no depths of cynicism to which corporations will stoop to maximize profit, externalize costs and minimize public accountability. Using these 'campaigns' as part of a PR blitz campaign, they have no qualms or sense of irony about PR spinning the world about how they are experimenting with 'carbon capture and storage' while simultaneously lobbying the federal government to open up more of the Arctic to dangerous oil and gas exploration or pushing to dramatically expand their tar sands operations. It is estimated that despite the enormous environmental risks taken by companies offshore drilling in the Arctic, their efforts if fully developed would add an additional 520 million tonnes of $CO2_2$ a year to global emissions by 2020, and as much as the entire national emission budget of Canada, and 1,200 million tonnes by 2030 (Voorhar and Myllyvirta 2013). If oil production plans for the Alberta's tar sands are realized, output will triple from 1.5 to 4.5 million barrels a day by 2035, adding 706 million tonnes of CO2 to global emissions a year (Voorhar and Myllyvirta 2013). Voorhar and Myllyvirta (2013) point out that by 2020, the tar sands expansion will add annual emissions of 420 million tonnes of CO2, equal to those of Saudi Arabia.

The expectation incorporated within advertising like "Fueling the Change" is that the 'consumer' will not do the basic math or invest in the background research to see how completely disingenuous any claim to 'change the world' for the better is. An active citizen on the other hand might just 'do the math'. They might ask basic questions like how could Shell's profit, increase by 54 % from 2010 to 2011 while its oil and natural gas production decreased by 3 % during the same time period? (Weiss et al. 2012). A critical citizen might ask as Weiss et al. (2012) do, why the economic theory that suggests companies will produce more of a good if its price is higher, or if it receives public subsidy, doesn't apply to oil companies. As Weiss points out, in 2011 the U.S. experienced the highest oil price adjusted for inflation since 1864, while subsidizing the five largest oil companies to the tune of $2 billion, despite the fact that their production yield was lower than in 2010. These corporations 'made a record high $137 billion in profits in 2011—up 75 % from 2010—and have made more than $1 trillion in profits from 2001 through 2011' (Weiss et al. 2012, p. 2). The major oil players according to Weiss spent "$1.6 million on campaign contributions and $65.7 million on lobbying efforts. For every $1 spent on lobbying in Washington, the big five received $30 worth of tax breaks" (Weiss et al. 2012, p. 1).

Conclusion

Above all, science and technological education must help young people make sense of their world, to help them think critically as citizens about how to enact change through science, technology and political engagement to shape a sustainable alternative to the neoliberal vision that pushes us ever closer towards a dystopian ecocidal future.

The 'occupy' movement with its diverse participants and diffuse objectives galvanized attention on growing inequality and neoliberal exploitation of disenfranchised groups. If science and technological education is to provide young people with both the means and critical knowledge to transform human economies into more sustainable and humane forms, it must embrace the vital bridge that activism builds between classroom practice and the messy world, this includes a critical examination of corporate activism. As Dinan and Miller warn:

> The PR industry is not some free-floating pustule on the surface of a globalizing world, but the cutting edge of corporate power in its campaign to stifle democracy. What is needed is the exposure of the PR industry and a series of measures to bring it and the corporations for which it acts to heel. Otherwise democratic politics is finished. (Dinan and Miller 2007, p. 18)

Science and technological education for activism begins with disclosure, by disclosing the real ecological and social costs of material consumption by the planet's affluent and the social and political systems that operate to keep these costs hidden. When youth begin to see what has been both purposefully and unintentionally hidden from their consciousness, they can begin a path to change the world for the better.

References

Abramowitz, M., & Mufson, S. (2007). Papers detail industry's role in Cheney's energy report. *Washington Post*. Accessed at http://www.washingtonpost.com/wpdyn/content/article/2007/07/17/AR2007071701987_pf.html

Aman, A. C. J. (2004). *The democracy deficit: Taming globalization through law reform*. New York: New York University Press.

Bagley, K. (2012a). Scientists Decry Canadian undermining of climate science. *Bloomberg News*. Accessed at http://www.bloomberg.com/news/2012-11-27/scientists-decry-canadian-undermining-of-climate-science.html

Berezow, A. B., & Campbell, H. (2012). *Science left behind: Feel good fallacies and the rise of the anti-scientific left*. New York: Public Affairs.

Boyce, D., Lewis, M. R., & Worm, B. (2010, July 29). Global phytoplankton decline over the past century. *Nature, 466*, 591–596. doi:10.1038/nature09268.

Brock, D., Rabin-Havt, A., Matters, M., & Media Matters for America. (2012). *The Fox effect: How Roger Ailes turned a network into a propaganda machine*. New York: Random House.

Bromwich, D. H., Nicolas, J. P., Monaghan, A. J., Lazzara, M. A., Keller, L. M., Weidner, G. A., & Wilson, A. B. (2012). Central West Antarctica among the most rapidly warming regions on Earth. *Nature Geoscience*. Advance online publication. doi:10.1038/ngeo1671

Buis, A. (2013). *Study finds severe climate jeopardizing Amazon Forest*. Accessed at http://www.nasa.gov/topics/earth/features/earth20130117.html

Burck, J., Hermwille, L., & Krings, L. (2013). *The Climate Change Performance Index 2013*. http://germanwatch.org/en/5698

Canadian Association of Oil Producers (CAPP). (2012a, November 29). The oil sands are tremendous for Ontario-advertisement. *Globe and Mail*, p. A3.

Canadian Association of Oil Producers (CAPP). (2012b). *Energy in action – What a difference a day makes*. Calgary: CAPP. http://www.youtube.com/watchfeature=player_embedded&v=iEiPEO5iBkc#!. Accessed 30 March 2013.

Cayley-Daoust, D., & Girard, R. (2012). *Big oil's oily grasp: The making of Canada as a Petro-State and how oil money is corrupting Canadian politics*. Edmonton: Polaris Institute.

Climate Action Network Canada. (2012). *Dirty oil diplomacy: The Canadian government's global push to sell the tar sands*. Ottawa: Climate Action Network Canada. Accessed at http://climateactionnetwork.ca/

Corbella, L. (2011, July 20). Industry sponsorship of ministerial conference throws gasoline on a simmering fire. *Calgary Herald*.

De Souza, M. (2012). It's official: Harper government withdraws from Kyoto climate agreement. *Canada.com*. Accessed at http://o.canada.com/2012/12/14/its-official-harper-government-withdraws-from-kyoto-climate-agreement/

De'ath, G., Fabricius, K. E., Sweatman, H., & Puotinen, M. (2012). The 27 year decline of coral cover on the Great Barrier Reef and its causes. *Proceedings of the National Academy of Sciences*. doi:10.1073/pnas.1208909109.

Dechene, P. (2012). Harper's war on science: An interview with Thomas Duck. *Prairie Dog*. Available http://www.prairiedogmag.com/?p=42551

Dinan, W., & Miller, D. (2007). Public relations and the subversion of democracy. In W. Dinan & D. Miller (Eds.), *Thinker faker spinner spy: Corporate PR and the assault on democracy* (pp. 11–20). London: Pluto Press.

Donaghy, T., Freeman, J., Grifo, F., Kaufman, K., Maassarani, T., & Shultz, L. (2007). *Atmosphere of pressure: Political interference in federal climate science*. Cambridge, MA: Union of Concerned Scientists Government Accountability Project.

Feldman, L., Maibach, E. W., Roser-Renouf, C., & Leiserowitz, A. (2011). Climate on cable: The nature and impact of global warming coverage on fox news, CNN, and MSNBC. *The International Journal of Press/Politics*, 1–29. doi:10.1177/1940161211425410.

Fischetti, M. (2012, December 22). Arctic sea ice hits record low. *Scientific American*. Accessed at http://www.scientificamerican.com/article.cfm?id=greene-arctic-sea-ice-hits-record-low&print=true

Gailus, J. (2012). *Little black lies corporate & political spin in the global war for oil*. Victoria: RMB.
Gardiner, S. M. (2011). *A perfect moral storm: The ethical tragedy of climate change*. Toronto: Oxford University Press.
Giroux, H. (2013). The new extremism and politics of distraction in the age of austerity. *Truthout*. http://www.truth-out.org/opinion/item/13998-the-new-extremism-and-politics-of-distraction-in-the-age-of-austerity?tmpl=component&print=1
Grant, J. (2011). *Denying science*. New York: Prometheus Books.
Greenpeace USA. (2012). *ExxonSecrets.org*. Washington: Greenpeace. http://www.greenpeace.org/usa/en/campaigns/global-warming-and-energy/exxon-secrets/
Hansen, J. (2013). *18 top climate scientists call on president Obama to reject Keystone XL*. Oakland: The International Research Institute for Climate and Society & 350.org. Accessed at http://www.350.org/en/about/blogs/18-top-climate-scientists-call-president-obama-reject-keystone-xl
Haring, N., & Douglas, N. (2012). *Economists and the powerful: Convenient theories distorted facts ample rewards*. London: Anthem Press.
Harper, T. (2011, November 17). For conservatives, contrary positions are treasonous. *Toronto Star*. Accessed at http://www.thestar.com/news/canada/2011/11/17/tim_harper_for_conservatives_contrary_positions_are_treasonous.html
Henderson, M. (2012). *The geek manifesto*. London: Bantam Press.
Holmes, P. (2012). *Global pr industry up eight percent to $10bn*. Retrieved December 12, 2012. Accessed at http://globalpragencies.com/
Hsu, S.-L. (2011). *The case for a carbon tax getting past our hang-ups to effective climate policy*. Washington, DC: Island Press.
International Energy Agency (IEA). (2012). *Coal's share of global energy mix to continue rising, with coal closing in on oil as world's top energy source by 2017*. Paris: International Energy Agency. Available http://www.iea.org/newsroomandevents/pressreleases/2012/december/name,34441,en.html
Kovach, B., & Rosenstiel, T. (2010). *Blur: How to know what's true in the age of information overload*. London: Bloomsbury.
Maher, S. (2011, December 2). Stephen Harper's PR obsession is fostering paranoia and paralysis in public service. *The Province*.
McChesney, R. W. (2000). *Rich media, poor democracy*. New York: The New Press.
McDonald, B. (2012). *Closure of polar lab a blow to Canada's scientific reputation*. Retrieved March 1, 2012. http://www.cbc.ca/news/technology/story/2012/03/01/f-vp-bob-mcdonald-pearl-closure.html
Milbank, D., & Blum, J. (2005, November 16). Document says oil chiefs met with Cheney task force. *Washington Post*. http://www.washingtonpost.com/wp-dyn/content/article/2005/11/15/AR2005111501842_pf.html
Morello, L. (2013). 2012 Proves warmest year ever in U.S. *Scientific American*. Accessed at http://www.scientificamerican.com/article.cfm?id=2012-proves-warmest-year-ever-in-us
Nadeau, C. (2010). *Rogue in power: Why Stephen Harper is remaking Canada by stealth*. Toronto: James Lorimar.
National Round Table on the Environment and the Economy. (2012). *Reality check: The state of climate progress in Canada*. Ottawa: National Round Table on the Environment and the Economy.
Nature Editorial. (2012). Death of evidence. *Nature, 487*, 271–272.
OECD. (2012). *OECD environmental outlook to 2050*. Paris: OECD Publishing. Accessed at http://dx.doi.org/10.1787/9789264122246-en
Oil Change International. (2013). *Dirty energy money*. http://dirtyenergymoney.com/. Accessed 3 Jan 2013.
Oreskes, N., & Conway, E. M. (2010). *Merchants of doubt: How a handful of scientists obscured the truth on issues from tobacco smoke to global warming*. New York: Bloomsbury.

Orlowski, P. (2011). *Teaching about hegemony race, class and democracy in the 21st century*. Dordrecht: Springer.
Orlowski, J. (Writer). (2013). Chasing Ice. Available http://www.chasingice.com/about-the-film/the-team/. In P. Dupre Pesmen & J. Aronson (Producer): Submarine Deluxe.
Otto, S. L. (2011). *Fool me twice: Fighting the assault on science in America*. New York: Rodale.
Payton, L. (2012). Radicals working against oilsands, Ottawa says. *CBC News*. http://www.cbc.ca/news/politics/story/2012/01/09/pol-joe-oliver-radical-groups.html. Accessed 9 Jan 2012.
Perkel, C. (2012). Noted scientists warn Stephen Harper against Fisheries Act changes. *Toronto Star*. http://www.thestar.com/news/canada/2012/03/22/noted_scientists_warn_stephen_harper_against_fisheries_act_changes.html. Accessed 22 Mar 2012.
Polczer, S. (2010). Oilsands ad comparing tailings to yogurt to be modified. *Postmedia Global News*. http://www.globaltvbc.com/oilsands+ad+comparing+tailings+to+yogurt+to+be+modified/89836/story.html. Accessed 30 Nov 2010.
Powell, J. L. (2011). *The inquisition of climate science*. New York: Columbia University Press.
Reynolds, C. (2012, October 31). Why are schools brainwashing our children? *Macleans*. Available http://www.macleans.ca/news/canada/why-are-schools-brainwashing-our-children/
Ricketts, A. (2012). *The activists' handbook*. London: Zed Books.
Rose, M. (2012, May 24). Global CO2 emissions hit record in 2011 led by China: IEA. *Reuters*.
Rothkopf, D. (2012). *Power, Inc.: The epic rivalry between big business and government – And the reckoning that lies ahead*. New York: Viking.
Ryckewaert, L. (2011). PM Harper takes communications strategy to new level. *The Hill Times*. http://www.hilltimes.com/news/2011/11/21/pm-harper-takes-communications-strategy-to-new-level/28868. Accessed 21 Nov 2011.
Scoffield, H., & Ditchburn, J. (2012, May 15). Tories admit to closing enviro research group because they disliked results. The Canadian Press. http://thetyee.ca/Blogs/TheHook/Environment/2012/05/15/NRTEE/
Shell Canada. (2012). *Energy 101*. http://www.shell.ca/. Accessed 10 Nov 2012.
Stromberg, S. (2010, December 15). Fox News on climate: Ignorant, manipulative – or both? *Washington Post*.
Toronto Star Editorial. (2002, September 29). Undermining Kyoto. *Toronto Star*.
Union of Concerned Scientists. (2007). *Smoke, mirrors & hot air how ExxonMobil uses big tobacco's tactics to manufacture uncertainty on climate science*. Cambridge, MA: Union of Concerned Scientists.
Union of Concerned Scientists. (2012). *A climate of corporate control how corporations have influenced the U.S. dialogue on climate science and policy*. Washington, DC: Union of Concerned Scientists. Accessed at www.ucsusa.org/corporateclimate
United Nations Environment Programme (UNEP). (2011). *Environmental assessment of Ogoniland*. Nairobi: United Nations Environment Programme. Available www.unep.org/nigeria
United Nations Environment Programme (UNEP). (2012). *The emissions gap report 2012 a UNEP synthesis report*. Nairobi: United Nations Environment Programme (UNEP). Accessed at http://www.unep.org/publications/ebooks/emissionsgap2012/
Vidal, J. (2011, August 10). UN report on the Ogoniland oil spills could be catalyst for change. *Guardian*. Available http://www.theguardian.com/global-development/poverty-matters/2011/aug/10/un-nigeria-ogoniland-oil-spills
Voorhar, R., & Myllyvirta, L. (2013). *Point of no return: The massive climate threats we must avoid. greenpeace.org*. Amsterdam: Greenpeace International.
Weiss, D. J., Weidman, J., & Leber, R. (2012). *Big oil's banner year: Higher prices, record profits, less oil*. Accessed at http://thinkprogress.org/climate/2012/02/08/421061/big-oil-higher-prices-record-profits-less-oil/
Welsch, E. (2012, May 29). Shell to expand Athabasca oil-sands output. *The Wall Street Journal*. Available http://online.wsj.com/news/articles/SB10001424052702303807404577435032862701706
Worldwide Fund for Nature (WWF). (2012). *Living planet report 2012*. Gland: WWF International, Zoological Society of London, Global Footprint Network.

Chapter 19
Joining Up and Scaling Up: Analyzing Resistance to Canada's "Dirty Oil"

Randolph Haluza-DeLay and Angela V. Carter

Abstract Canada's energy superpower ambitions depend on the continuation and expansion of Alberta's tar sands industry, but this industry comes at a cost of extensive environmental, political and economic impacts. While dissent is difficult in the new Canadian petrostate, a growing civil society movement is resisting tar sands production, its ecological and social implications, and the petro-capitalist political culture legitimating the industry. This chapter analyzes the discursive and action-oriented strategies of four kinds of social movement organizations leading the critique and opposition of the tar sands: Aboriginal, environmental, religious, and labour organizations. While these often local or provincially oriented organizations make valuable contributions to the tar sands debate, we argue the movement began to have political impact when broader cross-organizational coalitions were formed among unlikely allies and when the movement crossed local and provincial boundaries to the national and international level. One important success of this "joining up" and "scaling up" strategy was the creation of critical consensus around future oilsands developments and the seeds of a post-carbon development approach. Dominant political and industry actors were largely able to overlook the movement until a diverse and influential set of social movement actors began collaborating and shifting these local struggles transnationally. Yet while the anti-oilsands movement triggered a reaction by political elites, it was primarily rhetorical and reactionary – translating the movement into real social, environmental and political change remains a challenge.

Keywords Alberta's tar sands • Social movement organizations • Environmental organizations • Labour • Aboriginal communities • Religious groups

---

R. Haluza-DeLay, Ph.D. (✉)
Sociology, The King's University College, 9125 50th St., Edmonton, AB 76B 2H3, Canada
e-mail: Randy.Haluza-Delay@kingsu.ca

A.V. Carter, Ph.D.
Department of Political Science, University of Waterloo, Hagey Hall, Room 311,
200 University Avenue West, Waterloo, ON, N2L 3G1, Canada

L. Bencze and S. Alsop (eds.), *Activist Science and Technology Education*,
Cultural Studies of Science Education 9, DOI 10.1007/978-94-007-4360-1_19,

Canada's energy superpower ambitions hinge on the expansion of oilsands extraction. The scale of this industry is extreme: 170 billion barrels of reserves lie under 140,000 km^2 in Northern Alberta; currently more than a hundred projects daily produce 1.7 million barrels with production projections aiming as high as 3.7 million barrels per day by 2024. The industry's magnitude is paralleled both by its enormous environment and social impacts, given the dire consequences for land, freshwater, climate, air, and wildlife, as well as by its adverse political and economic impacts (Nikiforuk 2008). While dissent is difficult in this new Canadian petrostate, a growing civil society movement is resisting oilsands production, its ecological and social implications, and the petro-capitalist political culture legitimating the industry.

In this chapter, we describe the strategies (both discursive and action-oriented) of the four kinds of social movement organizations leading the critique and opposition of the oil sands: Aboriginal, environmental, religious, and labour organizations. While these often local or provincially oriented organizations make valuable contributions to the oilsands debate, we argue the movement began to have political impact when broader cross-organizational coalitions were formed among unlikely allies and when the movement crossed local and provincial boundaries and extended to the national and international level. One important success of these "joining up" and "scaling up" strategies were the creation of critical attention to future oilsands developments and the seeds of a post-carbon development approach. Dominant political and industry actors were largely able to overlook the movement until a diverse and influential set of social movement actors began collaborating and shifting these local struggles transnationally. Yet while the anti-oilsands movement triggered a reaction by political elites, it was primarily rhetorical and reactionary – translating the movement into real social, environmental and political change remains a challenge.

This analysis is rooted in social movement theory on political opportunity structures (Meyer and Minkoff 2004) and draws on a variety of data regarding social movement actions and counter-responses. Our approach resembles what Paul Gellert and Jon Shefner (2009) call 'structural fieldwork' wherein 'the deep familiarity with people and locales' offers analytic traction on the political-economic world-system. We independently conducted research from 2008–2012, involving participant-observation and interviews of key figures and citizens related to the oilsands and oilsands-opposition movements. Our research programmes draw also on participatory action research, content analysis of media responses to social movements' actions, discourse analysis of these actors' public statements, and focus groups with regular Albertans (Carter 2011; Haluza-DeLay et al. 2013; Haluza-DeLay and Berezan 2013; Kowalsky and Haluza-DeLay 2013; Le Billon and Carter 2012).

Activism Against the Oilsands

Contrary to the stereotypical notion that Alberta lacks vibrant political debate, there is significant local resistance to *status quo* oilsands development. New allies have joined and expanded the movement, groups have experimented with strategies, and alternatives to the carbon economy have emerged. Aboriginal, environmental, religious, and labour organizations have primarily led the opposition to the oilsands using a variety of discursive strategies and contentious actions.

Aboriginal Activism

Aboriginal communities downstream of oilsands developments, or in the path of pipelines to carry gas to the projects or to transport bitumen from them, have been longstanding sources of resistance. While seeking to benefit economically and socially from oilsands extraction, First Nations have protested the projects' environmental health impacts, their degradation of water and air quality, how they have increased the toxicity of subsistence food such as fish and game, as well as how the projects limit aboriginal peoples' access to traditional lands. Aboriginal communities have been increasingly disappointed in government and industry inaction on environmental issues in the region. A case in point is the withdrawal of the Athabasca Chipewyan First Nation and the Mikisew Cree First Nation from the Cumulative Effects Management Association (CEMA) in 2006 and 2007 respectively, in protest against the committee's lack of meaningful progress. The aboriginal groups pointed not only to the legitimation functions of the government-appointed body, but also to its failure to bring about effective environmental regulation of the resource sector. The Mikisew Cree First Nation explained its decision in this way:

> Membership of the Mikisew Cree in the CEMA organization was often cited and publicly stated by both government and industry as them fulfilling their constitutional duty to consult with the Mikisew Cree. The Mikisew Cree has also stated that despite numerous years of studies and meetings, CEMA has been too slow in developing tangible environmental limits for managing the impacts of oilsands projects. (Mikisew Cree First Nation 2007, p. 18)

The possibility of health impacts from oilsands development has gained the most public attention for First Nations. Although communities had been asking for baseline health studies since the 1990s, widespread attention to health problems was raised only in 2003 by physician Dr. John O'Connor who noted what he thought was too high a rate of rare cancers in Fort Chipewyan (Loyie 2009). O'Connor was later disciplined by Alberta College of Physicians and Surgeons for poor practice and raising 'undue alarm' – all charges of which he was later cleared (Loyie 2009). Questions about the reliability of health reviews conducted by both federal and provincial government ministries have generated growing distrust within the communities. As reported by Hanneke Brooymans (2009), Steve Courtorelle, Mikisew Cree First Nations councillor, stated: 'We just don't have enough faith in either government, and I just know they're going to try to protect their interests with continuing to develop the oilsands.' Eventually, the regional health authority commissioned an independent study which, when released in November 2007, proved the existence of levels of toxic and carcinogenic substances (such as mercury and arsenic) in fish and soil downstream of oilsands development that could harm Fort Chipewyan residents (Timoney 2007; Timoney and Lee 2009).

Despite the scientific evidence and the testimony of First Nations members regarding deleterious changes to the region's water, fish, game, and human health, the provincial government has downplayed or denied health risks and has not taken action to stop toxic emissions into the Athabasca River. Provincially, the issue has largely been constructed as one of health risks to First Nations communities, while the social justice (or environmental racism) dimension of the conflict has received little attention. At a 'Justice for Fort Chipewyan' rally held on the steps of the Alberta legislature March 1, 2008 – just a few months after the release of the Timoney study – speakers referred to an extensive range of related issues from environmental impacts and the problem of democratic deficits in the province to the infringement of treaty rights and the lack of local development benefits. However, when the news media reported on the rally, CBC TV, CTV, and the *Edmonton Journal* mentioned only concerns about 'health.' This framing of the issue is consistent with Leith Deacon and Jamie Baxter's (2009) research showing that mainstream media routinely ignore equity and justice framing in favour of health themes, generalized conflict, or 'typical' (i.e., 'nature') environmental topics. Such media framing greatly reduces the possibility for presenting diagnostic frames that have the potential for alternative trajectories for society. While health concerns may be received by the general public with sympathy, demands for justice challenge class and race privileges and the prosperity of the white settler population compared to the marginalization of Aboriginal communities.

Having experienced the unresponsiveness of health and environmental authorities, some Aboriginal communities have exhibited increasingly strong opposition to oilsands operations. While industry officials continue to tout the benefits of oilsands development for Aboriginal peoples in the region, first nations like the Athabasca Cree First Nation (ACFN) have sought greater returns for their people. Aboriginal communities have also begun court cases against the provincial or federal governments for failure to consult and for infringements on traditional lands or treaty rights. Providing an important legal basis, the Mikisew Cree First Nation won a landmark Supreme Court of Canada case in 2005 requiring the 'duty to consult' regarding land management in Aboriginal people's traditional use territories. This victory may have given confidence to some Aboriginal peoples to pursue this strategy to protect their rights and their livelihoods. However, other groups have been less successful. The Lubicon Cree have not been able to obtain recognition of their land claim and are therefore unable to make use of the 'duty to consult' constitutional provision. The Lubicon Lake Indian Nation, in collaboration with Amnesty International and other groups, has opposed the installation of pipelines through their non-ceded territory. For the Lubicons this is the latest instalment in a longstanding failure on the part of Canadian governments to negotiate their land claim, while both forest and oil and gas exploitation dissect their territory (Ominayak and Thomas 2009).

Environmental Activism

Numerous ENGOs are involved in oilsands mobilization. Some are nationally organized – including the Pembina Institute, Sierra Club of Canada, and Greenpeace – while others are provincial or regional, such as the Fort McMurray Environmental Association, Keepers of the Athabasca, or Toxics Watch Society of Alberta.

Environmental activists indicate that their initial strategies vis-à-vis corporations and government were more collaborative than later ones, often consisting of recommendations to 'tweak' industrial practices, specific projects, or policy regimes as well as participation in government-initiated consultation processes. However, by 2008, several ENGOs followed the two First Nations out of CEMA. This ENGO withdrawal followed other experiences of government consultation – such as the Special Places 2000 and Boreal Forest Conservation Strategy in the 1990s – that were later assessed by ENGOs as a pattern of wasted effort. Commenting on the Boreal Forest Conservation Strategy process, Richard Schneider of the Alberta Centre for Boreal Research wrote: 'Through the lens of 5 years of hindsight it now seems clear that orchestrating meaningful change was never the government's intent' (2002, n.p.). Environmentalists interviewed by Colette Fluet and Naomi Krogman (2009) expressed similar frustrations about the North East Slopes Land Use Strategy process. In their research on sour gas consultations, Jeffrey Masuda, Tara McGee and Theresa Garvin go so far as to say that 'public engagement' in Alberta is utilized 'less as a tool for promoting democratic consensus and more as means to legitimate particular forms of governance that privilege narrowly defined economic goals at the expense of citizen rights and values' (2008, p. 359). These experiences indicate that the institutional and other opportunities for meaningful political engagement on environmental policy have been very limited, and led to changes in ENGO strategies: ENGOs engaged in more direct action, sought broader coalitions with other citizen groups, and moved toward activism external to the province as a way of putting pressure on internal provincial regulation.

More frequent use of court cases and media-friendly direct action is the first significant strategic shift by ENGOs. For example, in April 2009, *Ecojustice* appealed to reconvene review hearings on approvals for two of Shell's oilsands projects (Jackpine and Muskeg River) after Shell broke written agreements with a coalition of major ENGOs to reduce greenhouse gases. The judicial strategy is expensive and has had mixed success as governments have not always acceded to judicial decisions, further demonstrating the closed political opportunity structure. When it became known than 500 ducks (later tallied at 1,600) had died on a Syncrude tailings pond in April 2008, the provincial and federal environment ministries laid charges only after *Ecojustice* launch a private prosecution against the company (*Ecojustice* 2009). Syncrude was found guilty of failing to have an effective deterrent system in June 2010.

Taking a less conventional approach, Greenpeace has made headlines by direct actions aimed at drawing national and international attention to the environmental consequences of the oilsands operations. The Greenpeace office in Alberta opened in 2007, making it one of the newer additions to the environmental organizational field, but it quickly became a central figure. Among Greenpeace's actions was the unfurling of a banner at one of Premier Stelmach's fundraising dinners which read: '$telmach: the best premier oil money can buy.' It also created a mock tourism website and video (*Experience an Oilsands Vacation*) to satirize the provincial tourism campaigns that highlight the natural beauty of the province.[1] Occupations of a Suncor bitumen conveyor and Shell's Albian mine in fall 2009 caused temporary, partial stoppages of operations at these sites.

Religiously-Based Activism

Religious groups are also involved in criticism of the oilsands, approaching the issue in a very different way and involving a different constituency than environmental and Aboriginal groups.[2] The impact of faith groups is broader than the relatively small proportion of the population that regularly participates in religious services. Participants in any faith group are heterogeneous compared to the members of other movement organizations; religious individuals, even if members of the same faith community, have diverse relationships to the oil economy (McKeon 2010).

In recent years religious groups have revisited environmental issues in light of their theologies of creation-care, stewardship and eco-justice (Gottlieb 2006). Two specific insertions of religious actors in the oilsands debate in 2009 drew media attention: a Roman Catholic report on the oilsands, and a week-long tour of the oilsands by church leaders. But for the most part, faith groups carry out study groups, workshops and other less noticeable events on the issue (Chetkovich and Kunreuther 2006), and these education techniques often focus on individuated lifestyle actions rather than socio-structural analysis.

The list of organizations calling for an oilsands development moratorium includes several religious groups, most notably KAIROS. A national organization with affiliated local chapters in many Canadian cities, KAIROS is an ecumenical coalition of ten of Canada's Christian denominations. KAIROS coordinates international development programmes through partnerships with organizations in 21 countries, runs campaigns on numerous social issues related to 'human rights,

[1] This 1-min video can be found by name on youtube.com

[2] Such actions have been primarily from Christian groups so far. The Edmonton Interfaith Association proposed a multi-faith climate change meetings but this initiative seems to have dissipated. However, there is evidence that broaching an issue in one faith context can influence other faith groups to broach the issue from within their own traditions (Gottlieb 2006).

justice and peace, human development and ecological justice' (KAIROS 2011), and provides educational material for congregational use. Its climate change campaign has specifically linked the issue with global poverty.[3]

KAIROS organized a 'Delegation to the Alberta Oilsands' with ten Canadian representatives accompanied by two indigenous representatives and one member each of Oil Watch Nigeria and Acción Ecológica-Ecuador in May 2009. Intended to be a 'listening tour' so that church leaders could have direct experience with which to engage their own denominational constituency on the issue, the tour began in Edmonton and headed north to Fort McMurray and Fort Chipewyan where it met with industry, government, and Aboriginal leaders. There was considerable publicity (Canada West Foundation 2009). A video and teaching material have now been produced and are in use in churches across the country (KAIROS 2011). Local KAIROS chapters have regularly included oilsands-related sessions in their annual conferences, but indicate that they were somewhat blindsided by the national office organizing a well-publicized tour by such prominent figures. KAIROS-Edmonton members report some intra-church discomfort with fellow congregants who are employed in or politically supportive of oilsands development.

A similar high-profile and contentious intervention into the oilsands issue occurred earlier in 2009. Roman Catholic Bishop Luc Bouchard, whose Diocese of St. Paul includes the Fort McMurray area, released an extensively researched pastoral letter titled *The Integrity of Creation and the Athabasca Oilsands* (Bouchard 2009). The document presented scriptural and theological reasons for viewing the safeguarding of the natural environment as a religious obligation, and summarized the environmental effects of the oilsands. Bouchard concluded that the extent and type of oilsands development 'cannot be morally justified.'

Bishop Bouchard's pastoral letter was widely read outside of the Roman Catholic community and created a firestorm of response. In a speech to over 100 people at a Roman Catholic conference called 'Living Faithfully in Oil Country' in Edmonton in February 2010, Bouchard said he was amazed at the quantity of responses and their tone (Warnica 2010). Among the most common sorts of negative responses were along the lines of 'Churches should stick to morality, and what they know about, and 'I'll stay out of your God business if you stay out of my oilsands business.' Bouchard pointed out that these responses indicate a prevalent societal belief that issues of economics are above moral comment and that the oilsands should only be dealt with by technical experts. According to Bouchard, industry and government representatives responded to his letter 'like they were reading from the same page.' Exemplifying what Martin Hajer (2009) has called technocratic closure, government and industry representatives described their technical efforts to reduce the oilsands'

[3] An apparent thorn in the side of the Harper government, KAIROS' 35-year relationship with the Canadian International Development Agency was abruptly terminated in November 2009, with conflicting explanations that relate to KAIROS' advocacy on Palestinian oppression, climate change and the oilsands.

environmental impact. They ignored the moral questions Bouchard raised as if it were already settled that oilsands development should proceed as it has.

Religiously-based activism provides a powerful counter-narrative of the dominant socio-political culture as it gives primacy to moral questions about 'what we ought to do.' Religious discourse raises moral concerns and duties within a framework of legitimacy that remains outside of the instrumental rationality of the hegemonic order, providing 'autonomous spaces' only partially controlled by the socio-political hegemony (Billings 1990). The result has been a widening of the legitimacy of questioning and opposing the oilsands.

Labour Activism

Labour movements have found themselves in a conflicted position, benefitting from petro-expansion yet struggling to articulate a 'green work' and justly sustainable alternative. Labour's position has typically been that environmental issues need addressing but in a way that reduces impacts on workers. Especially on climate change, the Canadian Labour Congress (CLC) has repeatedly endorsed ratification of the Kyoto Protocol or other initiatives to reduce greenhouse gas emissions, while advocating 'Just Transition' strategies.[4] Such strategies would include reskilling programmes for workers displaced from affected industries and compensation for reduced wages or lost jobs in specific sectors. In Alberta, one of the largest unions in the oilsands – the Communications, Energy and Paperworkers (2000) – advocated a just transition strategy and supported ratification of the Kyoto Protocol.[5] More specific to the oilsands industry, the Alberta Federation of Labour (AFL) produced the report *Lost down the pipeline* (AFL 2009), part of the *Will the Oilsands Be Used to Build a Brighter Future?* Campaign in which the AFL's opposition to the oilsands focused on ensuring value-added jobs from refining oil in the province, rather than transporting it elsewhere.

While labour organizations in Alberta have supported the transition to an ecologically sustainable model of development, the diverse interests and political positions among the unionized workforce make strong and unified commitments very difficult to negotiate (Spencer 1995). One significant characteristic of Alberta's labour force is the polarisation between the well-paid, mostly male workforce in the resource extraction and petrochemical sectors, on the one hand, and a poorly paid, largely female, service sector, on the other hand (Phillips 2010).

As a result of these internal divisions and other factors, union opposition to the oilsands *status quo* has primarily been about the manner in which the oilsands are

---

[4] CLC has produced several documents and statements on the topic of sustainability and a just transition. See www.canadianlabour.ca/issues/green-jobs and www.canadianlabour.ca/news-room/publications/just-transition-workers-during-environmental-change

[5] The CEP Policy 915 for a "just transition" in a global climate change world was adopted in 2000.

developed, and the distribution of costs and benefits (AFL 2008). However, as the above example suggests, even on issues of job creation and royalty rates the AFL does not command unified support.

Labour organizations remain involved in this debate, but it is unclear to what extent organized labour's activism or policy reports influence individual workers' perspectives on the oilsands. Labour activism has focused more on maintaining value-added jobs from the refining of bitumen, opposing the exportation of pipeline supply jobs, increasing royalties as a "fair-share" for the public, advocating that oilsands development be done in an environmentally and socially responsible manner, and, to a lesser degree, using oilsands revenue to move Alberta to a post-oil economy.

Growing the Movement (Horizontally and Vertically)

The anti-oilsands movement led by these four major kinds of organizations – aboriginal, environmental, religiously-based, and labour – was clearly a vibrant one that made important contributions to the debate on the oilsands as separate movements with distinct issues and strategies. However, we argue the movement began to have broader political traction when cross-organizational coalitions were formed and the movement "scaled up" to join with groups working at the national and international level.

Joining Up: Creating Coalitions Among Unlikely Allies

The political impact of these four groups was further strengthened through the coalitions among these organizations. For example, ENGOs in Alberta have been joined by organizations not primarily focused on environmental issues, such as Public Interest Alberta (usually focused on protecting public services), churches, and unions. Further, ENGOs like the Sierra Club have frequently drawn on scientific research and expertise, most notably from policy institutes such as the Pembina Institute (a unique hybrid of energy consulting firm and environmental nonprofit organization), and the Parkland Institute (a progressive research institute at the University of Alberta) which has produced several reports related to energy security and the royalty regime. ENGOs have also actively partnered with Aboriginal organizations such as the Athabasca Cree and Fort Chipewyan First Nations. The idea of environmental justice brought together Aboriginal groups, who use this concept as a way of describing the inequitable impacts of the oilsands, alongside ENGOs who adopted the term, as well as religious organizations who have long supported Aboriginal and other groups (Lind and Mihevc 1994) through a similar frame.

Unions and labour federations have also partnered with the other types of movement organizations. Several labour organizations signed the call for a moratorium on oilsands expansion. Less directly, unions have funded the research of the Parkland Institute, such as reports on the oilsands royalties. In April 2009, the AFL partnered with Greenpeace and Sierra Club to produce the report *Green Jobs: It's Time to Build Alberta's Future* (Thompson 2009). This report argued for policies to encourage 'high-quality jobs... that will shift our economy toward greater sustainability' (2) focusing on three sectors: energy efficiency, transit and high speed rail, and renewable energy. Further, it noted that of Alberta's 56 economic sectors, oil and gas extraction created the fewest jobs per dollar spent. Yet while labour organizations have partnered with environmental organizations, the level of support that such initiatives garner from workers is not clear. For example, when the provincial government was considering an increase in royalty rates for oil and gas producers, employers were able to mobilize workers to oppose such increases (Byfield 2007). The AFL as an organization supported the royalties increase.

Scaling Up: Crossing Borders

The resources and impact of these oppositional groups was further enhanced by scale-shifting – the mobilization of opposition to the oilsands in sites far from Alberta, within Canada, across the North American continent, and worldwide.

Recognizing the limitations of localized actions, Northern Alberta Aboriginal groups have sought allies who have partnered in taking the fight to Ottawa, corporate offices in the United States, the United Kingdom, and European parliaments. Similarly, opposition to the oilsands by environmental organizations has escalated in recent years and expanded in both geographic scale and intensity. Canadian-wide opposition to the oilsands grew from this movement in Alberta. Coalitions with high diversity now extend from local and provincial organizers to national, continental and transnational scales.

One manifestation of globalization is the linking of communities of resistance in Canadian and American communities impacted by the oilsands projects via the transportation or refining of bitumen, the transportation of "mega-loads" headed for northwestern Alberta, or by more far-reaching environment impacts such as climate change. Anti-pipeline groups have sprung up from Quebec to the American Midwest, bringing attention to the web of pipelines that bisect the continent. For example, in the summer of 2009, 'Dirty Oilsands,' an international network of social, environmental, aboriginal and research organizations, tried but failed to win U.S. government rejection of approval for the construction of the Alberta Clipper pipeline. Demonstrating tighter continental integration of the energy sector, the pipeline was intended to transport bitumen from the oilsands to Wisconsin to provide feedstock for American refineries. Transborder activism along pipeline routes is part of the coalition's broader agenda to raise awareness about how Alberta's oilsands projects are a threat to the development of an alternative energy economy in the U.S. This networking has provoked counter-organizing by corporations and the Alberta and Canadian governments. For example, in British Columbia there is strong opposition to pipelines and tanker traffic transporting bitumen (seen, for example, in the 'Oilsands Free B.C.' campaign). Coastal First Nations and local groups have organized to oppose the Northern Gateway Pipeline to transport crude oil from near Edmonton to the Port of Kitimat.

National NGOs with issue foci not limited to environmental concerns have also become active against the oilsands. One example is the Polaris Institute, which coordinates the 'Tar Sands Watch' program. Polaris is particularly interested in the political influence of the corporate sector and claims that the oilsands do not represent future 'energy security.' Polaris argues that the North American Free Trade Agreement may be an impediment to environmental protection as oil from Alberta has become a transborder commodity (Clarke 2008; McCallum 2006; Laxer 2008). Another example of national NGO activism is the Council of Canadians (CC), which also links the oilsands with other issues such as climate change, energy security, water, and opposition to free trade. CC president Maude Barlow has consistently called the oilsands 'Canada's Mordor' (Arrowsmith 2008; *The Star Phoenix* 2008).

Opposition in the United States had early successes by getting the U.S. Conference of Mayors to question cross-border importation of Alberta's unconventional oil for climate change reasons, leading to a motion, passed in 2008, to reduce the use of fuel from the oilsands due to its high greenhouse gas intensity (De Souza 2008). In response, Canadian oil producers launched a website, www.canadasoilsands.ca, to 'encourage dialogue' (Cattaneo 2008). Transnational groups like Oilwatch or Oil Change International specialize on this issue and comprise coalitions of indigenous solidarity, environmental, religious, and other social movement actors.

Another manifestation of transnational cultural politics and networks is demonstrated in recent American environmental groups' comparison of the oilsands to the popular movie *Avatar*. This movie represents the collision of a technological civilization's vision of progress with the cosmology and needs of an indigenous people. The 'avaTARsands' campaign draws parallels with Alberta, targeting American citizens and policy makers. A full page ad in American national newspapers and a website have gained the attention of politicians and industry (Hussey 2010). Invitations for *Avatar* producer James Cameron to visit northern Alberta were issued by indigenous and environmental groups as well as by Premier Stelmach (Bennett 2010). The Canadian Association of Petroleum Producers even responded with a special section on its website against 'claims that 'Pandora's unobtanium mining is Alberta's tar sands." Comparisons of the oilsands with the fictional planet, Pandora, have engendered increasing recognition among Albertans that the oilsands status quo is viewed beyond the province's borders as being problematic (Haluza-DeLay et al. 2013).

Apart from government lobbying and public education campaigns, transnational movement organizations have targeted the profits of key corporations directly. Rainforest Action Network (RAN) positions its strategy within a larger inter-organizational struggle to weaken the legal, fiscal, and political 'pillars' of oilsands extraction. In March 2009, RAN joined with members of the Lubicon Lake Indian Nation to lobby the Royal Bank of Canada – the Canadian bank providing the greatest amount of financing to the oilsands – to withdraw support from oilsands projects (Barclay 2009). Ethical investors have been targeted by Aboriginal activists who associate the oilsands with indigenous rights abuses. Canadian and American native activists have partnered in the development of the Indigenous Environmental Network, which draws attention to the human rights abuses, corruption, and violence often associated with oil extraction (Watts 2005; Zalik 2009).

As the Canadian and Albertan governments have resisted calls for the reduction of greenhouse gas emissions or better oilsands regulation, European groups have become increasingly active in opposing oilsands projects. All the actions reported in the oilsands section of the Greenpeace Canada website, from January to June 2010 took place in Europe, many with corporate targets and indigenous partners. Europe is a major source of investment in the oilsands, so activism in Norway, France, and the U.K. has sought to pressure investors in these countries. Since fall 2008, numerous actions occurred in which environmental, human rights and First Nations activists, financial institutions and oilsands companies interacted in an increasingly public manner. The oilsands have become perceived as a risky investment by financial markets – for climate change, long-term financial, and other reasons (Church Investors Group 2008; Crooks 2008). Scientists of international repute have also become involved. For example, American climate scientist James Hansen published an editorial in the Norwegian news media requesting disinvestment of Statoil from the oilsands in the spring of 2010. At nearly the same time, Alberta water scientist David Schindler and two Aboriginal Greenpeace activists held a series of meetings with Norwegian government and industry executives presenting a similar case.

Other transnational actions explicitly linked the commodity chain of the oilsands, which produces more emissions than conventional petro-production, the crisis of climate change and the challenges of peak oil. Indigenous, environmental, and other social movement organizations have increasingly called climate change – and the contribution of the oilsands – a form of 'climate injustice' (Angus 2010). During international climate change negotiations Canada has repeatedly been given 'fossil' awards by social movement groups.

Building Cross-Organizational, Cross-Border Consensus

One substantial outcome of this coalition building and scaling up was the solidification of a broad agreement on the need for a moratorium on new oilsands projects. While much of the movement action has been oppositional, since the expansion of the movement through coalitions and the transnational shift, there have also been efforts to develop alternatives to the existing development and regulation regime of the oilsands. Albertan and Canadian social movement actors have begun to create the broad outlines of an equitable and ecologically sustainable post-carbon economic model. The shift in movement strategy toward a leadership role in advancing an alternative hegemony began with calls for a moratorium on oilsands project approvals on the cusp of the recent financial crisis.

In 2008, as expansion of oilsands operations accelerated and the cumulative effects of resource development became clearer, over 70 Aboriginal, environmental, labour, and church groups, along with a variety of other social groups and research institutes, signed the call for a temporary moratorium on new projects.[6] The document was evidence of a broad-based understanding among unlikely allies – it demonstrated "homegrown" concern over the social and environmental impacts of the oilsands that stood at odds with government and industry representations of those opposing the developments as "outsiders." The diversity of the list showed that oilsands opposition was not a fringe movement but one of concern to many Albertans. The call for a moratorium also signalled the extent to which the issue had crossed the Albertan border to become a national issue.

Similarly, trying to seize an opportunity in the recession, social movement organizations began to conceptualize how to 'green' the economy. The 2009 report *Green Jobs: It's Time to Build Alberta's Future* is just one example of this effort. Canadian movement actors pointed to Ontario, which had increased investments in renewable energy technology.

Also during this period, opposition to oilsands projects spread across the border to the U.S. and to Europe just as awareness grew about the link between oilsands developments and climate change, and as key leaders (including American President Obama) expressed interest in taking action on the latter issue. Oilsands opponents began to position the reliance on the oilsands as an impediment to development of an American green energy economy.

Networked social movement actors have continued to assert their proposals for transition to a post-carbon economy. Earth Day 2010 saw the announcement of a nation-wide Green Economy Network. Founding organizations included 11 labour organizations, 5 environmental groups, the Canadian Federation of Students, KAIROS, Council of Canadians, Polaris Institute and the B.C.-based Columbia Institute (which promotes democracy and sustainability). In May 2010, Environmental Defence Canada and the United Steelworkers produced a report on the lost opportunities to invest in green jobs nationally and have allied as 'Blue-Green Canada.'[7] Even the Conference Board of Canada has produced several reports recently on the economic potential of such green economic initiatives as technologies to mitigate climate change (Conference Board of Canada 2010). Finally, the Pembina Institute has release numerous reports on economic and energy restructuring to reduce reliance on oil and gas and begin a shift towards a less carbon-intensive economy.

---

[6] Signatories to the 2008 moratorium on new oilsands development, *No New Approvals*, were originally accessed in November 2008 through a website that is no longer active. Similar information can be found at http://nonewoilsands.wordpress.com

[7] This report and others are available at www.bluegreencanada.ca

Despite these efforts, the critics of Alberta's reliance on intensive hydrocarbon extraction met a wall of resistance. An initial flurry of media interest met the *Green Jobs* report but it received no attention from the Alberta government. In a private conversation, an employee of the Alberta Public Affairs Bureau (a branch of the Executive Council) said the lack of a response by the department indicated the government did not notice the report.[8] The Alberta government continued its approvals of new oilsands projects apace, while launching repeated public relations campaigns in defence of oilsands expansion.

Since AFL was a partner in the *Green Jobs* report, its apparent absence in a green economy (RePower Alberta) campaign the next year was striking and may indicate labour's ambiguous relationship to oilsands employment. Dave Berry (2010) reported that Greenpeace's direct actions soured many labour members toward ENGO-led campaigns. And while the 'RePower' branding could imply empowering democratic engagement in the province, groups that would help produce this dimension (like Public Interest Alberta) were notably missing from the campaign.

Ordinary Albertans are aware of the problems related to the oilsands. Several polls show that a majority of Albertans agree that the government is not doing a good job of managing the environmental problems associated with oil extraction, and that a majority of Albertans want to see stronger measures to reduce greenhouse gas emissions. The Pembina Institute polled 500 Albertans in spring 2007 and asserted that an 'Overwhelming Majority of Albertans Support a Pause on New Oil Sands Approvals' (Pembina Institute 2007). Then a December 2008 poll released by Climate Action Network, Greenpeace Canada, Pembina Institute, and The United Church of Canada found that even in an economic recession, Canadians wanted governments to take action on climate change (Pembina Institute 2008). Cambridge Strategies, a conservative political strategy firm, released results of a poll of over a thousand Albertans in July 2010 that indicated the public questioned the effectiveness of Alberta government's management of the oilsands. As the consultant summarized,

> Our study demonstrated a serious disconnect between what the public thinks should be happening to address these concerns and what is in fact happening. So far, government and oil sands industry have failed to align their actions with the public's values. This is resulting in a growing sense that the public's trust is being betrayed. The clear message is that Albertans want to prosper from their oil sands development. But they also want a policy approach that is long-term, comprehensive, and integrated around environmental, social, political, and economic concerns (Chapman 2010).

There is an increasing problematization of the oilsands *status quo* on the part of the public. But this awareness is coupled with cynicism about the possibilities for changing this *status quo*, as well as poor awareness of alternatives to the hydrocarbon extraction-based economy. According to focus group participants in one study,

---

[8] Thompson commented in an email interview that the report has served as something of a model for other provinces or labour organizations, and that he continues to get requests for speaking engagements from the report. Subsequent meetings involved ENGOs, green business leaders, and labour but they were sporadic.

the environmental movement has been critical of oilsands development, yet has not offered convincing alternatives (Haluza-DeLay et al. 2013). Clearly, the studies of economic alternatives are not reaching the public, indicating that impediments to green transition remain entrenched in the institutions (media, political, educational) and political culture of the province. The challenge for movement actors is to translate the increasing legitimacy of their criticisms of the oilsands into solid support for a post-carbon model of development and broader citizen engagement in policy decisions.

Assessing the Movement's Strengths and Challenges

The anti-oilsands movement has made several important effective strategic choices. First, with political opportunity structures in the province generally closed to oilsands opponents, opponents of the existing level of oilsands development and regulation formed coalitions that heightened the impact of the organizations working in isolation. At the same time, they chose other venues outside provincial borders to mobilize pressure on the Alberta and Canadian governments. This is consistent with research that finds that when the domestic opportunities are closed, transnational activity becomes more likely (Poloni-Staudinger 2008; Van Der Heijden 2006). Opportunities for effective opposition were instead found in the United States and Europe, and an expanding set of actors became part of the organizational field of oilsands movement activism. Overall, this represents a significant shift from localized action to local actions that are plugged into an international network.

Largely through global coalition building and linkages among oilsands, pipeline risks, climate change, and social justice struggles, the opposition to Alberta's model of development has succeeded in a short period of time in putting the provincial and federal governments on the defensive. News reports and provincial and industry campaigns began to indicate in late 2007 that 'Alberta [was] losing the public relations war when it comes to the tar sands' (Yaffe 2008.) International campaigns have increased critical public attention to the issue in Alberta and Canada, and provoked shifts in corporate and governmental discourse as they struggle to defend the industry through counter-narratives.

A second trend that emerges is that provincial culture has become a site of more intense political struggle as various actors vie for popular support and the discursive terrain has shifted. Criticisms of the oilsands have become more legitimate – particularly as they pertain to the government's management of environmental harms. The more influential role of actors linked to global environmental and climate justice movements may be seen to be altering the discursive opportunity structure in Alberta and Canada. By discursive opportunity structure, we mean the meaning-making practices and institutions of a particular society (Gamson 2004; McCammon et al. 2007; Snow 2008). Stated in Gramscian terms, there has been some success in the 'war of position' to delegitimize the hegemony of the oil complex (Gramsci 1971; Golding 1992). The success of movements inside the province can perhaps be measured by the public awareness of the problematic nature of the oilsands, as well as the relatively higher level of trust that ENGOs command compared to government and industry (which are seen as too intimate) (Johnston et al. 2006; Thomson 2010).

Accelerating government public relations campaigns is one sign of an increasingly inhospitable public reception to the environmental and social problems associated with the oilsands. It is, of course, premature to predict how shifts in the discursive opportunity structure will impact public policy or alter power relations. The challenge for social movement actors is to transform these perceptions into solid support for a post-carbon model of development.

However, overlaying these strengths of the social movements opposing the oilsands are the interrelated political-economic and discursive challenges. Compared to its opponents, the anti-oilsands social movement is poorly resourced. This is a significant weakness in the movement's ability to shift public policy and promote environmental citizenship. Industry resources are comparatively huge, with staff and money to implement sophisticated public relations campaigns to counter the opposition. In addition, media outlets are not necessarily receptive to strong critiques of the oilsands, hence the opposition groups' difficulty in communicating alternatives.

Further, while a critique of the oilsands royalty regime and environmental management *status quo* seems to have gained acceptance among the public, it is unclear whether support exists for more extensive reforms of the province's economic model of development. The public will is divided: environmental protection is desired, but so, too, is a strong economy, and these are often represented as being opposing goals. In the absence of a compelling green economic programme, this conflict will not be easily overcome. Yet even the creation of a viable green economy alternative might be fundamentally problematic. To date, proposals for reform advocate a movement away from a hydrocarbon-based economy toward a less carbon-intensive capitalist economy through 'green jobs.' Yet as critics of ecological modernization have well argued, a "greener" capitalism may not adequately resolve the ecological crisis inherent in capitalism's need for incessant growth and consumption (Gould et al. 2008). Nor are these proposals definite solutions to the inequities of development experienced by indigenous and other local communities.

Also at issue is the recent powerful backlash against opponents to the oilsands. Both the Albertan and Canadian governments have attempted to criminalize opposition to the oil sands, comparing it with terrorist activity that posed a security threat (Le Billon and Carter 2012). Most recent evidence of this at the federal level included the Natural Resources Minister criticizing "foreign"-funded "radical groups," including environmental groups who threatened to slow the Northern Gateway pipeline review process (Payton 2012). By spring 2012, a counter-terrorism unit was in place in Alberta, justified by "factors such as [...] a strong economy supported by the province's natural resources and the need to protect critical infrastructure" (Royal Canadian Mounted Police 2012).

Conclusion

Environmental, labour, Aboriginal, and religious organizations have led the opposition to the tar sands using a variety of discursive and action-oriented strategies. Aboriginal groups have drawn attention to the issue by framing the oilsands as representing a health risk, by working through courts, and by presenting the oilsands as an issue of indigenous rights and justice. Environmental groups have moved away from working in consultative bodies and toward civil disobedience, using evocative description of the oilsands (such as 'Canada's Mordor') and linking the oilsands to climate change. Religious actors have engaged in educational events and challenged capital with the moral dimensions of its actions. Labour groups have primarily sought to propose transition to a renewable energy economy.

The local and then national anti-oilsands movement has recognized that international public opinion exerts greater pressure on the government of Alberta with regard to regulation of the oilsands than local campaigns targeting public opinion in the province. The strength of these initially separate efforts was enhanced by the formation of multiple coalitions across these groups as well as connections with movements and organizations crossing the Albertan border. One important result of this expansion was the creation of specific policy recommendations with broad-based support (the call for a moratorium), as well as proposals for a post-carbon model of development.

The Albertan and Canadian governments and industry have been disquieted by the movement's coalition building and scale shifting, particularly in their use of strategies such as the 'dirty oil' campaigns in the United States and Europe. Hence the government and industry attempts to re-legitimize oilsands development while de-legitimizing opponents to the oilsands. While the anti-oilsands movement has garnered considerable attention and roused important debate, to date even this broad-based, multi-scalar opposition has not moved the Canadian or Albertan governments. Both the country and province remain committed to the carbon energy superpower aspiration, an ambition fundamentally dependent on continued – and expanded – oilsands extraction.

Acknowledgement Haluza-DeLay's research has been supported by The King's University College Internal Research Committee, and Carter's research has been supported by the Social Sciences and Humanities Research Council.

References

Alberta Federation of Labour. (2008). *Black gold clear vision: A proposed policy framework for the Alberta oil sands*. Edmonton: Alberta Federation of Labour.

Alberta Federation of Labour. (2009). *Lost down the pipeline*. Edmonton: Alberta Federation of Labour.

Angus, I. (Ed.). (2010). *The global fight for climate justice*. Halifax: Fernwood.

Arrowsmith, L. (Oct 31, 2008). Alberta's oilsands like mordor: Maude Barlow. *The Star*.

Barclay, B. (2009). *Financing global warming: Canadian Banks and fossil fuels*. San Francisco: Rainforest Action Network.

Bennett, D. (2010, April 21). Alberta hopes James Cameron's oilsands comments won't open a Pandora's box. *Canadian Press*.

Berry, D. (Jun, 2010). The disobedient Albertans. *Alberta Views*.

Billings, D. B. (1990). Religion as opposition: A Gramscian analysis. *American Journal of Sociology, 96*(1), 1–31.

Bouchard, B. L. (2009, January 25). *A pastoral letter on the integrity of creation and the Athabasca oil sands*. Accessed at http://www.dioceseofstpaul.ca/index.php?option=com_contentandtask=viewandid=135

Brooymans, H. (2009, February 26). Cancer downstream from oilsands merits further study: Alta. Gov't. *Canwest News Service*.

Byfield, M. (Oct, 2007). Gil McGowan: A union vision for the future oilpatch. *Oil and Gas Enquirer*.

Canada West Foundation. (2009, July). *Media monitoring report*. Calgary: Canada West Foundation.

Carter, A. (2011). *Environmental policy in a petro-state: The resource curse and political ecology in Canada's oil frontier*. Ph.D. Dissertation, Government Department, Cornell University, Ithaca, NY.

Cattaneo, C. (2008, June 23). Oil sands producers start web site to encourage 'dialogue'. *Financial Post*.

Chapman, K. (2010). Where the oil sand owners stand. *The Mark* [cited January 2 2011]. Available from http://pioneers.themarknews.com/articles/1932-where-the-oilsand-owners-stand/#.UyyJvYXPbSg

Chetkovich, C., & Kunreuther, F. (2006). *From the ground up: Grassroots organizations making social change*. Ithaca: Cornell University Press.

Church Investors Group. (2008). *Unconventional oil: Tar and shale sands: A briefing for church investors*. London: Church Investors Group.

Clarke, T. (2008). *Tar sands showdown: Canada and the new politics of oil in an age of climate change*. Ottawa: Canadian Centre for Policy Alternatives.

Communications, Energy and Paperworkers Union of Canada. (2000). Policy 915: Just transition to a sustainable economy in energy. *Montreal* [cited January 2, 2011]. Available from http://www.cep.ca/docs/en/policy-915-e.pdf

Conference Board of Canada. (2010). *The economic and employment impacts of climate-related technology investments*. Accessed at http://www.conferenceboard.ca/documents.aspx?did=3586

Crooks, E. (2008, September 15). Investors warned of risk to oil sands plans. *Financial Times*.

De Souza, M. (2008, June 23). U.S. mayors point green finger at Alberta oil sands. *Canwest News Service*.

Deacon, L., & Baxter, J. (2009). Framing environmental inequity in Canada: A content analysis of daily print news media. In J. Agyeman, P. Cole, R. Haluza-DeLay, & P. O'Riley (Eds.), *Speaking for ourselves: Environmental justice in Canada*. Vancouver: University of British Columbia Press.

Ecojustice. (2009). *Syncrude cannot duck from charges in death of 500 waterfowl*. Accessed at http://www.ecojustice.ca/media-centre/press-releases/syncrude-cannot-duck-from-charges-in-death-of-500-waterfowl/

Fluet, C., & Krogman, N. (2009). The limits of integrated resource management in Alberta for aboriginal and environmental groups: The northern east slopes sustainable resource and environmental management strategy. In L. Adkin (Ed.), *Environmental conflict and democracy in Canada*. Vancouver: University of British Columbia Press.

Gamson, W. A. (2004). Bystanders, public opinion and the media. In D. A. Snow, S. A. Soule, & H. Kriesi (Eds.), *The Blackwell companion to social movements*. London: Blackwell.

Gellert, P. K., & Shefner, J. (2009). People, place, and time: How structural fieldwork helps world-systems analysis. *Journal of World-Systems Research, XV*(2), 193–218.

Golding, S. (1992). *Gramsci's democratic theory: Contributions to a post-liberal democracy*. Toronto: University of Toronto Press.

Gottlieb, R. S. (2006). *A greener faith: Religious environmentalism and our planet's future*. Oxford: Oxford University Press.

Gould, K. A., Pellow, D. A., & Schnaiberg, A. (2008). *The treadmill of production: Injustice and unsustainability in the global economy*. London: Paradigm Publishers.

Gramsci, A. (1971). *Selections from the prison notebooks of Antonio Gramsci*. London: Lawrence & Wishart.

Hajer, M. A. (2009). Ecological modernisation as cultural politics. In A. Mol, D. Sonnenfeld, & G. Spaargaren (Eds.), *The ecological modernisation reader: Environmental reform in theory and practice*. London: Routledge.

Haluza-DeLay, R., & Berezan, R. (2013). Permaculture in the city: Ecological habitus and the distributed ecovillage. In J. Lockyear & J. Veteto (Eds.), *Localizing environmental anthropology: Bioregionalism, permaculture, and ecovillage design for a sustainable future*. New York: Berghan Books.

Haluza-DeLay, R. B., Ferber, M. P., & Wiebe-Neufeld, T. (2013). Watching Avatar from 'AvaTar Sands' land. In B. Taylor (Ed.), *Avatar and nature spirituality*. Waterloo: Wilfred Laurier University Press.

Hussey, A. (2010, March 4). Seen as Canadian oilsands allegory, environmentalists want Oscar for Avatar. *Canwest News Service*.

Johnston, W. A., Krahn, H., & Harrison, T. (2006). Democracy, political institutions, and trust: The limits of current electoral reform proposals. *Canadian Journal of Sociology, 31*(2), 165–182.

KAIROS: Faithful Action for Justice. (2011). Accessed at http://www.kairoscanada.org/sustainability/the-tar-sands/kairos-response-to-the-tar-sands-3/

Kowalsky, N., & Haluza-DeLay, R. (2013). Homo energeticus: Technological rationality in the Alberta tar sands. In H. M. Jeronimo, J. L. Garcia, & C. Mitcham (Eds.), *Jacques Ellul and the technological society in the 21st century* (pp. 159–175). Berlin: Springer.

Laxer, G. (2008). *Freezing in the dark: Why Canada needs strategic petroleum reserves*. Edmonton: Parkland Institute/Polaris Institute.

Le Billon, P., & Carter, A. (2012). Securing Alberta's tar sands: Resistance and criminalization on a new energy frontier. In M. Schnurr & L. Swatuk (Eds.), *Natural resources and social conflict: Towards critical environmental security* (pp. 170–192). London: Palgrave MacMillan.

Lind, C., & Mihevc, J. (Eds.). (1994). *Coalitions for justice*. Ottawa: Novalis.

Loyie, F. (2009, November 7). Doctor cleared over suggested link between cancer & oilsands. *Edmonton Journal*.

Masuda, J., McGee, T., & Garvin, T. (2008). Power, knowledge, and public engagement: Constructing 'citizenship' in Alberta's industrial heartland. *Journal of Environmental Policy and Planning, 10*(4), 359–380.

McCallum, H. (2006). *Fuelling fortress America: A report on the Athabasca tar sands and U.S. demands for Canada's energy*. Ottawa: Canadian Centre for Policy Alternatives.

McCammon, H. J., Newman, H., Muse, C., & Terrell, T. (2007). Movement framing and discursive opportunity structures: The political successes of the U.S. women's Jury movements. *American Sociological Review, 72*(5), 725–749.

McKeon, B. (May 24, 2010). Base Energy Policy on Christian Values. *Western Catholic Reporter*, p. 15.

Meyer, D., & Minkoff, D. (2004). Conceptualizing political opportunity. *Social Forces, 82*(4), 1457–1492.

Mikisew Cree First Nation. (2007). *Response to the Multi-Stakeholder Committee Phase II proposed options for strategies and actions and submission to the Government of Alberta for the oil sands strategy*. Fort Chipewyan: Mikisew Cree First Nation.

Nikiforuk, A. (2008). *Tar sands: Dirty oil and the future of a continent*. Vancouver: Greystone Books.

No New Approvals for Tar Sands Developments. (2008). [cited November 3, 2008]. Accessed at www.nonewapprovals.ca

Ominayak, B., & Thomas, K. (2009). These are Lubicon lands: A first nation forced to step into the regulatory gap. In J. Agyeman, P. Cole, R. Haluza-DeLay, & P. O'Riley (Eds.), *Speaking for ourselves: Environmental justice in Canada*. Vancouver: University of British Columbia Press.

Payton, L. (2012, Januray 9). Radicals working against the oilsands, Ottawa, says. *CBC News*. http://www.cbc.ca/news/politics/radicals-working-against-oilsands-ottawa-says-1.1148310

Pembina Institute. (2007). *Backgrounder: Albertans' perceptions of oil sands development: Poll Part 1: Pace and scale of oil sands development* [cited January 2, 2011]. Accessed at http://pubs.pembina.org/reports/GHGPoll_Env_2007_MediaBG.pdf

Pembina Institute. (2008). *Poll: Canadians want action on global warming despite economic downturn*. Accessed at http://www.pembina.org/media-release/1736

Phillips, S. (2010). *Women's equality a long way off in Alberta: Alberta most unequal province in Canada* (Fact Sheet). Edmonton: Parkland Institute. Accessed at http://parklandinstitute.ca/research/summary/womens_equality_a_long_way_off_in_alberta/

Poloni-Staudinger, L. M. (2008). The domestic opportunity structure and supranational activity: An explanation of environmental group activity at the European Union Level. *European Union Politics, 9*(4), 531–58.

Royal Canadian Mounted Police. (2012). *Creation of an RCMP-led INSET in Alberta* [press release]. http://www.rcmp-grc.gc.ca/news-nouvelles/2012/06-06-inset-eisn-eng.htm

Schneider, R. (2002). *Alternative futures: Alberta's boreal forest at the crossroads*. Edmonton: Federation of Alberta Naturalists.

Snow, D. (2008). Elaborating the discursive contexts of framing: Discursive fields and spaces. *Studies in Symbolic Interaction, 30*, 3–28.

Spencer, B. (1995). Old and new social movements as learning sites: Greening labor unions and unionizing the greens. *Adult Education Quarterly, 46*(1), 31–42.

The Star Phoenix. (Nov 1, 2008). UN water advisor calls Alberta oilsands 'mordor'. *Canwest News Service*.

Thompson, D. (2009). *Green jobs: It's time to build Alberta's future*. Edmonton: Alberta Federation of Labour, Greenpeace and Sierra Club.

Thomson, G. (Aug 4, 2010). Greenpeace antics give pause for thought: Public increasingly in favour of greening of the oilsands. So why not do it?. *Edmonton Journal*.

Timoney, K. (2007). *A study of water and sediment quality as related to public health issues, Fort Chipewyan, Alberta*. Fort Chipewyan: Nunee Health Board Society.

Timoney, K., & Lee, P. (2009). Does the Alberta tar sands industry pollute? The scientific evidence. *The Open Conservation Biology Journal, 3*, 65–81.

Van Der Heijden, H.-A. (2006). Globalization, environmental movements, and international political opportunity structures. *Organization and Environment, 19*(1), 28–45.

Warnica, R. (2010, February 28). Bishop assesses fallout from letter. *Edmonton Journal*.

Watts, M. J. (2005). Righteous oil? Human rights, the oil complex and corporate social responsibility. *Annual Review of Environment and Resources, 30*(1), 373–407.

Yaffe, B. (Nov 25, 2008). 'In fact, every Canadian has a stake in this'; Alberta oilsands industry fights a public relations war in advance of a new energy policy in the U.S. *Vancouver Sun*.

Zalik, A. (2009). Zones of exclusion: Offshore extraction, the contestation of space and physical displacement in the Nigerian delta and the Mexican gulf. *Antipode, 41*(3), 557–582.

Part III
Elementary and Secondary Education

Larry Bencze and Steve Alsop

Preamble

In this Part, educational researchers and teachers provide examples of practice in elementary and secondary schools that challenge students to consider critical issues associated with fields of science and technology that may lead to actions to bring about a better world. Most of cases pertain to school science in elementary or secondary schools, while one deals with an environmental education course and two others relate to after-school programmes that engage youth. As may be apparent from the word cloud shown at above, in addition to emphases on activism, the authors here focus on issues of social justice and/or environmental wellbeing—often with students' perspectives prioritized.

The Chapters

Chapter 20
We Got Involved and We Got to Fix It!: Action-Oriented School Science

Erin Sperling, Terry Wilkinson, and Larry Bencze

Abstract The body of literature connecting science education and citizenship is growing, through the lens, for example, of science, technology, society and environment (STSE) education. The case study highlighted here uses the STEPWISE framework to explore ways in which students in a seventh-grade science class in an urban centre in Ontario, Canada used studies of waste management to engage in active citizenship. In our observations and analyses of their action projects, we suggest that students formed new connections between science education and citizenship. Through personal changes they appeared to experience through the projects, it seems that they gained recognition of the positive impact that an individual can have on the well-being of self, society, and environment. Factors influencing their personal changes, including changes in their science literacy and self-efficacy beliefs, as well as particular influences of their teacher, indicate directions for possible expansion studies and implementations of interactions between science and citizenship education.

Keywords Science education • Action project • Citizenship education • Middle school • STEPWISE

This chapter uses material previously published in a special issue (volume 10, issue 3) of the Canadian Journal of Science, Mathematics and Technology Education and it appears here with the permission of the copyright holder, the Ontario Institute for Studies in Education.

E. Sperling (✉) • L. Bencze
Ontario Institute for Studies in Education, University of Toronto, Bloor St W 252, Toronto, ON M5S 1V6, Canada
e-mail: erin.sperling@utoronto.ca; larry.bencze@utoronto.ca

T. Wilkinson
Faculty of Education, York University, Toronto, ON, Canada
e-mail: terry_wilkinson@edu.yorku.ca

L. Bencze and S. Alsop (eds.), *Activist Science and Technology Education*, Cultural Studies of Science Education 9, DOI 10.1007/978-94-007-4360-1_20,

Introduction

> … who asks the child for his opinion or consent?
> Who is likely to take note of any advice or
> approval from such a naïve being?
> What can a child possibly have to say? (Janusz Korczak)

This chapter explores the intersection of science and technology literacy and citizenship education involving environmental action. It highlights a case study within a publicly-funded curriculum development and research project, known as 'STEPWISE' (Science and Technology Education Promoting Wellbeing for Individuals Societies and Environments) (elaborated in: Bencze and Alsop 2009). A teacher followed the work of her seventh-grade students as they implemented personal change projects (Action Projects) related to waste reduction and management. The lead author (ES) worked with her to evaluate the outcomes. Through analyses of this case, we document and analyze possibilities for citizenship education outcomes in a science class and factors that may have contributed to such outcomes.

Theoretical Background

Citizenship Education

'Citizenship education' and 'science education' have often been treated as separate entities, each involving separate school subjects—such as 'Civics,' as separate from 'Biology.' There are contested visions of citizenship education, defined principally by divergent concerns about objectives and implementations. Regarding objectives, the literature points to several ideologies, such as the desire for 'nation-building,' for 'future economic sustainability,' and for 'social and/or political change' (Westheimer and Kahne 2004). Citizenship education may include issues of policy, identity, political structures, and action for change. Regarding the question of implementation, solely considering formal education models, there are *integrated*, *separated* and *overarching* (i.e., whole school/board culture) methods. In Ontario, for example, a discrete half-course is offered in grade ten as 'Civics.' Beyond that, forays into citizenship are either delivered piecemeal in social studies, or left to the teacher's discretion—which is likely to be influenced by a series of factors, including his or her comfort with the topic.

A component of citizenship education is *activism*, succinctly described by Cathie Holden and Nick Clough (1998) as an activity that "involves reflecting on values, assisting children to acquire the skills necessary for taking action and ultimately providing opportunities for them to become involved as active citizens" (p. 14). They reference the Convention on the Rights of the Child (United Nations 1989), Articles 12 and 13, in which children have rights to seek and impart information, to express their thoughts and feelings, to have these listened to, and to participate in decisions affecting them.

Like other areas of education, the learning of citizenship, of social responsibility, of policies and protocols, of nation and neighbour in the world, are not to be decontextualized and disconnected. Rather, the lessons should involve the context beyond the classroom. Ideally, as noted by Wilfred Carr (1998), formal education should occur through a curriculum for democracy—providing students with a critical engagement with issues in society. As Martin Ashley (1998) acknowledges, dissociation of knowledge from experience is 'undemocratic' while its coherence and interconnectedness leads to global citizen identity, which connects to particular notions of scientific literacy.

Scientific Literacy

Since about the mid-1990s, jurisdictions have been promoting 'scientific literacy' for all citizens through school science (Hodson 2008). Initially, the concept of scientific literacy was grounded in the ideology of expert/scientist and 'truth' production. In recent years, however, this concept has grown to include a means of addressing economic, social and environmental issues in preparation for responsible citizenship (Hodson 2003, 2008). Due to increased inequity in social and environmental justice globally, advantaged citizens cannot afford to be passive, buying into popular consumerist ideology at the detriment of human and ecological wellbeing. Access to resources is inherently political (Simms 2009). Hodson (1994) first referred to the political nature of issues-based scientific literacy in a four-level framework. Each level moves through a degree of 'STSE' (Science, Technology, Society, Environment) sophistication of student attainment: Level 1, appreciating the societal impact on scientific change; Level 2, recognition of stakeholders in scientific decisions and the link to wealth and power; Level 3, development of individual views and establishing value positions; and Level 4, preparation for and taking of action (p. 85).

This four-level classification could be referred to as critical-responsible scientific literacy, because of its orientation toward the actor taking a stand on issues, which lead to responsible action based on his or her values. This politicized version of science literacy amounts to a means of evaluating the degree of engagement with issues, which is used in the analysis below.

Intersection and Integration

The possibility of integrating citizenship education across the curricula is appealing for many reasons, and especially so in science education. A critical-responsible interpretation of issues-based scientific literacy is important as citizens feel pressing needs to demand and implement actions in their societies, with regard to, for example, the environment, consumer safety and energy consumption. As educators and citizens become more connected to the notion that "Life in the 21st century is irrefutably associated with science and technology, and formal education should help students prepare for active participation in modern democracies" (Sadler et al. 2007, p. 373).

Scientific literacy in its relationship to citizen-development is mainly expressed as STSE within formal science education in the Ontario context, as STSE focuses on issues in society that are impacted by science and technology and vice versa, for example. Students' interactions with each other and their teacher, in the science classroom, their school and community allow for accumulation of social and cultural capital, through the acquisition of skills, relationships and networks of knowledge, which can break down barriers, especially when combined with action (Hodson 2008). STSE education is a means towards promotion and engagement of citizens who may make informed decisions toward a sustainable future (Davies 2004). The goal of activism through science is not necessarily to develop justice-oriented citizens but to instill a sense of engagement that may lead to justice orientation through participation. A core assumption of the participatory citizen is that in order "to solve social problems and improve society, citizens must actively participate and take leadership positions within established systems and community structures" (Westheimer and Kahne 2004, p. 266). Engaged youth may become engaged adults, who are socially minded, scientifically knowledgeable citizens (Cargo et al. 2003); not simply "armchair critics" (Hodson 2003, p. 657). This chapter explores a case of this interaction in action.

Research Context and Methodology

Research Context

This is a case study of students in a seventh grade Science class, in which the teacher[1] encouraged the students to take action to promote personal, social and/or environmental wellbeing around the topic of waste management. The students had full choice of whether they would: work alone or with a team; focus their efforts at school, home or in the community; and, involve their families. They documented their 8-week projects through learning logs, classroom discussions and their culminating presentations. Once the Action Project was introduced, the students developed and followed their plans—mainly independently—using a SMART plan guideline (Specific, Measurable, Attainable, Realistic and Timely).[2]

[1] The teacher, 'Theresa,' was a 23-year veteran with a Visual Arts background and training in design and technology education (e.g., solving design problems and basic woodworking).

[2] The teacher's 'S.M.A.R.T. Plan' guideline for making a personal change was based on a widely used goal-setting framework developed by Jan O'Neill and Anne Conzemius (2006).

Table 20.1 Intersecting Hodson (1994) With Westheimer and Kahne (2004)

	Hodson (1994)			
Westheimer and Kahne (2004)	Level 1	Level 2	Level 3	Level 4
Personally responsible			X	X
Participatory (action oriented)				X
Justice oriented		X	X	X

The data from this study were collected from February to July, 2008, including three individual meetings with the teacher, five interactive classroom visits, and two group interviews with students. The teacher in the study was a member of an action-research group with the STEPWISE project, which addresses components of science education, STSE and sociopolitical action.

Data Collection and Analyses

To explore our above-mentioned research goals, data-collection and analyses relating to this case study had both naturalistic and rationalistic characteristics (Guba and Lincoln 1988). From a rationalistic perspective, we planned to look for evidence of students' orientation towards 'citizenship.' Our judgments were influenced by the typology proposed by Westheimer and Kahne (2004); that is, *Personally Responsible*, *Participatory* and *Justice-Oriented* Citizenships. We felt that this typology has some congruency with Hodson's four levels of STSE commitment, as summarized in Table 20.1.

In terms of factors influencing students' tendencies towards using their science and technology education for citizenship purposes, our planning and analyses were undoubtedly influenced by fundamental principles of the STEPWISE curriculum and instruction framework (Bencze and Alsop 2009). This framework arranges teaching and learning in ways that may provide all students with useful cultural capital (e.g., information about specific STSE issues), including that derived from their self-determined investigations and projects, so that they might 'spend' some of their capital on the common good.

While we had some pre-conceived notions about citizenship and factors possibly influencing it, our planning and analyses had a significant naturalistic character. Although Theresa had been involved with the STEPWISE project for 2 years and, accordingly, possibly influenced by its principles, we continually urged her and other members to make instructional decisions on the basis of various contextual variables associated with their work. Congruent with our belief that knowledge, thinking and doing are situated, we used qualitative, ethnographic research methods (e.g. Hammersley and Atkinson 1990)—allowing conclusions to emerge from studies of events.

Fig. 20.1 Reused Coke Cans™—Community-member engaged in waste diversion

Given our epistemological and methodological perspectives, data collected included:

- Semi-structured interviews: Six students were interviewed three times in small groups (2–6) for 20–45 min. From the interviews, data included interview transcripts with the teacher and the students, collected before, during and after the Action Project process. Students were asked open-ended questions, such as: What did you think about the Action Project? How did people get involved in your projects? What did you learn from this project? The teacher was also interviewed formally three times.
- Artefacts of Student Activities: Data included samples of students' learning logs, with student-written reflections on the progress of their Action Projects. As well, images were captured to represent students' actions (see Figs. 20.1, 20.2 and 20.3 for examples).
- Observational Records: The observational data included six sets of 50-min classroom visit field notes, and recordings of student presentations to their class of their Action Project progress and outcomes.

For analyses, constant comparative methods based on constructivist grounded theory (Charmaz 2000) were used. This is an abductive-deductive dialectic process, in which data were carefully reviewed and coded (as abductive acts). Codes were then grouped into categories and themes relating to knowing about and performing citizenship through science, which was evaluated through references to data (as deductive acts). These processes were repeated until 'saturation' of themes

Fig. 20.2 Conserving water poster—Sibling collaboration, and graphs of water conservation over time

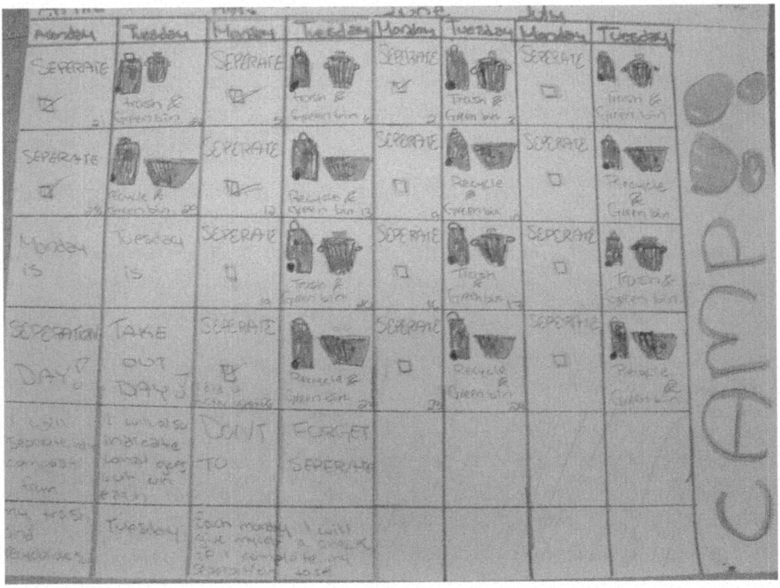

Fig. 20.3 Garbage separation calendar, an organized display of planning and data collection tracking

was achieved. Checks with the participants were conducted to ensure 'trustworthiness' of claims, each of which was based on at least three corroborating data sources (i.e., semi-structured interview transcripts, researcher observations, student project artifacts)—for 'triangulation' purposes (Lincoln and Guba 2000).

Results and Discussion

As a result of our analyses, it was apparent to us that students' Action Projects had some effects on their personal orientations towards knowledge and citizenship. Some students, for example, re-directed much of their personal and home waste away from the landfills, reduced their water usage and encouraged their neighbors to stop littering. In the sub-sections that follow, we provide support for this claim, along with discussions of factors apparently influencing students' orientations towards citizenship.

Students' Citizenship Orientations

As described above, we drew upon an intersection of Hodson's (1994) schema for levels of commitment to STSE issues and the typology for citizenship provided by Westheimer and Kahne (2004) to judge students' apparent orientations towards science-based citizenship. Our analyses led us to two aspects of such an orientation; that is, indications of students': (a) tendency toward 'active (participatory) citizenship'; and (b) commitment towards maintaining their changes beyond the completion of the Project. Each of these aspects is discussed, with examples, below.

- Tendencies Towards 'Active Citizenship': The term active, or participatory, citizenship relates to an indication by actors that they are involved in community life, that they have acknowledged the positive and negative possibilities of their actions. Many of the students indicated this through their projects. Student B, for instance, reflecting on her completed Action Project, stated:

  > It helped the environment, but it also helped us. It teaches discipline and to set goals, and you also get to record it down. And it does teach you to be responsible (Interview, June 6, 2008).

  They measured their success by setting goals for changes they would make in their waste management practices, and seeing them through. The students also felt rewarded in their changes when others became involved, and validated their efforts, as Student D stated:

  > I liked it because I knew that I wasn't the only one who was fulfilling my Action Plan. I was getting other people to help as well, and save, maybe not conserve more, [but] at least try and get involved or make the effort (Interview, June 6, 2008).

This statement also raises questions about voice, responsibility and impact, which are key to critical-responsible engagements with citizenship education (Roth and Désautels 2004). Recognition of responsibility is shown by the awareness of municipal laws and practices that were negotiated in the development of their Project, as well as setting and meeting goals, both socially and environmentally. When students recognized that they were making an impact not only on the environment but also on the people around them, they were often given validation of their voice and actions, and thus expressed that they felt they were participating as citizens in a broader unit than simply their classroom—as a family member, as a school participant and as a community member.

- Commitment Longevity: A sense of students' apparent commitments to citizenship was gathered from indications that they intended to maintain changes to their personal practices beyond the duration of the Action Project. Certainly, the interpretation of their responses is speculative, but their honesty with the project is also evident. Many statements, such as the following, indicated obvious intention:

  > I'm going to continue my showers to reduce the minutes... I'll still have the posters up to remind people in my family ... [but] I won't record it (Student E, May 30 2008).

  The lack of desire to continue recording waste management-related outcomes after the completion of the Action Project was consistent among students. While they described in great detail the methods they employed to record their data, rarely was it an activity that the students indicated enjoying or wishing to continue. However, it was observed in artefacts, such as posters, interviews and class presentations, that they expressed gratification by observing change in family members' actions which was in turn supportive to the students' maintenance of their personal change. This recognition, directly or indirectly, may create a community of support and practice (Wenger 1998). The participation of the students in their respective projects is often legitimated by the actors within their community—family, school, neighbourhood, etc.

Overall, there was mainly third and fourth level attainment under Hodson's (2003) prescription for STSE literacy. The students were able to establish their value positions based on the actions they designed (Level 3); e.g., organic matter *should go* in the green bin for compost. They also prepared and took action (Level 4); e.g., created a poster for one's family to redirect organic waste, and observed the organic waste diversion by measuring the volume in each bin per week. Thus the students in the case study experienced an intersection of science education and citizenship education through the personal change that took place in their tendencies toward Westheimer and Kahne's (2004) personally responsible and action-oriented citizenship (taking responsibility and initiating projects), and their commitment to the actions they had started as part of an individual project (highly positive and likely). The students experienced a form of democratic education that allowed for expression of their own critical analysis and a shared community of support, which reinforced their tendency and commitment through the process and outcomes.

Factors Influencing Citizenship Orientations

Students' orientations towards citizenship are, undoubtedly, influenced by many factors—apparently including the nature of the Action Project assignment, but likely also involving factors inherent to each child (e.g., a general feeling of self-efficacy). Our analyses suggested that at least three—not necessarily independent—general factors seemed to strongly influence students' orientations towards citizenship. The three factors, students' personal self-efficacy, influences from the teacher and science literacy are discussed in some detail, with examples, below.

- Personal Self-efficacy: Students who appeared to us to have the strongest citizenship leanings seemed to have a strong sense of *self-efficacy*, which Albert Bandura (1997) defines as a "belief in one's capabilities to organize and execute the courses of action required to produce given attainments" (p. 3).

  Several students encountered a positive response to their Projects from family or community members, which reinforced their role as leaders. For example, one student used posters to advertise a discount at her parents' restaurant for patrons who brought in representations of reusing their garbage. Figure 20.1 shows a model motorcycle made of Coke™ cans that was brought in by one of the restaurant patrons due to the prompting of the student. She expressed that the whole experience went well, and she exhibited self-efficacy by her engagement with the public and her ability to rally her family and immediate community to act on reusing waste items.

  Another student described a situation in which she caused stress for her young sister by explaining water usage in relation to the global freshwater shortage. When her sister began crying, she realized the implications of her Project, and had to negotiate a way of helping her sister to feel good, so, in Student E's words: "I talked to her and I made that poster with her [about water usage]... and she wrote her name on it too. She was really happy she pitched in" (Student E, Interview May 30, 2008). After six weeks, her sister's showers had reduced by more than half the time, as shown in Fig. 20.2, and Student E was very pleased. In several cases, the students recognized their abilities to influence their family and friends around an issue that was important to them.

  In the classroom, the teacher encouraged them to share with their classmates their progress and pitfalls, as well as offer suggestions to others when new problems developed. Early on in the project, one student exclaimed that "trying to get them to recycle is really hard," when some of the members in her blended family refused to support her efforts. Later on, however, she declared that separating the food containers was no longer an issue because she had "trained" her family (Theresa, Final interview, July 7, 2008). For the students, group discussions—how to begin and carry through with conversations of persuasion, breaking old and creating new personal routines, and making compromises with other family members—were, for the most part, enjoyable experiences. For the teacher, the students' telling and retelling of stories were engaging and provided evidence of their evolving forms of learning.

Wenger's (1998) interpretation of *learning in practice* is helpful in understanding the development of self-efficacy as a factor. We consider that what people learn is "not a static subject matter but the very process of being engaged in, and participating in developing, an ongoing practice" (p. 95). Significant learning, in Wenger's (1998) view, is:

> [w]hat changes our ability to engage in practice, the understanding of why we engage in it, and the resources we have at our disposal to do so. This kind of learning is not just a mental process—such as neurological memory, information processing in the brain, or mechanical habituation—though mental processes are surely involved. Such learning has to do with the development of our practices and our ability to negotiate meaning. It is not just the acquisition of memories, habits, and skills, but the formation of an identity. Our experience and our membership inform each other, pull each other, transform each other. We create ways of participating in a practice in the very process of contributing to making that practice what it is. (p. 96)

The idea of taking ownership for their learning was empowering to the students. Student D noted that the process by which the Action Project was implemented and tracked in school improved her sense of self-efficacy, independence and agency when she said, "We all got to show our point of view on things. If we had a problem with something, we got a chance to fix it and we got a chance to *make our own plan how to fix it*" (Interview, June 6, 2008, emphasis added). In this short quotation, the student proudly expressed how she and her fellow classmates "*got a chance to*" address issues of their choosing, by carrying out a plan of their "*own*" design and reporting on how well it turned out. There was some evidence of metacognitive understanding of how the Action Project might affect the students and increase their sense of self-efficacy. Additionally, Student E, referring to the action component of the project, commented, "When you do it hands on, you get your own opinion and perspective, and like maybe 5 years away, you might actually remember what you did" (Interview, June 6, 2008).

This is supported by the work of Cargo et al. (2003), as they found that "an empowering environment was characterized by adults acting to create a social context for youth to take responsibility for their [quality of life] issues" (p. s70). In this case, the quality of life is improved by the sense of reward from their own project work and by the less direct impact of diverting waste from the system.

- Teacher Influence: Certainly a significant feature of this project was its open-ended nature. Indeed, the young people responded enthusiastically to their teacher's invitation to "*show our point of view on things*" and to find their own solution for a problem that was personally relevant. Recalling Korczak's questions posed at the outset of this chapter—"*who asks the child for his opinion or consent?*" and "*What can a child possibly have to say?*"—it would seem that the teacher's willingness to give her students an opportunity to devise and independently carry out their activities was also a significant factor of influence. By recognizing and validating their capability for making a positive change in their own and their family's conservation or recycling habits, Theresa positively reinforced her students' sense of agency. In his work on community-based learning, Geoff Fagan (1996) viewed the role of the educator as one that facilitates and supports people—who are central to sustainability practice—to make their own choices about their future. In considering what action-based education might mean for a local community, he proposed,

> Education is about confidence—about saying..., 'You are able and clever, you do understand, you are perfectly competent and able to engage.' Education is also about risk—about creating the conditions in which people can engage and learn without being castrated in the attempt. It is about the knowledge they own and share, the knowledge they generate and update, and the knowledge that has purpose, direction, meaning and location. (p. 147)

In the process of 'creating the conditions,' the teacher also reported that she encountered a number of challenges—not the least of which was providing an adequate amount of class time for discussion and writing. Given the relatively small number of science periods allotted in the school timetable, she described her dilemma as one of continual trade-offs, referring to the tension she felt trying to achieve a balance between teaching more curriculum content and utilizing the time for students to reflect more deeply on a topic of personal and social/environmental importance. She justified her decision to grant more time for the Action Projects based on her students' enthusiasm and commitment to their learning, which aligned with her own views regarding the importance of learning about how daily practices of consumption impact on the quality of life and on the environment. Due to time constraints, she determined that it would be unreasonable to expect that all content expectations in the science curriculum could be adequately covered.[3] She, therefore, reconciled her decision as a matter of choosing 'quality over quantity.' Theresa dedicated much time to planning, preparing, and assessing the work of her students. Throughout the entire unit, she tracked each of her students' progress by collecting, reading, and providing written as well as oral feedback. In an attempt to connect students' learning with ministry expectations for report card evaluation, she searched for Expectations concerning basic concepts and skills of inquiry, design, and communication relevant to environmental issues that were scattered throughout the different strands and topics of the grade seven Science and Technology curriculum (MoE 1998).[4] Although she described this "search and rescue" exercise as both frustrating and time-consuming, she was committed to the project. Not unlike her students, Theresa was engaged in "dimensions of practice that are affected by significant learning" (Wenger 1998, p. 95). In the process of developing report card comments, she recounted having to reconcile what Wenger might refer to as "conflicting interpretations of what the enterprise is about" (Ibid.). In one interview, she explained,

[3] At the time, the 1998 Ontario Ministry of Education (MoE) curriculum for Science and Technology was still in use. MoE expectations describe "the knowledge and skills the students are expected to develop and demonstrate ... on which their achievement is assessed" (1998, p. 6). In grade seven, there were approximately 130 expectations listed for the five strands.

[4] The 1998 Science and Technology curriculum expectations were organized into five major areas or *strands*: Life Systems, Matter and Materials, Energy and Control, Structures and Mechanisms, and Earth and Space Systems.

> ... [b]ecause this was a fairly big part of their third term activities, ... I crafted a sentence that said something like .... "developed and carried out their own Action Plan." ... it actually was directly reported in the Report Card based on [the students'] activity with this project. I ran into a little trouble with the Vice Principal around the wording [because] ...it didn't fit the rubric format with the descriptors. So we had a little philosophical argument about that ... but I felt quite good about it because when I looked at some of the [report card] comments [written by]... some of the other teachers, ... most of it is focused on knowledge learning, ...[which is] an expectation about "understands heat and particle theory." (Theresa, Final interview, July 7, 2008)

In reflecting on the benefits of their engagement with environmental Action Plans—which far exceeded learning basic facts of science—the teacher expressed her unexpected delight for the enthusiastic way in which most of her students carried out their self-directed projects. In some cases, she was surprised by the degree of influence 12-year old children had on other family members, and their ability to change home routines to new waste management practices. It is important to note that through this project, Theresa also expressed that she felt that the positive outcomes of the project had increased her confidence as a science teacher.

The role of the teacher is both clear and vague, in just the same way that we may be aware of the official curriculum and the informal curriculum. Depending on one's educational philosophy, a teacher facilitates transmission, transaction or transformation, given the day, the lesson or the needs of the class at any moment (Miller 1999). In this case, the role of teacher as facilitator allowed for a transformative process for her students in their personal change (Westheimer and Kahne 2004; Holden and Clough 1998). Her ability to let the students direct their projects, to guide them with her questioning but mainly to provide the framework within which they may explore their own empowerment, is very valuable from a social constructivist perspective. For example, she encouraged the students to develop their Action Project based on their own needs or the needs of their family with regards to waste reduction and management. With the SMART plan she laid the groundwork for scaffolding which could help the students to guide themselves through the process, and she supported them in learning from the challenges and successes of their projects.

- Science Literacy: Another sub-theme of the data relates to scientific literacy. Using the STEPWISE model of science education, two categories for contextual knowledge became apparent in this case: *Products education*, i.e., understanding science laws and theories, such has how and why waste product components are disposed of or reused; and *Skills education*, such as knowing how to create a data table or communicate information to an audience.

The experience of products education occurred in multiple contexts, as the students connected their projects with their knowledge of waste management, as the constructivist and affective nature of the Action Project would suggest (Hodson 2003). The students designed their projects from the needs they identified from observations in their personal lives and/or communities. Because they had ownership of the project, they took responsibility, for the most part, for finding out the information that would help them to be successful. This, an enactment of secondary research, helped to inform and deepen their scientific literacy. For example, one of the students was curious about how a garburator, or garbage disposal unit, worked in her sink. She researched gray water disposal and treatment in the city and shared the information with the class. It followed that their contextual knowledge enabled the students to feel justified in their decisions for personal and community change (Carr 1998); while, reciprocally, their change decisions helped to enhance their science literacy (Ratcliffe and Grace 2003).

The students commented on the skills they gained from having to design their actions, record results and present outcomes. They recognized that procedural practices, such as measuring changes in water consumption or waste disposal, while having the possibility for human error, are important to the validity of the project and the presentation of the outcomes. Student B commented:

> It can help us in the long run when we have a plan for work or in our careers or in high school. And also, for just regular projects. Your hypothesis needs to be specific and it makes everything a lot easier if you know what you're doing and if you follow those rules. So, this project does benefit us! (Interview, April 16, 2008).

The data supported their skills acquisition: through the presentation of their results, they used sophisticated tracking methods and presentation methods, such as found in Figs. 20.2 and 20.3, with charts and graphs that were clear, as required in the Ontario curriculum (MoE 2007).

There are, undoubtedly, various ways of explaining why these three factors, self-efficacy, teacher, and scientific literacy, may have influenced students' orientations towards use of their science literacy for citizenship purposes. A framework that appears to work well is *knowledge duality theory* as presented by Wenger (1998). In this view, deep commitment to an idea, issue, approach, etc. may arise when people are engaged in personally relevant reciprocal interactions between phenomena of the world and representations of them. This seemed to occur, for example, when a student's study of personal water usage led her to co-develop a poster with her sister who, in turn, felt more comfortable about reducing her shower water usage (refer to Fig. 20.2).

Summary and Conclusion: Broader Social and Educational Importance

In Ontario, the location of the case study reported here, increased emphasis has been placed on STSE education in the latest version of the official Science curriculum. As part of this emphasis, students are required, for example, to develop "courses of practical action to deal with problems relating to science, technology, society, and the environment" (MoE 2007, p. 24). This mandate, among other new learning expectations, appears to reflect a particular spirit of citizenship education, at times addressed in subjects other than Science in schools, in that it promotes 'nation-building,' for 'future economic sustainability,' and for 'social and/or political change' (Westheimer and Kahne 2004). Furthermore, since the time of this study, a policy for environmental education in the province, called *Acting Today, Shaping Tomorrow,* was released; and, it demands that teachers and students take up and engage in opportunities for action on environmental issues (MoE 2009).

The case study reported here suggests some compelling starting points for assessing the value of the intersection of citizenship education and science education in support of the Ontario curriculum and policy documents. Perhaps usefully explained in terms of knowledge duality theory (e.g., Wenger 1998), students' orientations towards citizenship seemed to be enhanced when they had opportunities and resources (intellectual and physical) enabling them to engage in personally-meaningful reciprocal relationships between phenomena of the world (e.g., waste) and representations of them (e.g., artistic uses of material often treated as 'garbage'). Of course, we also have seen the importance of the teacher in facilitating these understandings.

Through engagement in participatory citizenship activities, students may feel both a greater desire for and affinity to science literacy in its many facets. Furthermore, engagement in science activities with positive support and outcomes may propel students, who from a global perspective have access to a relatively large proportion of the world's cultural and economic capital, to greater self-efficacy for acts of individual, social and environmental wellbeing. For a world facing multiple issues of social and environmental inequity, these are outcomes that need to be supported in the curriculum, allowing for and engaging resources to support individual and community transformation, and active citizenship with and through STSE issues-based science literacy, as offered by the STEPWISE framework.

References

Ashley, M. (1998). Economics, environment and the loss of innocence. In C. Holden & N. Clough (Eds.), *Children as citizens: Education for participation* (pp. 171–182). London: Jessica Kingsley Publishers.

Bandura, A. (1997). *Self-efficacy: The exercise of control.* New York: W. H. Freeman.

Bencze, J. L., & Alsop, S. (2009). A critical and creative inquiry into school science inquiry. In W.-M. Roth & K. Tobin (Eds.), *The world of science education: North America* (pp. 27–47). Rotterdam: Sense.

Cargo, M., Grams, G. D., Ottoson, J. M., Ward, P., & Green, L. (2003). Empowerment as fostering positive youth development and citizenship. *American Journal of Health Behavior, 27*(S1), S66–S79.

Carr, W. (1998). The curriculum in and for a democratic society. *Curriculum Studies, 6*(3), 323–340.

Charmaz, K. (2000). Grounded theory: Objectivist and constructivist methods. In N. K. Denzin & Y. S. Lincoln (Eds.), *Handbook of qualitative research* (pp. 509–535). Thousand Oaks: Sage.

Davies, I. (2004). Science and citizenship education. *International Journal of Science Education, 26*(14), 1751–1763.

Fagan, G. (1996). Community-based learning. In J. Huckle & S. Sterling (Eds.), *Education for sustainability* (pp. 136–148). London: Earthscan.

Guba, E. G., & Lincoln, Y. S. (1988). Naturalistic and rationalistic enquiry. In J. P. Keeves (Ed.), *Educational research, methodology and measurement: An international handbook* (pp. 81–85). London: Pergamon Press.

Hammersley, J., & Atkinson, P. (1990). *Ethnography: Principles in practice*. London: Routledge.

Hodson, D. (1994). Seeking directions for change: The personalisation and politicisation of science education. *Curriculum Studies, 2*, 71–98.

Hodson, D. (2003). Time for action: Science education for an alternative future. *International Journal of Science Education, 25*(6), 645–670.

Hodson, D. (2008). *Towards scientific literacy: A teachers' guide to the history, philosophy and sociology of science*. Rotterdam: Sense.

Holden, C., & Clough, N. (1998). The child carried on the back does not know the length of the road. In C. Holden & N. Clough (Eds.), *Children as citizens: Education for participation* (pp. 13–29). London: Jessica Kingsley Publishers.

Lincoln, Y. S., & Guba, E. G. (2000). Paradigmatic controversies, contradictions, and emerging confluences. In N. K. Denzin & Y. S. Lincoln (Eds.), *Handbook of qualitative research* (pp. 163–188). Thousand Oaks: Sage.

Miller, J. (1999). Making connections through holistic learning. *Educational Leadership, 56*(4), 46–48.

Ministry of Education [MoE]. (1998). *Ontario curriculum, Grades 1–8: Science and technology*. Toronto: Queen's Printer for Ontario.

Ministry of Education [MoE]. (2007). *Ontario curriculum, Grades 1–8: Science and technology*. Toronto: Queen's Printer for Ontario.

Ministry of Education [MoE]. (2009). *Acting today shaping tomorrow*. http://www.edu.gov.on.ca/eng/teachers/enviroed/ShapeTomorrow.pdf. Accessed 28 Mar 2013.

O'Neill, J., & Conzemius, A. (2006). *The power of SMART goals: Using goals to improve student learning*. Bloomington: Solution Tree.

Ratcliffe, M., & Grace, M. (2003). *Science education for citizenship: Teaching socio-scientific issues*. Philadelphia: Open University Press.

Roth, W.-M., & Désautels, J. (2004). Educating for citizenship: Reappraising the role of science education. *Canadian Journal of Science, Mathematics and Technology Education, 4*(2), 149–168.

Sadler, T. D., Barab, S. A., & Scott, B. (2007). What do students gain by engaging in socioscientific inquiry? *Research in Science Education, 37*(4), 371–391.

Simms, A. (2009). *Ecological debt* (2nd ed.). London: Pluto Books.

United Nations. (1989). *Convention on the rights of the child*. Retrieved May 15, 2010. www.unicef.org/magic/briefing/uncorc.html

Wenger, E. (1998). *Communities of practice*. Cambridge: Cambridge University Press.

Westheimer, J., & Kahne, J. (2004). What kind of citizen? The politics of educating for democracy. *American Educational Research Journal, 41*(2), 237–269.

Chapter 21
Undermining Neo-liberal Orthodoxies in School Science: Telling the Story of Aluminium

Ralph Levinson

Abstract Increased emphasis in jurisdictional science curricula of the interplay between science and society highlights many aspects of social justice but veils them with political neutrality. I argue that science curricula, while espousing progressive values, promote dependent consumerism and implicit nods towards corporatism and neoliberalism, ignoring many of the global problems that follow in the wake of these ideologies. Through a description of the extraction, manufacture and distribution, of a global commodity, aluminium, I identify the intrinsic interplay between science and society and the impossibility of disengaging core substantive science, as formulated through a Vision I depiction of science literacy, from its social, political and economic consequences. There are clear pedagogic implications which map the linkages between various actants in an overall interlocking narrative structure.

Keywords Neoliberalism • Science curriculum • Scientific literacy • Aluminium • Social justice • Interlocking narratives

Introduction

At the end of Premier League soccer matches in England the coach and a selection of players are lined up along a stretch of wall (I use this phrase with unintended irony) to give their verdict on the game to reporters. This is never any old stretch of wall. It is a colourful array of corporate logos complementing the sponsored shirts all players have to wear (replicas of which are often manufactured through sweatshops in Bangladesh and China) with the curiously meaningless names of the

---

R. Levinson (✉)
Institute of Education, University of London, London, UK
e-mail: Ralph.Levinson@ioe.ac.uk

L. Bencze and S. Alsop (eds.), *Activist Science and Technology Education*,
Cultural Studies of Science Education 9, DOI 10.1007/978-94-007-4360-1_21,

sponsors such as Aon and Etihad outflanking the club logos, reinforced by the constant flashing of advertising around the margins of the pitch.

The official sponsors of the 2012 London Olympic Games, included Coca Cola, McDonalds, BP, Procter and Gamble, Atos Origin, Adidas, ArcelorMittal, Cadbury, Deloitte. Despite the generosity of these conglomerates in helping to fund the Games (and incidentally preventing other less wealthy or charitable agencies from advertising their logos www.vanityfair.com/culture/2012/06/international-olympic-committee-london-summer-olympics), they are an odd bunch to be associated with a Global Activity designed to promote health, wellbeing and global communality. Products from Coca Cola, McDonald's and Cadbury contain very high proportions of sugar, salt and saturated fats contributing toward poor dental health, high blood pressure and obesity hence putting strain on health services: 'The promotion of foods that undermine healthy choices, such as energy-dense, nutrient-poor foods ... represents a major threat to healthy lifestyle choices' (Branca et al. 2007, p. 113).

BP, whose logo symbolises sunrise, photosynthesis and all matters green, was responsible for the Gulf of Mexico oil spill in 2010. Procter and Gamble, manufacturers of disposable nappies, have been implicated in the destruction of the Indonesian rainforest in the extraction of palm oil (van Gelder 2004). In 2010 ArcelorMittal, a global steel-producing company, was heavily fined for running a price-fixing cartel www.fin24.com/Companies/ArcelorMittal-slapped-with-price-fixing-fine-20100630. Just before the Games, *War on Want* exposed challenges to Adidas in its use of sweatshop conditions for workers in its supplier countries www.waronwant.org/olympics-home. Atos Origin won a UK government contract to assess disabled people for the validity of their benefits through tests that discriminated against disability www.guardian.co.uk/society/video/2011/may/11/disability-protest-atos-origin-video. Deloitte have been implicated in helping wealthy people to avoid paying UK taxes.

The Games were fraught with problems just before the start – the private security company, G4S, handed millions of pounds from public money to hire personnel, didn't do what it was supposed to do despite all the technology to hand – and the government had to draft in the army at the last moment to secure the Games. The soldiers did more than that: they were brought in civilian uniform to occupy empty seats aired on the media at the Games reserved for corporate sponsors who couldn't be bothered turning up, despite intense demand from the public for seats. The Games were described euphorically by the media as a fantastic success.

One of the heroes, associated with the success of the Games is the Mayor of London, Boris Johnson, a crafty, right-wing politician masquerading as a populist buffoon, known adoringly to his many fans as 'Boris'. Writing in *The Daily Telegraph,* a U.K. newspaper associated with right-of-centre politics, he declaimed on Climate Change with his usual chutzpah, gall, or conniving deception: 'We human beings have become so blind with conceit and self-love that we genuinely believe that the fate of the planet is in our hands' (Johnson 2013).

Why start an article on the science curriculum and its associated pedagogy with an exposure of global corporations flying their brands for the Olympics, and professional sport more generally? This is because the Olympic Games was seen as an excellent context in the U.K. to teach science in the curriculum. One of the STEM challenges was for 'a team of pupils that can help design an Adidas glove for canoeing, sailing or rowing' (www.sciencelearningcentres.org.uk/centres/westmidlands/news/19808?searchterm=olympic%20games (my italics)), others look at a range of concepts such as aerodynamics of the javelin, measuring heart rates and the nature of measurements. Another, *Olympic Alloys,* looks at the metals and alloys used in the construction of the stadium and sports equipment (Oliver 2012). There was little mention of relating high sugar and fat-loaded diets of the sponsors to sport and health, the relationship between the chemistry of the manufacture of steel and goings on the world market aimed at enrichment of the few and impoverishment of the many, the economic pathway between the exploitation of palm oil and its role in detergents, nappies and other manufactured goods for corporate profit and the effect on the rain forests. In other words there were no critiques of the relationship between social, economic, political and ethical productions and the representations of sport. One exception was the prompted investigations in *Practical Action*, entitled *Winners and Losers at the Olympics*, to encourage students to research Fair Trade practice (http://www.nationalstemcentre.org.uk/dl/a81d132b0afd81d885508273591e6c981b0a7b58/3790-Fair_Trade.pdf).

Wrapped up in sports activities which are firmly associated with social and personal wellbeing are its exact opposites, commodified and represented into a counter-reality – or a hyper-reality (Baudrillard 1994) – so destructive to the public good and yet crafted through symbolic manipulation to appear benevolent and healthy.

A common aphorism aimed at debunking the politicisation of the science curriculum is that there is nothing wrong with the science and technology; it's the applications that they are put to which can be the problem. These are Mertonian and academic science (Ziman 1984) perspectives which have science controlled by agreed virtuous norms, communalism, universalism , disinterestedness, originality and scepticism. What science produces emerges from 'justifiable' values. What happens after that is the problem. That outdated social representation of science has been replaced by more contemporary descriptors such as post-academic science (Ziman 2000), post-normal science (Funtowicz and Ravetz 1993), mode 2 science (Gibbons et al. 1994) which sees science practice as enmeshed with social, economic and political drivers. Regardless, the Mertonian norms persist in school science curricula.

Controlling the Science Curriculum

In a long article over-viewing different discourses through scientific literacy, Roberts (2007) distinguishes between Vision I and Vision II approaches. These provide a functionally useful but ultimately problematic delineation. Vision I prioritises the core substantive ideas of science, and at the most, social applications of these concepts become an add-on in teaching and learning. Vision II starts with issues or social problems to be addressed and science becomes one of a range of domains of knowledge brought in to seek solutions. Vision II approaches have become more prevalent in science curricula over the past few decades. One reason is to strive to show the relationship between science and its applications to society in a science for all policy (DES 1985; AAAS 1995). Not unconnected with this in an increasingly accountable society is the need to have a decision-making public informed by scientific knowledge and how scientific knowledge is made, i.e. a scientifically literate populace, through STEM activities (http://sciencepolicyforall.wordpress.com/2013/03/19/stem-education-the-value-of-a-scientifically-literate-population/). However, as I have indicated above, and as I hope to demonstrate below, these formulations often come with a covertly neo-liberal view of society.

Callon (1999) posits three roles for citizens with regard to science and technology; that is, as (i) dependent consumers; (ii) interactive negotiators; or (iii) co-constructors of knowledge. The position of the National Science Standards, for example, is to prepare students predominantly as dependent consumers:

> The overarching goal of our framework for K-12 science education is to ensure that by the end of 12th grade, *all* students have some appreciation of the beauty and wonder of science; possess sufficient knowledge of science and engineering to engage in public discussions on related issues; are careful consumers of scientific and technological information related to their everyday lives; are able to continue to learn about science outside school; and have the skills to enter careers of their choice, including (but not limited to) careers in science, engineering, and technology. (NRC 2012, p. 1)

Within science education, school students are positioned as 'future citizens' in a reactive and dependent role in relation to science and technology such as in the Nuffield 2000 report (Millar and Osborne 1998), which is driven by:

> . . . a sense of a growing disparity between the science education provided in our schools and the needs and interests of the young people who will be our future citizen . . . the rapid pace of technological change and the globalisation of the marketplace have resulted in a need for individuals who have a broad general education, good communication skills, adaptability and a commitment to lifelong learning. (p. 2001)

The role of the school here appears to be one of enculturation, to instruct pupils in the skills and knowledge and values of what it means to be part of a contemporary global market. Being responsive to the needs of a global neoliberal economy (Smith 2011) is implied in the phrase 'the rapid pace and the globalisation of the marketplace have resulted in a need for individuals . . .'. The future citizen might be a decision-maker but one constrained by, and subject to, the rationalities of global economics. In such a socio-political context of schooling, students are given certain dilemmas, aimed at learning critical thinking and argumentation skills (Lee 2007). Substantive scientific concepts become a resource through which to respond to socio-scientific issues, and how students as future citizens use such knowledge in decision-making can be evaluated in terms of socio-scientific reasoning skills (Sadler et al. 2011). Ostensibly progressive, these formulations promote individualist consumerist discourses, which cohere smoothly with strongly capitalist agendas (Bencze and Alsop 2009).

Since this article is concerned with the inter-relation of science and society my arguments will be framed predominantly within the socio-political sphere.

But it would be helpful, nonetheless, to explore aspects of the Vision I approach. The reason I want to do this is because a common argument against a science and society approach is that the rigour of academic science is sacrificed at the altar of social consciousness or the excitement of contemporary affairs (Donnelly 2004); for example, an interview in *The Guardian* newspaper with Tim Oates, an educational advisor to the U.K. Conservative government, commenting on a review of the National Curriculum of England: ‘We have believed that we need to keep the National Curriculum up to date with topical issues but oxidation and gravity don’t date ... we are taking it back to the core stuff. The curriculum has become narrowly instrumentalist.’ (www.guardian.co.uk/education/2011/jun/12/climate-change-curriculum-government-adviser).

In epistemological terms science describes Nature which involves unravelling the hidden, the abstract and the theoretical (Haack 1996; Wolpert 1992), ethics is axiological, what we ought to do for social and individual betterment. Conflating descriptive and normative depictions is epistemologically untenable as Hall (1999) argues:

> It is widely recognised that “is” statements in science cannot be turned into the “ought” statements of moral discourse ... The domains of scientific and moral discourse are fundamentally different; they have different core concepts ..., different procedural ground rules and different tests for truth ... To apply science’s empirical test for truth within the moral domain would turn morality into pragmatism. (p. 15)

The distinction made between fact and value is derived from David Hume’s naturalistic fallacy, which aims to demonstrate that an ‘ought’ statement cannot be deduced from an ‘is’ statement. But facts and values are deeply entangled both in thought and in common discourse, and the dichotomy falls apart on diverse grounds. Pragmatically, imagine a classroom where the teacher is explaining the critical mass required for the fission of Uranium-235, and students then bring up the question of the moral responsibility of dropping a nuclear bomb on densely-populated areas. Retaining the fact-value dichotomy, according to Hall (2004), the teacher has limited options. One is to politely point out to the students that this is a science lesson and such questions have no place in science. Or to say ‘I am explaining a scientific process; your question is an ethical one and therefore it requires a different way of thinking about things which I have to make explicit to you before we can discuss this matter.’ Regardless of the rather odd approach and the implicit suppression of moral imagination, it begs a number of questions: the topic of critical mass is not only discussed in school just because it is a scientific concept; it appears in curricula precisely because nuclear power is of contemporary relevance.

Putnam (2002) has pointed out that the fact-value dichotomy, a mainspring of logical positivism and pre-war economic thinking, was used to maintain the status quo and advise against wealth redistribution because such policy formulations were based on subjective value judgements.

One distinction drawn by the logical positivists between fact and value was that scientific statements are empirically verifiable and value-judgements are not.

As Putnam (1993) points out this position is untenable. Scientific statements, as empirically verifiable sentences, which follow from the logical positivist claim, do not hold up. 'Newton's entire theory of gravity, for example, does not *in and of itself* (i.e. in the absence of suitable 'auxiliary hypotheses') imply any testable predictions whatsoever' (p. 143). Putnam (2002), drawing on Dewey, maintains the central importance of intelligent reflection, the intellectual tools of science, philosophy and the arts in trying to resolve common problems, adapt to new solutions, where 'changing one's values is not only a legitimate way of solving a problem, but frequently the only way of solving a problem' (p. 98). Facts in concert with values are not only instruments in attaining an end but are a means subject to change and reconsideration.

Donnelly (2002) defines the key characteristics of natural science by their ontic categories, the explanatory entities which explain phenomena -electron charge clouds, thermodynamic equations and causation. Unlike the humanities, Science is, therefore, instrumental; enabling prediction and control, which go beyond any values attributed towards its procedures. The reviewing of scientific papers, the ethical constraints, the processes of the scientific community are contingent upon, but not intrinsic to, these ontic categories, the 'potentialities of the material world are not to be altered by any number of social values, though of course such values may well influence which possibilities are realised' (Donnelly 2002, p. 138). The implications for the science curriculum are that attempts to humanise science or place it in a social and ethical context result in the 'replacement of education in science with curricula in what might be loosely called the political sociology of science' (Donnelly 2002, p. 147).

In 1999, public concern grew about the use of genetically modified foods in the U.K. when a television programme reported on findings of a researcher that rats fed genetically modified raw potato developed a dysfunction of their intestines. Thomas (2000) argues that in order for the public to follow this debate and understand why the experiments were contested they would have to understand the complexity of statistical analysis and 'the subtleties of interpretation, spanning arcane fields as disparate as lectin chemistry and intestinal histology. . .' (p. 140), procedures and terminology beyond that of many teachers let alone school pupils. It follows from this that it would be best to concentrate on what can be taught well about the natural world in school and leave scientific decision-making to the experts. Such an approach is often termed the 'deficit' approach although it oversimplifies the realpolitik between diverse public interests and science.

While arguments such as those deployed by Thomas focus on the erudite, specialised and abstract nature of knowledge emanating from scientific research, others argue that science knowledge is often irrelevant or needs to be recontextualised for public decision-making. (Chapman 1991; Dawson 2000). As Dawson (2000) writes on his study of Ovine Johne's Disease in a farming community in South Australia, decisions were based on 'economic and political reasoning' (p. 127) and the amount of scientific content knowledge needed by citizen-participants was minimal. In terms of use in supporting decision-making in real life scientific issues the core concepts in the science curriculum are often inert (Ryder 2001).

Commenting on the use of scientific knowledge in technology, Layton (1993) refers to science as 'cathedral, quarry or company store' (p. 54), hence the cathedral is the source of the natural laws which bounds the technology; the quarry focuses on what science is useful to the project rather than its form or structure; the company stores operate as workshops where the products of scientific knowledge are remodelled to meet the demands of technology.

The Nuffield 2000 (Millar and Osborne 1998) and NRC (2012) depictions are one version of school science, society and citizenship. If we envisage the socio-epistemic relationship between school science and society as a continuum then at one end the curriculum perspective would be that the science curriculum simply isn't the place to deal with social and political issues because of their complexity (e.g. Thomas 2000). Pedagogically this is represented in Simonneaux's metaphoric contrast of heating up or cooling down socio-scientific issues where teaching at the cool end, Vision I approaches predominate and if socio-scientific issues are taught they are simulated or fictionalised to promote the teaching of core substantive scientific concepts (Simonneaux 2013).

At the other end of the spectrum is an 'activist' approach which aims to empower school students to make changes that they see as addressing the manifest injustices in a profit-driven society (Bencze and Carter 2011). Schools thus need to be transformed from non-democratic simulacra of society, pipelines for academic scientists, to resources that provide opportunities for social action (Roth 2009). Whereas the 'future citizen' might role-play whether change is desirable and how they might achieve such change, activist students draw on relevant knowledge and skills to enact change.

The shift towards engagement of the science curriculum with social issues has been viewed broadly in the science education literature as progressive in removing the science curriculum from its authoritarian clothes (Fensham 1997). But curricula discourse, while taking care to be balanced and non-indoctrinatory, has a very clear economic purpose, that the place of science on the curriculum leads to the flow of personnel which provide the skills businesses and employers need, and the creation of scientists and technologists who enhance the nation's wealth to keep it competitive, for example, with the BRIC economies (CBI 2012). Regardless of the global divisions intrinsic in such formulations, the ultimate aim is economic progress, despite the nods to democratic processes. While it seems desirable that a knowledge of science by its citizenry underpins a nation's democratic processes there is no substantive evidence or research to support this claim. Indeed a strong counter to democratic claims is the dominance by expert science in national deliberations on techno-scientific projects (Jasanoff 2003; Hendricks 2006).

In contrast is an approach that embraces diverse social and epistemological positions such as critical pedagogy, feminism, post-colonialism, anti-capitalism, post-modernism and race theory, and starts with a radical critique of society broadly under a science for social justice banner. Emanating from such approaches are precisely the kinds of programmes which aim to expose injustice and the neoliberal agenda such as the STEPWISE project (Bencze and Carter 2011) and Roth and Lee's (2002) ethnographic study of a community to remedy the pollution of a local watercourse. Some of these works shine a light on narrowly positivist and scientistic agendas but they remain marginalised by the hegemonic emphasis on managerialistic qualities in the school sector such as 'value-added', 'quality', 'excellence' (Ball 2003).

The point I want to argue is that emancipatory knowledge for improvement of human wellbeing can be understood and become all the more powerful precisely because it is deepened through interrogation of substantive scientific knowledge. In formulating such a relationship I want to look at the example of the manufacture and uses of aluminium as a means of reconfiguring scientific knowledge and its links to social justice in the curriculum.

The Story of the Manufacture and Use of Aluminium

One of the stories that receives little attention in jurisdictional science curricula is the sourcing of those materials that are indispensable for life in industrial and post-industrial economies: the semi-conductor materials for our cell phones and computers, the metals essential for structures such as those of the Olympic Games, containers and transports; the fossil fuels that burn and drive electricity supplies; the food that sustains populations; the precious gems often mined in inhumane working conditions. In the latter case, in an article entitled 'Digging for diamonds and democracy', the BBC correspondent explains how young children in the Democratic Republic of Congo clandestinely work the diamond mines – guarded heavily by armed soldiers to prevent theft of property – while the diamond companies make profits of $30 million per month (news.bbc.co.uk/1/hi/world/africa/5213528.stm). In August 2012, 34 striking miners were shot dead by police at the Marikana platinum mine in South Africa, exacerbated by falling platinum prices on the world markets, union struggles for living wages, infighting between unions, and the government's anxieties about investment opportunities in South Africa (http://www.guardian.co.uk/world/2012/aug/17/south-african-police-shot-miners?intcmp=239).

In all these cases the relay of goods from source to refinement involves the amassing of large amounts of profit at the high end of the production line and exploitation and oppression at the low end. The world's most valuable resources are found in Asia, Africa, Russia and Latin America and enjoyed in Western Europe and North America.

To look further into the ways in which scientific knowledge and issues of social justice ineluctably interact I want to take the case of Earth's most abundant metal, aluminium. In everyday use aluminium metal is electroplated with a layer of aluminium oxide which is impermeable to most materials such as water, air, salts. As a result, aluminium, a light, refractory, pliable metal is extremely durable and has many economic uses ranging from incorporation in aircraft to containers for fizzy drinks, sealed capsules for medicines and cooking meats and fish in ovens. Aluminium alloys are used for example in the construction of bikes used in the Olympic Games, in the design of javelins they are used in combination with carbon fibres to provide rigidity and low density, for dinghies in the sailing boat competitions, particularly useful for resisting water corrosion (Oliver 2012). Anecdotally Napoleon was said to have used aluminium cutlery for State occasions. Because aluminium was so difficult to extract in a relatively pure form it had a higher value than gold. (acswebcontent.acs.org/landmarks/landmarks/al/revolution.html).

Since aluminium is such an economically important material its extraction and production are usually found in school science curricula to emphasize its range of uses, and one of the chemical principles of its production, reduction through electrolysis. What is omitted from curriculum and examination specifications in the story of the production of aluminium is perhaps illuminative of the socio-political context in which decisions on content are made.

Sourcing, manufacture and purification of aluminium, and hence its potential social and physical impacts, can only be understood in a socio-scientific context. Aluminium is an element found high up in Group III of the Periodic Table. This means that the atomic volumes of the atoms of which it is composed are relatively small and that they have three electrons in their outer shell. In turn this means that aluminium atoms are weakly electropositive and through electron transfer, and the resulting electrostatic attraction, will combine with non-metals, hence they are found in raw materials as aluminium oxide, where the aluminium is combined with that abundant and reactive gas, oxygen. The first problem, therefore, is to remove this aluminium oxide (Al_2O_3) in relatively pure form, known as alumina, from the rocky raw material, bauxite, in which it is encrusted. As well as alumina, bauxite contains other mineral deposits, sand, clay, iron oxide, and traces of heavy metals, some of which are radioactive. This removal is a tricky process because the aluminium oxide can only be separated through dissolution in a caustic alkali, concentrated sodium hydroxide, NaOH (also used as a drain cleaner).

Deducing from its position in the Periodic Table the oxide of aluminium is amphoteric, i.e. it has both alkaline and acidic properties. The oxides of those elements to the left of aluminium in the Periodic Table generally have basic or alkaline oxides of a high pH while those oxides of non-metals to the right have an acidic or low pH. Al_2O_3 although nominally a metal oxide is acidic enough to react and be dissolved in an alkali. While this manages to separate off the alumina from the bauxite, the residue is a mixture of undissolved material including the insoluble iron oxide (Fe_2O_3) which gives the residue toxic sludge a red colour (deriving from the Fe^{+++} ion most commonly recognised in rust), known as ‘red mud’. The amount of red mud left behind is at least equal in quantity to the amount of alumina produced, and usually more than double the quantity. Because of the high pH of the residue due to the alkalinity of the caustic soda, and some small amounts of toxic heavy metals found in bauxite, the red mud destroys animal and plant life with which it comes in contact, and if not properly controlled can infiltrate the water table.

Dealing with the red mud poses scientific, technical, environmental and hence economic, social and health problems. To prevent toxic exposure, the red mud is kept in thick-walled holding reservoirs which take up vast areas of land space. To staunch leaks, some manufacturers try to squeeze the water out of the red mud.

This process requires huge amounts of energy where the mud is put under high pressure and then heated to evaporate the water. To ensure as little spoilage as possible, high publicly accountable regulation must be supported by a strong infra-structure.

Just after mid-day on October 12th 2010, the retaining wall of the red mud reservoir of the Ajka aluminium plant in Hungary collapsed, releasing over one million cubic meters of red sludge over an area of about 50 km^2 killing ten people and injuring about 150 others. It turned out to be the 'greatest environmental crisis ever of Hungary and of the whole region' (Javor 2011). The main factors relating to the cause of the damage were deemed to lie at the door of the Hungarian government's privatisation of the industry just as the price of aluminium started to rise in the 1990s, the consequent lack of environmental monitoring, the lack of dry processes for safe disposal and failure of health and safety monitoring (Javor 2011).

Not all is bad news in the red mud waste industry however. "Where there's muck there's brass" – as the common British euphemism has it. Red mud left over from the Jamaican aluminium smelters has been found to harbour high concentrations of rare earth elements used in the manufacture of products such as hybrid cars, smart phones and DVDs. In early 2013 a multinational, Nippon Metal Products http://jamaica-gleaner.com/gleaner/20130213/business/business2.html started to explore the potential of red mud for concentrating heavy metals for commercial use.

The case of red mud, however, is only a small part of the natural and social impact necessitated by aluminium's chemistry and economic value. In terms of space I have left out the economic and environmental impacts of the mining of the bauxite, and the shipping of large quantities of bauxite for processing to parts of the world where labour is relatively cheap.

Once the alumina has been purified the aluminium metal has to be extracted from it. Because of the immense electrostatic attraction between the aluminium and oxygen atoms in the alumina, there is no other way to separate the aluminium than by freeing the atoms from their tightly bound solid state and then reducing the aluminium ions through the flow of charge by electrolysis. This can only be carried out in a liquid aqueous or molten medium. Elemental aluminium cannot be liberated by electrolysis in aqueous solution because hydrogen gas is produced preferentially as the electrons reduce the hydrogen ions from the water. To melt the alumina requires a temperature of 2072C which is simply too high a technical and economic challenge.

The Role of Cryolite

Since it is technically impossible to electrolyse aluminium oxide directly in the pure molten state and chemically impossible in aqueous solution, another method is needed. Hall and Heroult discovered in 1886 that aluminium oxide can be dissolved in the chemically inert mineral cryolite (Na_3AlF_6) which has a much lower melting temperature (1 012C) than alumina. The chemical advantage here is that Al^{+++} ions are then generated in the ionic state ready for electrolysis. Cryolite is thus central to the aluminium production process. It is mentioned in Peter Høeg's novel *Miss Smilla's Feeling for Snow* featuring a German attempt to gain control of the rare natural source of this material in Greenland in WWII. So vital was cryolite for the war effort (in terms of the use of aluminium for aircraft manufacture) of both the Axis and Allied Powers that the U.S. annexed Greenland as a Protectorate during the war in order to secure the Ivittuut mine at Arsuk Fjord in southern Greenland. This had the Planet's richest and purest seam of cryolite. Cryolite was mined here from 1854 before the quarry fell into disuse in 1987, the reserves of cryolite having been worked to extinction. Mining of cryolite in Ivittuut has left its mark, particularly on the local fishing industry where lead levels, caused by leaching into local waters through blasting of the mine, have accumulated in local blue mussels to such an extent that 'it is recommended not to eat blue mussels from this area' (NERI 2010, p. 30). While lead and zinc levels have decreased since the mine was abandoned, they still remain at significant levels of pollution.

Because there are no longer exploitable natural sources of cryolite, it is now manufactured synthetically using Hydrogen Fluoride. This process carries its own problems and possibilities but that is yet another story.

Electrolysis

With the aluminium ions dissolved in cryolite the solution is ready for the smelter, and it is the process of electrolysis that is usually included in all textbooks and school examinations. However, the challenges and problems which precede and arise from this process are barely alluded to. The chemistry behind the reduction of the Aluminium ions in the smelter is represented by a simple, rather unobtrusive equation

$$Al^{3+} + 3e \rightarrow Al \tag{21.1}$$

The chemistry of the reaction, however, is considerably more complex than this equation suggests. Simply stated it means when an ion of aluminium reacts with three electrons one atom of aluminium is discharged, or when a certain quantity of aluminium (measured in an amount called moles) is liberated three times as many moles of charge are needed to discharge a mole of aluminium atoms.

The current global annual production of aluminium is about 40 million tonnes which consumes about 600 billion kilowatt hours of electrical energy. This means that 3 % of the world's energy supply goes into the production of aluminium http://wordpress.mrreid.org/2011/07/15/electricity-consumption-in-the-production-of-aluminium/. For a number of reasons – the rising costs of fossil fuels and reduction of carbon pollution – hydroelectricity is now the main means of supplying energy to smelters. At first, and as acknowledged by many school textbooks, this looks like a 'clean' option. Rio Tinto Alcan, one of the major producers of aluminium asserts that '84 per cent of the power we use in primary aluminium comes from clean hydropower.' http://www.riotintoalcan.com/ENG/ourapproach/360_sustainable_development.asp

But an elementary understanding of the principles of hydroelectricity gives an insight into the environmental depredation involved. Water has to fall from a great height on to a turbine whose movement transfers electric charge to the smelter. For water to fall from a great height two natural features are essential: mountains and rivers. As well as the physical changes to the landscape, when a power station and dams are built, the heat outputs raise the temperature of surrounding water-courses and damage local wildlife. Hydroelectric power stations for aluminium smelters are usually located in wildernesses; however as an Icelandic report (2009) on the cost-benefit analysis of locating smelters in Iceland suggests:

> Obviously, great care must be taken when making an economic assessment of the environment so as to avoid errors in the result. It must also be borne in mind that the value of the environment is always assessed indirectly in terms of other goods expressed on a monetary scale. This may result in environmental quality being overvalued, compared with other goods since, among other reasons, the existence value of other goods sold on the market is not included in their valuation. (Institute of Economic Studies, University of Iceland 2009, p. 62)

The aesthetic nature and wonder of wildernesses become a commodity to be weighed against other commodities. But that is only the siting of the power plant. The smelter itself is a huge steel vat which acts as a cathode, the site at which aluminium ions are reduced to aluminium. The smelter is charged with alumina dissolved in cryolite but chemically the main reactants are the aluminium and oxide ions dissolved in the inert (in terms of the electrolysis) cryolite. The aluminium ions are discharged (collected and fashioned as ingots and bars) at the base of the steel vat which acts as the cathode reducing the aluminium ions as in eq. 21.1. At the anode, in the high temperature of the smelter, the oxygen is discharged at the carbon anode. The oxygen reacts with the carbon generating large volumes of the greenhouse gas CO_2. As the anode is worn away it is replaced with fresh stocks of carbon. The cryolite, Na_3AlF_6, also combines with the carbon forming perfluorocarbons (PFC) in the atmosphere, which in high enough concentrations are potent greenhouse gases. Depending on the political and regulative mechanisms the discharges of CO_2 and PFCs vary enormously.

The Waste Collectors of Rio de Janeiro

In cities in Brazil you will see people, on the margins of society, old and young including small children, male and female, collecting rubbish from the streets usually in small handcarts. Until quite recently such people operated in a haphazard and unofficial way, picking up mainly aluminium cans and taking them to festering dumps where they were paid a small amount in proportion to the number of drink cans collected. Brazil is one of the main growing economies in the world, part of the

BRIC group and has one of the largest discrepancies between rich and poor in the world both between different regions and within small areas of cities. As a result there is a great deal of profligacy and dumping of rubbish without the necessary capital or infra-structure to deal with the waste. It is through the enterprise, initiative and life-needs of the destitute waste-pickers that Brazilian towns and cities have not themselves become overrun with rubbish.

There are many social and political opportunities and problems associated with this process. These include enhancing the sustainability of waste recycling, easing the burden on landfill sites, providing occupational outlets and sources of income for unemployed people. Recycling reduces the impacts from the manufacture of primary aluminium from bauxite as a raw material. Problems come with the scramble for profits and infighting that breaks out among gangs of collectors. In order to reduce squabbling and street fights some local government authorities supported the establishment of collectors' co-operatives; these help to enhance human rights such as entitlements to living wages, decent sanitation conditions and basic housing. On the other hand, in the case of Rio de Janeiro, this seeming move towards both collaboration and independence further subjugated the waste-pickers 'to the interests of buyers and middle men' (Pinto and Scarlet Do Carmo 2012, p. 403).

Inter-relationships between diverse actors in the production and recycling of commodities such as aluminium involves winners and losers. But these inter-relationships are always dynamic and fluid. For example, increase of recycling of aluminium in a huge economy such as Brazil is likely to benefit those involved in recycling systems as a whole and might reduce the need for building of aluminium smelters and hydro-electric dams in areas of outstanding natural beauty. In fact, use of secondary aluminium in the automotive industry has grown enormously (The Aluminium Association 2011). However, reduction in the production of primary aluminium is a trade off for decrease in employment opportunities and national taxes levied both on large corporations (should they pay up), and on workers employed in the manufacture of aluminium together with subsidiary activities such as leisure facilities, holidays, which accompany salaried employment and so forth. Even with the gain of rights of the Brazilian waste-pickers there are still tensions and forms of exploitation. In Fig. 21.1, I attempt to capture some of the interconnections between the different agencies and how shifts in the viability of one agency effects the entire nature of the system in the production and uses of aluminium.

Pedagogy

Figure 21.1 demonstrates only a segment of the related processes and agents involved in the manufacture of aluminium. The system extends infinitely and there are subsystems within the system or network, the links between the different parts are the components of a story which can be told from any perspective, what I

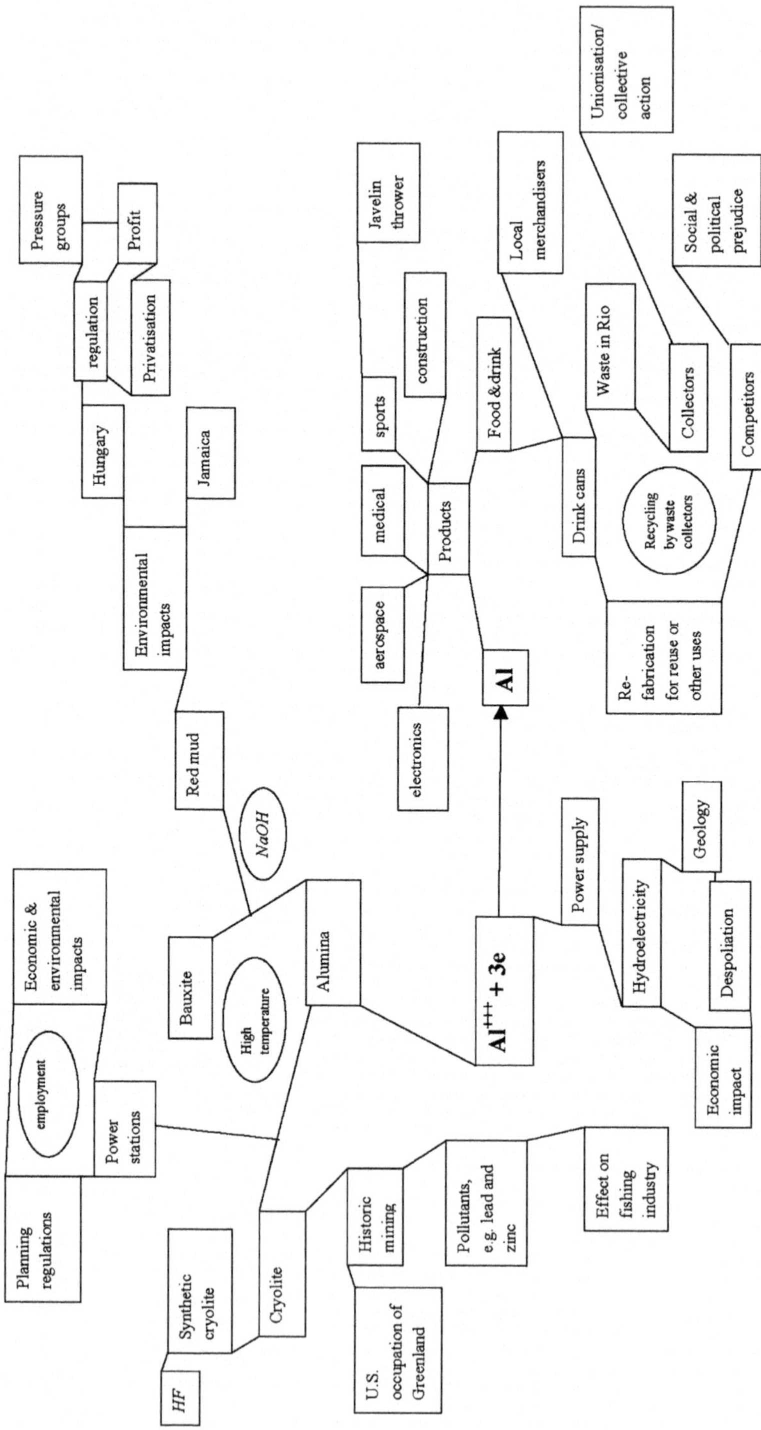

Fig. 21.1 Components of the interlocking narratives in the manufacture of aluminium

have previously called 'interlocking narratives' (Levinson 2009). Nor is this system or any other related system static, it is always fluid. But what I want to show is that the struggle for social justice among Brazil's waste collectors has effects on others, and is mediated by events such as political control, corruption, power struggles, media and other symbolic actors. Actors and actants here are not necessarily human although only human agents can be ascribed intentionality. Stories could be told in any number of ways, from the perspective of an aluminium ion, a villager affected by the red mud overspill, an executive from one of the global mining companies, or a rubbish picker in Rio.

Knowing the equation for the reduction of aluminium only has meaning within a larger scientific and social systemic framework. The 'cathedral' of science as represented by the equation of the reduction of aluminium, is inhabited by people and objects positioned in a variety of relationships with each other and the fabric of the 'cathedral' itself. The project of humanising science is not necessarily one that detracts from understanding the materiality of science, or indeed the way science is practised and made. To comprehend why the javelin-thrower at the Olympic Games is connected to corporate acquisitions of red mud waste in Jamaica, or to the struggle for the recycling economy in Brazil, is to appreciate that the manufacture of aluminium acts as a mediator through these events. Knowledge of the science deepens understanding of the interconnections between events and people, and understanding of the systemic network deepens knowledge of the science and the problems such knowledge generates. Students can change, challenge, or develop new narratives at any point in this figure.

To expose these inter-relationships demands a distinct pedagogy, seeing relationships between entities and events in narrative terms, a refusal to decontextualise and to enhance transparency. Narrative affirms agency and provides means of interpreting experience (Bruner 1996). Different agents involved in the manufacture and recycling of aluminium can tell their stories through means which reveal injustices (Nussbaum 1997) and express conflicts and synergies (Levinson 2009). Figure 21.1 acts as a map to initiate such narratives, to express the complexities of science-society interactions and the ways in which the effects of corporate action can be captured and critiqued.

References

American Association for the Advancement of Science (AAAS). (1995). *Project 2061. Science for all Americans summary*. Washington, DC: American Association for the Advancement of Science.

Ball, S. J. (2003). The teacher's soul and the terrors of performativity. *Journal of Education Policy, 18*(2), 215–228.

Baudrillard, J. (1994). *Simulacra and simulation*. Ann Arbor: University of Michigan Press.

Bencze, J. L., & Alsop, S. (2009). Anti-capitalist/pro-communitarian S&T education. *Journal for Activist Science and Technology Education, 1*(1), 65–84.

Bencze, J. L., & Carter, L. (2011). Globalizing students acting for the common good. *Journal of Research in Science Teaching, 48*(6), 648–669.

Branca, F., Nikogosian, H., & Lobstein, T. (Eds.). (2007). *The challenge of obesity in the WHO European Region and the strategies for response*. Copenhagen: WHO Office for Europe.

Bruner, J. (1996). *The culture of education*. Cambridge, MA: Harvard University Press.

Callon, M. (1999). The role of lay people in the production and dissemination of scientific knowledge. *Science, Technology and Society, 4*(1), 81–94.

CBI. (2012). *Learning to grow: What employers need from education and skills*. London: CBI/Pearson.

Chapman, B. (1991). Overselling science education in the 1980s. *School Science Review, 72*(260), 47–63.

Dawson, C. (2000). Selling snake oil: Must science educators continue to promise what they can't deliver? *Melbourne Studies in Education, 41*(2), 121–132.

Department of Education and Science (DES). (1985). *Science 5–16: A statement of policy*. London: Her Majesty's Stationery office.

Donnelly, J. (2002). Instrumentality, hermeneutics and the place of science in the school curriculum. *Science and Education, 11*(2), 135–153.

Donnelly, J. (2004). Humanising science education. *Science Education, 88*(5), 762–784.

Fensham, P. J. (1997). School science and its problems with scientific literacy. In R. Levinson & J. Thomas (Eds.), *Science today: Problem or crisis?* (pp. 119–136). London: Routledge.

Funtowicz, S. O., & Ravetz, J. (1993). Science for the post-normal age. *Futures, 25*, 739–755.

Gibbons, M., Limoges, C., Nowotny, H., Schwarzman, S., Scott, P., & Trow, M. (1994). *The new production of knowledge: The dynamics of science and research in contemporary societies*. London: Sage.

Haack, S. (1996). Science as social? – Yes and no. In L. H. Nelson & J. Nelson (Eds.), *Feminism, science and the philosophy of science* (pp. 79–93). Dordrecht: Kluwer Academic.

Hall, E. (1999). Science education and social responsibility. *School Science Review, 81*(295), 14–16.

Hall, E. (2004). Science and ethics: Give them a break. *School Science Review, 86*(315), 25–28.

Hendricks, C. M. (2006). When the forum meets interest politics: Strategic uses of public deliberation. *Politics and Society, 34*, 571–602.

Institute of Economic Studies, University of Iceland. (2009). *The effect of power-intensive industrial developments on the Icelandic economy*. University of Iceland. Available at http://eng.atvinnuvegaraduneyti.is/media/Rafraen_afgreidsla/HHI-skyrslan-a-ensku.pdf

Jasanoff, S. (2003). Technologies of humility: Citizen participation in governing science. *Minerva, 41*, 223–244.

Javor, B. (2011). *The Kolontár Report: Causes and lessons from the red mud disaster*. Budapest: University of Technology and Economics.

Johnson, B. (2013, January 20). It's snowing, and it really feels like the start of a mini ice age. *The Daily Telegraph*.

Layton, D. (1993). *Technology's challenge to science education*. Buckingham: Open University Press.

Lee, Y. (2007). Developing decision-making skills for socio-scientific issues. *Journal of Biological Education, 41*(4), 170–177.

Levinson, R. (2009). The manufacture of aluminium and the rubbish-pickers of Rio: Building interlocking narratives. *School Science Review, 90*(333), 119–124.

Millar, R., & Osborne, J. (1998). *Beyond 2000: Science education for the future*. London: King's College.

National Environmental Research Institute (NERI). (2010). *Environmental monitoring at the cryolite mine in Ivittuut, South Greenland, in 2010*. Aarhus: Aarhus University.

National Research Council [NRC]. (2012). *A framework for K-12 science education. Practices, crosscutting concepts, and core ideas*. Washington, DC: National Academies Press.

Nussbaum, M. (1997). *Cultivating humanity*. Cambridge, MA: Harvard University Press.

Oliver, R. (2012). Olympic alloys. *Catalyst, 22*(4), 21–22.
Pinto, R. F., & Scarlet Do Carmo, M. (2012). Fighting poverty and protecting the urban environment: The refuse collectors of Rio de Janeiro, Brazil. *Public Administration and Development, 32*(4–5), 402–415.
Putnam, H. (1993). Objectivity and the science-ethics distinction. In M. Nussbaum & A. Sen (Eds.), *The quality of life* (pp. 143–157). Oxford: Oxford University Press.
Putnam, H. (2002). *The collapse of the fact/value dichotomy*. Cambridge, MA: Harvard University Press.
Roberts, D. A. (2007). Scientific literacy/science literacy. In S. K. Abell & N. G. Lederman (Eds.), *Handbook of research on science education* (pp. 729–780). Mahwah: Lawrence Erlbaum.
Roth, W.-M. (2009). Taking a stand(point). In W.-M. Roth (Ed.), *Science education from people for people* (pp. 1–14). New York/London: Routledge.
Roth, W.-M., & Lee, S. (2002). Scientific literacy as collective praxis. *Public Understanding of Science, 11*, 33–56.
Ryder, J. (2001). Identifying science understanding for functional scientific literacy. *Studies in Science Education, 36*, 1–44.
Sadler, T. D., Klosterman, M. L., & Topcu, M. S. (2011). Learning science content and socio-scientific reasoning through classroom explorations of global climate change. In T. D. Sadler (Ed.), *Socio-scientific issues in the classroom: Teaching, learning and research* (pp. 45–78). Dordrecht: Springer.
Simonneaux, L. (2013). Questions socialement vives and socioscientific issues: New trends of research to meet the training needs of post-modern society. In C. Bruguière, A. Tiberghien, & P. Clément (Eds.), *9th ESERA conference selected contributions. Topics and trends in current science education* (pp. 37–54). Dordrecht: Springer.
Smith, D. V. (2011). One brief, shining moment? The impact of neoliberalism on science curriculum in the compulsory years of schooling. *International Journal of Science Education, 33*, 1273–1288.
The Aluminium Association. (2011). *Aluminium: the element of sustainability*. The Aluminium Association. Available at http://aluminum.org/Content/ContentFolders/Miscellaneous/Aluminum_The_Element_of_Sustainability.pdf
Thomas, J. (2000). Using current controversies in the classroom: Opportunities and concerns. *Melbourne Studies in Education, 41*(2), 133–144.
Van Gelder, J. W. (2004). *Greasy palms: European buyers of Indonesian palm oil*. London: Friends of the Earth.
Wolpert, L. (1992). *The unnatural nature of science*. London: Faber and Faber.
Ziman, J. (1984). *An introduction to science studies*. Cambridge: Cambridge University Press.
Ziman, J. (2000). *Real science: What it is and what it means*. Cambridge: Cambridge University Press.

Chapter 22
Preparing Students for Self-Directed Research-Informed Actions on Socioscientific Issues

Mirjan Krstovic

Abstract The central idea of this chapter is about engaging youth in research-informed action (RiA) projects to address various socio-scientific issues (SSIs) that affect individual, social and ecological well-being. Given that there is a growing concern around the world about many of the negative aspects of SSIs, with climate change being probably the most pressing issue of the twenty-first century, I have decided to transform my science curriculum in a way that will allow my students to examine critical socio-scientific issues in the classroom, propose practical courses of actions and implement these actions in their community to address the issues they examined. In this chapter I reflect on my experiences as a teacher guiding my Grade 10 'Academic' science students through three RiA projects with the last project being mostly directed by the students. The use of social correlational studies are discussed as a type of open-ended science inquiry that has been shown to motivate students' socio-political actions. I include various examples of action-ready materials that students developed to address issues concerning household chemicals, recycling in the school cafeteria, modes of transportation, water consumption, subliminal messages in advertising, consumption of multivitamins and many others. I believe that students around the world can be empowered to become agents of positive social change and that teachers and science educators can become the facilitators of issues-based, action-oriented science curriculum helping their students develop expertise and confidence for self-directed research-informed activism.

Keywords Research-informed action (RiA) projects • Socio-scientific issues (SSIs) • Correlational studies • Action-ready materials • Activism • Issues-based • Action-oriented curriculum

---

M. Krstovic (✉)
Erindale Secondary School, Peel District School Board, Mississauga, ON, Canada
e-mail: mirjan.krstovic@peelsb.com

L. Bencze and S. Alsop (eds.), *Activist Science and Technology Education*, Cultural Studies of Science Education 9, DOI 10.1007/978-94-007-4360-1_22,

Introduction

The world that we live in today has undergone rapid changes – in both good and bad ways – compared to the time when I was a high school student in the mid to late 1990s. We have some better technologies today and access to information is at the tip of our fingers. However, our world still faces increasingly complex social and environmental problems – from growing gaps between rich and poor, alarming levels of air pollution and catastrophic effects of climate change, to lack of universal access to education, especially among women in 'emerging' countries. These problems seem to be left to adults to solve with little consideration as to what our youth in schools can do to be part of the solution. As a high school science teacher of 9 years, I have mostly noticed that science education has remained virtually unchanged since I was in high school almost 20 years ago. It still tends to focus instruction "almost exclusively on the well-established products of science [e.g., laws and theories] and cookbook approaches to laboratory exercises, using authoritarian teaching modes" (Bell 2006, p. 430). I think that many of us are all too familiar with this traditional 'content only' approach to science education. However, over the past decade, a number of science educators have called for a much more politicized, issues-based and action-oriented science education – education that empowers youth to start to address some complex and contentious social and environmental issues through socio-political activism. Hodson (1998) states that the central goal of such an approach is to "equip students with the capacity and commitment to take appropriate, responsible and effective action on matters of social, economic, environmental and moral-ethical concern" (p. 4). Given the highly controversial nature of many socio-scientific issues[1] (SSIs) and the possible threat they pose to the wellbeing of individuals, societies and the environments, it makes sense that we need a much more active citizenry to address these issues.

Over the last few years, I turned the spotlight on the socio-scientific issues in order to build a more *balanced* science curriculum in which students would learn about various SSIs and start to address these issues through socio-political actions. I was intrigued by the possibilities of this approach to science education! Prior to starting to transform my practice, I wondered what my classroom would look like, and feel like, if I let students voice their opinions on critical SSIs, if I gradually handed over control of learning to students so that they conduct their own research and direct their own socio-scientific actions; and if the entire learning process is more *customizable* to students' interests and their emotional, aesthetic and spiritual needs. This chapter is about my journey towards a more issues-based, action-oriented, science curriculum that engages students in self-directed, research-informed activist (RiA) projects. I describe how students in my classes became more independent in conducting their research and directing their actions. I share

[1] In Ontario, Canada, the context for the research reported here, SSI education is addressed in a domain of learning known as 'STSE' (Science, Technology, Society & Environment) education.

several examples of students' socio-scientific actions on various SSIs. I also argue that RiA projects can motivate students to learn science and the complex interactions among science, technology, society and environment, while also developing their twenty-first century skills and myriad character traits that should prepare them for life.

How and Why I Started Building an Issues-Based, Action-Oriented Curriculum

I began to think more critically about the purpose of science education and my role as a twenty-first century science teacher when I started my graduate studies in 2010 at the Ontario Institute for Studies in Education (OISE) at the University of Toronto. In the spring of 2011, I took a course in the history, philosophy and sociology of science for which the instructor was Larry Bencze. As I was conducting my research in this course for a final paper about a reform in science education, I came across the STEPWISE framework[2] developed by Larry. STEPWISE is the acronym for Science & Technology Education Promoting Well-being for Individuals, Societies and Environments. STEPWISE offers an approach for STSE education that enables students to engage in student-directed, open-ended, primary (e.g., experiments and correlational studies) and secondary (e.g., Internet searched) research as bases for developing and implementing plans of action to address SSIs. In the past, I rarely addressed SSIs or asked students to propose any practical courses of action, let alone *take* socio-political action on SSIs. I knew that something was missing in the courses that I was teaching. I did not feel that there was any real, long-lasting and tangible connection to the world outside of the classroom, despite my enthusiasm for science and the many innovative teaching strategies that I learned at various professional development workshops. Although I had designed a number of creative, cross-curricular projects that engaged students in exploring connections between science, technology, society and the environment, they were largely teacher-directed and rarely involved students in deep exploration of contemporary SSIs with the intention of taking appropriate actions to address these issues. Therefore, I wanted to implement the STEPWISE instructional framework to promote a broader and more socio-politically engaged science education for students. Moreover, since the first goal of the revised secondary science program in Ontario is to relate science to technology, society and the environment (STSE), it gave me even more reason to put contemporary SSIs at the forefront of my teaching.

After completing the course in the history, philosophy and sociology of science, I initially asked Larry if he could facilitate a study of progress in my students' views of the nature of science (NoS), since that was the focus of my reform

[2] For a fuller description of STEPWISE, please visit: www.stepwiser.ca

paper for the course. However, I agreed that it would be better if Larry acted as a researcher-facilitator of my efforts to implement the STEPWISE framework with my tenth-grade 'Academic[3]' science class, starting in the fall of 2011, and that students' NoS views would relate to their actions on SSIs. Over the last year and a half, the focus of our work within the STEPWISE instructional framework has been on increasing students' expertise and motivation for self-directing research-informed actions on socio-scientific issues. It wasn't until I started implementing the STEPWISE framework that my thinking and practice started to change radically along with my students' learning.

Another reason why it seemed so natural for me to embrace a system of science education that encourages students to take appropriate research-informed actions on SSIs is of a personal nature. As a young boy living in the former socialist/communist Yugoslavia, I saw the collapse of a once prosperous nation. I lived through the initial stages of the 1992–1996 civil war and spent nearly a year in a refugee camp. Along with other kids in the community, I walked door-to-door collecting anti-war petitions from neighbours and family members to send to government officials. Hundreds of scribbled signatures gave me hope that they, the children of Bosnia and Herzegovina, could make a difference. Motivated by the anti-war youth solidarity campaign, I felt compelled to act for the first time on an issue that would affect the rest of my life. And even though my petitions made no difference in ending the war, this act made me feel united with other children, despite their ethnic and religious differences. My dedication to improve the wellbeing of individual, societies and the environments has followed me into adulthood and has become the central goal of my science education reform.

My First Teacher-Guided Research-Informed Action Project

I started my journey into research-informed actions on SSIs in 2011–2012 school year. I was one of a dozen science teachers in one of the largest and most culturally and ethnically diverse high schools in my district. That year, my school population was close to 1,700 students with four sections of Grade 10 Academic Science per semester. It was not until the middle of October of 2011 that I introduced my tenth-grade Academic-level students to several SSIs related to the Chemistry unit. The list of topics was limited to issues related to household chemical cleaners, oil spills, acid rain and cigarette smoking. The students selected their own groups and picked one of the four issues. They expressed what they knew about the issue by

[3] In Ontario, the jurisdiction of the research reported here, students are able to take courses at different levels of sophistication. Those labeled 'academic' are organized for students likely to pursue university-level courses in the subject area. Meanwhile, 'applied' courses are designed for students likely to enter the workforce or college after high school.

brainstorming their ideas on a placemat.[4] I provided several thought-provoking cartoons about each of their issues to inspire the students to think critically. Finally, I showed several student-developed YouTube™ videos as examples of actions that youth have taken to raise awareness about controversial issues such as combustion of fossil fuels, consumption of fast-food and the garbage dump created by 'drive-thru' restaurants. After watching these youth activist videos, the students were asked to discuss the research that was conducted to produce the videos as well as how effective each video was at creating positive change. These apprenticeship activities prepared students for their first research-informed STSE action project. While guiding the students through their first RiA project, I also taught the important concepts (e.g., the law of conservation of mass, types of chemical reactions and balancing chemical equations) and whenever possible, related them to their SSIs.

At the same time that I was preparing my students for their first research-informed action (RiA) project, the Occupy movement that started in New York City's Wall Street financial district was on the rise around the globe, including in Toronto. I felt that it was important for me to attend Occupy Toronto to show my solidarity with the kindred spirits who opposed social and economic inequality, environmental degradation and corporate influence on democracy. I took pictures of protestors and video interviews with various groups, including one of the First Nations (Indigenous People) activist groups who spoke against the 'Keystone' Pipeline Project and its impact on the environment. I brought these artifacts back to the classroom to inspire my students to become interested in issues-based and action-oriented projects. Two photos that I took, one of a quote by Gandhi "FIRST THEY IGNORE YOU, THEN THEY LAUGH AT YOU, THEN THEY FIGHT YOU, THEN YOU WIN" and another photo depicting the quote "I want an economy based on the principles of SUSTAIN-ability not GROWTH" caught students' attention (see photos below). My presentation fired the kids up making them feel that they could make a difference with the issues they were studying (Fig. 22.1).

For the first teacher-guided RiA project, the students conducted secondary research and developed ready-to-use action materials; however, they did not conduct any primary research (i.e., experiments or correlational studies) until the second teacher-guided RiA project in the Climate Change unit. Two examples of students' actions from the first RiA project are worth mentioning. One group of girls developed a YouTube™ video to address potentially hazardous household chemical cleaners.[5] They leveraged the power of social media to share their learning and bring awareness to their issue. The girls were very proud when they noticed that their video attracted attention by hundreds of viewers of whom several

[4] A 'placemat' is an instructional tactic that involves students in writing individual ideas in sections (e.g., quadrants) and then, after sharing, writing common ideas in the middle of the placemat.

[5] http://www.youtube.com/watch?v=nK5X0rCxzlA

Fig. 22.1 Photos taken at the Occupy Toronto movement in October, 2011

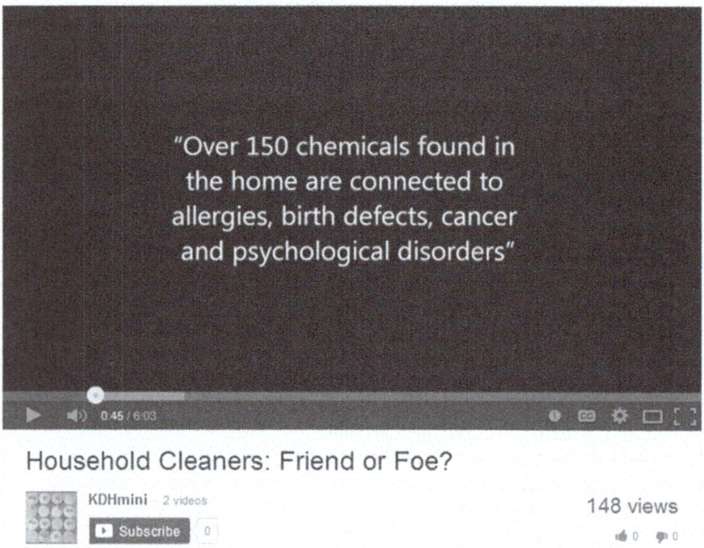

Fig. 22.2 YouTube™ video excerpt from a student group's action on hazardous household chemicals

left encouraging comments, like the teacher from Australia who wrote the following comment (Fig. 22.2):

> This is an excellent video. There has been a growth in the number of potentially toxic chemicals in the house. Non-toxic cleaners can do as good a job but don't leave nasties behind to harm us. Well done! (Teacher, NSW, Australia, October 22, 2011).

Another group of girls who researched the effects of oil spills on society and the environment created several attractive posters which they took to the local community centre and asked the manager for the permission to post them in the main foyer. They also presented their project to cafeteria workers at the community centre, parents who brought their kids to the hockey practice and their peers who came to the community centre to exercise or play their favorite sport. The girls developed deep attachment to the issue as they addressed it with the public and this was evident during their oral presentation to the class. Not only did this group of students learn about oil spills, they also learned how to communicate to different groups in their community and how to solve problems that posed a challenge to their action plan, such as asking authorities for permission to post their activist posters in various places.

At the time when I was introducing my students to research-informed action projects in science, the other three sections of the tenth-grade academic science were following more traditional modes of learning. At the school where I had started my teaching career and where I dedicated seven amazing years, I had the reputation of experimenting with different instructional methods and trying out new ideas. This meant that I also led various professional development workshop sessions at the school. As long as I shared my experiences and showed how it was benefiting my students, most of my colleagues were fully supportive of my efforts to engage students in research-informed activism. In addition, the principal of the school and her administrative team were highly supportive of the STEPWISE instructional framework. The principal shared my belief in creating stronger partnerships with teachers and university researchers/professors. They both consider research in education to be an important guide in practice and often talked about how to make research more accessible to all the teachers at the school. I have always considered my classroom as a 'laboratory' where I was not only cultivating students' inquiry habits of mind, but also my own. Having Larry act as a coach/mentor as well as a researcher/facilitator ensured that I was implementing the STEPWISE framework effectively – nothing beats an instructional coach when implementing a novel idea. I believe that this partnership approach to teaching and learning has been one of the most important factors in the success of my students.

Use of Correlational Studies to Motivate Students' Socio-political Actions

There is an apparent bias in science education towards experimentation and away from social (correlational) studies (Bencze 1996). The curriculum guidelines, as well as science textbooks, emphasize the experimental nature of science through the 'scientific method' (Bencze 1996; Gott and Duggan 1995). However, experiments are not always ideal when students are asked to explore issues that relate science to technology, society and environment. Sometimes, experiments are difficult, expensive and unethical (Bencze 1996). Therefore, the focus of the second teacher-guided research-informed action project on SSIs in the Climate Change unit was on developing students' confidence and expertise in designing and conducting correlational studies; and using students' results and conclusions to inform socioscientific activism. The students first learned how to conduct correlational studies through several apprenticeship activities (e.g., a small scale study to determine if there is a correlation between gender and involvement in extracurricular activities) before I guided them through their first, large scale correlational study involving about 250 students at the school.

The students worked in groups to create a survey that would investigate relationships between gender and/or age to everyday actions such as recycling, water consumption, dietary lifestyles, modes of transportation and uses of electronic devices. The class decided which survey questions to take from each group and a master class survey was created. Each student collected data from ten peers in the school and, over 260-mi periods, the students worked collaboratively to summarize the data and identify correlations while all along I guided the students through secondary research to learn more about how each of the issues is related to climate change. Each group decided to focus their research and actions on one particular issue (e.g., modes of transportation) that interested them the most.

Although the goal in the first semester was to get the students to completely *self-direct* their research and their actions, by the time they would do their third RiA project, this did not happen as planned, given the late start into the RiA projects and the fact that it was my first time implementing the STEPWISE framework. Nonetheless, the entire learning experience was positive as most students generally achieved outcomes comparable to Hodson's (2011) criteria for evaluating SSI expertise; that is, *understanding* of SSI issues; awareness of *power relations* in SSIs; establishment of *personal SSI positions*; and, *sociopolitical actions* on SSIs. In terms of the variations of their actions to address the issues that affect climate change, most students targeted their peers and a few groups targeted the authority figures in the school (e.g., the principal and her administration). This made sense, given that they were using the results of their correlational study done at the school to motivate and direct their socioscientific actions. For example, two groups of students decided to tackle the recycling issue at the school after finding out that an alarming percentage of boys and girls rarely ever recycled after eating lunch in the cafeteria. One group decided to design a series of creative, eye-catching posters and place them in and outside of the school cafeteria (see Fig. 22.3) in addition to developing and handing out brochures to their peers to encourage more recycling. The other group wrote a letter to the principal asking for her permission to relocate the recycling bins in the cafeteria so that they are more accessible to students. In addition, they made several morning announcements to remind their peers to put recyclable materials in the appropriate bins. During an interview with Larry, a student from the first group said that

> … in about a week we saw some progress; we saw more people recycle. One thing that helped us was our close interaction with another group; it was 'Zoe's' group. Their group wrote a letter [to the principal] saying that we want more recycling bins. … .Both of us [the two groups] were communicating. . .if this letter got accepted, our posters would have some effect. ('Sam', Interview, April 24, 2012).

Fig. 22.3 Group of students holding their recycling posters during a presentation to the class

In addition to evaluating their actions by observing the progress in the recycling habits of their peers, the students from these two groups demonstrated some understanding of the nature of collaborative knowledge creation. The fact that the two groups had the same goal (i.e., to improve recycling at the school), but a different approach, naturally led them to communicate with each other, thus helping them develop a more effective collaborative approach leading to better understanding of *what* they are doing, and *why* they are doing it. The students believed that this contributed to the success of their actions. When Larry asked a few members from these two groups if there was anything else that they would like to add about their RiA projects, one student said:

> This whole idea of taking action at an early age is a good idea and lots of schools should participate in this because not only did we learn about the aspects of science that we are suppose and that are part of the curriculum in Grade 10, we made a change, so you actually feel like you did something while learning. . .like, the term comes again '*street smarts*,' [emphasis added] not only book smarts. ('Shan', Interview, April 24, 2012).

The concept of '*street smarts*' that 'Shan' was referring to was mentioned in one of the earlier interviews by another student. It may relate in some ways to practical intelligence and experiential education. As 'Shan' put it in the interview: "For street smarts, you're actually experiencing something. . .experience always stays with you. . .book smarts might go away after you're done the subject ('Shan', Interview, April 24, 2012). My colleagues and I have explored the concept of street smarts further. We found that research-informed student activism on SSIs can lead to moderate increases of street smarts in science students Phillips et al. (2013)). The students' social (correlations) studies influences on their socio-scientific actions and the idea that research-informed student activism increases students' street smarts were both accepted for presentation at the 2013 AERA (*American Education Research Association*) Annual Meeting in San Francisco.

My experiences from the first semester motivated me to continue to learn more about issues-based and action-oriented approach to science education. It was rewarding to see the students learning to become part of the solution to society's problems and that I, as a teacher, can play a role in increasing civic participation.

I was eager to implement the RiA projects again in the second semester; however, this time around, I was determined to get the students to self-direct their final RiA project from start to finish.

Developing Students' Expertise and Confidence for Self-Directed RiA Project

I started the Climate Change RiA project almost immediately in the second semester. Similar to the first semester, I led the students through their first and second RiA projects by helping them develop their expertise and confidence in conducting primary and secondary research. In addition, there was a lot of modeling of activism through YouTube™ videos (e.g., an '*Edds World*' video on climate change[6] and *The Story of Stuff* video[7]) and personal videos made by me (e.g., UN protest during his 2012 March Break in New York City – see still photos in Fig. 22.4). Discussions pertaining to variations in actions ensued following each video with a stronger emphasis on actions targeting powerful actors (e.g., governments and corporations), since most actions in the first semester were local and targeted the students at the school.

By the time the students conducted the second RiA project in the Light and Optics unit, the process was already becoming more student-directed in terms of topics, methods, conclusions and the use of findings to direct their socio-scientific actions. Students were transitioning more effectively from teacher-directed to student-directed RiA activities as my efficacy in facilitating these projects increased over time. I was confident that by the third RiA project the students would be fully self-directed.

Fig. 22.4 Still images from video recording of two different protests in front of the United Nations Headquarters in New York City (March, 2012)

[6] http://www.youtube.com/watch?v=uvqU_L5PZtk

[7] http://www.storyofstuff.org/

Fig. 22.5 Students' new water bottle label to encourage more recycling of plastic bottles

From the first two teacher-guided RiA projects in the second semester in the Climate Change and Light and Optics units a few notable actions stand out. In order to address lack of recycling of plastic water bottles, one group developed new water bottle labels (see Fig. 22.5). They proposed the idea that companies should use these new labels as it may encourage more people to recycle or reuse the bottles. I challenged this group of students by asking them why they didn't advocate for a city-wide ban on single-use plastic water bottles. They responded by noting that they may be fighting a very powerful group (i.e., the corporations) and that their idea may be a gateway into a future in which single-use plastic water bottles will be banned forever.

One group of boys developed an RSA-style YouTube™ video[8] to address the issue of transportation and climate change after learning that a large percentage of boys and girls (with slightly more girls) get to school by a car (see still image in Fig. 22.6). They leveraged the power of social media (e.g., YouTube™) to disseminate their findings and recommendations.

In order to raise awareness about the water consumption issue, one group developed a brochure (see Fig. 22.7) in which they included both their secondary research and the data from their original study performed at the school. The students handed out the brochures to their peers. They also left samples of brochures in the library, the main office and they even went to the feeder (local elementary) schools to drop of a few samples.

At this stage, it is difficult to ascertain if their actions had any real positive change; however, the students developed significant understanding of power relations associated with socio-scientific issues and relatively sophisticated use of social media. With the exception of only a few groups, most students did not study the extent to which their actions affected change; however, during the 2 h reflection conference with the students on research-informed STSE action projects following their first two teacher-guided RiA projects, the students were engaged in thinking about and discussing the possible impact of their actions.

The *semiotics of advertising* was an issue in the Light and Optics unit that drew a lot of attention from students at the school. One group studied the intent of advertising and how semiotic representations in various ads (e.g., 'cool,' 'sexy,' 'powerful,' etc.) send powerful messages that overlook the possible personal, social and environmental problems resulting from using certain products and services

[8] http://www.youtube.com/watch?v=GCL1By923zk

Fig. 22.6 Still image from an RSA-style YouTube™ video developed by students to address the issue of transportation and climate change

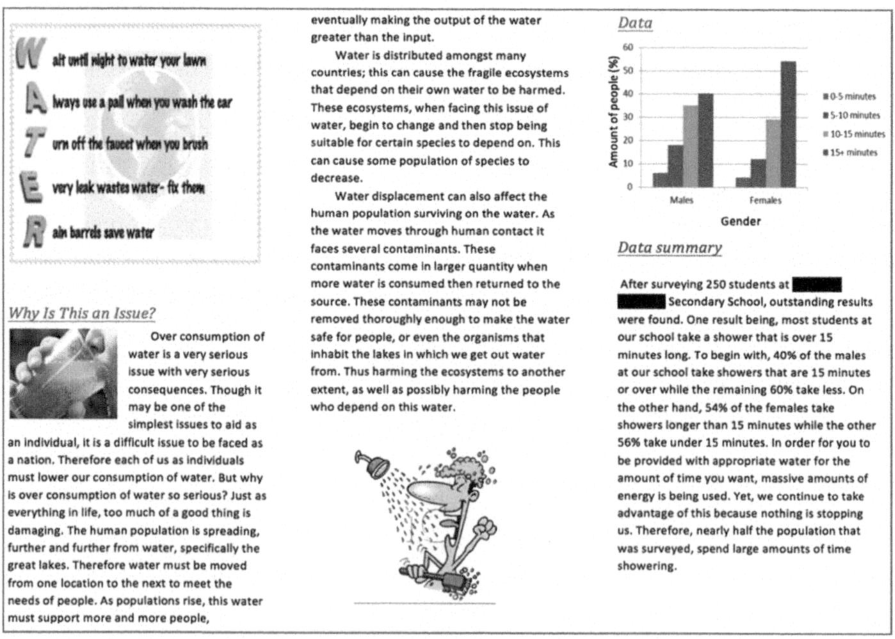

W ait until night to water your lawn
A lways use a pail when you wash the car
T urn off the faucet when you brush
E very leak wastes water- fix them
R ain barrels save water

Why Is This an Issue?

Over consumption of water is a very serious issue with very serious consequences. Though it may be one of the simplest issues to aid as an individual, it is a difficult issue to be faced as a nation. Therefore each of us as individuals must lower our consumption of water. But why is over consumption of water so serious? Just as everything in life, too much of a good thing is damaging. The human population is spreading, further and further from water, specifically the great lakes. Therefore water must be moved from one location to the next to meet the needs of people. As populations rise, this water must support more and more people, eventually making the output of the water greater than the input.

Water is distributed amongst many countries; this can cause the fragile ecosystems that depend on their own water to be harmed. These ecosystems, when facing this issue of water, begin to change and then stop being suitable for certain species to depend on. This can cause some population of species to decrease.

Water displacement can also affect the human population surviving on the water. As the water moves through human contact it faces several contaminants. These contaminants come in larger quantity when more water is consumed then returned to the source. These contaminants may not be removed thoroughly enough to make the water safe for people, or even the organisms that inhabit the lakes in which we get out water from. Thus harming the ecosystems to another extent, as well as possibly harming the people who depend on this water.

Data

Data summary

After surveying 250 students at [redacted] Secondary School, outstanding results were found. One result being, most students at our school take a shower that is over 15 minutes long. To begin with, 40% of the males at our school take showers that are 15 minutes or over while the remaining 60% take less. On the other hand, 54% of the females take showers longer than 15 minutes while the other 56% take under 15 minutes. In order for you to be provided with appropriate water for the amount of time you want, massive amounts of energy is being used. Yet, we continue to take advantage of this because nothing is stopping us. Therefore, nearly half the population that was surveyed, spend large amounts of time showering.

Fig. 22.7 Excerpt from a student group's action on Climate Change

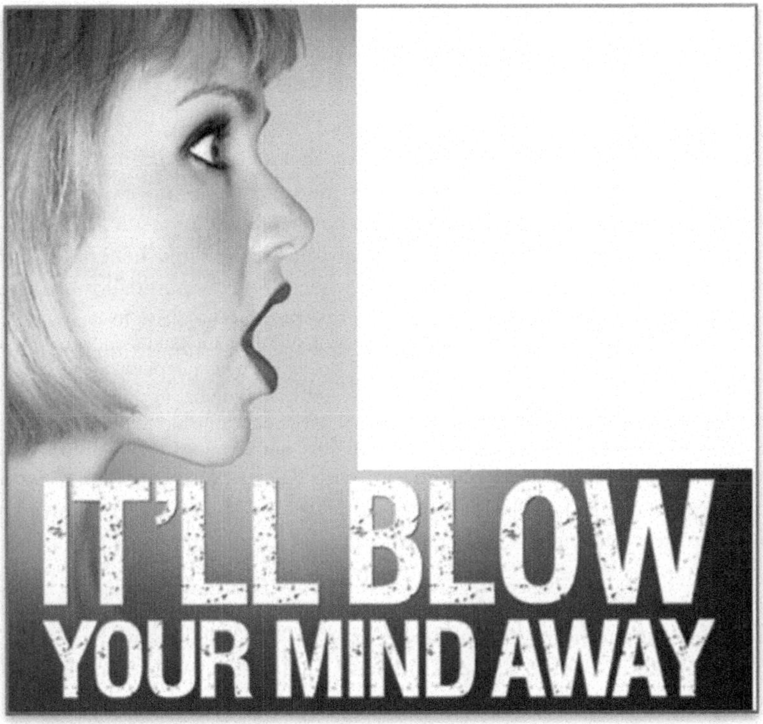

Fig. 22.8 An example of a student-modified ad to study the semiotics of advertising

depicted in the ads. For example, the students modified an ad by Burger King™ (see photo in Fig. 22.7) and conducted a study to learn how their peers at the school perceived this ad and if they knew what product was advertised. By removing the company's seven incher sandwich and leaving the photo of a woman with her mouth wide open; and a statement "*It'll blow your mind away,*" the students explored the extent to which various ads can occlude the health risks and environmental concerns associated with production, transportation and consumption of fast foods. As for the science, the students explored the anatomy of the eye (e.g., cones versus rods) and how the eye works at detecting colour; and why yellow, which is used in many ads, is the most eye-catching colour. For their action, the students decided to raise awareness (*educational activism*) by informing their peers about the possible negative impacts of ads during the lunch periods while documenting their conversations on the video, which they later showed to the class during their oral presentation. The group was deeply engaged and they also engaged their peers in meaningful conversations around the impact of advertising on youth. Supported by the *knowledge duality theory* (Wenger 1998), the students seemed to develop deep commitments to action(s) when they were personally engaged in reciprocal relations between *phenomena* (e.g., peers' perceptions of various ads and the associated consumer products) and *representations* of them (e.g., survey data and graphs) (Fig. 22.8).

Common Concerns and Considerations with Issues-Based, Action-Oriented, Curriculum

The secondary science curriculum is 'crammed' and contains a raft of demanding concepts that are to be assessed. This means that teachers often struggle to find the time to properly address the most important instructional expectations. In a conversation with one of my colleagues who has been teaching science for 22 years in six different schools, he stated that he is aware of the many benefits of infusing SSIs into the science curriculum. However, he is unsure about how to *properly* incorporate SSIs into his practice using *effective* pedagogy, how to manage groups of kids in various *open-ended*, and often 'messy,' learning situations, how to develop equitable assessment and evaluation practices, how to handle parents/guardians who prefer traditional learning styles for various reasons, and how to ensure that kids are still learning the content knowledge and developing the skills that may be needed for deeper engagement with SSIs ('Andy', personal communication, February 26, 2013). Therefore, it appears that the most common concerns when it comes to addressing SSIs is the shortage of curriculum time and a perceived lack of teacher expertise and confidence in conducting a largely divergent classroom environment as opposed to an environment where there are single correct answers. The same findings have been reported by Alsop and Pedretti (2001) and their suggestion is that "in spite of these challenges, we cannot lose sight of STSE education as a critical part of what we do with our students in science education" (p. 205).

One of the questions that I have been asked by my colleagues is if RiA projects compromise the learning of important concepts in science, given that they initially require a significant investment of time and energy. An important finding that I share with colleagues is that my students performed just as well, and in some cases better, on the unit tests and the final exam when compared to students taking the same course who did not engage in research-informed action projects. This important finding suggests that students' mastery of concepts may not be affected by their engagement in RiA. Au contraire, they may be performing even better!

As teachers become more skilled over the years, they also become more *efficient* at teaching the content knowledge. Efficiency is all about getting the most with the least effort (and in less time). When it comes to teaching science knowledge, efficiency means teaching the concepts well the first time around (i.e., having a high success rate the first time without having to re-teach it because you felt that kids didn't get it). This obviously requires a rich repertoire of teaching tactics and strategies for all sorts of concepts, which takes time and energy to develop. Also, if science concepts are presented in real life context such as SSIs, and if the teachers' pedagogical approaches allow students to make their own connections between content knowledge and various context, then the students may be more successful at mastering the content, or at least they may feel more motivated to study the content knowledge.

RiA projects are not only excellent activities to engage students in contemporary socio-scientific issues, but they also help students hone twenty-first century skills, which include critical thinking, collaboration, communication and creativity. In addition, RiA projects have the potential to develop various character traits such as: responsibility, care, courage, generosity and imagination, which are an integral part of many schools' and school Boards' mission statements and core values. I believe that *learning science* (i.e., development of cognitive skills by mastering content knowledge) should be balanced with *doing science* (e.g., learning to perform social (correlation) studies and controlled experiments) and *learning about science* (e.g., learning about socio-scientific issues and complex interactions among science, technology, society and environment). Hodson (1998) defined the multi-dimensionality of critical scientific literacy in term of these three major elements. I see these three elements as the critical components of a *holistic* science education for which the purpose is to develop students' skills in the cognitive, affective and psychomotor learning domains.

Reflecting on the Nature of RiA with Students

At the end of the first two teacher-guided RiA projects in the second semester, the students participated in a 2-h reflection activity. The students first reflected and evaluated their previous projects in small groups and then as a class they discussed the *nature* of RiA by considering the extent to which they felt that they developed: (i) sufficient understanding of the controversy involved in various SSIs; (ii) confidence in conducting primary and secondary research; and (iii) expertise in using research to direct their actions and target their actions towards people in power. The purpose of this exercise was to prepare the students for the final *student-directed* STSE action project in the Biology unit by getting them to think about what it means to prepare for and engage in responsible actions that are informed by their research. This metacognitive step appeared to have contributed to students' successes in self-directing research-informed actions on socio-scientific issues (Bencze and Krstovic 2013). It has been argued in the past that such epistemic reflections can motivate social agency (Damsa et al. 2010).

Most students understood that many of the SSIs that they explored in the Climate Change and the Light and Optics units are highly controversial: electricity usage, water consumption, dietary choices, modes of transportation, electronic waste disposal, health hazards associated with laser eye surgery, subliminal messages in advertising, privacy invasion and other SSIs that affect the wellbeing of individuals, societies and environments. They also commented on the extent to which human emotions (e.g., vs. logic) influence RiA projects. One student stated that: "in order to help people, you need to have some compassion" ('Jay', Reflection Conference Interview, May 24, 2012). Another student said that: "I think it depends on what you're trying to do, like what topic it is because for my group we [focused on] privacy and that really was based on how people felt about being watched, or videotaped, so it was emotional ('Tina', Reflection Conference Interview, May 24, 2012). One's personality, family background, culture and life experiences were also discussed as some of the factors that influenced how students perceived various SSIs.

In terms of the primary research, most students understood the difference between (social) correlational studies and experiments; and the fact that correlational studies need a large sample size to be more valid. The students identified ethics as the major reason for why investigators conduct studies as opposed to experiments. It was more difficult for students to identify other reasons for why correlational studies are preferred over experiments, such as logistics and the fact that when studying age as a variable it is difficult, or impossible, to force organisms to age. As for the secondary research, most students, if not all, were able to identify sources (i.e., on-line sites) that were more trustworthy and reliable than others. One student noted that the corporate sites are biased and that they only present content that is in favour of the products and/or services that are sold by the company. When students were asked why one might take action(s) on various SSIs, one boy, whose group tackled the issue of water consumption in the Climate Change unit, stated that:

> . . . it's not a just a research project, it's an action project. The research is the base of it - the actual project itself is to take the action. . . .It's our responsibility to take action. Us as a people, we need it [water] to survive so we have to take the action in order to benefit the society and the world itself. ('Chris', Reflection Conference Interview, May 24, 2012)

The use of the words 'action,' 'responsibility,' and 'benefit to society and the world' speaks about this particular student's understanding of how the RiA projects are intended to take students further by becoming radical agents of change. Another student concluded this part of the discussion by stating that "if you are really moved by what you are studying, then you will act on it!" ('Mona', Reflection Conference Interview, May 24, 2012). This further supports the idea that the research-informed action projects involve human emotions and that emotional attachments to SSIs can motivate sociopolitical actions.

The final, student-directed RiA project was introduced to the students at the end of this 2 h activity. The class read through the following list of possible SSIs that students could explore in the Biology unit: effects of energy drinks, consumption of snack foods, impacts of smoking, long-terms effects of ultrasounds, consumption of multivitamins, controversies surrounding vaccines (e.g., flu shots) and the possible negative effects of sunscreens. Students were also able to add their own SSIs to this list with my approval. One group chose GM foods as the focus for their RiA project. Even before they selected their SSIs, the students already started to become engaged in a debate around certain SSIs (e.g., energy drinks) based on their preconceived ideas about the issues. This excitement was a good sign that they would enjoy this project.

Students' Self-Directed Research-Informed Actions on SSIs

With 4 week left in the course, the students had to start their secondary research on their SSIs immediately. The first step was to get the students to complete a rough outline of their project in which they: (i) describe the STSE issue in their words; (ii) state what they already know about the issue; (iii) list what they would need to learn; (iv) propose an idea for the primary research (correlational study or an experiment); and (v) state at least three possible actions to address the issue with an explanation for each action. Based on the students' preliminary responses, I developed a Wiki template with six guiding questions[9] for the secondary research that students had to complete by a certain deadline. The Wiki template provided an on-line collaborative environment for the students and it was easy for me to track and assess the contributions to secondary research by each group member. In addition, I was able to provide *immediate* feedback to students by leaving comments on their Wiki pages as they progressed through their final self-directed RiA projects. My role in the final self-directed RiA projects was to facilitate students' progress, act as a critic, consultant and a learning resource.

After completing their secondary research, the groups started on the primary research with correlational studies being the popular choice for their method of inquiry. It was not surprising that all groups decided to conduct correlational studies for their final self-directed RiA project since they were most comfortable with these types of investigations, even though experiments were performed in class and were suggested as an option for the primary research.

By the end of the final self-directed RiA project, all student groups were generally successful in conducting such autonomous projects. Compared to projects these students completed earlier in the course and those completed by students in the first semester, their actions were more *diverse* (e.g., pamphlets and YouTube™ videos), *broader* (e.g., beyond the school) and more often *aimed at powerful actors* (e.g., a drug company). The group that explored the issues surrounding consumption of multivitamins took several actions to address this SSI: (i) they developed informative brochures that they gave out to their peers at the school; (ii) they visited the local Shoppers Drug Mart™ and gave out brochures to the shoppers in the multivitamin aisle; (iii) they wrote a note on Centrum's™ Facebook™ page stating what they have learned about some of the negative effects of multivitamins (later Centrum™ removed the students' note from their wall); and (iv) they developed a YouTube™ video[10] to raise more awareness about some of the negative effects of multivitamins while suggesting healthier alternatives such as fruits and vegetables.

A group of three girls studying the effects of energy drinks on the wellbeing of individuals, societies and environments devised the largest social (correlational)

---

[9] http://mrkrstovic.wetpaint.com/page/Multivitamins

[10] http://www.youtube.com/watch?v=3Z6a2F7fh6o

study in which over 100 students per grade participated. The girls found that significantly more Grade 11 and Grade 12 students consume energy drinks compared to the junior students. The group correlated higher level of consumption of energy drinks amongst seniors to their busier academic lives and less hours of sleep per night (relative to juniors). In terms of their sociopolitical actions, the girls sent letters to Rockstar™ and Monster™ energy drink companies as well as Health Canada (federal government department) stating that our government should ban the sale of energy drinks to minors; and they collected petitions from their peers, which they sent to both Health Canada and the energy drink companies along with the letters.

Other student groups engaged in a variety of sociopolitical actions and often more than just one per group. The group of students who selected GM foods as their SSI wrote a letter to Greenpeace™, an activist organization that opposes genetic modification of foods, expressing their solidarity with this organization while sharing some of their primary research. They also developed an RSA-style YouTube™ video to disseminate their research and raise awareness. A different group of enthusiastic boys engaged the whole class in a discussion about the effects of snack food consumption on the well-being of individuals, societies and the environments drawing special attention to the influence of advertizing in consumption of snack foods (e.g., Dorito™ ads during Super Bowl™ commercials). They showed the posters that they had created and put inside the large supermarkets, such as Freshco™ and Food Basics™ in order it alert people to the potential dangers of snack foods. The last two groups took very different actions to address their SSIs – one group developed an education brochure to raise awareness about some possible negative effects of ultrasounds and another group sent a petition to the local member of the provincial parliament to ban smoking in one of our city's largest parks. I was especially impressed with the last action, as I envision a future when all public parks will become smoke-free.

Even though the actions that students took may not have a significant social or environmental impact, the students took a lot of pride in learning how to be young activists. In her letter to an online site aimed at expectant mothers, 'Nancy,' started with: "I am a student and a young activist. . . ." (June 19, 2012). Many students seemed to identify themselves as activists. Hodson (2014, pp. 67–98) argued that "the likelihood of students becoming active citizens in later life is increased substantially by encouraging them to take action *now* (in school), by providing opportunities for them to do so, and by providing detailed examples of successful actions and interventions engaged in by others." I would also argue that engagement in socio-political actions by students in school is contagious. The more the boundary between classroom and life is blurred, the more the students become aware of how their education can have an impact on what goes on both inside and outside of their school. Once they are made aware of this through their research-informed action projects, they want to continue learning and engaging in transformative socio-political actions. And this is very empowering for the students and the teachers who embrace issues-based and action-oriented science curriculum.

Concluding Thoughts

In this chapter, I sought to provide an overview of how I guided my tenth grade 'Academic' students through research-informed action projects and how I ultimately led them to *self-direct* their own research and actions on various socio-scientific issues during the second semester. By providing various examples of student actions, I want to demonstrate how science education can be transformed from "passive, technical, and apolitical orientation that is reflective of most students' school-based experiences to an active, critical, and politicized life-long endeavor that transcends the boundaries of classrooms and schools" (Kyle 1996, p. 1).

I continue to build a more issues-based and action-oriented science curriculum. My students continue to amaze and inspire me with their ideas, their passion and their commitment to make a difference in the world. Teachers hold the key to unlocking tremendous student potential. They are the torches lighting the way to a brighter future and I feel that it is an honour to be on this personal and professional journey. While this is the end of the chapter, I feel that I am only at the beginning of what may become the best journey in the world of science education.

References

Alsop, S., & Pedretti, E. (2001). Science, technology and society. In S. Alsop & K. Hicks (Eds.), *Teaching science: A handbook for primary and secondary school teachers* (pp. 193–205). London: Kogan.

Bell, R. L. (2006). Perusing Pandora's box. In L. B. Flick & N. G. Lederman (Eds.), *Scientific inquiry and nature of science: Implications for teaching, learning, and teacher education* (pp. 427–446). Dordrecht: Springer.

Bencze, J. L. (1996). Correlational studies in school science: Breaking the science-experiment-certainty connection. *School Science Review, 78*(282), 95–101.

Bencze, J. L., & Krstovic, M. (2013, April). *Students' social studies influences on their socioscientific actions*. Paper presented at the American Education Research Association, San Francisco, CA.

Damsa, C. I., Kirschner, P. A., Andriessen, J. E. B., Erkens, G., & Sins, P. H. M. (2010). Shared epistemic agency: An empirical study of an emergent construct. *Journal of the Learning Sciences, 19*(2), 143–186.

Gott, R., & Duggan, S. (1995). *Investigative work in science curriculum*. Buckingham: Open University Press.

Hodson, D. (1998). In pursuit of literacy. In *Teaching and learning science: Towards a personalized approach* (pp. 1–8). Philadelphia: Open University Press.

Hodson, D. (2011). *Looking to the future: Building a curriculum for social activism*. Rotterdam: Sense.

Hodson, D. (2014). Becoming part of the solution: Learning about activism, learning through activism, learning from activism. In J. L. Bencze & S. Alsop (Eds.), *Activist science and technology education* (pp. 67–98). Dordrecht: Springer.

Kyle, W. C. (1996). Editorial: The importance of investing in human resources. *Journal of Research in Science Teaching, 33*, 1–4.

Phillips, C., Krstovic, M., & Bencze, J. L. (2013, April). *Can research informed student activism promote the use of street smarts to navigate between the academic realm and everyday life?* Paper presented at the American Education Research Association, San Francisco, CA.

Wenger, E. (1998). *Communities of practice*. Cambridge: Cambridge University Press.

Chapter 23
Activism in Science and Environmental Education: Renewing Conceptions About Science Among Students When Considering Socioscientific Issues

Barbara Bader and Yves Laberge

Abstract This chapter's aim is to highlight the importance of critical pedagogy for activism in science and environmental education, notably an «education for awareness» regarding the dominant ideologies and the instrumental rationality related to climate change. We tried to apply some of these principles at high school level, with French speaking students in Quebec, to question and enhance their conceptions of the nature of science, by inviting them to document issues on climate change, including uncertainties, controversies and research practices. We also present some results about their civic engagement on climate change.

Keywords Activism • Citizen science education • Environmental education • Critical pedagogy • Education for awareness • Instrumental rationality • Nature of science • Climate change issues • Social justice • High school students

Introduction

Introducing activism in science education classrooms can be seen as provocative or debatable. Any approach in that sense must be made intelligently and with nuance in order to be legitimized. Defending an authentic citizenship engagement by specific groups and coherently prioritizing goals of social justice, however, seems

B. Bader (✉)
Chaire de leadership en enseignement des sciences et développement durable, Université Laval, Quebec City, QC, Canada
e-mail: Barbara.Bader@fse.ulaval.ca

Y. Laberge
Centre de Recherche en Éducation et Formation Relatives à l'Environnement et l'Écocitoyenneté, Montreal, QC, Canada
e-mail: ylaberge@uottawa.ca

L. Bencze and S. Alsop (eds.), *Activist Science and Technology Education*, Cultural Studies of Science Education 9, DOI 10.1007/978-94-007-4360-1_23,

to be valorized in the field of environmental education and in the domain of sustainable education. Politicizing science education can only be credible, on the other hand, if the goals of promoting activism in the science classroom are clearly exposed right from the start with respect to students' and teachers' positions, in order to maintain dialogue and praxis, which are basic principles within critical pedagogy.

Overall, however, linking science education and politics can be seen as necessary and constructive when justifying a politicization of science education, drawing from Ulrich Beck's (1992, 2001) 'risk society,' which considers environmental issues as indicators and consequences of our industrial relationship with nature. This seems especially appropriate, given that fields of technoscience often are used as an unfair productivist system that appears to generate significant risk for public health and environments.

Defining Activism

Following Steve Alsop and Larry Bencze (2010) and the special issue of the *Canadian Journal of Science, Mathematics and Technology Education* they edited, we take with them the concern for a science education engaged in concrete politicized actions, reflecting aims of social justice; that is, to seek reductions in social inequalities because of our capitalist system and, in that way, also trying to protect nature. With them, we draw on Maxine Greene (1995), who proposes to advocate and explore 'norm-governed situations,' in which "students discover what it is to experience a sense of obligation and responsibility, whether they derive that sense from their own experiences of caring and being cared for or from their intuitions and conceptions of justice and equity" (p. 66, cited in Alsop and Bencze 2010, p. 178).

We illustrate later in this chapter how these concerns guide our work with young people completing high school in Quebec. Whenever it is seen from the outside and especially from English-speaking provinces within Canada, teaching in Quebec implies specific approaches that refer to specific models and traditions, not only because programs are developed in the French language, but also because it is apparent that teachers in Quebec rely on some unique values and perspectives in terms of nature, environment, governance, and politics.

Note that 'activism in science education' implies in-depth analyses of our culture and its effects on youth (Giroux 2000, 2003), trying to resist a 'dominant ideology' characterized by: (1) a conception of nature as a 'material resource to be exploited,' which some authors also find in the United Nations way of defining sustainable development (Sauvé et al. 2003); (2) belief in the progress of science and technology to analyze environmental issues and sustainable development and to propose solutions, giving knowledge and science experts a special status; and (3) promotion of a globalized system of industrialization that creates risks (Beck 1992, 2001), in a context where the world of high finance seems to have monopolized the power to decide on the use of public funds (George 2010) at the expense of the common good. This ideology is also characterized by an *instrumental rationality*, which often frames our ways to define and address environmental issues (Lash et al. 1996).

'Instrumental rationality' is conceived following Scott Lash, Bronislaw Szerszynski and Brian Wynne (1996, p. 1) as the tendency to translate environmental issues into "authoritative scientific and policy vocabularies," which could best be described as "epistemologically 'realist', positivistic, disembedded, technological and cognitivist", and where the "Horkheimer and Adorno's dire warnings of a dialectic of enlightenment in which reason would metamorphose into technology is nowhere more profoundly confirmed than in 'man's' domination and instrumentation of nature – including human nature." According to this ideology, Szerszynski, Lash and Wynne argue that the "[c]olonization of nature is being achieved through technology, machinery and computers, as well as through a range of expert-systems – especially capitalist management and the administrative apparatus of the state" (1996, p. 3). 'Managing Planet earth' by technocratic expertise' (1996, p. 4) minimises concerns and awareness related to global equity, justice and basic human rights whenever considering environmental issues. Following this ideology, these issues are mostly addressed by "deterministic social-scientific predictions of human inputs to the climate system (see the Intergovernmental Panel on Climate Change 1996) and the equally deterministic predictions of human impacts of thus predicted climate change", without any serious political and cultural analyses of the environmental problems (p. 4).

If activism in science education means social change, it should address some of these considerations, in reflexive, critical and constructive ways.

Activism in Environmental Education and Education for Sustainability

To engage students in concrete actions in connection with citizen groups is an avenue that is favored by environmental education and education for sustainable development in different contexts (Ardoin et al. 2012; Driskell and Chawla 2009). It is also part of the educational practices in science education (Calabrese Barton and Tan 2010; Roth 2003).

The North American Association for Environmental Education advocates, for example, for knowledge, dispositions and skills to be included in an 'environmental culture,' which especially contributes to "[c]itizen participation and action strategies – forms of citizen participation, action, and community service intended to preserve or improve the environment. The 'action strategies' include restoration projects; consumer and economic action; effective communication strategies; political action; and collaborative solution seeking" (North American Association for Environmental Education 2011, pp. 3–5).

In the same way, David Driskell and Louise Chawla (2009) strongly advocate for positive experiences in childhood and adolescence, as well as a real commitment in conservation actions to protect nature. For example, they led a project that took part in the Growing Up in Cities program and engaged children and adolescents as co-researchers in their surroundings to document ways in which they use and value the places where they live, and take action to correct the situation when necessary. As they say, this kind of project is valuable for different reasons:

> GUiC (. . .) then involves them in analyzing their findings, determining strategies for taking action on their priorities, and working in partnership with adults to realize at least some of their ideas. (. . .) Trough exploration and discovery in the places where they live, young people begin to understand 'nature' as an important aspect of their local area and their daily lives, rather than something that exists only outside the city or on television programs. By acting on tangible issues within their control, and understanding that the condition of the environment is connected with issues such as livelihoods or violence, young people are «learning by doing». In the process they have articulated their concern for natural settings in their area, questioned their own role in local environmental conditions, and developed the skills and confidence to be successful stewards on their local environment. (p. 97)

It is, thus, working towards a school culture of democratic and civic engagement; such as what we have tried to promote in a popular area of Quebec City, a program that encourages high school students to diagnose a social or environmental problem in their neighborhood, and propose a solution strategy, while being supported by some players in their school and neighborhood (Bader et al. 2010). By doing so, we share many of the objectives found in the conceptual framework for activism in science education proposed by Calabrese Barton and Tan (2010). Indeed, we pursue several different purposes, including: trying to give sense to school learning by relating it to real concerns, educating students about social issues that affect them, empowering and involving them in concrete actions that may improve situations, as well as introducing them to democratic ways of operating in society to protect the environment and strengthen solidarity of activist networks.

In Quebec, the school network Brundtland Green Schools (Réseau des Écoles vertes Brundtland[1]) supports educational initiatives that strengthen values of democracy, solidarity, ecology and peace. This network celebrates its 20th anniversary this year. It calls for civic engagement of students and taking action on issues of sustainable development according to principles of critical pedagogy (principles renamed the 'pedagogy of hope'). Through provision of educational kits, for example, teachers and schools are confirmed and strengthened in their efforts to engage students in action for the common good. Indeed, it is apparent that most of the time these projects are designed to engage students in a process of social transformation, more or less, ranging from the establishment of a recycling system in a school, strengthening everyday actions seen as 'good for the environment,' to raising awareness of a group facing a social or an environmental problem, or restoring a place for people by political pressure.

[1] http://www.evb.csq.qc.net/

This program prioritizes 'learning by doing' and demonstrates marked concern for student engagement in actions, sometimes promoting social justice. This is considered legitimate in environmental education and education for sustainable development (Sauvé and Orellana 2008). It emphasizes critical pedagogy, following authors such as Paulo Freire (2013) and Henry Giroux (1988).

Several authors refer to the Earth Charter,[2] as an inspiring document in environmental education (EE) and education for sustainable development (ESD) because of its concern for social and environmental justice (Corcoran and Osano 2009). As critical pedagogy takes on significant importance in environmental education and sustainable development education, let's consider some of its important characteristics and then specify why they seem important and how to apply them in science education.

At the Heart of Critical Pedagogy Are the Ideas of 'Education for Awareness,' Praxis and Dialogue

It should be understood that Paulo Freire, Henry Giroux and others (e.g., Michael Apple 2002), combine micro-social and macro-social analysis for a better understanding of challenges to which the education system should respond. Freire did not see the development of human beings in a deterministic way, but situates people in contexts that influence them. He also recognized that human beings are free and have the ability, particularly through education, to emancipate and to identify constraints that violate their freedom and try to work with others to try to change things.

Critical pedagogy also insists on reflexively naming the ideologies that orientate our worldviews. To promote social interaction in order to reduce social inequalities, we need indeed to identify some of the strongest ideologies that reinforce social inequalities. But, in order to implement some 'praxis,' critical thinking should also include actions. This proposal is legitimate, but makes sense only as tested in specific contexts. It is not effective if it remains 'only' theoretical.

Critical pedagogy also underlines the importance of dialogue, based on the recognition of the dignity and respect of each other, avoiding various forms of violence more or less symbolic (Freire 2013). This does not mean we do not recognize the teacher's authority, but this authority is linked to his/her competence, based on clear and shared objectives. The same authority can be given to students or adults as they develop knowledge and take a stand on environmental issues of importance to them.

These principles are to be used to guide educational practices in environmental education and citizen science education in order to avoid an 'educational agitation' that does not lead to actions for social change, changes in our worldviews and in our relationship to nature and to others.

---

[2] http://www.earthcharterinaction.org/content/

A Particular Concern to Make School and Science Courses Meaningful for Students

In an effort to root our pedagogical practices in analyses of their socio-cultural contexts, Henry Giroux (2000, 2003) considers how culture and media influence the way young people define their identity. He focuses on media discourse analyses to illustrate how they reinforce a commercial consumer culture into which many children and young people are socialized. He highlights, as well, the lack of spaces where young people learn to develop their sense of initiative, commitment ('agency'), solidarity values and social relations for more cooperation. For Giroux, it seems the (corporate) media promote a culture that is increasingly 'private' (Kellner 2001, p. 225), with the consequences of strengthening individualism and the market economy, and more political indifference of youth. As specified in the introduction of the FreeChild Project reading list, for Henry Giroux, academia is impossibly far away for too many young people today. Scholar, author, and professor Henry Giroux demands more. For more than 25 years he has called for academia to be more accessible, more relevant, and more meaningful to all young people, particularly low-income and youth of color. This advocacy for young people, democracy, and social change has constantly grown: throughout the 1980s he challenged teachers and school leaders to transform learning into meaning; in the 1990s he took a critical eye towards popular culture and its effects on young people and society. His analysis of the war against youth is unparalleled; his call for action, reflection, and responsibility to, for, and with young people is vital. Henry Giroux is an ally to all people fighting for the radical notion that democracy is more than a perfectionist, idyllic utopia: it is an action, an authority, and a requirement for our future. Giroux' writings leads many young people and adults down a path to understanding, creating, and challenging that future together, today.

These principles of critical pedagogy need to be applied in science education and especially, as Bruno Latour (1999) said, in order to "reintroduce sciences into democracy".

Critical Pedagogy and Science Education

To use these principles of critical pedagogy in some form of activism in science education, it seems important to specify what in our cultural context seems problematic about our relationship to nature and to others. In particular, it seems that the instrumental rationality that characterizes our way of defining environmental issues should be explained and illustrated with teachers and students. The point would be not to discredit it, but to allow a form of reflexivity and critical approach to our usual way of dealing with issues. The systemic overvaluation of certain areas of knowledge, at the expense of ethical, social, political and cultural analysis, associated with instrumental rationality, seems to be a problem and tends to extend the issues it should address (Bader 2011). Accordingly, we are motivated to propose ways to renew current dominant conceptions of sciences, emphasizing more contextualized and up-to-date propositions about the nature of science (Richard and Bader 2010). In this work, we particularly promote attention to environmental issues that take into account socio-cultural contexts.

It seems clear that our society is highly individualistic and over-consuming. Related to that, our worldviews seem to prioritize instrumental rationality, which apparently results in overvaluing the technological expertise to deal with our relationship to nature and to others at the expense of political analysis, ethical and/or cultural values. Discussions tend not to include deep political analyses of issues and seem to reinforce citizen disengagement. By contrast, following the work of Henry Giroux, at the heart of our concerns in science education, we would like to value young people and their cultures, what drives them and motivates them. Accordingly, we should work with them to promote 'self-awareness,' defining their socio-cultural context, its characteristics, in order to base their social engagement on a thorough analysis of the causes of current environmental issues. Such an approach would favor 'praxis'; that is to say, a reflection in action that combines 'critical analyses' and 'practical confrontation,' open to change. Hence the importance of an emerging pedagogy, whoever are our interlocutors (Bader et al. 2010).

To engage students in science learning, we stand in line with authors in science education advocating a renewal of how school science is conceived, considering current research practices. In this view, science is seen as contextualized and the actors, the research practices and projects that give them meaning are included in science courses (Richard and Bader 2010). It is also important for us to present limitations, uncertainties and indeterminacies that are part of science knowledge (Wynne 1987), so as not to overvalue the position of experts when science addresses environmental issues, and to invite youth and the public to appropriate these social, political and cultural issues (Bader 2011).

Drawing from 'education for awareness' and to promote a critical science education, led us to questioning and actualizing the current conception of the nature of science with students and, notably, ways in which scientists portray climate change issues. We can draw on Fumiyo Kagawa and David Selby (2010), who insist that:

> At the moment the educational response to climate change has not been of this kind ... The curriculum focus has been on imparting the science, but less often wrestling with the ethics of global warming. The exhortation has been for personal change of a reformist (rather than transformative) nature. There has been an overarching absorption with the technological fix with a focus on reducing carbon emissions by the educational institution in question as well as by society at large... Climate change education learning experiences have, thus, tended to be confined by «business as usual» parameters. There has been minimal recognition of the need to engage learners in openly debating and discussing the roots, personal meanings, and societal implications of climate change scenarios that are likely to play out during their lifetimes, and what needs to be done and achieved of a transformative nature by way of mitigation. The academy has tended to fiddle while Rome begins to burn. (p. 5)

It is, therefore, central to ask for a thorough diagnosis of the state of our society to understand what is wrong and then to educate in ways favoring political action aiming to reduce inequalities and restore a respectful relationship to other and to nature.

It seems important to explain how some instrumental rationality operates to identify and resolve environmental issues. Students can, for example, be initiated into a reflexive and critical epistemological analysis of the work of the Intergovernmental Panel on Climate Change (IPCC) (2007). Such an approach might involve designing an educational activity for secondary science classes on climate change with two main goals. First, based on a summary of the main conclusions of the IPCC (2007) we propose to discuss the general orientation of the IPCC's way of defining and analyzing climate change issues, highlighting how some instrumental rationality is at work and how it does not include the political, ethical and cultural concerns.

To be more aware and reflexive on what is considered here as a dominant epistemology when scientists deal with climate change and, at the same time, trying to invite students to understand research practices, being conscious of this dominant epistemology and more aware of social justice concerns, seem of importance if we want to educate citizens and to form future scientists capable of introducing considerations for the common-good in their work. Another important element to contextualize science and environmental education is to follow Ulrich Beck (1992, 2001) in order to understand how industrial society sees and uses nature. It seems relevant to consider how industrialization includes life itself in a capitalist system that relies on science and technology innovation, copyright, and production. Traditionally, science has enjoyed considerable public authority. However, in recent years and especially since industrial risks became an important part of the game, raising doubts related to the legitimization of technocracy to manage nature has increased. Such doubt seems to apply to much work in the sciences and fields of technology, as they produce new knowledge, new technologies and new products.

According to Beck (2001), risks sensed by members of societies help to legitimize capitalism and its logics. He argues that industrial societies take advantage of the risks that they produce because these new risks need to be addressed and resolved by new technologies and new research, in a never ending spiral of risks and research, thus creating commercialization and profits, but only for a minority of actors. Beck's idea of a 'risk society,' however, allows us to picture and conceptualize environmental issues as *political* issues. If one tries to eliminate the causes of these risks while considering them completely being part of the industrializing process, then the public debate, citizens and polities must be considered as an integral part of the analysis of environmental issues, which are issues of science in society. For example, as pointed by Beck (2001, p. 57), finding a level of lead in breast milk is not in itself the definition of danger. We must add a causal link so that danger appears to be the product of a certain type of industrialization, way of defining that 'manufactured' problem, which may then lead to a political analysis mobilizing a network of actors.

Illustration on the Issue of Climate Change at the Secondary School in Quebec

In our work, we rely on the implementation of interdisciplinary science projects related to climate change and study relationships to scientific knowledge of secondary school students and their ways of dealing with climate change as citizens (Bader et al. 2013).[3] We seek to actualize students' conceptions about sciences in the context of regular science classes and give some autonomy to students. Six groups of students were involved in one secondary school in Quebec City. These 16 year-old students had to answer two questions: "What should we do to face the prospect of climate change?" and "Can Science tell us how to act?"

The interdisciplinary approach tested in the course of science and technology spanned seven (7) teaching periods of 75 min each. It has mobilized two science teachers and nearly 200 students enrolled in the International Education Program (IEP) took part.[4]

We introduce in this teaching process some aspects of the social construction of sciences (Richard and Bader 2010): elements of debate among researchers, controversial aspects, but also the importance of peer review and publication of results, as well as the strengthening of researchers' credibility when being members of a recognized institution or research center (Latour 2001).

The entire process of education was introduced in the classroom listening to a seem-like debate between researchers who did not share the same research priorities on the issue of climate change. The first insisted on pursuing basic research to better understand the climate, while the latter wanted to focus on adaptation to climate change, especially when poor people are at risk. The two researchers were presented as being affiliated with renowned institutions; referred to papers published in leading scientific journals. They worked in collaboration with other researchers and studded the issue for several years.

Each team of students had to choose a theme related to climate change that they saw as relevant. The topics that the teams have treated were for example:

- The controversy over the governance of the Northwest Passage (studying of satellite images/studying ice cores)
- The use of fossil and meteorological data
- The relationship between climate change and natural disasters
- The homogenization of climate data

[3] This research was made possible with funding from the Research Council Social Sciences and Humanities (SSHRC, B. Bader and C. Lapointe, 2008–2011) and the Quebec Fund for Research on Society and Culture (FQRSC, G. Therriault, 2008–2010). Some of the results presented here have been adapted from the paper: Bader, B., Jeziorski, A. & G. Therriault (2013). Conception des sciences et d'un agir responsable des élèves face aux changements climatiques. In Jean Simonneaux et Bernard Calmettes (Eds.). *Les Dossiers des Sciences de l'Éducation*, 29. Theme issues on *'Les sciences et les crises contemporaines'*, 15–32.

[4] These students are in the (IEP) international education program, whose official name is the program of junior high (MYP), and the responsibility of the International Baccalaureate Organization (http://www.ibo.org). This program is based on three main principles: intercultural awareness, global education and communication.

- The involvement of developing countries in the Kyoto Protocol
- The causes and consequences of desertification
- The decline in salmon populations: climate change or overfishing?
- Biofuels as a solution to climate change?
- The Australian animals in danger: caused by tourism or climate change?

They then had to describe how researchers work to address their issue, to specify some existing knowledge and identify uncertainties that remain, whether this was relevant. They had to refer to at least two science disciplines and discuss their proposal with invited guests that had some special expertise on climate change or with engaged citizens. Here, one of the two guests was a master degree student in physical Geography. The other one had a master degree in Physic. Both of them were studying climate change. Students finally answered the two initial questions based on their research in a text and presented their findings to their colleagues.

At the end of the process, we conducted 12 interviews with volunteers. These interviews (each with ~~two or~~ three students) allowed us to question thirty-six 16 year-old students regarding their understanding of science and climate change, and also regarding their ways to engage themselves as citizens dealing with the issue of climate change. As just said, each team had to document how scientists worked on the topic they chose to deepen on Climate Change. A 'research practice' was defined here as the methods – usually empirical but not only – allowing researchers to collect data. This idea of 'research practice' was a problem for students at the beginning of the course. They did not understand what it was about and, accordingly, it was necessary for teachers to explain it. A set of pre-selected texts was provided to them and some teams searched on the internet for articles on how some research were conducted and some data were obtained. During the interviews, several students said they had been very interested in documenting this aspect of the subject and that they appreciated knowing more about how researchers could manage to produce new knowledge. Of course, this is only a first introduction to the work of data collection, but this first exploration of the work of researchers seemed of interest for many students and it has encouraged us to strengthen this aspect of the approach in the similar project that we are currently undertaking to enrich the current conception of sciences (Weinstein 2008).

Students also noted that this approach made them realize that scientists do not always agree when studying the same subject. This idea was new to them and we found in their comments the same discomfort often documented elsewhere when facing the possibility that two researchers studying the same subject don't come to the same conclusions (e.g., Bader 2003; Driver et al. 1996).

Types of Civic Engagement

One of the interview questions focused on the ways in which students consider the scope of their actions as citizens facing climate change, and knowing that some uncertainties remain. Table 23.1 shows the main categories related to the theme of civic engagement.

Table 23.1 Types of civic engagement

Categories	Number of Interviews	Number of statements
Making everyday actions for the environment	9	20
They don't know on what knowledge their actions are based	8	12
They wonder about the relevance of their actions	5	8
They don't expect to know everything before engaging themselves	7	8
Climate change requires further research	6	8
They consider not having a lot of power	5	6

Many students refer mainly to everyday actions such as recycling, the use of bicycle, bus, or composting. They also mentioned 12 times, and in eight of the 12 interviews, that they do not understand a part of the science underlying their actions but that, despite this lack of science knowledge, they act as stated above because these actions cannot, in their eyes, have major negative impacts on the environment. While some of these students question the relevance of their actions or consider not having a lot of power to protect the environment, others point out that we should not expect to know everything before acting. Students mentioned eight times in the 12 interviews, the need for scientists to produce further research in order to guide citizens towards the reduction of climate change. While mentioning the remaining uncertainties or that there are controversies, students still place trust in 'science.' Some even compare it to a 'religion' to which they adhere (Interview 03). Other students, while recognizing some limits of their scientific knowledge, do not want to deepen and sometimes base their actions on incorrect knowledge without knowing it (Interview 04). Others choose a sort of precautionary principle (e.g., "We must act now in spite of the uncertainties"; interviews 01 and 05). Meanwhile, for some, further research is necessary (which will undoubtedly lead to well-established knowledge to guide actions accordingly). To consider the closure of the debates and controversies among scientists through more research that leads inevitably to the truth is a well-known position, characteristic of the current understanding of science and a realistic conception of knowledge that students have used here (Bader 2003; Potter 1996).

Overall, our research suggests that this type of teaching interested students and enabled them to develop critical thinking and become more aware of different aspects of this complex issue – since most of them took into account research practices and controversial elements that were new to them in their answers to the two initial questions. But at the same time, they emphasized during the interviews that they act on climate change by focusing on everyday activities, without always being able to define the impact or the science underpinning them, believing that the sum of small actions will ultimately have an impact on the issue at hand.

It should be noted here that students have widened their knowledge on climate change issues during their history and citizenship education lessons. They learned measures taken by the Quebec provincial government and society on the issue of climate change. At the end of the history class, several teams recognized that the provincial government and Quebecois society took several actions and created new policies in order to deal with this issue, but students also underlined that it is still possible to do more. Students also agreed with the principle that Quebec provincial government had responsibility for addressing climate change. Additionally, the involvement of young people and citizens is considered important to them, both at school and within their communities.

Although their interest in the issue of climate change was low at the beginning of this course, this way of teaching science as being situated in a specific context, inviting students to specify more concretely how researchers work and illuminating areas of uncertainty and controversy, seems to have mobilized the majority of students interviewed. Addressing the issue by allowing each team to choose a theme to deepen and supporting them in their research seemed to be useful and constructive. Our results, even if they do not allow us to better understand views of the minority of students who were not really interested in this work, undoubtedly point to an interest in this type of educational approach. Considerations about the uncertainties that still surround some knowledge about climate change, about the controversies remaining, and about ways of doing science on climate change have been mobilized in the discourses collected. Note, however, that students incorporated these new considerations in their usual, empiricist and realist conceptions of science, while promoting science expertise as a legitimate source of knowledge on climate change.

Conclusion

The heart of our work is to try to ensure that the way we teach science interests and engages students (Bader et Sauvé 2011). It is also to assure a more reflexive and critical way of considering the consequences of the epistemological mainstream when dealing with environmental issues (Bader 2008). It should also allow students to better understand what it means to do research on an everyday basis, so that when the sustainable development issues and environmental polemic debates mobilize experts and citizens, these young people will have developed the possibility of considering themselves as stakeholders within a reflection about the development of scientific knowledge. Of course, the example reported here is only a first step in this direction. Nevertheless, improvements to their understanding of science while reflecting on ways to engage as citizens on the issue of climate change may ultimately lead to more reflexive and critical thinking about the meaning and scope of their actions.

Our work based on critical pedagogy, which we have briefly presented above, is yet to be completed. In our ongoing research, students completing high school will be invited to participate in an educational project on climate change, in science, history and citizenship education. Reflective and critical analyses of some conclusions of the IPCC report of 2007 will be included to illustrate how researchers are working on the issue of climate change. We would like to introduce some considerations about this 'instrumental rationality,' which seems to engender many problems. In that vein, we will try to present the issue of climate change more related to social justice and a more democratic way of sharing power. The ways students engage as citizens will be analyzed more explicitly in the context of the over-consuming society in which we live. Of course all of these propositions will be discussed with the science and history teachers.

This way, we hope to better take into account Paulo Freire's (1998) proposition as he insists on the importance of 'education for awareness' for the students so that they become 'subjects, rather than objects, of the world.' As Freire says, this is done by teaching students to think democratically and to continually question and make meaning from (critically view) everything they learn:

> Our relationship with the learners demands that we respect them and demands equally that we be aware of the concrete conditions of their world, the conditions that shape them. To try to know the reality that our students live is a task that the educational practice imposes on us: Without this, we have no access to the way they think, so only with great difficulty can we perceive what and how they know. (p. 58)

References

Alsop, S., & Bencze, J. L. (2010). Introduction to the special issue on activism: SMT education in the claws of the hegemon. *Canadian Journal of Science, Mathematics, and Technology Education, 10*(3), 177–196.

Apple, M. (2002). *Power, meaning and identity: Essays in critical educational studies*. New York: Peter Lang (Counterpoints).

Ardoin, N. M., Clark, C. R., & Kelsey, E. (2012). An exploration of future trends in environmental education research. *Environmental Education Research, 18*, 1–22.

Bader, B. (2003). Interprétation d'une controverse scientifique: stratégies argumentatives d'adolescentes et d'adolescents québécois. *Revue Canadienne de l'enseignement des sciences, des mathématiques et de la technologie, 3*(2), 231–250.

Bader, B. (2008). Education to climate change. In J. Golson, P. Geoffrey, & S. George (Eds.), *Encyclopedia of global warming and climate change* (pp. 354–357). Thousand Oaks: Sage. Collection Sage reference.

Bader, B. (2011). Éducation à l'environnement dans une société du risque: la conception des sciences privilégiée de l'éducation au développement durable. In B. Bader & L. Sauvé (Eds.), *Éducation, environnement et développement durable. Vers une écocitoyenneté critique* (pp. 223–250). Québec: Presses de l'Université Laval.

Bader, B., & Sauvé, L. (Eds.). (2011). *Éducation, environnement et développement durable. Vers une écocitoyenneté critique*. Québec: Presses de l'Université Laval.

Bader, B., Horman, J., & Lapointe, C. (2010). Fostering community and civic engagement in low income multicultural schools through transformative leadership. *Exceptional Education International, 20*(2), 25–37. Theme issues on "Educational leaders and student diversity: Leading the way to inclusion".

Bader, B., Therriault, G., & Morin, E. (2013). *Rapports aux savoirs scientifiques et formes d'engagement écocitoyen des jeunes. Communication sur invitation – 3ème séminaire du Centr'ERE – Premières fenêtres sur un paysage théorique: Identités, engagements, écocitoyenneté*. 15 Mars 2013, Université du Québec à Montréal. http://www.centrere.uqam.ca/evenements/115-seminaires-centr-ere.html. Accessed 2 Mar 2013.

Beck, U. (1992). *Risk society: Towards a new modernity*. London: Sage.
Beck, U. (2001). *La société du risque. Sur la voie d'une autre modernité*. Paris: Aubier.
Calabrese Barton, A., & Tan, E. (2010). "It changed our lives": Activism, science and greening the community. *Canadian Journal of Science, Mathematics, and Technology Education, 10*(3), 207–222.
Corcoran, P. B., & Osano, P. M. (Eds.). (2009). *Young people, education and sustainable development. Exploring principles, perspectives, and praxis*. Wageningen: The Center for environmental and sustainability Education of Florida Gulf Coast University/Wageningen Academic Publishers.
Driskell, D., & Chawla, L. (2009). Learning by doing: Education for sustainable development through place-based action research. In P. B. Corcoran & P. M. Ozano (Eds.), *Young people, education, and sustainable development. Exploring principles, perspectives and praxis*. Wageningen: The Center for environmental and sustainability Education of Florida Gulf Coast University/Wageningen Academic Publishers.
Driver, R., Leach, J., Millar, R., & Scott, P. (1996). *Young people's images of science*. Bristol: Open University Press.
Freire, P. (1998). *Teachers as cultural workers – Letters to those who dare teach* (D. Macedo, D. Koike, & A. Oliveira, Trans.). Boulder: Westview Press.
Freire, P. (2013). *Pédagogie de l'autonomie*. Paris: Érès.
George, S. (2010). *Whose crisis, whose future?: Towards a greener, fairer, richer world*. Cambridge: Polity.
Giroux, H. (1988). *Teachers as intellectuals: Toward a critical pedagogy of learning* (Critical studies in education series). Granby: Bergin & Garvey.
Giroux, H. (2000). *Stealing innocence. Youth, corporate power and the politics of culture*. New York: St. Martin's Press.
Giroux, H. (2003). *The abandoned generation. Democracy beyond the culture of fear*. New York: Palgrave Macmillan.
Greene, M. (1995). *Releasing the imagination: Essays on education, the arts, and social change*. San Francisco: Jossey-Bass Publisher.
Intergovernmental Panel on Climate Change (IPCC). (1996). Houghton J. T., Meira Filho L. G., Lim, B., Treanton, K., Mamaty, I., Bonduki, Y., Griggs, D. J., & Callander, B. A. (Eds.), *Revised 1996 IPCC guidelines for national greenhouse inventories*. Paris: IPCC/OECD/IEA.
IPCC. (2007). *Climate change 2007: Synthesis report. Contribution of Working Groups I, II and III to the fourth assessment report of the Intergovernmental Panel on Climate Change* (104pp). Geneva: IPCC [Intergovernmental Panel on Climate Change].
Kagawa, F., & Selby, D. (Eds.). (2010). *Education and climate change. Living and learning in interesting times*. New York: Routledge.
Kellner, D. (2001). Critical pedagogy, cultural studies, and radical democracy at the turn of the millennium: Reflections on the work of Henry Giroux. *Cultural Studies ↔ Critical Methodologies, 1*(2), 220–239.
Lash, S., Szerszynski, B., & Wynne, B. (Eds.). (1996). *Risk, environment and modernity: Towards a new ecology*. London: Sage/TCS.
Latour, B. (1999). *Politiques de la nature. Comment faire entrer les sciences en démocratie*. Paris: La Découverte.
Latour, B. (2001). *Le métier de chercheur, regard d'un anthropologue*. Paris: INRA Éditions.
North American Association for Environmental Education. (2011). *Developing a framework for assessing environmental literacy*. Washington, DC: The North American Association for Environmental Education (NAAEE). Available online: http://www.naaee.net/what-is-ee
Potter, J. (1996). *Representing reality: Discourse, rhetoric and social construction*. London: Sage.
Richard, V., & Bader, B. (2010). Re-presenting the social construction of science in light of the propositions of Bruno Latour: For a renewal of the school conception of science in secondary schools. *Science Education, 94*(4), 743–759.

Roth, M. (2003). Scientific literacy as an emergent feature of collective human praxis. *Journal of Curriculum Studies, 35*(1), 9–23.

Sauvé, L., & Orellana, I. (2008). Conjuguer rigueur, équité, créativité et amour: l'exigence de la criticité en éducation relative à l'environnement. *Dans Éducation relative à l'environnement. Regards, Recherches, Réflexions,* (7), 7–20. Available online http://www.revue-ere.uqam.ca/PDF/Volume7/01_Editorial.pdf

Sauvé, L., Berryman, T., & Brunelle, R. (2003). Environnement et développement: la culture de la filière ONU. *Éducation relative à l'environnement. Regards, Recherches, Réflexions,* (4), 33–55. Available online http://www.refere.uqam.ca/FR/publications_articles.php#2007

Szerszynski, B., Lash, S., & Wynne, B. (1996). Introduction: Ecology, realism and the social sciences. In S. Lash, B. Szerszynski, & B. Wynne (Eds.), *Risk, environment and modernity: Towards a new ecology* (pp. 1–26). London: Sage/TCS.

Weinstein, M. (2008). Finding science in the school body: Reflections on transgressing the boundaries of science education and the social studies of science. *Science Education, 92*(3), 389–403.

Wynne, B. (1987). *Risk management and hazardous waste: Implementation and the dialectics of credibility*. London: Sage.

Chapter 24
Utilizing Social Media to Increase Student-Led Activism on STSE Issues

Brandon Zoras and Larry Bencze

Abstract Through Brandon Zoras' graduate work at OISE on urban boys and science education, he was most interested in papers written on STSE (Science, Technology, Society, and Environment) and social justice in science. Social justice is not always first associated with courses like science, but often is addressed in the social sciences. Nevertheless, many social justice issues are rooted within fields of science; and, ensuring students have some scientific literacy in this regard is critical so they can navigate and understand complex social justice issues. The work of Angela Calabrese Barton and colleagues (e.g., Barton AC, Teaching science for social justice. Teachers College Press, New York, 2003; Barton AC, Tan E, Can J Sci Math Technol Educ 10(3):207–222, 2010), Larry Bencze (e.g., Bencze L, STEPWISE: Science and technology education promoting wellbeing for individuals, societies and environments. Accessed at http://stepwiser.ca, 2013; Bencze L, Bowen M, Alsop S, Sci Educ 90(3):400–419, 2006), Christopher Emdin (e.g., Reality pedagogy and urban science education: Toward a comprehensive understanding of the urban science classroom. In: Fraser B, Tobin K, McRobbie C (eds) Second international handbook of science education. Springer, New York, pp 67–80, 2010; Int J Qual Stud Educ 24(3):285–301, 2011), Wanja Gitari (e.g., Can J Sci Math Technol Educ 9(4):262–275, 2009), and Erminia Pedretti and colleagues (e.g., Sci Educ 17(8):941–960, 2008), had inspired him to get students looking at their own communities for social justice issues that involved science. Focusing on urban education and also teaching in urban schools within Toronto, issues of social

B. Zoras, M.Ed. (✉)
Toronto District School Board (TDSB), 5050 Yonge Street, Toronto, ON M2N 5N8, Canada
e-mail: brandon.zoras@tdsb.on.ca

L. Bencze
Ontario Institute for Studies in Education, University of Toronto, Bloor St W 252, Toronto, ON M5S 1V6, Canada
e-mail: larry.bencze@utoronto.ca

L. Bencze and S. Alsop (eds.), *Activist Science and Technology Education*, Cultural Studies of Science Education 9, DOI 10.1007/978-94-007-4360-1_24,

justice, equity, and socioeconomic status are important factors to discuss. Having a better understanding of science can lead to better careers, understanding of health diagnosis and power. Through the semester-long apprenticeship and exposure to various types of technology and social media, students reported being able to better understand the STSE issues as well as learning activism strategies that can be applied within their lives. From simple self-advocacy next time they are faced with an issue, to being able to start their own action on an issue, they felt prepared to research and take action.

Keywords STSE • Social media • STEPWISE • Activism • Urban science education • Educational technology • Social justice

Introduction

Through Brandon Zoras' graduate work at OISE on urban boys and science education, he was most interested in papers written on STSE (Science, Technology, Society, and Environment) and social justice in science. Social justice is not always first associated with courses like science, but often is addressed in the social sciences. Nevertheless, many social justice issues are rooted within fields of science; and, ensuring students have some scientific literacy in this regard is critical so they can navigate and understand complex social justice issues. The work of Angela Calabrese Barton and colleagues (e.g., Barton and Tan 2010; Barton 2003), Larry Bencze (e.g., Bencze et al. 2006, 2013), Christopher Emdin (e.g., 2010, 2011), Wanja Gitari (e.g., 2009), and Erminia Pedretti and colleagues (e.g., 2008), had inspired him to get students looking at their own communities for social justice issues that involved science. Focusing on urban education and also teaching in urban schools within Toronto, issues of social justice, equity, and socioeconomic status are important factors to discuss. Having a better understanding of science can lead to better careers, understanding of health diagnosis and power. Activist Majora Carter's TED Talk[1] on "Greening the Ghetto" (Carter 2006) discusses social justice, ecological and economical issues in her Bronx neighbourhood. The power in her voice and descriptions of how she represents her community engages and hooks others to take action. Similarly, Chris Emdin[2] (2013) speaks on urban science education:

> Urban pubic education must open you up to the realities of their experience, people are making decisions about them without them. Make them realize their asthma cause is in their neighbourhood. To let them realize there is pollution in the environment, to let them understand that climate change is happening, not just habit of memorizing information, because memorizing information and being comfortable in that space, means that you are not being innovative, you will never fight, and to fight you need first have to feel the pain,

[1] Majora Carter – TED Talk http://www.ted.com/talks/majora_carter_s_tale_of_urban_renewal.html

[2] Chris Emdin TED Talk http://tedxtalks.ted.com/video/Empowering-Children-through-Urb

> and we removed the pain from their experience in the interest of making them feel better, but if you're making them feel better but not giving them a thing to fight for and new opportunity and future than what's the purpose?

The Ontario science curriculum (Ministry of Education [MoE] 2008) asks teachers to explore, assess and analyze these STSE issues with students. Zoras felt that just knowing about an issue is not enough to make a difference or change what students can actually see. Letting students have choice in issues and getting them to become a researcher and activist is crucial for change. Within an urban setting, Rowhea Elmesky and Ken Tobin (2005) suggest having students take a researcher role that will disrupt the unequal power hierarchies within society. This would require a shift in curriculum and pedagogy. Derek Hodson (2010) proposes removing the ideological oriented interests of particular groups and then understanding causes of social disadvantage and environmental degradation that then leads to social action that can make change in society. Zoras indicated that it was the STEPWISE framework of Bencze (Bencze and Carter 2011) that took him to the next step of incorporating student-led research-informed actions on STSE issues in his teaching.

Within the last 3 years, Zoras has tried various action-based projects within the grade 9 and 10 sciences, as well as senior chemistry and biology. This past semester, with a grade 10 science class, he completed a full apprenticeship using the STEPWISE framework.[3] Students were highly engaged in all aspects of the lessons, activities and projects. This was largely due to the fact that they were involved in action research and had the choice on topic. Many of the students decided on local issues that impact their own community but soon realized the connectivity was also global. Traditional projects, where students hand in posters and brochures, usually result in students throwing them out shortly after they are graded. With the past round of projects, students have continued working on them after the course has finished. Their own activism continued through the use of social media like Twitter™ and Facebook™, to technology like smart phone apps they are trying to develop.

Teaching with Technology

Today's classrooms vary from complete digital wonderlands of interactive whiteboards, digital projectors, tablets, WIFI, laptops and netbooks to those with little more than a single classroom computer. Funding from the school board, grants, pilot projects and partnerships with corporations often limit the amount of technology available. Regardless of resources, being creative with what technology you have and what you do with it is most important. Siddika Guzey and Gillian Roehrig (2012) found "internalization of the technology use comes from reflection and that teachers' use of technology in classroom instruction is constructed jointly by their technology, pedagogy, and content knowledge; beliefs; identity; and the

[3] For STEPWISE framework see www.stepwiser.ca

resources that are available to them" (p. 162). Teachers over the years have to modify existing lessons to integrate technology or create new lessons that integrate technology. It is the technology, pedagogy, and content knowledge (TPACK) explained by Matthew Koehler and Punya Mishra (2009) that is needed to integrate technology meaningfully into the classroom. There are no rules mandating every teacher to use technology in the classrooms although school boards promote technology use and some schools are known as technology schools.

Having students 'bring your own device' (BYOD) is a common practice within classrooms. BYOD allows easy access to resources and services, increased productivity and enhanced communication (Violino 2012). Zoras encourages students to use their own devices, during appropriate times, to search for information, take photos and videos, document, record, create content and publish. The comfort level of the teacher with these technologies varies along the software, apps, programs and sites that are constantly evolving. Digital technology can be challenging for teachers, since each technology can have multiple uses, is rapidly changing and how the technology works is often hidden from the user (Mishra and Koehler 2007). Educators can take a deep breath as experts exist right within the student population and they are able to troubleshot and fix most issues. Students come in often knowing how to use technology, software and social media, but require guidance adapting them to the context of school. By modeling technology and social media, with which they are familiar in a STSE and science context, students were able to use these tools in their actions.

Expressing STSE Issues with Technology

With the most recent Ontario science curriculum (MoE 2008), greater emphasis has been placed on STSE issues. Placing the STSE learning goal first among three, ahead of 'content' knowledge (e.g., laws and theories), in the curriculum has brought about this shift. Previous provincial curriculum placed this learning domain in the third position. STSE issues might be taught towards the end of a unit, if the teacher got to that point. The curriculum now allows a vast amount of choice in terms of which STSE issues should be taught. You can easily incorporate local issues or brand new issues that are trending currently. It is also critical to realize that so many issues overlap across science and can fluctuate with a variety of factors. "We conceive STSE education as a vast ocean of ideas, principles, and practices that overlap and intermingle one into the other" (Pedretti and Nazir 2011, p. 603). Students come into the classroom with many ideas of the world around them. The pre-conceived notions about STSE needs to be validated within the classroom setting. The first step of the STEPWISE framework suggests to brainstorm and express their pre-instructional ideas and opinions on STSE issues. Zoras's students had many great ideas but some conflicted with new knowledge they were about to learn. By validating their ideas and shifting their current conceptions of STSE issues together, the learning was more meaningful. This is inline with constructivist theory of constant updating of ideas based on new interactions and experience. Expressing one's pre-conceived notions allows a person to consider them in light of new information. This was achieved in the classroom through the use of STSE cards. Students were given a STSE topic on a card and within groups brainstorm what they knew about it before they learned. For example, cards would contain issues of stem cells, medical imaging, laser weapons, vaccines, or types of power generation. They brainstormed this on the class set of netbooks using SMART Ideas™ concept mapping software. This allowed them to interact, manipulate and express opinions on the various topics. Students then revisited the concept map after the unit to modify their initial ideas and compared to their previous version. It is important to model technology use within the STSE issues when taking students through the STEPWISE apprenticeship. By expressing STSE issues with the use of technology students are more engaged and guided towards their own projects.

A great starting point to many high school STSE topics is from *The Story of Stuff* [SoS][4] project. After the project's initial success with The Story of Stuff, with over 15 million views, new videos have been developed, such as the Story of: *Electronics*, *Bottled Water*, *Cosmetics* and *Change*. Another project, 94 Elements,[5] features films on a particular element and an STSE issue associated with it. One of the videos shot in Hindi is "Copper: Acid and Dust," showing young men in India using nitric acid to extract copper from electronic circuit boards (Paterson 2011). When Zoras showed his students this video, they were shocked that these young men are 'farming' our old circuit boards instead of farming crops. Many students in his class can speak Hindi and thought that they may direct their action towards people in their own community. Both of these STSE issues were expressed through video and a website, and they utilized social media to share their message. A second example of teaching STSE issues is using computer simulations. Explore Learning's Gizmos[6] is an excellent use of technology to explore, manipulate and test STSE issues. Simulations range from greenhouse effect, climate change, ecology, and photosynthesis, where students can change variables and see results. Students can explore videos, simulations, animations and other STSE issues from a wide variety of platforms. The issue not only has to be relevant to youth but engaging. To bring about change, the STSE issues need to be presented with some form of activism, action or reason.

Modeling Activism by Viewing Research-Informed Actions

The Internet paired with modern technology has allowed mobilization of activism through many different means. During class, many sample STSE issues were examined with ease, due to the increasingly connectedness technology brings. It was important to not just present STSE issues but focus on those that had an action component. By nature, documentaries are actions based on research

[4] Story of Stuff (2008) – www.storyofstuff.org

[5] 94 Elements – www.94elements.com

[6] Explore Learning Gizmos – www.explorelearning.com

meant to expose a situation. A great documentary we examined was *Blood In The Mobile.*[7] Our dependence on mobile phones has fueled civil conflict, child labour, and environmental degradation due to mining of minerals required by the phone manufacturers. Frank Piasecki Poulsen explores the Congo and documents the conditions that are linked to the supply chain of major mobile phone companies. After watching the film, many students felt guilty and surprised that their own phone may contain traces of blood minerals. The point was not to guilt the students but to place them in the web as a consumer that fuels production. The Toronto Zoo has taken its own action on the same issue of blood minerals by recycling cell phones through their *Phone Apes Program*[8] to protect gorilla habitat in the Congo. Both of these programs are large actions in which students could take part, but they felt their own action wouldn't be so large-scale. Accordingly, actions that students could easily achieve were also important to show. A spoken word piece titled *Come Clean For Congo*, performed by Micah Bournes, clearly communicates about blood minerals and brings attention to the issue through a YouTube™ video.[9] This is an action that a student could create to address an STSE issue. The three previous examples utilize technology to share their message and elicit a response. The responses often match or exceed the technology used by branching out to response videos, various social media, blogs or websites.

Practicing Research-Informed Actions

Having been shown STSE issues and actions, students were eagerly waiting to jump off the starting block on their own actions. Since students have been exposed to and influenced by the actions of others, such as advertisements and opinions, they had no problem forming their own actions. Zoras found students are very good at locating secondary research; but, when required to do their own primary research, they needed more support. He wasn't surprised by this, since students are not often given the opportunity within school science to become primary researchers. Students were introduced to primary research through correlational studies. They were given a reaction time experiment they were to conduct within the class by dropping the reaction timer and seeing how long it took for another student to catch it. As a group, they decided on factors that may influence the reaction time and carried out their own correlational study. They were able to

[7] See http://bloodinthemobile.org for movie trailer, background and resources.

[8] More information on Toronto Zoo Phone Apes – www.torontozoo.com/conservation/PhoneApes.asp

[9] http://youtu.be/406TLCNksM8

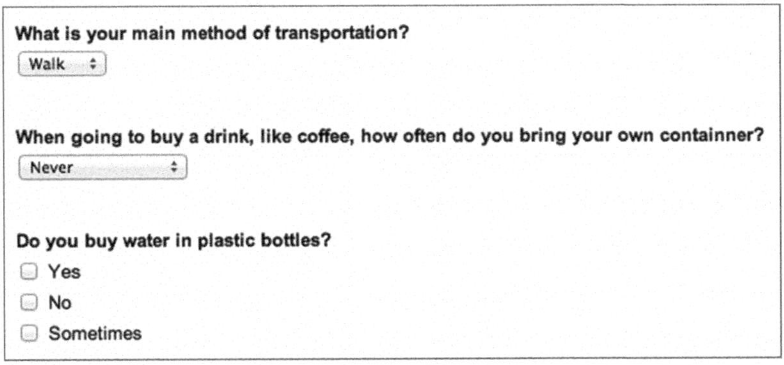

Fig. 24.1 Google Forms™ student survey

graph and form conclusions based on their own data. By preparing them to start their own correlational study based on their secondary research of the STSE issue, students were put into groups. They were able to collaborate within groups virtually using a wiki by Wikispaces.[10] This platform allowed a digital collaborative workspace for the project. They brainstormed forms of actions they could take, examined who were the stakeholders, which audience they wanted to target and by what means they wanted to convey their message. Students were also provided with other examples of correlational studies to help them choose their own variables. They created their own correlational studies by collecting data using Google Forms™, which is a tool within Google Drive™.[11] Students were able to create a survey digitally (see Fig. 24.1 above), which was then distributed over email, Twitter™ and Facebook™ for their peers to answer. Within Google Drive™ students were able to download the data (see Fig. 24.2 below) and also graph the results directly in the program. All their results were then published to the wiki for the class to see.

Having learned transferable skills, such as survey development, distributing surveys, collecting data, graphing, and then posting results, students were able to start on their own projects. Students began creating a class survey on climate change with modifications learned about their first round of surveying. They wanted a more valid survey than the first time, so wanted a larger sample size. Using social media, students shared their survey with friends and also went class to class with mobile netbooks to collect over 100 students worth of data.

[10] Wikispaces™ – www.wikispaces.com

[11] Google Drive™ – http://drive.google.com

Fig. 24.2 Google Spreadsheet™ student survey results

D	E	F
What is your main method of transportation?	When going to buy a drink, like coffee, how often do you bring your own containner?	Do you buy water in plastic bottles?
Walk	Most of the time	No
Walk	Never	Yes
Car	Never	Sometimes
Bus	Not very often	Sometimes
Walk	Never	Sometimes
Bus	Never	Yes

Digital Activism Through Student-Led Research-Informed Action Projects

Getting the class to this stage required building the foundation throughout the semester by guided practice. At this point, students have expressed opinions on STSE issues, seen examples of research informed actions, and guided through research informed actions. The class was now able to conduct their own student led research informed actions based on STSE issues. More and more, corporations are moving to the digital realm to reach consumers. Some companies have complete departments dedicated to social media. Everyday, average people can take to social media to review products, complain about service and get answers from these companies. As a result, businesses have devoted lots of resources to respond and also build their own digital identity. Activists have turned to social media to rally, provide information and get others to join the cause. A negative aspect of the ease created by using technology in activism is so-called ‘slacktivism.’ Some people may “like” a page, “share” a cause, or “tweet” support but these are surface measures that may not produce real change. Often, people feel they are supporting and helping, but no real action towards the cause is taking place. Students were cognizant of this when creating their own actions as they have sometimes self-identified as being ‘slacktivists.’

The students in Zoras’s class took to Twitter™, Facebook™, YouTube™, email, wikis and more to share their research informed actions. Many wanted answers from major companies, others were out to expose the truth and some wanted to promote greener lifestyles with their own products and services. The following are examples of actual student work grouped by type of technology used. They accessed these technologies from smart phones, tablets, computers from school, library and home.

Twitter™ – twitter.com

Twitter™ has the ability to send out a 140 character message (a tweet) to millions of users worldwide (Twitter 2013). Most companies have set up Twitter™ accounts with followers in the hundreds of thousands. A customer can send a quick message and get a response within seconds. Twitter™ also is a valuable tool to gather information on needs and opinions, as well as engaging communities (Hagman 2012). Students took to twitter to inform peers, seek out new followers with hashtags, and get answers from companies. Students created accounts that enabled them to connect to other users through their phones, tablets, and computers. They easily were able to find similar users, obtain followers and start trends using hashtags. Hashtags allow users to track messages, for example, #recycle allows users to tag a message with this hashtag and allow other users to find them (Twitter 2013).

One group's action was dedicated to getting people to use re-usable shopping bags. They would tweet out facts about the harmful effects of plastic bags, tips to remembering your re-usable bag, and getting companies to offer them (see Fig. 24.3 below). Another group strived to get customers to support companies that used recycled plastic in their bottles. They would tweet facts and also contact companies asking them to switch to a better recyclable bottle.

Students were successful in gathering a following of people, using hashtags to connect with other users and getting the attention of companies. They felt that by contacting companies and getting them to offer bags, change their policies and

Fig. 24.3 Sample student tweets

Fig. 24.4 Response from the Real Canadian Super Store™

make it easier for consumers, they would see change. They reached out to large companies like Coca Cola™, Loblaws™, and Walmart™, and got responses back on that very day (See Fig. 24.4). This instant gratification is what many of the youth are used to with technology today. The thought of them sending a letter was foreign to many of the students. They explained it doesn't make sense to type it out, print it, put a stamp on it, go to the mailbox and then wait 6–8 weeks for a response. They achieved the same results as a letter but nearly instantaneously.

YouTube™ – youtube.com

Posting a video and sharing it with a large audience has never been easier than with YouTube™. Users can easily produce, edit, record and post videos onto the Internet right from the YouTube™ app on their phone or from their computer. The videos can be posted and shared via various social media partners like Facebook™ and Twitter™ or hosted directly on YouTube™. Many companies are turning to YouTube™ to post videos about their products and services, as well as to pay for advertising on others' videos.

Students took to YouTube™ in a variety of ways. Some commented on the videos made by others eagerly waiting for a response, while others took control by creating their own videos. Students were able to upload videos straight from their phones, while others filmed and edited videos on a computer. The students used their primary research from the survey to come up with demographic information for their target audience and the message they wanted to send. Some examples of videos ranged from candid interviews asking students in the school if they knew the

Fig. 24.5 Recycle rap lyrics
[a]Timmies refers to a large chain of coffee shops called Tim Hortons™ that uses a wax liner in their cups which makes them unrecyclable in city recycling

'2 Cupz' – I Recycle (based on a song by 2 Chainz)

I can't recycle my Timmies[a] cup
Wax on wax off, get a discount when I bring my own cup
'2 Cupz', I be drinking my water
When I finish my water
I bring it to the recycle centre
Trying to make the world better

{Chorus}

I recycle, ya I recycle
I recycle, ya I recycle
I recycle, ya I recycle

meaning of GOOS (Good On One Side) paper to creating their own music videos. Others took to making commercials, while some made documentary style videos. One pair of students decided to use a beat from a popular hip-hop song but changed the lyrics (Fig. 24.5) to convey the importance of recycling in a music video. The students were given choice in this part but asked to think of the ways that would best reach the audience they were targeting. Having the students apply previous knowledge regarding videos or learning how to do this the first time is beneficial as it is a transferable skill they can use for other classes or work. Students were able to sign in to track views, demographics and share their videos. They could then reflect upon the extent to which their style of video was successful in reaching their intended audience.

Facebook™ – facebook.com

Facebook™ relies on networks of people making connections with each other. Advertisers use people in their own network to endorse products, hold online contests and share information about upcoming sales. Growing from a small network within colleges and universities, Facebook™ has extended publicly and has a large audience for advertisements. They were able to tap into the power of word-of-mouth, but on a digital platform. Students created 'Pages' within Facebook™, in which they posted facts, videos, and resources on their topic to followers. They were able to quickly build a network of friends and like-minded individuals on their topic. Some posted to their friends and others made new accounts, enabling them to reach out to new audiences. The page could easily be edited by the group and allow for posts and contributions from anyone with an account. This fluid page can be updated and tailored to the users who are following it. A group decided to use their page as a collection of recycling information. They took pictures around the school of recyclable items in the garbage, asking students to place them in the proper recycling bins. They posted recycling tips, videos, how to pack litter-less lunches, and facts. The group would also 'share' and 'like' similar content and have the same done for them. This would spread the content to the new users' group of friends and extend to friends of friends. The students really enjoyed this aspect, as nearly all students had a Facebook™ account and could rely on pre-developed networks of friends to support their cause and become involved. They were also able to extend beyond their own networks and make new connections. Many groups would share and support each other's projects by putting their social media actions on their page.

Wikispaces™ – wikispaces.com

Wikis are great tools to create, engage in collaboration and publish information to the Internet. Many wikis allow for integration of social media and embedding elements – such as videos – from other sites. Wikis are fluid, in the sense that content can be updated as often as required. Although this is good for keeping up-to-date information, it does need to be validated to be considered a credible source (Gooding 2008). Larusson and Alterman (2009) give several examples of the benefits of wikis:

> The plasticity of wikis is conducive to customizing, preformatting, or scaffolding the online interaction among the students. The malleability of wikis means teachers and students can further adapt the application after its initial deployment. The non-hierarchical control structure enables students to "own" and "control" their workspaces. (p. 398)

All these factors make wikis a great tool for teachers and students to allow collaboration. Wikispaces™[12] has great wikis for educational purposes that are free, easy to use and offer privacy controls. Educators also have the luxury of the ad-free version with no pop-ups or banners. Students can access wikis from school or home on a variety of devices. Wikispaces™ allows the teacher to track changes by students, reverse edits, control members and edit the privacy of the wiki.

The entire STEPWISE project for the semester was set up on Wikispaces™. This allowed for documentation of the process and collaboration with the students throughout the semester. Zoras was able to post videos, images, documents and embed other HTML items for students. They were then able to download, reply back and upload their own content to the wiki. They collaborated in groups and were able to learn the tools quickly. This allowed them for their final action to post the results and, in some cases, to use wikis as their action. Furthermore, Zoras was able to share, get support and feedback from Larry Bencze on activities and student work. This allowed Zoras to grow as an educator and collaborate online with a professor currently researching in this field.

[12] For more information on Wikispaces for educators see http://www.wikispaces.com/content/teacher

Student Perspectives

Students reported this apprenticeship program as fun, interesting and important to solving STSE issues. Many of the students appreciated the integration of social media into the projects. They felt that they were already comfortable with using social media personally and liked being able to adapt their skills for school use. A study by Yun-Ke Chang (2010) found "...the wiki was easy to use to some extent, and it helped to develop students' abilities in connecting new knowledge with their personal experience in online learning environment" (p. 300). They quickly began to recognize that advertisements have infiltrated their personal social media and sometimes without them even knowing it. Often by clicking "Like" on a certain product, it would appear to another friend as an endorsement for the product or service. By giving them a better understanding of how social media is being used by advertisers they were able to effectively use these tactics to spread their activist messages. Having students also create content instead of consuming was highly engaging. Students reported that they enjoyed researching and learning about the topics because it would require them to teach others about it. They were then able to choose how to best convey their message across a variety of platforms. Many of them chose social media, as they felt this would reach the greatest number of people and be most effective. For those students who were not personally into social media, they became familiar with a variety of sites and learned how to navigate those safely. It also allowed Zoras to grow as an educator, as students showed him new ways to use social media that he has not used in a classroom setting; examples including: Pinterest™ and Instagram™.[13] Many of them included their own social media into their own projects. One student used Instagram™, a social photo sharing site, to post photos of recycling. They felt that it was a site they used already and could apply it to the project. This is very important to teach transferable skills and also to encourage students to be resourceful. Students have already reported using social media in other class projects due to trying it first with this project. Ultimately, students will continue to use and practise these skills in post-secondary education and on into their careers. Many careers require keeping up-to-date with social media. Zoras's class also felt they were better able to understand advertising within social media, as they are often the target of commercial promotions. One student felt he couldn't escape the ads on social media, since he feels subjected to bombardment of ads on every site or device he uses. Students were able to see that all of these practices can be applied to science and felt they were making an actual change. All of the students walked away proud of their projects knowing they had impacted peers, companies and government in some way.

[13] Pinterest – http://pinterest.com and Instagram – http://instagram.com

Closing

Incorporating technology into student-led research-informed action leads to critical connections within STSE and social justice education. It allows students to be activists and shape their own learning by giving choice in the process and products. This is central to the idea of differentiated instruction. "Learning science means being able to engage in practices that are informed by thoughtful reflection on scientific ideas and principals and the ways these ideas and principals shape our understanding of the world" (Basu and Barton 2009). Having this activist approach is beneficial for all students taking the science course, whether or not they choose to continue in the sciences. With urgency, we must recognize that personal, social and environmental problems are embedded within science as well. It is suggested that these problems have become so serious we need students to become citizen activists and take personal and social action to combat these issues, such as, climate change (Bencze and Sperling 2012). There is the responsibility of educators to help youth make connections between equality, civil rights, and environmental justice and they can have an impact if they get involved both socially and ecologically (Alsop and Bencze 2010). Through the semester-long apprenticeship and exposure to various types of technology and social media, students reported being able to better understand the STSE issues as well as learning activism strategies that can be applied within their lives. From simple self-advocacy next time they are faced with an issue, to being able to start their own action on an issue, they felt prepared to research and take action. Brandon Zoras would like to thank his fall 2012 grade 10 science students for their hard work and great results, as well as support from Larry Bencze for meeting, sharing and modeling STEPWISE.

References

Alsop, S., & Bencze, J. L. (2010). Introduction to the special issue on activism: SMT education in the claws of the Hegemon. *Canadian Journal of Science, Mathematics and Technology Education, 10*(3), 177–196.

Barton, A. C. (2003). *Teaching science for social justice*. New York: Teachers College Press.

Barton, A. C., & Tan, E. (2010). "It changed our lives": Activism, science, and greening the community. *Canadian Journal of Science, Mathematics, and Technology Education, 10*(3), 207–222.

Basu, S. J., & Barton, A. C. (2009). Critical physics agency: Further unraveling the intersections of subject matter knowledge, learning, and taking action. *Cultural Studies of Science Education, 4*(2), 387–392.

Bencze, L. (2013). *STEPWISE: Science and technology education promoting wellbeing for individuals, societies and environments*. Accessed at http://stepwiser.ca

Bencze, L., & Carter, L. (2011). Globalizing students acting for the common good. *Journal of Research in Science Teaching, 48*(6), 648–669.

Bencze, J. L., & Sperling, E. R. (2012). Student-teachers as advocates for student-led research-informed socioscientific activism. *Canadian Journal of Science, Mathematics & Technology Education, 12*(1), 62–85.

Bencze, L., Bowen, M., & Alsop, S. (2006). Teachers' tendencies to promote student-led science projects: Associations with their views about science. *Science Education, 90*(3), 400–419.
Carter, M., (2006). *Majora Carter: Greening the ghetto* [Video file]. Accessed at http://www.ted.com/talks/majora_carter_s_tale_of_urban_renewal.html
Chang, Y. K., Morales-Arroyo, M. A., Than, H., Tun, Z., & Wang, Z. (2010). Collaborative learning in wikis. *Education for Information, 28*(2), 291–303.
Elmesky, R., & Tobin, K. (2005). Expanding our understandings of urban science education by expanding the roles of students as researchers. *Journal of Research in Science Teaching, 42*(7), 807–828.
Emdin, C. (2010). Reality pedagogy and urban science education: Toward a comprehensive understanding of the urban science classroom. In B. Fraser, K. Tobin, & C. McRobbie (Eds.), *Second international handbook of science education* (pp. 67–80). New York: Springer.
Emdin, C. (2011). Citizenship and social justice in urban science education. *International Journal of Qualitative Studies in Education, 24*(3), 285–301.
Emdin, C. (2013). *Empowering children through urban education: Christopher Emdin at TEDxColumbiaSIPA* [Video file]. Accessed at http://tedxtalks.ted.com/video/Empowering-Children-through-Urb
Gitari, W. (2009). Exploring the understanding and common use of scientific and everyday knowledge and students' agency in a Kenyan High School. *Canadian Journal of Science, Mathematics, and Technology Education, 9*(4), 262–275.
Gooding, J. (2008). Web 2.0: A vehicle for transforming education. *International Journal of Information and Communication Technology Education (IJICTE), 4*(2), 44–53.
Guzey, S. S., & Roehrig, G. H. (2012). Integrating educational technology into the secondary science teaching. *Contemporary Issues in Technology and Teacher Education, 12*(2), 162–183.
Hagman, J. (2012). Joining the twitter conversation. *Public Services Quarterly, 8*(1), 78–85.
Hodson, D. (2010). Science education as a call to action. *Canadian Journal of Science, Mathematics, and Technology Education, 10*(3), 197–206.
Koehler, M. J., & Mishra, P. (2009). What is technological pedagogical content knowledge? *Contemporary Issues in Technology and Teacher Education, 9*(1), 60–70.
Larusson, J. A., & Alterman, R. (2009). Wikis to support the "collaborative" part of collaborative learning. *International Journal of Computer-Supported Collaborative Learning, 4*(4), 371–402.
Mishra, P., & Koehler, M. J. (2007). Technological pedagogical content knowledge (TPCK): Confronting the wicked problems of teaching with technology. *Society for Information Technology & Teacher Education International Conference, 2007*(1), 2214–2226.
Ontario Ministry of Education. (2008). *The Ontario curriculum grades 9 and 10: Science.* Accessed at http://www.edu.gov.on.ca/eng/curriculum/secondary/2009science11_12.pdf
Paterson, M. (2011). *Copper – Acid and Dust.* Accessed at http://www.94elements.com/elements/29/
Pedretti, E., & Nazir, J. (2011). Currents in STSE education: Mapping a complex field, 40 years on. *Science Education, 95*(4), 601–626.
Pedretti, E. G., Bencze, L., Hewitt, J., Romkey, L., & Jivraj, A. (2008). Promoting issues-based STSE perspectives in science teacher education: Problems of identity and ideology. *Science & Education, 17*(8), 941–960.
Story of Stuff. (2008). *The Story of Stuff project.* Accessed at http://storyofstuff.org/about/about-the-project/
Twitter. (2013). *Help centre: Twitter basics.* Accessed at https://support.twitter.com/groups/31-twitter-basics
Violino, B. (2012). Education in your hand. *Community College Journal, 83*(1), 38–41.

Chapter 25
Developing an 'Activist Mentality' in an Environmental Science Course

Erica N. Blatt

Abstract This chapter presents a framework for understanding the goals of the Environmental Science course, including promoting an activist mentality, through Kempton and Holland's (Identity and sustained environmental practice. In: Clayton S, Opotow S (eds) Identity and the natural environment: the psychological significance of nature. MIT Press, Cambridge, MA, pp. 317–341, 2003) stages of environmental identity development. Subsequently, the concept of the 'zone of proximal identity development' (Polman JL, Revista de Educación, 353:129–155, 2010) is introduced as a useful theoretical notion in conceptualizing how teachers can encourage activism within the Environmental Science course. Excerpts from teacher and student interviews from a recent ethnographic study in an Environmental Science high school course are shared in the discussion of the goals for the course and in the presentation of several activities which encourage student activism. The categories of environmental action described as 'civic action' and 'cultural reform' are used to further analyze the various types of activities that may be enacted in an Environmental Science course in the teaching of environmental issues. Additionally, this chapter considers how we can employ the objectives of the social reconstructivist movement of the early twentieth century as a guide for promoting activism in our schools, specifically to counter the current narrative of learning and schooling advocated by the 'education reform' movement. Finally, several teaching strategies that can be used in an Environmental Science course (or on a school-wide level) are recommended in order to extend the goals of the course beyond environmental awareness to a level where scaffolding for environmental activism becomes a central component.

Keywords Controversial issues • Curriculum • Environmental activism • Environmental identity • Environmental science • Role of schooling

---

E.N. Blatt (✉)
School of Education, College of Staten Island, City University of New York,
2800 Victory Boulevard, Building 3S-208, New York, NY 10314, USA
e-mail: Erica.Blatt@csi.cuny.edu

L. Bencze and S. Alsop (eds.), *Activist Science and Technology Education*,
Cultural Studies of Science Education 9, DOI 10.1007/978-94-007-4360-1_25,

> You get the knowledge, but in order for the knowledge to really have legs, you have to have a mentality that believes that there's something that can actually be done. You have to-, it's that activist mentality, and by that I don't mean that the person has to necessarily become an environmental activist, but I think that a person can believe in the activist ability that they have as a single person: "I can change the stuff that I do enough to suit the parameters that I'm going to define for myself as environmentally responsible," and so somebody begins to act in a way that they feel like they have changed their own little paradigm, their way of being in the world has shifted, even if it's only slightly (Mrs. P,[1] Environmental Science teacher, 7/01/09)

Introduction

Given that Environmental Science high school courses have grown in prominence in recent years (Edelson 2007), this chapter aims to further the conversation regarding the goals for this course, including the development of what Mrs. P calls an 'activist mentality.' Willett Kempton and Dorothy Holland's (2003) framework for the development of environmental identity among environmental activists is presented, and the relationship between this framework and the goals for the Environmental Science course is explored in depth. Subsequently, the concept of the 'zone of proximal identity development' (Polman 2010) is introduced as a useful theoretical notion in understanding how teachers can encourage activism within the Environmental Science course. In addition, this chapter considers how the current narrative of learning and schooling is at odds with the promotion of activism amongst our students, and how we can utilize the goals of the social reconstructivist movement of the early twentieth century as a guide for promoting activism in our schools. The discussion incorporates examples from a recent ethnographic study in an Environmental Science high school course that utilized a sociocultural approach to investigate the teacher's goals for the Environmental Science course, how these were enacted through various activities, and how the activities affected the environmental identity and environmentally-related behaviors of the students in the class. Finally, the chapter outlines several teaching strategies that can be utilized in an Environmental Science course in order to extend its goals beyond environmental awareness to a level where scaffolding for environmental activism becomes a central component of the course.

[1] Mrs. P (pseudonym) is an Environmental Science teacher who participated in a recent ethnographic study of students' environmental identity and behavioral change as they participated in an Environmental Science course (see Blatt 2010).

The Stages of Environmental Identity Development

The term 'ecological identity' came to the fore as it was used by Mitchell Thomashow in his book *Ecological Identity: Becoming a Reflective Environmentalist* (1995), in which he states that "ecological identity refers to all the different ways people construe themselves in relationship to the earth as manifested in personality, values, actions, and sense of self" (p. 3). Building upon this description, and incorporating several fields, including deep ecology, environmental ethics, ecopsychology, environmental education research, and history of science, Erica Blatt (2012a) outlines various relational views of the environment, including viewing oneself as a part of nature; as damaging to nature; as superior to nature; as separate from but connected with nature; as a protector of nature, etc. In accordance with the understanding that individuals possess multiple situational identities that shift over time (Stryker 2004), it is likely that our environmental (or ecological) identity is subject to change as a result of cultural influences and our experiences with the environment.

Various aspects of environmental identity are explored in Clayton and Opotow's (2003) *Identity and the Natural Environment: The Psychological Significance of Nature*, highlighting several recent research studies involving environmental activists or professionals. For example, Kempton and Holland (2003) investigated the process of environmental identity development by conducting 'identity interviews' with 159 members of representative types of environmental organizations, where they investigated each participant's history of involvement with the environmental movement. Utilizing the information gained in these interviews, Kempton and Holland constructed a framework for the development of environmental identity.

According to this framework, there are three interrelated aspects of environmental identity formation (Kempton and Holland 2003). As individuals move through these aspects of development, Kempton and Holland refer to these identity processes as 'reformulations.' The first stage of development involves a new awareness of environmental issues, whereby an understanding of environmental threats becomes more salient. Environmental activists describe this stage of increased *salience* as becoming 'aware' or 'waking up' as a result of direct experience with local environmental destruction or a connection with a larger environmental issue. The second type of reformulation occurs as an individual gains a sense of *empowerment*, or a belief that one's actions, either individual or as a member of a group, can have an impact. Kempton and Holland (2003) found that this sense of empowerment is often acquired by taking on a role where action becomes part of one's environmental involvement. The third type of reformulation occurs as one further engages in environmental practices and becomes active in the environmental movement. At this stage of *activism*, knowledge about how to be an effective activist may be gained through mentorship and connections with other active members of the environmental community.

Activism: Individual vs. Collective Action

In Kempton and Holland's description of the activism stage of environmental identity development, they describe an important component as the mentorship of more experienced others with common values and imply involvement with a network of others working for a similar cause. Interestingly, in the opening quote, Mrs. P suggests "that a person can believe in the activist ability that they have as a single person," which raises the question as to whether activism can represent either an individual or a collective pursuit. While Mrs. P focuses on the small acts that individual students can take, a sociocultural perspective considers "all individuals as dialectically interconnected with the collective, and each presupposes the other (i.e., individual/collective). This implies that no individual can act independently of the collective and individual actions become material resources that structure collective agency" (Tobin 2007, p. 7). Dewey (1916) suggested a similar framework for considering this relationship:

> A being connected with other beings cannot perform his own activities without taking the activities of others into account. For they are the indispensable conditions of the realization of his tendencies. When he moves he stirs them and reciprocally. We might as well try to imagine a business man doing business, buying and selling, all by himself, as to conceive it possible to define the activities of an individual in terms of his isolated actions. (Dewey 1916, p. 14)

Therefore, in viewing all individual action as dialectically connected with the collective, it seems to be a false dichotomy to view environmental action as *either* individual or as part of a collective, as all actions are carried out by individuals acting within a social network.

A more useful frame for viewing various types of environmental action is suggested by Kempton and Holland (2003), as they describe the categories of 'civic action' and 'cultural reform.' The first of these includes actions aimed at reforming government policies, institutional practices, or corporate behavior, including membership in environmental groups, petitioning local government, or grassroots organizing. Cultural reform includes actions that are directly attempting to counter consumer culture, including carrying out environmentally-friendly practices in an attempt to be role models for others, and direct efforts to encourage others to reform their own practices. In the discussion below, these categories of environmental action are used to discuss the various types of activities that may be enacted in an Environmental Science course.

Goals for the Environmental Science Course

Following the opening quotation made by Mrs. P, she goes on to state:

> I see my goal as simply raising awareness of the student and empowering them to understand that they can make a difference, and I think that if I can make them believe that they can make a difference, even if it's a tiny difference, if they can make a difference, then that will inspire an activist mentality and leave them open to learning more in the future about what they can do. (Interview 3, 7/01/09)

While Mrs. P states that she sees the goals of the Environmental Science course as going beyond increasing students' environmental awareness, this may not be the case for all Environmental Science teachers, as there are many different approaches to teaching the Environmental Science course for a variety of reasons. First, there is a lack of established standards at the national or state level for the Environmental Science course. Environmental Science teachers may consult the *National Science Education Standards*, which includes several environmentally-related content standards for grades 9–12, including the topics of "population growth, natural resources, environmental quality, natural and human-induced hazards, and science and technology in local, national, and global challenges" (National Research Council 1996). Missing from these standards, however, is guidance on the role of high school science courses in teaching action or advocacy skills. Similarly, the course description for the College Board's Advanced Placement Environmental Science course is focused on teaching students about environmental systems and issues (The College Board 2010), rather than acting or caring for the environment.

In contrast, the North American Association for Environmental Education's *Excellence in Environmental Education: Guidelines for Learning (Pre-K-12)* (revised 2004) contain very specific suggestions for what should be taught at the 9–12th grade level, divided into the categories of:

Strand 1: Questioning, Analysis, and Interpretation Skills
Strand 2: Knowledge of Environmental Processes and Systems
Strand 3: Skills for Understanding and Addressing Environmental Issues
Strand 4: Personal and Civic Responsibility

In these guidelines, only Strand 2 is focused on environmental knowledge and awareness, while Strands 1, 3, and 4 emphasize development of both analysis and action skills, as well as a sense of both personal and collective responsibility for helping the environment. Many Environmental Science teachers are not aware of the *NAAEE Guidelines for Learning*, however, and they are therefore underutilized in the development of course curricula. The differences between the *National Science Education Standards* and the *NAAEE Guidelines for Learning* in including guidelines that go beyond teaching environmental knowledge into the categories of responsibility and action provide an indication of the diversity in approaches that exist at the classroom level.

In addition to a lack of established standards for the course, teachers also have differing levels of comfort with teaching both controversial issues, and what might be perceived as advocacy skills. Controversial issues have been defined by Stradling (1985) as "those issues on which our society is clearly divided and significant groups within society advocate conflicting explanations or solutions based on alternative values" (p. 9). Within the Environmental Science course, many issues can be considered controversial, including energy issues, climate change, genetic engineering, industrial farming, land use and development, resource depletion (logging, fishing, mining), species extinction/loss of biodiversity, etc. Unique pedagogical issues arise when attempting to teach students about controversial issues because the issues often involve differing interpretations of 'scientific facts' depending on one's values (Oulton et al. 2004b). Therefore, if controversial issues are included in the Environmental Science curriculum, then the teacher must contend with how best to present these issues which involve conflicting values (environmental, consumer-materialist, etc.). Since it is also probable that the teacher has strong opinions on the issues, this presents further challenges in considering best practices in presenting issues of environmental consequence to the students.

Concerns raised during focus groups of primary and secondary teachers in England who were teaching controversial issues included the need to 'stick to the facts,' respect the values of students, teachers, schools, and the community, and also present a balanced view of the issues (Oulton et al. 2004a). These teachers' concerns indicate that in teaching controversial issues teachers need to take into account several factors beyond what is necessary in the teaching of 'non-controversial' scientific knowledge, which teachers may or may not feel adequately prepared to do.

Research indicates that teachers may also be concerned about the 'proper' role of a teacher, and may feel stymied in including environmental action projects in the curriculum as a result. For example, Barrett (2006) documents the beliefs of an outdoor education teacher who is concerned that he will be perceived as a 'social engineer' if he includes projects that are interpreted as promoting environmental advocacy. Stevenson (2007) suggests that these difficulties arise as a result of the discrepancy between the traditional view of 'school knowledge as storage for future use' and the 'function of knowledge in environmental education [as] immediate use for the social value of a sustainable and emancipated quality of life' (p. 147). This incongruity between the traditional role of schooling and an advocacy-based approach to Environmental Science courses makes it difficult for teachers to decide upon appropriate educational goals for the course.

The Role of Schools in Promoting Activism: The Social Reconstructivist Model

In the age of accountability and testing, schools have become absorbed in the narrative that learning is that which can be demonstrated on a standardized exam (Ravitch 2010). A brief look back at the education reform movements of the 1930s indicates that a similar narrative was being pushed during that period in the name of 'efficiency.' A counter movement, called social reconstuctivism, led by George S. Counts arose in opposition to the standardization of education at that time. Notably, Counts' statements are remarkably relevant to the state of education today, as he argues that "the feverish and uncritical fashioning of tests in terms of the existing curriculum and in the name of efficiency has undoubtedly served to fasten upon the schools an archaic program of instruction and a false theory of the nature of learning" (cited in Kliebard 2004, p. 158). Replacing the word 'efficiency' with 'accountability' transforms this statement to a twenty-first century context. Dewey (1928) also recognized that the emphasis on achievement standards, data collection, and precise measurement at that time was "ignoring the social impact of education" (cited in Kliebard 2004, p. 161). The participants in the social reconstructivist movement realized that the prevailing narrative of schooling was established to perpetuate the status quo and preserve existing social conditions, and in response, called for schools "to address ongoing social and economic problems by raising up a new generation critically attuned to the defects of the social system and prepared to do something about it" (Kliebard 2004, p. 157). Similarly, in today's schools, teaching to create an 'activist mentality' requires a very different narrative regarding our view of the role of schooling and the purpose of education.

In order to shift the culture of schools away from a view of learning as 'value added' on standardized tests, it may be necessary to adopt a similar narrative to that of the social reconstructivists as a guiding vision. In this alternative narrative, learning is viewed as student understanding of the current issues in society, as well as preparation for 'doing something about it.' In the next section, we will explore some examples of what a modern-day social reconstructivist model of education might look like in the Environmental Science classroom.

What Does a Social Reconstructivist Model Look Like in Practice?

Blatt (2010) considers the impact of various environmental activities in an Environmental Science course on students' environmental identity and environmentally-related behaviors. During interviews, students express various reactions to these activities, which are informative in considering what a "social reconstructivist" Environmental Science curriculum would look like. In this section, several activities and student reactions are presented, followed by a discussion about whether these activities promote activism (or fall short of doing so). In the final section, further suggestions are provided for activities that may be utilized in the Environmental Science course (and other science courses as well) to encourage student activism.

The setting of this study, Mrs. P's class, was an elective Environmental Science course for 9th through 12th graders in a large suburban high school in New Hampshire. Many students reported enrolling in the course because they viewed it as an easy alternative to chemistry, which several had previously failed. There were 17 students in the class, all of whom participated at varying levels of involvement in the Blatt (2010) ethnographic study. Ten students and the teacher were interviewed three times during the semester, and excerpts from four of these interviews are shared below. The main activities in the course included an on-line ecological footprint, inventory of everything the students own, videos of several controversial environmental issues (including population issues, logging, industrial farming, and the Exxon Valdez™ oil spill), a mock town hall meeting of a local wetlands development issue, a class debate on oil drilling in the Arctic National Wildlife Refuge, a field trip to the local landfill, weekly collection efforts in the school-wide recycling program, and tree coring outdoors on the school grounds.

In addition to these activities, the teacher often engaged the students in a Socratic dialogue type of discussion regarding the various issues in an attempt to help them both connect emotionally with the issues and think critically about them. In the section below, interview segments are presented in which students or the teacher are discussing their reactions to various activities in the course. These excerpts have been chosen because they demonstrate varying levels of environmental awareness, empowerment, and activism (Kempton and Holland's stages of environmental identity development), and include responses to activities in both the categories of 'cultural reform' and 'civic action.' For further analysis of students' reactions to the activities in the course, see Blatt (2012a).

Student Reactions to Environmental Activities

The Ecological Footprint

In Mrs. P's class, students participated in a series of activities that were intended to lead to reflection of their consumptive values, including an ecological footprint activity, writing an inventory of everything they owned, and visiting the local landfill. The ecological footprint was a survey about consumptive behaviors that students filled out on-line, which resulted in a 'measure' of how many Earths would be needed if all the humans on the planet lived at the consumptive level of the student. A student named Kat (pseudonym), when asked during an interview about her reaction to the ecological footprint activity, states (E is the researcher):

> K: Yes, well, I was surprised because Americans take up at least 7 Earths, and then I was sad because I'm one person, and for everybody to live like me, you would need like 3.15 Earths.
> E: Right, yeah, it's pretty shocking, huh?
> K: Yeah, I think I started to cry in class.
> E: Oh, so does that make you want you to change anything you're doing, or what does that make you think about?
> K: Economically I really can't change anything because of the pressure.
> E: Yeah, so does that make you feel, sort of like frustrated, or-?
> K: Yeah, because I'm stuck between a rock and a hard place. I can't change, but, you know, I want to. (Interview 1, 2/12/09)

Kat's reaction indicates increased environmental awareness regarding the impact of her behaviors on the environment. While Kat's initial response to this activity is frustration at not perceiving what she is able to do to change her behavior, by the end of the semester Kat indicates that she has found several ways to do so on a small scale, including turning off the water when brushing her teeth, showering less, etc. (Interview 3, 6/10/09). This particular activity seems to have 'awakened' Kat, who already had a strong environmental background, leading her to look for ways she could adapt her behavior further.

School-Wide Recycling Program

During the Environmental Science course, students participated weekly in the school-wide recycling program, as this class was responsible for emptying the recycling bins throughout the school, which they did during class time. One day, Mrs. P suggested that they remove all the paper from the bins that had only been used on one side, and keep it in a scrap paper pile in the class. Shortly after this exercise, a student named Allan stated during an interview:

> I think that we need to use our resources less, and we need to use them more wisely. There are ways that we can cut out using a lot of stuff, like using recycled paper–that day that we pulled out all the paper [from the recycling bins] that had only been used on one side, and then using it. That's just wasteful, there are simple ways that you can cut down on the amount of resources that we use, and it doesn't take all that much. (Interview 2, 4/15/09)

Towards the end of the semester, another student, Greg explains how his recycling behavior has changed as a result of experiences in the course:

> G: I would say, recycling and stuff, I throw my paper definitely in the recycling bin, or opposed to I would always just go to the garbage or something. [Now] I make the extra walk, like Mr. G., all his recyclables, all his stuff is in the back of the room, when the garbage is literally three feet from my hand.
> E: Yeah, but you make the effort.
> G: But I do, I definitely do, like I don't just throw it away, I can recycle it. I've changed a lot that way, cause in the beginning of class, I was like, 'Oh yeah, recycling [with a negative, sarcastic tone]' but it really makes a difference, just a little one, but at least I'm going for it. I might be destructive here, so I'll try to help myself here to kind of counterbalance. (Interview 3, 6/3/09)

From these student responses, it is evident that they have gained a sense of empowerment that through their own actions they can make a difference, even if 'just a little one.' Awareness of the impact of their consumptive habits in connection with the experience of participation in the recycling program has led these students to take on new environmental behaviors.

Mock Town Hall Meeting

A third activity in the Environmental Science course was a mock town hall meeting where students were asked to take on various roles as stakeholders in a local wetlands development issue. The issue involved development of an affordable housing complex on a wetlands area commonly used by outdoor enthusiasts and hunters. As several students in the class lived in affordable housing and many spent time in the outdoors, this was an issue to which many students could relate. Students were assigned to groups of two to three students; the assigned roles included the developers, the town manager, social workers, birding enthusiasts, recreational users of the land, an engineer, etc. After 2 days of researching and preparing their positions, a mock town hall meeting was held where each of the stakeholders shared their position on the issue with the class. Mrs. P's comments regarding this activity are informative:

> I think it is real important the way that the town meeting went. Kids got an opportunity to experience sense of place. Their town was under an onslaught from a developer who was going to take away something that was possibly important in their town for a variety of different reasons. If they cared more about animals, animals were going to lose their habitat, or whatever it was, and through citizenship you can develop a sense of place, and this year, because of that exercise they had an opportunity to dabble in citizenship, and for somebody like Scott, he blossomed with that role that he had, and it changed the way the kids looked at him, and it changed his behavior in class as a result...I mean, now he's becoming this student, but I think that he understood how much power you can have...and there were a variety of kids that understood that they have the ability to control their world to a greater or lesser degree, and so even though it's not sense of place in the fru-fru sort of touchy-feely kind of way, there's something real about it for them that I think is equally important. (Interview 3, 6/5/09)

As Mrs. P indicates, the mock town hall meeting resulted not only in students developing an understanding of a local development issue, but it also gave students an opportunity to experience what it is like to 'dabble in citizenship' and have one's voice heard in this type of setting.

Promoting Activism?

In our discussion of the role of activism in the Environmental Science course, it is important to consider whether these activities are in fact promoting activism. Through each of these activities, the teacher is guiding students in the process of gaining environmental awareness, the first stage of environmental identity development. She is encouraging empowerment (stage two of environmental identity development) by providing students with the opportunity to participate in collective action for the environment, such as the school-wide recycling program. Finally, students are able to practice citizenship skills during the mock town hall meeting, as they gain an awareness of how local environmental issues may be addressed.

In the framework of Kempton and Holland's (2003) categories of action, the first two activities are promoting 'cultural reform' types of action, as defined by Kempton and Holland (2003), where students are encouraged to reflect upon their behaviors and adopt more environmentally-responsible ones. The third activity, the mock town hall meeting, is an example of 'civic action,' as students are learning about the process of becoming involved with an environmental issue at the local governmental level. As students engage in these activities during the course, it could be argued that they are not necessarily becoming environmental activists (as that process requires extended experience), but rather they are being provided with opportunities to develop an 'activist mentality' – they are being encouraged to reflect upon their behaviors and to question the status quo. They are also realizing that there are actions that can be taken to improve situations that they deem in need of improvement.

In discussing the relation of these activities to the goals for the Environmental Science course, a useful construct to consider is that of the 'zone of proximal identity development' (ZPID). Adapted from Vygotsky's (1978) 'zone of proximal development,' the term ZPID has been used by Polman (2010) to describe:

> the distance between the actual identity developmental level as determined by an individuals' past positionings and the level of potential identity development as determined through mutual negotiation of positioning and stance during actions associated with an identity, under adult guidance or in collaboration with peers. (p. 134)

Polman suggests that through support and scaffolding, individuals can be guided on a journey "along possible identity development pathways that the individual is capable of recognizing and willing to explore," and goes on to state that "not all these pathways will be taken a long distance but they are related enough to that individual's past understanding and identification to be imaginable and explorable" (p. 134). As we relate this concept to students in the Environmental Science course, a possible goal for the course then becomes helping students move further along in the development of their environmental identity – for some, this may mean strengthening the salience of their environmental identity through new awareness of environmental issues or impacts of their behavior; for others, this may mean helping them to realize that there are behaviors they can change, supporting their movement to a stage of empowerment through their own actions. Finally, we can provide students with opportunities to 'dabble in citizenship,' connect with others involved in environmental issues, and become role models for others through their own behavior. Since students will enter the class at all different levels of environmental identity development, the teacher's role can then be seen as providing a variety of activities designed to stretch students beyond their "already-achieved states" (Polman 2010, p. 134), encouraging their exploration of new levels of identity development.

What Else Can Teachers Do to Promote Activism Within the Environmental Science Course?

As we attempt to develop an 'activist mentality' in our students and scaffold their environmental identity development, there are many other useful teaching strategies that can help with this endeavor. Here are a few suggestions:

1. Work with students to develop a community-based project. This type of project will enhance students' understanding of local issues, while empowering them to realize that there are actions that can be taken to address the issues. An example is discussed in Janice Koch's *Science Stories* (2005) where a fifth-grade teacher decides to guide her students in a project to turn an abandoned lot that students pass on their way to school into a school/community garden. The students engage in writing letters to local officials to have the garbage and debris removed from the site, and gain permission to utilize the lot for the creation of

a garden. The students are then guided by the teacher in developing a plan for the garden and cultivating their own plants and vegetables.

2. Create an after-school environmental group for students. This type of after-school program can be guided by student interest, but the development of practical action-based projects should be encouraged to engage the students in meaningful activities that will promote sustainability. In a recent study of pre-service teachers' youth experiences in the outdoors (Blatt 2012b), one pre-service teacher writes about her experiences in this type of program:

> As a child in school, I remember going outside during Science class to learn about the nature that surrounded me: the animals, the hills, the trees, everything. I began to become more and more interested and even joined *an after school group* that my teacher had put together for those who were interested. We visited a nearby park, Blue Heron, and walked through the trees, examining and identifying everything that we saw. The amount that I learned could have only been surpassed by the fun that I had with my fellow classmates, as I knew I was doing my part to understand nature and even help to better it as we cleaned up the nearby area. The experiences that I had stand to be some of my greatest childhood memories and I will never forget my time at that park. (Nancy, Fall 2010)

   Notably, this student's experiences at a local park seemed to develop her feelings of stewardship, as the students participated in cleaning up the local environment. Through both classroom activities and after-school programs that bring students into the outdoors, teachers can help connect students' youth experiences in the outdoors with an understanding of the need to protect such areas through conservation and sustainability practices.

3. Develop school-wide projects, such as an energy audit, recycling initiatives, or fundraisers for sustainability projects (such as solar panels or a school garden). While Mrs. P had her students participate in the school-wide recycling program by collecting the recycling bins, this project fell short of reaching beyond the students in the class. The ideal school-wide project will involve multiple stakeholders in the school, so that current efforts can be both evaluated and improved upon. For example, after the activity where students collected all the pieces of paper that were only used on one side, they could have been encouraged to create a school-wide challenge to both increase the amount of paper recycled and, more specifically, the amount of paper on which both sides have been used. Whenever possible, students should be actively involved in the creation and implementation of these projects, so that they gain an understanding of the process, including the challenges and obstacles, to carrying out this type of project.
4. Encourage positive experiences in the outdoors. Research has indicated that environmental activists often refer to meaningful experiences in the outdoors when questioned about influences that led to their environmental activism. For example, Kals et al. (1999) investigated the relationship between the construct 'emotional affinity towards nature' and nature-protective behavior with a questionnaire study of 200 participants from the general population (of Germany) and 81 environmental activists. Results showed that emotional affinity towards nature is a powerful predictor of nature-protective behavior, and found that 39 % of emotional affinity toward nature traces back to present and past experiences in

natural environments. The authors suggest that environmental education programs should promote time spent in nature as a means of encouraging an emotional connection with the natural environment. Similarly, in the article, Caring for the Environment: Challenges from Notions of Caring (2007), Martin points out that many young people feel separate and distant from the natural world. Consequently, he suggests that at best these children can be taught to *care about* Nature, unless environmental education puts a high priority on direct personal contact so that students can learn to *care for* their local environment. According to Martin, this is best accomplished through activities where children can experience responsiveness to and from other components of Nature – whether this involves tree planting, working with animals, etc – developing a truly caring relation, where they begin to see themselves as part of the natural world rather than disconnected from it. Additionally, during science classes, students can be taken outside for fieldwork to build their connection with the local environment. Activities, such as water quality studies of local streams or tree studies calculating the CO_2 sequestration of local trees (Blatt 2013), can be incorporated into Environmental Science, Biology, or Chemistry coursework. Building this connection with the environment is a key step in guiding our students on a path towards becoming environmental stewards.

Conclusion

This chapter has presented a framework for understanding the goals of the Environmental Science course through both Kempton and Holland's (2003) stages of environmental identity development and Polman's (2010) notion of the 'zone of proximal identity development.' If we adopt a social reconstructivist model of education and take the third stage of environmental identity development – environmental activism – to be a long-range target for students in the Environmental Science course, and perhaps for all of our students in general, we can use the idea of the ZPID to understand how to move students along the path of environmental identity development from whatever "already-achieved state" they are at. We ought to consider the important role that the Environmental Science course can play in the effort to develop socially aware students that are willing to act for the environment, but also recognize the support that teachers may need in guiding students along this pathway.

References

Barrett, M. J. (2006). Education for the environment: Action competence, becoming, and story. *Environmental Education Research, 12*(3–4), 503–511.

Blatt, E. N. (2010). *An ethnographic investigation of the process of change in students' environmental identity and pro-environmental behavior in an Environmental Science course*. Doctoral dissertation. ProQuest Dissertations & Theses: Full Text (Publication No. AAT 3430780).

Blatt, E. N. (2012a). Exploring environmental identity and behavioral change in an Environmental Science course. *Cultural Studies of Science Education*. doi:10.1007/s11422-012-9459-2.

Blatt, E. N. (2012b). *Pre-service elementary teachers' outdoor experiences: How do these translate into beliefs about taking students outdoors?* Annual conference of the National Association for Research in Science Teaching, Indianapolis, IN.

Blatt, E. (2013). Local tree mapping: A collaborative, place-based activity integrating science, technology, math, and geography. *Science Activities: Classroom Projects and Curriculum Ideas, 50*(3), 99–109. doi:10.1080/00368121.2013.808165.

Clayton, S., & Opotow, S. (Eds.). (2003). *Identity and the natural environment: The psychological significance of nature*. Cambridge, MA: MIT Press.

Dewey, J. (1916). *Democracy and education: An introduction to the philosophy of education*. New York: Macmillan.

Edelson, D. (2007). Environmental science for all? Considering environmental science for inclusion in the high school core curriculum. *Science Educator, 16*(1), 42–56.

Kals, E., Schumacher, D., & Montada, L. (1999). Emotional affinity towards nature as a motivational basis to protect nature. *Environment and Behavior, 31*(2), 178–202.

Kempton, W., & Holland, D. C. (2003). Identity and sustained environmental practice. In S. Clayton & S. Opotow (Eds.), *Identity and the natural environment: The psychological significance of nature* (pp. 317–341). Cambridge, MA: MIT Press.

Kliebard, H. M. (2004). *The struggle for the American curriculum: 1893–1958* (3rd ed.). New York: RoutledgeFalmer.

Koch, J. (2005). *Science stories: Science methods for elementary and middle school teachers* (3rd ed.). Boston: Houghton Mifflin Company.

Martin, P. (2007). Caring for the environment: Challenges from notions of caring. *Australian Journal of Environmental Education, 23*, 57–64.

National Research Council. (1996). *National science education standards*. Washington, DC: National Academy Press.

North American Association for Environmental Education. (2004). *Excellence in environmental education: Guidelines for learning (Pre-K-12)*. Available at http://www.naaee.org/npeee/learner_guidelines.php

Oulton, C., Day, V., Dillon, J., & Grace, M. (2004a). Controversial issues – Teachers' attitudes and practices in the context of citizenship education. *Oxford Review of Education, 30*(4), 489–507.

Oulton, C., Dillon, J., & Grace, M. (2004b). Reconceptualizing the teaching of controversial issues. *International Journal of Science Education, 26*(4), 411–423.

Polman, J. L. (2010). The zone of proximal identity development in apprenticeship learning. *Revista de Educación, 353*, 129–155.

Ravitch, D. (2010). *The death and life of the great American school system: How testing and choice are undermining education*. New York: Basic Books.

Stevenson, R. B. (2007). Schooling and environmental education: Contradictions in purpose and practice. *Environmental Education Research, 13*(2), 139–153.

Stradling, R. (1985). Controversial issues in the curriculum. *Bulletin of Environmental Education, 170*, 9–13.

Stryker, S. (2004). Integrating emotion into identity theory. *Advances in Group Processes, 21*, 1–23.

The College Board. (2010). *College Board Advanced Placement Environmental Science course description*. Available at http://apcentral.collegeboard.com

Thomashow, M. (1995). *Ecological identity: Becoming a reflective environmentalist*. Cambridge, MA: MIT Press.

Tobin, K. (2007). *Collaborating with students to produce success in science*. Second International Conference on Science and Mathematics Education, Penang, South Korea.

Vygotsky, L. S. (1978). *Mind in society: The development of higher psychological processes*. Cambridge, MA: Harvard University Press.

Chapter 26
Responsible Stewards of the Earth: Narratives of Youth Activism in High School (Science)

Ashley S. Kerckhoff and Giuliano Reis

Abstract The present chapter explores significant factors influencing the successful implementation and development of youth environmental activism in high schools from the perspective of a group of teachers and students in Ontario (Canada). According to participants, the emergence and perceived long-term positive influence of a particular youth environmental action initiative in their school—an environmental club founded by a science teacher—depend largely on the existence of environmentally-motivated teachers and students who are committed to developing a (social, intellectual and action-oriented) venue for activism, are open to mentorship activities, and who believe in the possibility of making a positive difference in their communities through collective enterprises. Our discussion offers insight into how (science) teachers and students can work together with (in) different disciplines to support environmental youth activism in schools.

Keywords Youth activism • School clubs • Responsible steward • Environmental education • Science education

Introduction

In 2000, The United Nations' Earth Charter (UNESCO 2000) acknowledged that the world's societies and ecological systems are greatly interconnected and that everybody has the responsibility to care for the earth. In 2002, the international community recognized that sustainable development can—and should—be taught

---

A.S. Kerckhoff (✉)
Rainbow District School Board, City of Greater Sudbury, ON, Canada
e-mail: ashleykerckhoff@gmail.com

G. Reis
Faculty of Education, University of Ottawa, 145 Jean-Jacques Lussier, Ottawa, ON K1N 6N5, Canada
e-mail: preis@ie.ul.pt

L. Bencze and S. Alsop (eds.), *Activist Science and Technology Education*, Cultural Studies of Science Education 9, DOI 10.1007/978-94-007-4360-1_26,

at the heart of the education system in order to properly address the human-caused damage to the earth's ecological systems[1] (UNESCO 2007). In addition, the United Nations (UN) declared the years 2005–2014 as the Decade of Education for Sustainable Development (DESD). The aim of this resolution is to assist citizens in developing "the attitudes, skills, and knowledge to make informed decisions for the benefit of themselves and others" (UNESCO 2009, paragraph 1). Therefore, through these documents and initiatives, it becomes evident that there is an urge for the incorporation of an education for sustainable development (ESD) into the heart of school curricula worldwide in an attempt to boost society's commitment to act responsibly towards the health of the planet.

In line with the UN, the Ontario Ministry of Education states that society needs to understand how social and environmental complex systems interact in order for us to live sustainably (OME 2009a, b, c). In addition, it claims that one's preparation for a role as a responsible steward of the earth occurs in a large part through environmental education (EE) in schools, where a historical connection with science continues to exist (Campbell and Robottom 2004). Here, the term *responsible steward of the earth* is used broadly and it refers to those people who participate in environmental action/activism at various levels—e.g., from families who participate in shoreline clean-ups on scheduled weekends to the field activists who risk their lives to denounce illegal wildlife killing. More precisely, it entails the decisions, planning, implementation and reflection carried out by an individual (or group) with the deliberate aim to achieve a specific positive environmental outcome (Emmons 1997).

Despite official sanctioning, the daily experiences of teachers and students throughout the province with the Ministry-mandated EE expectations and recommendations may vary as they struggle to interpret and put them into practice. In other words, this (relative) pedagogical flexibility that teachers enjoy to deliver the curriculum in their own way contributes to making teaching unique and successful, but could also pose an obstacle for novice teachers who may require additional directive guidelines to conduct their daily classroom businesses. Therefore, research-informed pedagogical strategies offer more grounds on which teachers can make informed decisions on how to liven current EE-oriented (science) curricula in schools.

The present chapter draws on the narratives of teachers and youth from a local public high school in Ontario (Canada) to investigate those factors that contributed to their involvement in local environmental activism. In doing so, we hope that participants' stories of learning about/for/in the environment can inspire others in becoming more environmentally active in their own communities and thus respond to current environmental issues through local, national, or global actions (Barrett 2006). Otherwise, (science) curriculum documents in themselves are unlikely sufficient to effectively place EE at the foundation of the current education system (Renton and Butcher 2010).

[1] The terms *environmental* and *ecological* are used interchangeably here.

Youth Environmental Activism: The Potential Role of School (Science)

The UN's Tbilisi Declaration states that the world needs to strive for environmental activism in all people (Chawla and Cushing 2007). This, in turn, requires that people accept responsibility—both individually and collectively—to participate in solutions to current environmental issues to the point where they become capable of "influencing the actions of others" (Lester et al. 2006, p. 316). In this context, the commitment to action is a form of environmental activism pedagogy intended to tackle the societal roots of our current ecological issues rather than produce schoolwork that is disconnected from society (Roth 2010). For instance, these actions can take the form of involvement with a community project, participation in town meetings and even (out-of-)school clubs (Lauglo and Øia 2008).

On the other hand, it remains unclear the exact mechanisms through which one's environmental concerns translate into action (Kim et al. 2013), specifically when it comes to youth engagement (Olofsson and Ohman 2006). Indeed, many factors seem to interact and mediate one's decision to become a responsible steward of the earth. As an example of this complexity, Blanchet-Cohen (2008) offers a framework that details the different stages that people must experience on their journey towards environmental activism: (i) connectedness; (ii) engaging with the environment; (iii) questioning; (iv) belief in capacity; (v) taking a stance; and (vi) strategic action.

The first stage—*connectedness*—is reached through the development of a positive emotional link with the natural environment. That is, unless one feels affection for the environment the mere acquisition of knowledge about environmental issues won't necessarily lead to environmental action (Heimlich and Ardoin 2008). This emotional connection can be achieved through direct exposure to nature, where people become aware of the existence and the beauty of the natural world to the point where they are able to explore and feel comfortable with their surroundings (Lugg 2007). The second stage—*engaging with the environment*—refers to a more or less "universal tendency" (Sobel 2008, p. 39) that people have to learn about how natural systems work and also cultivate a passion to protect them. In addition, it helps to provide a safe space for exploring increasing autonomy from other influences, like family. Unfortunately, this stage may be vanishing altogether as modern children seem to experience reduced amounts of free unstructured play during their childhood (Pyle 2008). As for the third stage—*questioning*—it implies that people begin to question certain accepted norms in society that favour environmental destruction. For instance, the over-consumption behaviour patterns mainly adopted by the developed world in face of the existing global unequal distribution of resources (Urry 2010) are now confronted with a food justice movement, in which teachers and youth play an important role (Gottlieb and Joshi 2010). The fourth stage—*belief in capacity*—alludes to the hope that people have for the future of the planet, which also lets them perceive their role as catalysts that can bring about positive change (Farmer et al. 2007). Finally, the last two stages—*taking stance and strategic action*—require people to have the necessary skills to take action in/for/about the environment. While living through all these stages is not always cut-and-clear, they provide insight into some of the intricacies of those influential factors existing throughout various aspects of life and that increase the probability of people (including youth) to develop into responsible stewards of the earth.

Additionally, there are two other elements that seem indispensible for shaping the foundation of environmental activists: influential experiences and people (Gruenewald 2003). These two forces work in conjunction as well as independently throughout the lives of children and youth. For example, children may have independent influential experiences in (quasi-)natural settings, such as playing in the forest behind their house or the park nearby their apartment building; or there may be a person in their lives that is passionate about the environment and influences their attitudes (e.g., close relative). Alternatively, this influential person may have taken them places where they shared influential experiences. Notably, this individual can be a teacher as teachers are known to influence students in their EE learning (Duvall and Zint 2007). It has also been suggested that for teachers to adequately encourage the achievement of the goals of EE in their students they must hold a positive environmental attitude as well as be environmentally literate and active (Ernst 2007). Therefore, if the opportunity to develop a *connection with* and *understanding of* nature is lacking in home life, schools seem to hold the potential to expose students to other similar and valuable EE learning experiences (Barrett and Sutter 2006; OME 2007). In addition, although environmental concern among youth in the USA, for example, has been decreasing since the early 1990s (Wray-Lake et al. 2010), those who continue to participate in environmental action often do so out of their own *understanding of* and *authentic concern for* the well-being of the planet (Schusler et al. 2009).

In the school context, activism requires students to be involved at a personal level or else the disconnect between knowledge and action tends to widen (Reis and Roth 2010). Amongst the possible school-based initiatives that exist, environmental clubs remain a popular choice. On one hand, their practices can be greatly limited to basic activities—like recycling or water saving campaigns—for fear of confrontation arising from more controversial topics, such as targeting forestry practices of a local company in a logging dependent community (Dyment 2008). These types of school-based environmental initiatives tend also to foster a "resourcist attitude toward nature" (Howard 2008, p. 309), where any actions taken are for human benefit instead of in the interest of the health of the environment. On the other hand, we believe that environmental clubs have the potential to serve as a venue for supporting the attainment of the goals of EE in schools by the youth—and the study summarized and reported in the present chapter supports this claim.

Data: Collection and Analysis

In order to learn about what factors mediated participants' decision to become responsible stewards of the earth, we used narrative inquiry. Within this framework, participants' narratives are the context for making meaning of their school situations, which are temporarily continuous and socially interactive (Connelly and Clandinin 1990). Moreover, the temporality aspect of narratives also suggests that the stories told are consisted of significance (past), value (present), and intention (future) (Carr 1986).

During the individual semi-structured and open-ended interviews, eight students (from grades 11 and 12) and two teachers—Rachel Smith and Dawn Baker—in a town located in South-Central Ontario were invited to look retrospectively at some influential factors shaping their environmental activism through the school environmental club (or EcoAction).[2] Their stories, which lasted between 25 and 45 min, were audio-recorded and transcribed verbatim for posterior analysis. In the process, a number of themes emerged across participants' narratives through subsequent reads of the transcripts. These themes intertwine and therefore are not completely distinct from one another. Therefore, the differentiations that we make here are artificial and meant to clarify certain aspects of the narratives collected rather than to indicate any independent constitutions. The themes we found include: opportunities for networking and connecting with nature as well as challenges that made participants' actions difficult, like the initial lack of EE initiatives in school. On that note, and due to necessary space limitations imposed to this chapter, we decided to focus on those themes that related to the emergence and potential long-term positive influence of this particular youth environmental action initiative on participants. Our choice also reflects a growing interest on these topics as alternative ways of measuring the successful implementation of similar programs in schools worldwide (Reis and Iosif-Guimaraes 2012).

Participants: A Glimpse

High school teacher Dawn Baker is the developer and supervisor of EcoAction at the school in North Gate. According to her, the environment plays an important part in her life. Dawn grew up in the area where she now teaches, which helped her gain an appreciation of the local environment when she lived away. When she returned home to teach a few years ago, the need for an environmental club at the school became evident when students began asking her for this type of initiative. During her interview, Dawn also shared that the club is one of the reasons that she loves to go to work every day as she believes that her work with the club members has lasting impressions that go beyond the four walls of the classroom.

Rachel Smith and Dawn Baker began working at the high school on the same day. Prior to that, Rachel worked for several years in residential outdoor education

[2] Pseudonyms are used throughout to maintain confidentiality of participants.

Table 26.1 Pseudonyms, grade and gender of students interviewed along with a short descriptor of their identities associated with their club membership

Pseudonym	Grade	Gender	Descriptor
Greg	11	M	Impacted by the environment as a child
Heather	12	F	Motivated to maintain sport environments
Isabelle	12	F	Realized career path when joined the club
Laura	12	F	Has a love for summer camping
Liam	12	M	Driven by facts
Madison	12	F	Sees activism as a choice, not obligation
Samantha	12	F	Learns in the environment
Sarah	12	F	Is an environmental activism leader

programs, where youth would stay for an entire week to participate in various environmental activities. At some point in her teaching career Rachel decided that she wanted to work with students for a longer period of time, which led her to take her current job at the high school. The local natural environment has always been an important part of her life so much so that she found in EcoAction an opportunity to maintain her connection with environmental education and youth activism at the school while working in special education. She is currently a co-facilitator in the program.

There were eight students interviewed for this study who were members of EcoAction. They were mostly female students (75 %) and were mainly in their last year in High School (87.5 %). Their pseudonyms, grade, and gender are presented in the table above. In addition, we have used a short descriptor for each one to offer our readers a fragment of their club identities as they came through to us during the analysis of the interviews (Table 26.1).

EcoAction: Emergence and Potential Lasting Positive Impact

EcoAction is a student-driven action group that started at North Gate Secondary School about 7 years ago by Dawn Baker (science teacher and main teacher responsible for running the club) and Rachel Smith (special education teacher). It is now a youth action group of about 50 students that fluctuates from semester to semester and from beginning to end of year. The club is responsible for recycling and all the eco-initiatives and awareness campaigns that go on in the school. For instance, according to their website, they have introduced the importance of re-usable mugs in the school (which was reported by students as having been very successful in reducing the use of disposable coffee cups in the school), planted trees in the community, and hosted an Environmental Film Festival and an eco conference to hundreds of students. As part of their annual planning, they pick or target a few small goals to be accomplished during each school year. Consequently,

EcoAction and their members fit well the definition of responsible stewards of the earth that we adopt here. This was confirmed by one of the grade 12 students, Isabelle:

> I guess kind of 'cause we're trying to preserve the environment, and protect it so, and we're kind of charged ourselves with that responsibility. And we do take on like a lot of responsibility and I like, personally I'm a kind of person that feels responsible for like the actions of people and the effects that we have on the environment. I feel like it's our like my responsibility to protect the environment and where we live, so yeah [we are stewards of the earth].

EcoAction was born out of Dawn's initial aspiration to promote a venue for youth environmental activism in the school. As Rachel put it:

> I think you know clubs like this, you know they're small but important, I think um, you know the kids need an opportunity to be able to be activists and to be a voice on behalf of the environment.

Moreover, participation in the club has social networking benefits to members. According to Dawn:

> [EcoAction] gives [students] a place to belong and it's a way to make connections with other like-minded people. Uh, so I think that's a really important piece that they do take away from it you know it's like a home for them. A safe place for them to be and hang out.

The students' narratives also corroborated the perspectives offered by the two teachers above. For example: "So I think that's, it's been a really positive experience, and it's also definitely opened the doors to other experiences that I can have" (Sarah) and "we don't just hang out in [EcoAction], we do other stuff together too, so um, and we're kind of like we all like even when we're doing other stuff we're kind of environmentally conscious" (Isabelle). More so: students also expressed the ambition to positively influence other people's actions, similarly to what Dawn and Rachel did for them: "I think that it makes a huge difference that we are educating younger children who will then educate their children. It's just start a ripple effect" (Laura). This, in turn, points to the fact that the actions initiated and carried by Dawn and Rachel in the school through their involvement with the club did not go unnoticed by the students. That is, the presence and commitment of these two teachers are recognized as influential in the lives of students.

Even thought there was mention that the school was full of supportive people, including the administration and teachers with whom they have had courses, both Dawn and Rachel were specifically viewed as mentors and resourceful people regarding environmental action. For example, Heather demonstrated this perspective: "Obviously like the teachers in [EcoAction] are likely some of the most environmentally motivated people that I know in terms of teachers and mentors." In addition, both Dawn and Rachel acknowledged having the desire to see the long-term impact of the club on students' school experience. For instance, Dawn mentioned:

> But I'd love to see, like their goal, their dream. I'd love to see a solar panel on our roof. I'd love to see you know alternative energy being pumped through this building instead. And a deeper connection between the classes that are all using something together. So, you know, those are good goals. And the kids came up with those, and they seem to come up with those same goals year after year after year, regardless of who's in the club. Because the passion and vision is, you know, we need to do something bigger.

Once again, Dawn's vision for the future of the program and its long-term influence on students' lives was also shared by some of the students, like Sarah:

> Something that I would really love to do is eventually be able to implement strategies for communities abroad, um that are both like sustainable economically but also environmentally, so I'd really like to see environmentalism play a huge role in how that works in my career. Um, personally, I would really like to keep striving towards being a local organic food consumer, and I would like I said, really love to have my own garden so I can be self-sustaining.

In sum, participants' interviews point for the existence of two main reasons that motivated and sustained participants' involvement in environmental youth activism at North Gate High School: the existence of a *venue for activism* (i.e., EcoAction) and the presence of *teacher mentors* who, in this case, had different academic backgrounds (i.e., Dawn in science and Rachel in special education). Therefore, both teachers and students spoke to their mutually importance to the success of the program: the teachers were determined to provide a venue for students who demonstrated a need and interest for such an initiative. That is, teachers were influential for students inasmuch as students were influential for teachers. In addition, both groups were keen on influencing other people: for teachers, their involvement in activism came out of their desire to impact students. As for students, they expressed a keen interest to influence younger people to whom they could serve as role models.

EcoAction: Some Implications for Youth (Science) Environmental Activism

Overall, the vision and goals of EE from the Ontario Ministry of Education's perspective is to encourage participation in environmental action, or to instil the idea of environmental activism into students' lives. These activists would display their environmental commitments through their actions to create a society that exists in harmony with earth's natural systems. However, what school-related experiences influence student and teacher's engagement with environmental activism? More so: how might the participants' stories of learning about/for/in the environment assist in the interpretation and implementation of EE-based curriculum practices in schools?

In brief, our study seems to indicate that there are three basic conditions that would need to be fulfilled in order for youth environmental activism programs to succeed in the context of schools (and which were not ranked in order of occurrence or significance by our participants): first, there needs to exist an opportunity for students and teachers to engage in environmental activism. Although there will exist difficulties in the initial implementation of initiatives of the like in schools, it is important for participants (i.e., teachers and students) to anticipate some of the benefits for the school community. These could range from a cleaner school to the sparking of students' interested in careers in environmental sciences. According to the participants' stories, the EcoAction club at North Gate Secondary School was the most relevant form of school-related experience that mediated their involvement with youth environmental activism. Therefore, as an organization, the environmental club is a—social, intellectual, and action-oriented—venue that provides a foundation for participants to develop their activism skills and work collectively towards local environmental causes of their choice.

Secondly, there needs to exist teachers—i.e., more than just one and not necessarily from science exclusively—who are committed to mentor students through their various activism projects along the year. These projects could take the form of action research projects, where both mentors and mentees would work together in a more democratic and respectful (harmonious) relationship. Although in our study the teachers' influence in shaping youth to become responsible stewards of the earth occurred mainly through the creation of EcoAction as a venue for activism, participants never reported an obligation to participate in any of the decisions taken in relation to any of the environmentally-oriented activities performed in the school. Indeed, those involved with the club seemed to have developed ownership of it from the very beginning stages, which is an EE goal not easily achieved (Kennelly et al. 2008). On that note, some teachers might be discouraged from attempts to involve students in environmental action due to their own actual or perceived lack of environmental knowledge. In this case, as Dawn Baker indicated, it is not the teacher's job to be the expert (in science or any other subject), but rather help students to become activists in a student-driven environment. The teachers are there to help students acquire the necessary skills and build the social networks required to carry out their actions. Put differently, teachers who are not entirely confident with their level of environmental knowledge could still play a role as a teacher-mentor by providing students with a venue for activism, where they can learn skills for social or environmental action as the result of their membership to those social networks that tend to develop naturally over time. Once people become involved in social networks related to environmental action, they can tap into the vast amount of information that is collectively distributed across the group.

Thirdly, participants need to want to be positively influential in the school. In other words, a strong appreciation for making a difference in the local community is a desirable trait in youth activists (of any age) (Barton and Tan 2010). For instance, Dawn and Rachel's involvement with environmental activism initially took place because of their desire to have a positive impact on students' environmental actions. Consequently, students now seem to have picked up on that discourse so that they too talk about teaching a younger generation about what they have learned and are doing as responsible stewards of the earth. Altogether, these three conditions not only can contribute to changing harmful practices to—either the natural or social aspects of—the environment (Boyes and Stanisstreet 2011), but they can also help to boost and reinforce members' self-image (identity) as catalysts for change (Carlsson and Sanders 2008).

The journey to becoming an environmental activist who is environmentally literate takes time—and this fact can be the same for teachers as it is for students.

Each person comes to experiences with different narratives that impact how they work through the experiential learning cycle and build their knowledge. Becoming involved in a social network and a venue for activism can have beneficial effects for developing responsible stewardship in both students and teachers. Also, while it is important for all teachers to be environmentally inspired (and inspiring), there needs to be a supportive administration to facilitate the survival of these types of initiatives. Otherwise, they might die when its champions leave (Reis and Iosif-Guimaraes 2012). The inclusion of parents, the use of media technology and funding for fieldtrips to provincial parks and outdoor education centres are equally critical elements to complement regular offered hands-on learning opportunities in the local natural environment. Finally, both pre- and in-service teachers need to receive professional development on the integration of an EE-based pedagogy through the curriculum (and beyond science-based courses). It is also possible that the program reported here was successful mostly because it already had highly motivated participants involved with it since the beginning. That is, it could be that participants had attained all six stages suggested by Blanchet-Cohen (2008) for becoming a responsible steward of the earth before EcoAction was even conceived. Whatever the case might be, the three conditions detailed above continue to provide important criteria for indicating the level of success achieved/projected for youth activism initiatives implemented and developed in schools.

We recognize that the factors discussed here might be more complex than they appeared to us, but that should not prevent future research to explore how they relate more specifically to one another and could apply to diverse educational settings. Ultimately, this chapter was meant to be a sharing of ideas in hopes that teachers and student become inspired to take the first steps on their journey to becoming responsible stewards of the earth.

References

Barrett, M. (2006). Education for the environment: Action competence, becoming, and story. *Environmental Education Research, 12*(3–4), 503–511.

Barrett, M., & Sutter, G. (2006). A youth forum on sustainability meets the human factor: Challenging cultural narratives in schools and museums. *Canadian Journal of Science, Mathematics, and Technology Education, 6*(1), 9–23.

Barton, A., & Tan, E. (2010). "It changed our lives": Activism, science, and greening the community. *Canadian Journal of Science, Mathematics, and Technology Education, 10*(3), 207–222.

Blanchet-Cohen, N. (2008). Taking a stance: Child agency across the dimensions of early adolescents' environmental involvement. *Environmental Education Research, 14*(3), 257–272.

Boyes, E., & Stanisstreet, M. (2011). Environmental education for behaviour change: Which actions should be targeted? *International Journal of Science Education, 11*, 1–24.

Campbell, C., & Robottom, I. (2004). Environmental education: Appropriate vehicle for teaching science? *Teaching Science, 50*(2), 18–23.

Carlsson, M., & Sanders, D. (2008). School councils as an arena for pupils' participation in collaborative environmental education projects. In A. Reid et al. (Eds.), *Participation and learning: Perspectives on education and the environment, health and sustainability* (pp. 321–337). New York: Springer.

Carr, D. (1986). *Time, narrative, and history*. Bloomington: Indiana University Press.

Chawla, L., & Cushing, D. (2007). Education for strategic environmental behavior. *Environmental Education Research, 13*(4), 437–452.

Connelly, M., & Clandinin, J. (1990). Stories of experience and narrative inquiry. *Educational Researcher, 19*(5), 2–14.

Duvall, J., & Zint, M. (2007). A review of research on the effectiveness of environmental education in promoting intergenerational learning. *Journal of Environmental Education, 38*(4), 14–24.

Dyment, J. (2008). Student participation in school ground greening initiatives in Canada: Reflections on research design decisions and key findings. In A. Reid et al. (Eds.), *Participation and learning: Perspectives on education and the environment, health and sustainability* (pp. 241–255). New York: Springer.

Emmons, K. (1997). Perspectives on environmental action: Reflection and revision through practical experience. *The Journal of Environmental Education, 29*(1), 34–44.

Ernst, J. (2007). Factors associated with K-12 teachers' use of environment-based education. *Journal of Environmental Education, 38*(3), 15–32.

Farmer, J., Knapp, D., & Benton, G. M. (2007). An elementary school environmental education field trip: Long-term effects on ecological and environmental knowledge and attitude development. *Journal of Environmental Education, 38*(3), 33–42.

Gottlieb, R., & Joshi, A. (2010). *Food justice*. Boston: The MIT Press.

Gruenewald, D. (2003). The best of both worlds: A critical pedagogy of place. *Educational Researcher, 32*(2), 3–12.

Heimlich, J., & Ardoin, N. (2008). Understanding behavior to understand behavior change: A literature review. *Environmental Education Research, 14*(3), 215–237.

Howard, P. (2008). Ecology, phenomenology, and culture: Developing a language for sustainability. *Diaspora, Indigenous, and Minority Education, 2*(4), 302–310.

Kennelly, J., Taylor, N., & Jenkins, K. (2008). Listening to teachers: Teacher and student roles in the New South Wales sustainable schools programme. *Environmental Education Research, 14*(1), 53–64.

Kim, S., Jeong, S.-H., & Hwang, Y. (2013). Predictors of pro-environmental behaviors of American and Korean students: The application of the theory of reasoned action and protection motivation theory. *Science Communication, 35*(2), 168–188.

Lauglo, J., & Øia, T. (2008). Education and civic engagement among Norwegian youth. *Policy Futures in Education, 6*(2), 203–223.

Lester, B. T., Ma, L., Lee, O., & Lambert, J. (2006). Social activism in elementary science education: A science, technology, and society approach to teaching global warming. *International Journal of Science Education, 28*(4), 315–339.

Lugg, A. (2007). Developing sustainability-literate citizens through outdoor learning: Possibilities for outdoor education in higher education. *Journal of Adventure Education and Outdoor Learning, 7*(2), 97–112.

Olofsson, A., & Ohman, S. (2006). General beliefs and environmental concern: Transatlantic comparisons. *Environment and Behaviour, 38*(6), 768–790.

OME (Ontario Ministry of Education). (2007). *Shaping our schools, shaping our future: Environmental education in Ontario schools*. Toronto: Queen's Printer for Ontario.

OME (Ontario Ministry of Education). (2009a). *Acting today, shaping tomorrow: A policy framework for environmental education in Ontario schools*. Toronto: Queen's Printer for Ontario.

OME (Ontario Ministry of Education). (2009b). *Environmental education, Grades 1–8: Scope and sequence of expectations*. Toronto: Queen's Printer for Ontario. Retrieved from http://www.edu.gov.on.ca/eng/curriculum/elementary/environ18curr.pdf. Accessed 10 May 2010.

OME (Ontario Ministry of Education). (2009c). *Environmental education, Grades 9–12: Scope and sequence of expectations*. Toronto: Queen's Printer for Ontario. Retrieved from http://www.edu.gov.on.ca/eng/curriculum/secondary/environ9to12curr.pdf. Accessed 10 May 2010.

Pyle, R. (2008). No child left inside: Nature study as a radical act. In D. A. Gruenewald & G. Smith (Eds.), *Place-based education in the global age: Local diversity* (pp. 155–172). New York: Taylor & Francis.

Reis, G., & Iosif-Guimaraes, R. (2012). The death and life of a school-based environmental education and communication program in Brazil: Rethinking educational leadership and ecological learning. *Applied Environmental Education and Communication, 11*(3–4), 123–132.

Reis, G., & Roth, W.-M. (2010). A feeling for the environment: Emotion talk in/for the pedagogy of public environmental education. *Journal of Environmental Education, 41*(2), 71–87.

Renton, Z., & Butcher, J. (2010). Securing a sustainable future for children and young people. *Children & Society, 24*, 160–166.

Roth, W.-M. (2010). Activism: A category for theorizing learning. *Canadian Journal of Science, Mathematics, and Technology Education, 10*(3), 278–291.

Schusler, T., Krasny, M., Peters, S., & Decker, D. (2009). Developing citizens and communities through youth environmental action. *Environmental Education Research, 15*(1), 111–127.

Sobel, D. (2008). *Childhood and nature: Design principles for educators*. Portland: Stenhouse Publishers.

UNESCO (United Nations Educational, Scientific and Cultural Organization). (2000). *The Earth charter initiative: Promoting change for a sustainable future*. Retrieved from http://www.unesco.org/education/tlsf/TLSF/theme_a/mod02/uncom02t05s01.htm. Accessed 5 Oct 2009.

UNESCO (United Nations Educational, Scientific and Cultural Organization). (2007). *The UN decade of education for sustainable development (DESD 2005–2014): The first two years*. Retrieved from http://unesdoc.unesco.org/images/0015/001540/154093e.pdf. Accessed 5 Oct 2009.

UNESCO (United Nations Educational, Scientific and Cultural Organization). (2009). *Education: Learning our way to sustainability*. Retrieved from http://www.unesco.org/en/esd/. Accessed 15 Oct 2009.

Urry, J. (2010). Consuming the planet to excess. *Theory, Culture, & Society, 27*(2–3), 191–212.

Wray-Lake, L., Flanagan, C., & Osgood, D. (2010). Examining trends in adolescent environmental attitudes, beliefs, and behaviours across three decades. *Environment and Behaviour, 42*(1), 61–85.

Chapter 27
Climate Change and Citizen Science: Early Reflections on Long-Term Ecological Monitoring Projects in Southern Ontario

Ana Maria Martinez and Steve Alsop

Abstract Citizen science projects have become quite popular of late but still retain some controversy. Within a particular Canadian context of declining governmental environmental monitoring, this chapter explores the ideal of enhanced civic participation and experience gained through a long-term citizen science project. We offer this as one possible expression of activism. Drawing on specific tree planting and long-term monitoring programs established by the Association for Canadian Educational Resources (Climate's Sake) as an illustration of citizen science, we argue that these programs offer opportunities for those involved to increase their knowledge of local ecologies, share concerns and potentially contest local forest policies and management approaches linked with climate change and biodiversity conservation. We conclude by highlighting some associated tensions and contradictions.

Keywords Citizen science • Environmental monitoring • Democratising science • New social movements

Introduction

In a recent 2013 budget, the Canadian government announced a further round of funding cuts to federal environmental programs that deal with environmental monitoring programs and assessments. A number of key environmental departments,

---

A.M. Martinez (✉)
Faculty of Education, York University, Toronto, Canada
e-mail: Ana_Martinez@edu.yorku.ca

S. Alsop
Faculty of Education and Department of Science and Technology Studies, York University, 4700 Keele Street, Toronto, ON M3J 1P3, Canada
e-mail: SAlsop@edu.yorku.ca

L. Bencze and S. Alsop (eds.), *Activist Science and Technology Education*, Cultural Studies of Science Education 9, DOI 10.1007/978-94-007-4360-1_27,

it was announced, are to be absorbed by other agencies or closed for good, and there are expected to be massive job losses for scientists in key areas including climate change, natural resources and fisheries (CBC 2011a, b; Public Service Alliance of Canada 2012; National Post 2013; *The Star* 2013). These cuts build on widespread discussions of closer monitoring of scientists in Canadian federal agencies in which media-relations officers vet governmental scientists work before they are able to speak to the public.[1] This has created considerable discomfort within particular governmental scientific communities to the extent that one of the most renowned peer-review science journals has very recently accused the Canadian's government of 'silencing its scientists' (CBC 2011a, b; Nature 2012). Moreover, as this edited collection goes to press a series of public demonstrations have occurred in major Canadian cities, a new national organisation ('Scientists for the Right to Know') has been established; and Chris Turner has published a high-profile text-length indictment entitled *The war on science: Muzzled scientists and wilful blindness in Stephen Harpers Canada* (Greystone Books).

It seems at the very least that being a governmental scientist in contemporary Canada is profoundly paradoxical. On the one hand, the role equates with controversy and job instability, especially for those who dare to affirm that climate change and environmental decline is happening on Canadian political soil; whilst on the other hand, citizens are told—through very expensive advertising campaigns[2]—that scientists are 'hard at work' on new technologies that will give us back our clean water, wetlands, and forests after mining or tar sands explorations; or finding ways to secure how food and technology are produced to cheaply feed seven billion plus people.

Perhaps this serves as just one illustration of the ways in which the status and positioning of scientists and scientific knowledge in some Western democracies is changing. After all it was not that long ago that the 'myths of modernity' (according to many high profile scholars) were such that science was seen as reliably speaking the truth of nature, whilst politics, on the other hand, was concerned with much more murky and subjective values and opinions. This is not to say that science and politics did not interact, but to narrate this interaction as predominately one in which science provided the 'robust facts' from which to justify political decisions, policy reforms and actions (see Latour 2004). Today, however, this seems to be more complicated and politicians no longer feel the immediate, or compulsion to court scientific expertise in ways in which they might have done in the past. Indeed, the question of the roles of scientific expertise within Western democracies has now become a subject of considerable exploration, contestations and extended speculations (see Brown 2009).

[1] See Dr. Kristina Miller's case on salmon genetics and how after publishing her research in a peer-reviewed journal, she was ordered not to speak to the media about her findings.

[2] In 2012, the Canadian Federal Government spent nine million dollars on a natural resources national campaign promoting development projects (CBC 2012).

Within this context, this chapter offers some reflections on the potential contributions, effects and associated tensions and contradictions of citizen science programs in schools, communities and democracies. We focus on a particular organisation. The Association for Canadian Educational Resources (ACER) is an organization whose central focus is climate change education. It has developed programs and worked with students, communities and scientists in Southern Ontario during the last 17 years. Through a case study of this organization, our chapter explores a widely held assumption that through civic participation in monitoring activities, citizens can potentially become more knowledgeable and responsive political agents of change within their own communities. ACER's programs have evolved as one grassroots response to create awareness of climate issues and to monitor changes in biodiversity within local forest ecosystems. These programs are based on the premise that learning through monitoring, participation and community-based scientific inquiry is a key part of democratic participation and enhanced civic social responsibility.

What Is Citizen Science?

The concept of citizen science is not new, although it has gained greater prominence lately, especially on projects where researchers are constrained by limited funding and seek the collaboration of citizens as a way to further their research (Cohn 2008; Silvertown 2009). Some recent high profile citizen science projects include 'Zooniverse,' 'the Christmas Bird Count' and 'Rink Watch'. Zooniverse attracted nearly 800,000 online participants due to its innovative and interactive approach in which participants can assist research projects that range from exploring the ocean floor to the surface of the moon (Zooniverse 2013). The 'Christmas Bird Count' organized by the National Audubon Society recently celebrated its 113th anniversary (2012). This count enlists tens of thousands of participants annually. The Audubon society produces a yearly report with regional summaries and featured articles from the data collected by these volunteers (Audubon 2013). 'RinkWatch,' is a very recent initiative developed by researchers at the Wilfried Laurier University, which tracks climate changes by collecting weather data at the local level. Launched in early 2013, this project attracted more than 500 participants within the first 2 weeks of operation. Evidence suggests that participants eagerly and enthusiastically monitored and logged weather conditions on their backyard ice rinks (CBC 2013a, b).

Citizen scientists might be seen as volunteers who go beyond the philanthropic notion of helping a cause, or as Cohn (2008) describes it, they become in a certain way 'field assistants [who collaborate] in scientific studies' (p. 193). The data collected by citizen scientists participating in such research projects (whether online or *in situ*) has evolved from being highly controversial, in terms of its precision, to be publically valued and validated by scientists and publics (pp. 193–194). Cohn argues that after 'trial and error' on several research studies, scientists realize that not only training citizen scientists was crucial in order to be able to use and rely on data collected, but also it was important to develop simplified methodologies or protocols that would make these collected data more reliable. Thus, citizen science projects offer support to scientists. In Cohn's examples, participants primarily act as data gatherers, and as a consequence more data can be collected and analysed. Participants are given opportunities to record observations and in so doing potentially gain some local appreciation of the phenomenon being monitored (for instance bird migration, or forest growth), as well as methods of data collection used (although these, as noted, are often simplified for increased reliability). However, as Mueller and colleagues (2012) critically note, there are some very real caveats here: projects seldom involve citizens actually watching scientists in action (the data is mostly collected in isolation) or developing their own research projects and protocols (the project questions and observation protocols are largely prescribed by scientists). In this regard, the relationships between the scientists and citizens are sharply hierarchical and as such enact particular centralised and centralising relationships and epistemic assumptions. However, perhaps what makes citizen science projects especially appealing for us is the ways that they aspire to participants' better understanding and changing civic roles in democratic society. By becoming aware of socio-scientific issues affecting local communities and the sciences behind them, it is genuinely hoped that citizens are more likely to become more closely engaged and active participants in making personal and political choices concerning these issues. In this way, citizen science potentially offers not only the prospect of being a enthusiastic participant, a data collector, but also to becoming an empowered citizen, hopefully with an increased capacity and motivation to voice and act upon their and others' concerns in democratic societies.

Some citizen science projects even have self-identified activist orientations. For instance, Ottinger (2010) provides an example of how citizens in Norco, Louisiana 'challenged the standard practices used by regulators for assessing air quality' (p. 245). In a practice he calls 'buckets of resistance,' communities living at the edge of a Shell chemical plant were concerned with the levels of toxics emitted into the air by the plant. The residents organized in order to independently assess the air quality. They measured and compared the results against state regulatory standards and their findings established that the state's standards were missing crucial evidence to completely assess the air quality in an area. The project was not without tensions. Although policy makers addressed citizens concerns by reinforcing regulatory and state practices, in the end, citizens' data was disregarded because of the perceived lack of standardized methodologies that were compatible to the one used by the state. Ottinger argues that even with the lack of such standards, the experience at Norco is critical for activism movements because 'citizen science [. . .] is appealing from the standpoint of projects [that] democratize science and science-based policy. Citizens' scientific efforts demonstrate that so-called "laypeople" can be meaningfully involved in knowledge and policy making; by revealing the values inherent in expert knowledge, citizen science [ultimately] bolsters arguments that they must be' (2010, pp. 264–265).

Silvertown argues that the concept of citizen scientist has been under-represented in peer-reviewed journals because it is relatively new and more importantly because the accuracy of the data collected has been repeatedly brought into question on scientific projects (2009, p. 470). There is an open question whether this lack of representation in the literature might be interpreted as entirely a consequence of perceived inaccuracies or might be linked in some way with a degree of indifference or mistrust of specific sources of knowledge—such as traditional or community-based knowledges. The "subjective probabilities" (Schneider 2000) within any type of scientific inquiry seem more tolerable when produced by "independent scientists" and associated rhetoric of objectivity, but more questionable when produced by local citizen scientists, perhaps.

Schneider draws our attention to the early Working Group of the Intergovernmental Panel on Climate Change (IPCC) who explicitly recognise complexities and relevancies of all sources of knowledge, and consequently introduced the concept "science as community" (2000, p. 109). Although idea of community science was proposed nearly two decades ago, contemporary citizen science projects still face a magnified level of scrutiny and scepticism. Riedlinger and Berkes (2001) argue that although 'environmental change[s] associated with variations in weather and climate has not gone unnoticed by communities that are experiencing changes first hand' policy makers have deliberatively decided to ignore the potential 'contributions that [local] knowledge [might have] to climate-change research.'

Schneider continues by arguing that a; 'sophisticated ability to discern who is more credible requires a citizen that is more than casually interested, but who is passionately involved'. This citizen, he continues by stating is one that 'can... figure out where the mainstream is and whose subjective probabilities to trust.' (2000, p. 111). An informed citizen, in Schneider's sense therefore is not just one that watches the news or is up to date with the latest local/global trends. An 'informed citizen' is one who can, through different sources of information, formulate knowledge opinions and then act upon those topics or causes accordingly. This is admirable. However, at the very least, this is likely to be a complex process and is unlikely to happen spontaneously.

Many citizen science programs argue for the benefits of active participation whether this is learning or contributing to the community or environment. Bohman, for instance, advocates the 'creation of deliberative situations and institutions in which those affected by [an issue] are able to make judgements about the credibility of experts [whether scientists or policy makers] and to influence the terms of their on-going cooperation with them' (1999, p. 592). Participation then, it seems, moves citizens in a way that enables them to make decisions with others (individuals and groups) based on 'shared' experiences. The significance of engaging citizens through direct experiences of participation has been proposed by a number of environmental education scholars. David Orr (2004), for example, recommends that in order to develop affection and a sense of responsibility towards our environment, a fundamental change is necessary in which communities 'adopt [local] streams or entire watersheds and make their full health an educational objective'. Orr argues that this is 'an [unexploited] opportunity to move education beyond the classroom and laboratory to the outdoors, from theory to application and from indifference to healing' (pp. 58–59). Many informal programs developed by Environmental Non Governmental Organizations (ENGOs) inspire to these types of teaching-learning experiences in their collaborations with schools and others in the wider public sphere.

There are inspiring instances of interested groups responding to their concerns of lack of policies in environmental education. In Ontario, one of the most widely publicised is "Greening the Way Ontario Learns" published in 2003 and prepared by Environmental Education Ontario (EEON) in collaboration with more than 40 ENGOs. This document presents the Ontario Government with a public strategy to address environmental and sustainable education. In its latest publication, the Ministry of Education of Ontario has stressed its commitment to an environmental education that 'must be defined locally to meet the differing environmental, social and economic conditions that exist in Ontario communities' (Ministry of Education of Ontario 2009, p. 4). This vision is one of inclusiveness that recognises the role that communities play in the education of school children and in which citizen scientists are most certainly a part.

With this brief introduction to citizen science, we now reflect on a particular project that the first author (Martinez) has been centrally involved in. The organisation concerned might be described as a "boundary organisation" insofar as it seeks to work with both scientists and citizens in planting and monitoring trees in response to Climate Change.

ACER: Understanding Change Through Long-Term Forest Monitoring

For the last 17 years a program has been monitoring forests and working with data produced by citizen scientists (ranging from elementary students to seniors). The Association of Canadian Educational Resources (ACER) is one of the oldest citizen science projects in Canada. In the mid 1990s, ACER established a partnership with scientists from the Ecological Monitoring and Assessment Network from Environment Canada and the Smithsonian Institution (SI). The latter created a community-based monitoring protocol called the 'Measurement and Assessment of Biodiversity (SI/MAB) protocol' to measure changes in the biodiversity of forest ecosystems. This initiative was tested and implemented by ACER and its partners in a project called the Niagara Escarpment Biosphere Reserve Studies (NEBRS). ACER's role was to provide the volunteer and community participation necessary to collect benchmark data for a three paired one-hectare plots in Southern Ontario. The data collected by ACER was analyzed and has been published and presented in different conferences in and outside Canada by both ACER and Environment Canada (Casselman 2008; Karsh et al. 2007).

This early partnership enabled ACER to start five additional programs in which the SI protocol continued to be the standard for data collection. Although the protocol has been tailored according to the needs of each program, the basis for measuring, monitoring and reporting is the same. The concept was to produce a simplified and straightforward methodology that could be used by non-experts, and that was, at the same time, reliable enough to be used and analyzed by climate change scientists. To date, ACER has worked with over 10,000 students and 2,000 adult volunteers through its monitoring programs.

ACER's focus is to create awareness—through hands-on experience—to the changes that are occurring in participants' own backyards. In other words, our experiences offer participants an opportunity to investigate the implications of policy makers' decisions through closely and systematically monitoring changes in local forest ecosystems that they are part of.

Different ACER programs have been designed to serve a specific purpose and target a different—although not exclusive—audience. Go Global is the name of a program linked with a one-hectare network of forest plots established in 1996 to assess biodiversity. It was implemented in partnership with Environment Canada and the SI. Through this program, nearly 15 plots have now been established in Southern Ontario in collaboration with landowners, parks and forest reserves. Within these plots, all mature trees are measured, tagged, coded and assessed for health (e.g. presence of invasive species) by ACER volunteers.

'Let's Plant, Measure and Mulch' is a 10-year biodiversity experimental plot established in 2002 at the Humber Arboretum in Toronto, in which 2,230 trees from 61 different species were planted by students and volunteers on a one-hectare research plot. Considered as an instance of 'unprecedented community planting' (Karsh and Casselman 2012) for its extent and level of effort, this site was designed so each one of the 28 quadrats would contain different numbers of hardwood, mixed wood, Carolinian, city street or forestry species as well as high, moderate or low biodiversity. ACER participants measure and monitor the growth of trees roughly once a year. The study, which initially sought to identify which species would be more likely to survive under changing climatic conditions, also identifies other aggravating problems affecting urban forests. Fragmentation due to urbanization has allowed certain populations to thrive (e.g. deer, invasive herbaceous plants, new insect infestations). Despite all the efforts from Humber and ACER's staff and volunteers to protect the trees and shrubs with tree guards, the existing deer population caused a total loss of 76.2 % of the trees in this experiment (Karsh and Casselman 2012). This gives an indication of some of the types of considerations that need to be taken into account within projects of this type.

Another program specifically designed for schools is entitled 'Measuring Our Resources was launched in 2004. This is built on a particular protocol to collect and monitor existing trees in schoolyards. The schools take responsibility to measure and map existing mature trees, whether these are located on the schoolyard, a park, or a forest nearby the school. The idea is to teach the students how and why it is necessary to track changes in existing trees and forests. The school's commitment is to send the data to ACER every year. In 2012, a new approach was developed and implemented by ACER (in partnership with a local municipal group 'Conservation Halton') to allow schools to measure a selected quadrat at the Rattlesnake Point Park in Milton, Ontario.

'Planting for Change (P4C)' initiated in 2008, is a program that creates an outdoor laboratory, where students can plant and monitor the growth of a suite of 5 different species from the same nursery. The same species are planted in each school, following a protocol, to be able to compare the results. In 2012, 21 schools had planted these 5 species strategically chosen by ACER's technical advisory committee, composed mainly by foresters. The program has been implemented in public, private, elementary and secondary schools. The schools' commitments with ACER is to report back every year on the data collected with the principle of sharing findings and the capacity to compare results with other schools. Data collected since 2008 is currently being analyzed to be published in a report in the Fall of 2013.

In 2012, a new program was designed to monitor newly planted trees through planting efforts developed by other organizations. The program called 'Tracking for Success' uses the same protocol as the one used for P4C, with the difference that a 10 % of the total of trees planted are measured. The idea with this project is to determine which species survive and 'succeed' in community or mass planting events. This program is a partnership between ACER and other organizations, which have been conducting reforestation efforts within their properties or through their own programs.

Reflecting on Experiences with ACER

ACER's approach to community-based measuring and monitoring has been scrutinized by other NGOs and funding organizations. One of the strengths of ACER's approach, we like to think, is the way that it has retained a focus on tracking local biodiversity changes and enabling citizens to measure these changes themselves by developing methodologies to collect data that can be easily understood and replicated anywhere. ACER's approachable monitoring methods allow participants, in principle, to compare what was happening in their backyards to other sites, and to raise questions, potentially enabling them to request and demand answers on issues affecting—in this case—forest ecosystems.

Funding organizations continue to question the need for continuous training when working with lay-communities. Although in our experiences there is a constant need for revisiting data protocols with experienced participants as well as inviting others to become involved. We find the participant turn over in these programs quite high as participation is influenced by a series of external factors, not least the availability of volunteers time because of other commitments. It takes time to educate people about the measuring equipment. The labour involved in actually taking these measurements is also time-intensive. In the case of ACER, my (Martinez) experience has demonstrated that a degree of data verification is necessary in the field, such that reliability is ensured. Although most participants in the field take their work responsibly and with a high level of commitment, there is always room for human error (e.g. lack of knowledge of the equipment, where units and models change from one tool to the next; confusion when reading or following instructions, etc.). These errors (or standard deviations in the data) need to be acknowledged and identified as part of running evolving programs, so the analysis and dependability of results will reflect a substantial level certainty. We constantly revisit our protocols with participants to ensure consistently high quality data.

As previously mentioned, Mueller and colleagues (2012) observe that it is often difficult to locate papers published by citizens scientists. It is the case that ACER tends to publish articles 'in-house' rather than in peer reviewed scientific journals. Indeed, we openly suspect that prestigious peer-reviewed journals might question the validity and reliability of the ACER data sets. Whilst ACER's intent is to secure high quality data that can be published in peer reviewed journals, it should also be recognised that ACER's role is more localised and ultimately responsive to the contexts of our participation (the communities and ecosystems that we work within). A success of ACER in this respect links, in part, to the way that it mobilises groups to engage in longer term monitoring. Long-term monitoring programs potentially enable participants to understand and possibly contest local ecological policies and management approaches concerning climate change or the conservation of local biodiversity. As Braschler (2009) puts it 'these projects are not just about the data: a key aim is raising the participants' awareness about the science behind the projects. These projects combine research and outreach, which can improve understanding of biodiversity while simultaneously raising awareness about the threats to it' (p. 103). ACER is particularly concerned about the quality of its data and therefore seeks to combine this with quality participation.

Some Emergent Tensions and Contradictions

Citizen Science programs have certainly thrived in the last couple of decades. Many Environmental Non-Governmental Organizations (ENGOs), for example, continue to play a central role in the creation of programs that allow citizens to participate more actively in scientific research projects. ENGOs mobilise volunteers as a way to complement their efforts to run programs, whether their focus is in education, awareness, or collecting data. There is some evidence, that data collected and vetted by non-specialists has been gaining credibility and influence in the last two decades to the point of affecting national policies. Throughout the years, ACER has recruited a vast number of researchers, students, teacher and community member to implement its projects. The number of volunteers willing to undertake potentially arduous workdays in the field or continuous annotations of observations has been encouragingly substantial. The willingness of participants has allowed organizations and institutions to execute projects that otherwise would be difficult to accomplish in the light of austerity measures, limited resources and staff.

Citing a number of germane examples, McCormick argues that 'citizens have taken research into their own hands, using it to frame issues, challenge corporate [and governmental] practices, and change policy' (2009, p. 34). She writes of the 'democratizing science movements', but then continues by arguing that 'creating new science is no longer always central [to these movements], making science accessible and politically relevant may be more so' (p. 35). By doing so, citizens could become actors of change and potentially influential stakeholders. Schneider (2000) adds that 'what environmental literacy can do is empower citizens to begin to pick a scientific signal out of the political noise that all too often paralyzes the policy process' (p. 119).

Citizen science projects clearly have much to offer, although they, of course, present a series of tensions and contradictions. For example, as Mueller and colleagues (2012) note, many projects are 'top-down', and as such participants have a somewhat peripheral and instrumental role. They collect data by following prescribed protocols for the most part. In this regard, there are questions of the types of practices that participants are experiencing, and the extent to which these meaningfully represent the complexities of scientific practices and knowledges, and the events being monitored. As Mueller and colleagues (2012) contend;

> Whereas citizen science gets its power from basking in the sun of science, there is much more to science that is often not acknowledged as part of the endeavour, such as cultural, ethical, political and spiritual (and virtual) studies (p5.).

With a series of persuasive examples, Mueller et al. argue that many citizen science projects are sometimes wanting because they fail to embrace dynamic and multidimensional notions of science and as a consequence can overlook important, axiomatic socio-ecological antecedents. In the case of ACER's tree planting and monitoring programs this raises important considerations and there are many open questions that call for our future investigations and careful critical reflections.

From the perspectives of the scientists involved, the ACER community monitoring of forest plots offers a way of collecting local and regional data. These data have been used in local, national and international policy discussions. As a high profile climate scientist noted at the recent ACER conference, climate models are no substitute for local data collected by dedicated community groups carefully and systematically monitoring adaptation in local tree plots (comprising in this case of carefully chosen mixed tree species). Macro-models of climate change are unable to include micro-environmental conditions (such as specific humidity, soil type, water drainage, and deer and insect population growth, for instance).

ACER has been able to bring climate scientists, school students and community groups together within these 'shared' contexts of concern and support actions associated with climate change. The tree planting and stewarding initiatives have demonstrably brought about significant environmental changes (new forest lots have become established, and in this regard, local green coverage has increased in some areas. It continues, of course, to be lost in others). We like to believe that participating in research projects of this type energises those involved. There is a sense of purpose and a strong desire to become more active. It is possible to see a difference. A local environmental context, in this regard, situates a global, often distant, pressing concern. It offers the possibility of acting, and also discussing the best way to apply pressure on regional governmental decision-making and planning. Of course, this is not without partialities. Those involved tend to be much more comfortable focusing on tree planting and monitoring rather than embracing sociocultural and economic structural changes persuasively advocated by many environmental and social justice educators. David Orr's (2004), for example, presses the need to change the drivers in society such that people and institutions become stewards of local streams and forests and are able to question and challenge decisions that concern their lives and those of their children; take responsibility for their actions and be proactive in their communities; be part of any cause that seeks to rebuild or restore their surroundings; foster environmental education within their own communities. Whilst all those involved with ACER's programmes inspire us, we are reflexively aware that these efforts seem to fall short of Orr's desires for fundamental social change.

Within the context of this book, we offer ACER's tree planting and monitoring as a form of activism. In a broader sense, citizen-science programs are to greater or lesser extent instruments of participation that to some degree blend scientific inquiry, enhanced agency and local participation. Such involvements, we posit aspire to 'offer opportunities for rich, diverse pedagogies of experience, participation, and empowerment in which students, teachers, researchers can through knowing and acting, come together to name the world and explore and reshape their place in it' (Alsop and Bencze 2010, p. 181). It might be argued that after participating 'first-hand' in monitoring projects, a form of power is given to those involved, they 'take-back' (although this seems a rather strong term) their communities by becoming more aware of the issues concerning their environment. If 'knowledge is power,' as many claim, then would it not be in the best interest of youth, communities and civic institutions to play a more active role in local monitoring?

Mueller and colleagues (2012) argue that 'democratizing science education is vital to fostering students' understanding of how science can be relevant to their lives and communities' (p. 1). Perhaps, ACER offers a context in which to begin to unpack these complex aspirations. There are inevitable partialities and trade offs. The simplified protocols involved in tree monitoring serve to arguably enhance reliability, but raise questions of validity and misrepresentation that ultimately distort the complexities of science (as Mueller and colleagues (2012) note). The aspiration to 'democratizing science' is far from self evident and raises complex questions, not least of which are questions pertaining to whose and what science as well as whose and what democracy? Central to such considerations are questions of representation. As Brown (2009) brings to our attention, science "stands for" (represents) phenomena in the natural world, whilst democracy "stands for" (represents) the voices, or will of the people (or something else). To embark upon an analysis of citizen science as a 'democratizing of science', or for that matter a 'democratizing of science education', as Mueller and colleagues (2012) suggest, inevitably calls into question representations and who speaks for the phenomena that they come to represent. Such an analysis ultimately emanates from a series of assumptions and commitments concerning whose voices ought to count, as well as when, and how, they get to count in representations and representatives of nature, culture or nature-cultures.

Acceptance, denial, or ignorance of issues such as climate change, won't make them disappear, especially now, while many governments retreat from international commitments that will have enormous implications for the lives of current and future generations. At the very least civic environmental monitoring programs offer the hope of disclosing local ecological changes within a local sphere of constant vision. They recognise the potential of short and longer-term community based environmental monitoring programs and protocols. The difficult question is what this might mean in light of the recent actions of the Canadian government in disbanding federal scientific monitoring. There is a loss of expertise here that needs to be recognised. It seems wilfully inadequate to couch these changes as a shift from centralised democratic representatives of science (governmental scientists) to community-based-representatives of science (citizen scientists). This is not an opportunity to replace one type of ideal of representative democracy with another,[3] but to be clear about the silencing of expertise at a time in which science seems to be fading as a political tour de force. It is also, once more, an opportunity to recognise the political power of shorter-term economic representations as the basis of Canadian governmental policy and decision-making.

References

Alsop, S., & Bencze, L. (2010). Editorial: Introduction to the special issue on activism: SMT education in the claws of hegemon. *Canadian Journal of Science, Mathematics, and Technology Education, 10*(3), 177–185.

Audubon. (2013). *Christmas Bird Count*. Retrieved February 20, 2013, from Birds: http://birds.audubon.org/christmas-bird-count

Bohman, J. (1999). Democracy as inquiry, inquiry as democratic: Pragmatism, social science, and the cognitive division of labor. *American Journal of Political Science, 43*(2), 590–607.

Braschler, B. (2009). Successfully implementing a citizen-scientist approach to insect monitoring in a resource-poor country. *BioScience, 59*(2), 103–104.

Brown, M. (2009). *Science in democracy: Expertise, institutions and representations*. Cambridge: MIT Press.

Casselman, A. (2008). SI/MAB paired forest monitoring plots in Canada. In A. Fenech, D. MacIver, & F. Dallmeier (Eds.), *Climate change and biodiversity in the Americas* (pp. 68–69). Toronto: Environment Canada.

[3] Some might compare this with a debate between John Dewey and Walter Lippmann concerning the complexity of the modern world. For Dewey the response was in the form of a great community, and educated public that could take on the issues of the day. In contrast, Lipman called for specialist representatives whose expertise stem beyond the local.

CBC. (2011a, August 9). *Federal government job cuts: The story so far*. Retrieved November 23, 2012, from CBC News. Politics: http://www.cbc.ca/news/politics/story/2011/08/05/pol-public-job-losses.html

CBC. (2011b, July 29). *Federal scientist unfairly silenced, union says*. Retrieved April 24, 2013, from CBC News: http://news.ca.msn.com/top-stories/federal-scientist-unfairly-silenced-union-says-3

CBC. (2012, November 15). *Resource development ads*. Retrieved December 7, 2012, from CBC Player: http://www.cbc.ca/player/News/Politics/Power+%26+Politics/ID/2305146747/

CBC. (2013a, January 23). *Outdoor rink climate change project gets hundreds of citizen scientists. Wilfrid Laurier University researchers recruit RinkWatch participants to record climate patterns*. Retrieved February 20, 2013, from Technology & Science: http://www.cbc.ca/news/technology/story/2013/01/23/pol-backyard-rink-monitors-climate-change.html

CBC. (2013b, March 21). *Budget offers little new on the environmental front*. Retrieved April 21, 2013, from CBC News, Politics: http://www.cbc.ca/news/politics/story/2013/03/21/pol-budget-2013-environment-energy.html

Cohn, J. (2008). Citizen science: Can volunteers do real research? *BioScience, 58*(3), 192–197.

EEON. (2003). In E. Houghton (Ed.), *Greening the way Ontario learns. A public strategic plan for environmental sustainability education* (127p.). Toronto: Education Ontario (EEON).

Karsh, M., & Casselman, A. (2012). *Monitoring ecological sustainability in the Humber Watershed: Examples of community and inter-agency partnership*. Mississauga: ACER.

Karsh, M., MacIver, D., Fenech, A., & Auld, H. (2007). *Climate-based predictions of forest biodiversity using Smithsonian's global earth observing network* (Occasional Paper 8). Adaptation and Impacts Research Division (AIRD). Environment Canada, publications.gc.ca/collections/collection_2012/ec/EN57_41_8_2007_eng.pdf

Latour, B. (2004). *Politics of nature: How to bring the sciences into democracy*. Cambridge: Harvard University Press.

McCormick, S. (2009). From "politico-scientists" to democratizing science movements. *Organization & Environment, 22*(1), 34–51.

Ministry of Education of Ontario. (2009). *Acting today, shaping tomorrow. A policy framework for environmental education in Ontario schools*. Toronto.

Mueller, M., Tippins, D., & Bryan, L. (2012). The future of citizen science. *Democracy & Education, 20*(1), 1–12.

National Post. (2013, March 17). *Federal budget cuts undermine environment Canada's mandate to enforce clean air regulations: emails*. Retrieved April 21, 2013, from Canadian Politics: http://news.nationalpost.com/2013/03/17/federal-budget-cuts-undermine-environment-canadas-mandate-to-enforce-clean-air-regulations-emails/

Nature. (2012). Frozen out. *Nature, 483*, 6.

Orr, D. (2004). *Earth in mind*. Washington, DC: Island Press.

Ottinger, G. (2010). Buckets of resistance: Standards and the effectiveness of citizen science. *Science, Technology and Human Values, 35*(2), 244–270.

Public Service Alliance of Canada. (2012, April 22). *Cuts to federal science and research programs – Backgrounder*. Retrieved November 23, 2012, from Third Choice Campaign: http://www.psac-afpc.com/issues/campaigns/3c/labs-e.shtml

Riedlingera, D., & Berkesa, F. (2001). Contributions of traditional knowledge to understanding climate change in the Canadian Arctic. *Polar Record, 37*(203), 315–328.

Schneider, S. (2000). Is the "citizen-scientist" an Oxymoron? In D. Kleinman (Ed.), *Science, technology, and democracy* (pp. 103–120). Albany: State University of New York Press.

Silvertown, J. (2009). A new dawn for citizen science. *Trends in Ecology and Evolution, 24*(9), 467–471.

The Star. (2013, March 21). *Federal budget 2013: Tories fold CIDA into foreign affairs department*. Retrieved April 21, 2013, from News Canada: http://www.thestar.com/news/canada/federalbudget/2013/03/21/federal_budget_2013_tories_fold_cida_into_foreign_affairs_department.html

Zooniverse. (2013). *Zooniverse. Real Science Online*. Retrieved February 20, 2013, from https://www.zooniverse.org

Chapter 28
"It Changed Our Lives": Activism, Science, and Greening the Club/Community

Angela Calabrese Barton and Edna Tan

Abstract Drawing upon critically oriented studies of science literacy and environmental justice we posit a framework for activism in science education. To make our case, we share a set of narratives on how the River City's Youth Club acquired a new green roof. Using these narratives we argue that the ways in which youth describe their accomplishments with respect to the roof reflects a range of subject positions that they carve out and take up over time. These subject positions reveal how activism is a generative process linked to "knowing" and "being" in ways that juxtapose everyday practices with those of science.

Keywords Youth clubs • Social justice • Urban studies • Environmental justice • Performativity • Critical science literacy • Narrative studies

Introduction

In the summer of 2009, the "River City Youth Club," a neighborhood youth organization, which serves a predominantly lower-income and African American population, had a new green roof installed on its facility. While the club's facility needed a new roof because the "old one leaked," club leaders sought out a *green*

"This chapter uses material previously published in a special issue (volume 10, issue 3) of the *Canadian Journal of Science, Mathematics and Technology Education* and it appears here with the permission of the copyright holder, the Ontario Institute for Studies in Education."

A. Calabrese Barton (✉)
Department of Teacher Education, Michigan State University, 327 Erickson Hall, East Lansing, MI 48824, USA
e-mail: acb@msu.edu

E. Tan
School of Education, University of North Carolina, Greensboro, 341H Curry Building, Greensboro, NC 27402, USA
e-mail: e_tan@uncg.edu

L. Bencze and S. Alsop (eds.), *Activist Science and Technology Education*, Cultural Studies of Science Education 9, DOI 10.1007/978-94-007-4360-1_28,

roof because it would be energy efficient, reduce their energy bills, and offset their carbon footprint. As Sarah,[1] one of the club leaders, explained:

> I had never heard of urban heat islands before. I did not realize that a building's roof could contribute to our own carbon footprint. Actually, I had never really thought about how the club could have its own carbon footprint. When the GET City youth started to raise these ideas, I began to question what approach we were taking to getting a new roof. We had to be more responsible.

When Sarah spoke these words, we were surprised. GET City [Green Energy Technologies in the City] is a year round program intended to engage youth in authentic investigations into green energy and the environment. As teachers, however, we had not anticipated that the program would have such a substantive impact on local practices – especially those that required substantial financial costs. While the youth had previously studied the impact that urban design has on climate change, *we never took green roofs up specifically as a target of investigation* until the club leaders had gone public with a plan to replace their leaky roof with a green roof. We take up two questions in this manuscript:

- What are the narratives that youth tell about how the club got its new green roof and their role in the process?
- How do the youth's narratives inform activism in science?

A Conceptual Framework for Activism in Science Education

Activism generally implies taking action to bring about change – socially, economically, politically, or environmentally. Often framed by ritualized activity (Robins 2006), activism is often re-presented as "spectacle" replete with police and protesters clashing in public spaces (Urrieta 2004). Such re-presentations remove activism from normal daily activity; they situate activism within ideological convictions rather than in the lived experiences of everyday people (Brodkin 2009). Further, activism has been framed narrowly as action taking, sidestepping the deeper significance of the ways in which such actions are deeply embedded in cultural knowledge and experience. Why and how one critically engages the subject in an effort to transform routine practices is both a reflection of one's subjective locations and one's daily effort to transform those locations. Below, we build a case for activism in science as both an identity and knowledge building project deeply rooted in everyday practice. To do so, we turn to critically oriented studies on science literacy and environmental justice.

[1] A pseudonym.

Environmental Justice

The dominant discourse of environmental justice of the past three decades has been to highlight who has the power to create and enact environmental policies and practices and their effects on oppressed communities (Hobson 2006). As the "anti-toxics" movement shows, environmental justice efforts have largely focused on challenging the polity (Teelucksingh 2002). This focus has been important for it has brought into the public discourse dialogs on *in*justices – in terms of the right to information, to healthy environs, and to a voice in broader policy and practice. However, such a discourse has left little room for incorporating how *youth* enact environmental justice (Stephens 1996). The marginalization of youth in environmental justice discourses is acute because youth do not, generally, possess the rights of "full members" of societies (e.g. neither allowed to vote nor considered experts who can make a change).

However, recent work on environmental justice fronts has paid more attention to the everyday enactment of environmental concerns – or the lived experiences of environmental *in*justice – among low-income communities and communities of color (Brodkin 2009). In recent work, environmental justice is located not only in anti-toxic movements, but also in how the *boundaries* of environmental issues are framed across race and class. Studies on access to quality housing (Molina 2000), venues for participation in urban planning (Teelucksingh 2002), and the design and dissemination of stoves for heating and cooking (Subramaniam 2000), for example, all challenge the form and function of environmental justice discourse and activity globally. This re-framing expands the discourse from rights-based to the performative, with its attention on the "everyday actions and representations" individuals take up, which often fall "under the radar" of rights-based environmental justice (Hobson 2006, p. 673).

The move towards the performative is crucial to understanding youth practices with respect to environmental justice, specifically, and activism in science more generally. Performative lenses highlight the agency of individuals and collectives when they take actions to re-inscribe everyday spaces and activities with meaning: a process that re-formulates environmental justice discourse in terms that acknowledge and value youth participation in local, situated environmental concerns. Such a reformulation may help to advance critically-oriented place-based efforts in environmental education intended to make matter how and why youth are positioned within the environment in the learning of science (Bowers 2002). This reframing is important because it positions African American and low-income youth as activists even when their actions fall outside what is normally constitutive as activist (Cohen 2006).

Critical Science Literacy

Activism involves a critical engagement with the subject. Individuals or collectives take action because they believe in something as the good or moral thing to do. Yet, values are not always a recognized part of the discourse or practices of science, at least in schools and other formal learning environs.[2] The inherent challenge here is that activism in science has been accepted in the discourses of teaching/learning science only in so far as individuals can take action on science-related topics, rather than through their actions transform the daily practice of doing science.[3]

The reforms of the past two decades have been premised on functional science literacy. The term functional is meant to imply that individuals gain the knowledge, skills, and habits of mind of science necessary for "personal decision making, participation in civic and cultural affairs, and economic productivity" (NRC 1996, p. 22; see also Ryder 2001). However, functional science literacy attends to participation in the world as it is now, without explicit critical attention to how or why scientific ways of knowing or being might bring about a more just world for individuals or communities while being transformed by broader and more diverse participation.

While such a stance has advanced the debate around why learn science, it also leaves uncontested what it means to function in society. Critical science literacy, while embracing the broader notion that individuals ought to have facility with the big ideas and practices of science, also privileges critical engagement with text, ideas, and ways of knowing and being that frame the discourse and practice of science. Teaching and learning practices often represent science in its final form, yielding descriptions of content that appear complete and stable rather than as knowledge-in-the-making. Critical engagement with the text of science deprivilege the authority of text and teacher, thereby expanding opportunities to more fully define and situate scientific problems, describing methods, and posing limitations to knowledge claims. An important part of this framing is in acknowledging the social networks that facilitate and constrain individuals as they seek to perform the necessary tasks and sociocognitive work of science. "Lone individuals do not solve problems, but rather problem solving is embedded in a social network that collectively performs necessary tasks and cognitive work" (Nasir and Hand 2008, p. 144). At the same time, such social networks can legitimize or delegitimize the knowledge, experiences, identities and practices one brings to doing such work.

Returning to Activism in Science

Bringing critically oriented stances on science literacy and environmental justice together provide us with a powerful way of framing activism in science. With

---

[2] It is important to note that feminist, multicultural and queer perspectives on science do take on the relationship between scientific knowledge and practices and values.

[3] Weinstein's discussions of street medics and guinea pigs are excellent examples of activism transforming the daily practice of science (see Weinstein 2006, 2008a). However, he, too, notes that such a stance is divorced from the discourse and practice of school science (see Weinstein 2008b).

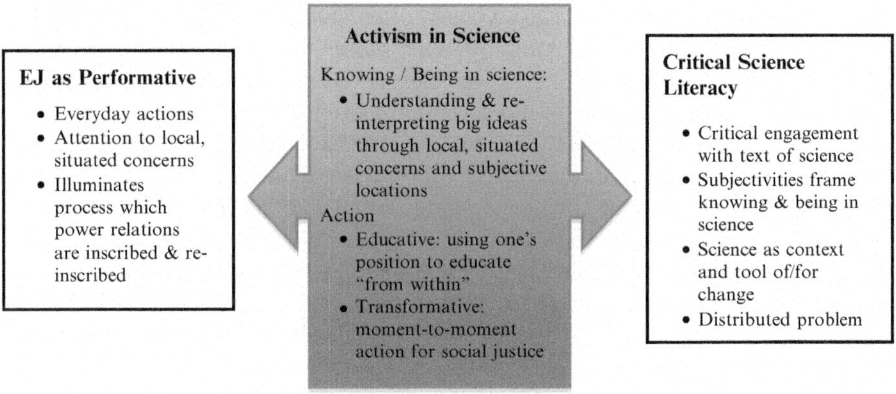

Fig. 28.1 A framework for activism in science education

attention paid to the role of the everyday in the critical engagement of science, activism in science incorporates knowing and being in science (in terms of understanding and re-interpreting big ideas through local, situated concerns and subjective locations), and taking action. Activism thus privileges two forms of action: the *educative*, where individuals or collectives seek to use their subject locations to educate others from within; and the *transformative*, where emphasis is placed on "moment-to-moment" (Urrieta 2004, p. 6) actions meant to work towards a just world one step at a time (see Fig. 28.1).

Activism in science is not limited to the environmental realm. However, we find power in drawing extensively from the environmental justice movement. It is in this movement that we witness the fight between ideology and lived experience (Brodkin 2009) as well as the struggle to integrate scientific knowledge and practice, power and positionality, economic and corporate concerns in a racialized, gendered and classed global society. The stakes are high (from personal health to global sustainability), and there is a global pattern of unjust practices. Reflecting upon how and why knowledge and action come together in ways attentive to these vast tensions can offer a broader model for activism in science.

Green Energy Technologies in the City: GET City

GET City is a year round program for youth ages 10–14 in River City, an economically depressed city in one of the most economically-depressed states in the US. Located at a local youth club, GET City began in the summer of 2007 with a cohort of 20 students who investigated whether their city, River City, contributed to the urban heat island phenomenon. Since then, the program expanded to reach about 30 students per year, and also to year-round programming. The club has been in existence in River City for over 40 years and is one of the more robust youth centers in the city serving over 250 children and youth per day.

The original goals of GET City were to offer youth an opportunity to engage with advanced information technology skills while learning about green energy issues. Over the 3 years of GET City we have strived to create authentic investigations rooted in local problems of global importance (e.g., Should river city build a new hybrid power plant?). However, over time, the youth have authored novel pathways for contributing to the local and national discourse and taking action on green energy and the environment (Calabrese Barton and Tan 2010). What was new in GET City was the collective authoring of new ways to push towards a deeply meaningful engagement with others on how and why caring for the environment matters in ways that were science rich and authentic to the situated needs of youth, most of whom live in the most economically depressed neighborhoods in the city.

Crafting Narratives

Our work in GET City is a part of a broader critical ethnographic and community-based effort to engage low-income youth in green energy technologies. In our work we have fieldnotes, video of weekly sessions, interviews, and a corpus of student work. However, because the "new green roof" was not an actual target of the GET City program and emerged as an artifact of broader community participation in the youth's work on green energy, we sought additional information. We purposefully set up additional experiences to fully capture different interpretations of "getting a new green roof." First, we asked youth to generate a map that showed the "critical events" and "experiences" that led to the new green roof. We conducted individual interviews, and asked the participant to tell us their story of how the club got its new roof. We prompted for what roles they believed GET City played along with their own personal contribution.

A grounded theory approach to data analysis yielded three salient coding trees: (a) what the roof signified; (b) steps, knowledge, and practices that led to the roof; and (c) youth positionings. Cutting across each of these three coding trees were ideas about the science youth drew upon. These coding trees were used to help structure the creation of a set of individual narratives on how the club got its new roof, of which we tell three below. Looking across the narratives we noted similarities in how the youth positioned themselves vis-a-vis science, community, and the roof process, along with what the roof signified for them.

Narrative Constructions of the New Green Roof

We share three narratives of how the club got its new green roof. In telling the narratives we seek to show that the ways in which youth describe these accomplishments and how the relative priorities they ascribe to them works to reify their identities as community science experts "who make a difference."

Fig. 28.2 Jana working on a graph

Jana's narrative: "It changes our lives." Jana is a vivacious 6th grader who attends the local elementary school adjacent to the Club (see Fig. 28.2). While small in stature, she exudes confidence. Jana joined GET City in the Fall 2008, in part because her older sister had participated in GET City the previous year and she was eager to participate in some of the activities and to gain access to the computers while learning more about the environment.

Jana described the green roof as "a *big* step ahead." She was careful to note that "just because you say green doesn't mean you have to change your whole life. You can still be you and still help the earth at the same time." Yet, Jana also pointed out that the new green roof "changed our lives." Jana is a matter of fact person. She has a deep thirst for knowledge and will often wade into complexity as she seeks a carefully thought out position (see Kissing 2010). Not prone to the dramatic, such a statement that the roof "changed our lives" carries weight. Jana was quick to point out that the roof impacted both the individuals who came to the club and the larger community. She explained that the roof's skylights brought much needed natural lighting in spaces in the club that are windowless. This mattered because "a lot of kids come to the club." She also noted that as a result of the roof, "our club has more money to run other programs:"

> The roof would help them to save a lot of money. Since in the summer a lot of kids come here and so that means a lot of energy and stuff to feed us and have the lights on. And when a lot of people come in one area, it gets hot. So when you have a green roof in the summer time it keeps it cool in the building and outside [on the rooftop] and when its cool like it is now, it keeps it warm and dry and stuff. The green roof is a big change.

Most importantly, Jana suggested that the new roof showed that even in their neighborhood a green roof was possible. That Jana described the new roof as a "big step" and as "real change" seems important given the backdrop of the current political climate where "change" is an important part of the lexicon, but frustration exists among those in her community suffering greatly from the current economic recession.

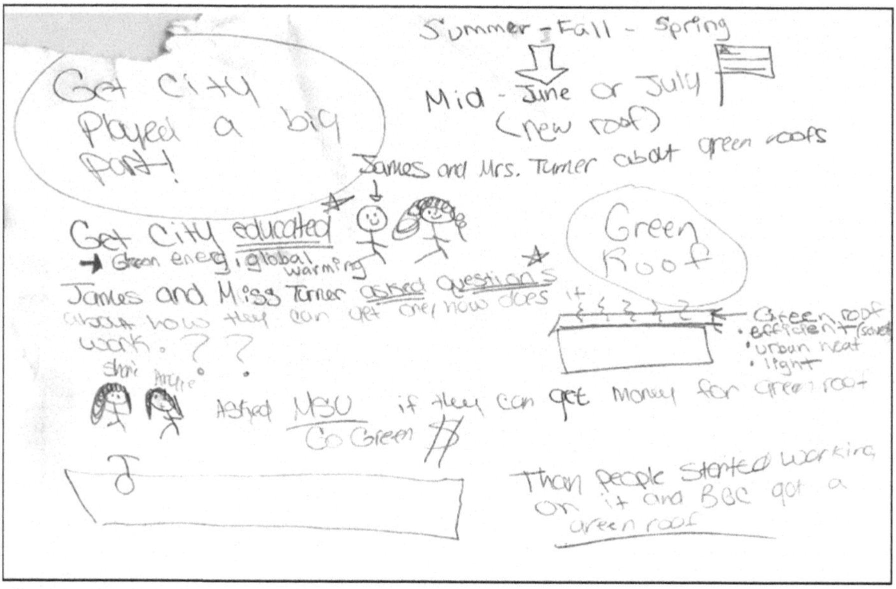

Fig. 28.3 Jana's map of "how the club got its new green roof"

Jana believed that the youth in GET City "played a big part" in how and why club leaders sought out a green roof. She placed primary emphasis on the role that GET City youth played in educating others on green energy issues, and getting others "talking about how to change people's lives." In her map of how the club got its new roof, Jana underlined twice the point that the youth "educated" the club leaders, and similarly underlined twice the fact that club leaders turned to youth for direction by "asking questions" (see Fig. 28.3). Jana viewed the youth as experts on the topic who carried a heavy responsibility for getting others to reflect and act upon their concerns. In fact, Jana credited their Public Service Campaign, a series of 30 and 60s digital shorts that the youth created earlier in the Fall 2008 on green energies, as the key event that got the green roof process started because they "got people talking."

Jana wove in two related science content lines in describing the value of the new roof. First, she pointed towards the role of cutting edge technology. She stated that "Back then there were a lot of ways to be green, but now there are newer ways to be green, and we are more educated and we are hard working kids who care." She was clear in her narrative that even a few years ago not enough was known about the connection between energy efficiency and global warming such that considering a green roof was a viable idea. Jana is clear that the youth's up-to-date knowledge of alternative energies and urban design directly parlayed into a new roof. Given that the club has one of the few green roofs in the city that contribute to LEED certification, her stance is not a surprise.

Second, Jana indicated that a detailed plan with "proof" would be needed to educate others about how to get a roof. The roof "is more efficient" and "you use less energy to heat and cool the building." The "less energy you use," the less we "contribute to global warming." The central concern for Jana was making this point evident to others in a carefully planned way: "What you have to do is to convince people. First of all, you have to have a plan and you have to stick to it and be determined. After you have that figured out, you get the proof, then make a video and then back it up with information, and then show it to the highest people in charge." What seems especially important here is how being knowledgeable in science can break down power relations in youth's efforts to gain access to resources to acquire something like a roof. Such power relations matter to Jana not only because of how they – as "club kids" – are positioned *without* resources but also because of their age. As she poignantly stated, "Most people say kids are too young and they can't really do anything. We are in 5th and 6th grade and we got Mr. J. and Ms. T. to get a new roof."

Cathy's Narrative: "They Had No Idea They Were Wasting Energy!"

Cathy is 13 and an 8th grade honors student who has participated in GET City weekly for 2.5 years. When we first met her she informed us that science was for "nerds." The only reason she joined GET City was because her mother made her. Yet, Cathy takes decisive ownership of the new green roof (Table 28.1).

Cathy's description of the process of getting a new green roof was rich with references to relevant scientific ideas and with descriptions of the importance of research and evidence. The new roof was most important because it would help to mitigate the urban heat island phenomenon and reduce the club's carbon footprint. To Cathy, green roofs are "energy efficient" and would "reflect rather than absorb the sun's heat." She also noted that the research helped them to make connections that she was not aware of before, such as "why roofs even matter [in climate change]."

Cathy highlighted four steps that led to the club's green roof: The urban heat island investigation, the youth-produced digital public service announcements and survey work on River City's official energy policies, the carbon footprint investigations, and investigations into green energy (see Fig. 28.4). Cutting across her description of each of these was the importance of doing research and getting evidence. In one 20-min conversation about her map, Cathy used the word evidence, research or data 12 times. In reference to green roofs she says, "We had evidence that showed that green roofs were better."

More importantly, however is *how* she used the terms research and evidence to position herself and her peers with powerful knowledge. Cathy pointedly illustrates

Table 28.1 What the new roof signifies

	Cathy	Jana	Janis
	"They had no idea"	"Big step" ahead	Saving the environment and money
Science	Mitigates the urban heat island phenomenon Carbon footprint Energy efficient and reflect sunlight Research and evidence	Prevent global warming More efficient and uses less energy to cool and heat the building Is possible because of scientific advances	Saves the electricity, the environment and money Provides natural lighting, which people enjoy
Social	Had a responsibility to teach peers and club leaders because "no one knew" Providing evidence and sharing ideas in relevant and fun ways Youth are leading the way	It changed our lives Convincing people in power Model for other youth and communities Club leading the way	If the club can make a change then others in the community can too A new roof costs a lot of money but the GC efforts were worth it
Personal	Dedication Hard work	Dedication Hard work	Dedication Hard work
Youth with similar narratives	Daniel, Patrice	Carla, DaShawn	Shernice, Kayla

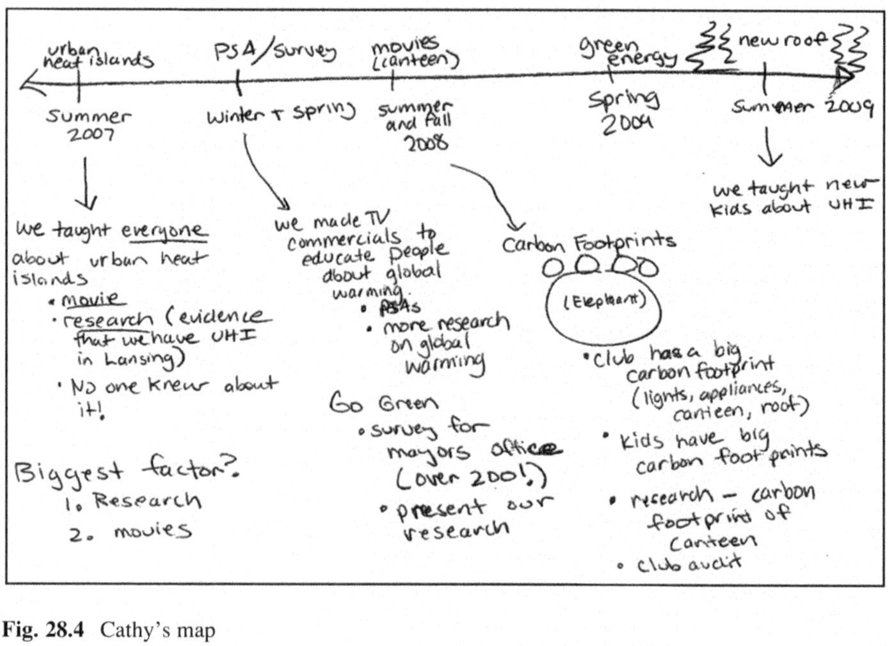

Fig. 28.4 Cathy's map

how research and evidence allow her to convince others in more powerful positions, such as the people with money or the club directors:

> I think research is a good thing too because then we can actually show like the board members and people who gave money like the things that GET City is doing.
>
> We figured out how our roof helps make urban heat islands. . .Well we had to like, cause [Sarah, one of the club directors] didn't know what it was either. So we had to tell her what it was. We had to show her all the things that we did and what we know about it and everything.

In fact, when asked what the most important factors in her mind were with respect to the club getting the new roof, she amended her map to include two factors: research and movies. She talked extensively about her movie making activities as an important component of conducting authentic research. She stated on three separate occasions in her interview that their digital products, such as their scientific documentaries and public service announcements, allowed them to organize and present their evidence in useful and fun ways:

> The movies were all about our research and what evidence we gathered. . .You have to show them [Club leaders] somehow. Like if we just wrote papers and stuff it would be just like school and stuff but I think that it was a fun way for [others] to learn so we need to do this for them."

While science played an important role in Cathy's description of the roof, it is in how her expertise positions her as an educator that seems to matter more in her narrative. Being an expert through her own research and dedication compelled her to take action with her peers because "no one else knew." In describing her work on carbon footprints she says, "Like we did those surveys about our carbon footprints and it proved that like a lot of us, it proved that a lot of our carbon footprints are big. And adding a green roof would put our carbon footprints low." She felt that this research had "a big impact cause they didn't know how much energy that they [the club] were wasting." Cathy was clear that the new roof was not the focus of their research but getting others to understand how their everyday practices contributed to climate change was. Figuring out that a roof could be an important step was part of the education process.

Cathy summed up the roof by saying that "personally I think that it gives me one more thing to say that I did to help the earth, you know?" (see Fig. 28.5). This comment reflects a deeper tension expressed in how Cathy positioned herself as both an expert and an activist. She stressed that African Americans are stereotyped in the media as people who do not care for the earth and who are not interested in using their knowledge and power to work for environmental causes. As she stated, "There's a stereotype. Because in the media and stuff they only show the negatives about African Americans. They don't show the positive and stuff." This comment deepens the meaning that the roof carries as a reflection of her seriousness and hard work.

For Cathy, the roof signified a great deal of seriousness and hard work and many hours of scientific research, ideals that run against the stereotypes held against African American youth. She also believed that because she had expertise that she

Fig. 28.5 Cathy posing in front of the Club's new green roof sign

deemed critical to the earth's survival, she was positioned as someone who *needed* to educate others. The roof signified "one step ahead" for one to build a bigger legacy in the community and for African American youth.

Janis' Narrative: "The green roof is the most important example of how we are make a difference experts!" Janis is a quiet 12 year old who attends the elementary school next door to the club. A gifted artist, Janis spends nearly all of her free time sketching images from pictures in magazines or photos. She attends GET City because it is "fun" and she can "make a difference" but she is also quick to note that she does not like science – it is for "geeks." She does however "love" GET City because it's a "different kind of science" where you can have fun and use computers (see Fig. 28.6).

Like Cathy, Janis also took decisive ownership over the new green roof. In fact, Janis explained that the green roof was "the most important example of how we are make-a-difference-experts." According to Janis, a make-a-difference-expert was someone who "knows what they are doing" and "how to make a difference":

> We know what we are doing. We know how to make a difference. [We know] how to save energy and how to convince other people of better ways to do things with electricity. That is one way that we are experts. The roof is probably the best example because we actually helped the club save money. They spent a lot of money getting the roof but now they have probably already saved enough to get that roof again. In the long run it saved money.

Fig. 28.6 Janis working on her skylight report

The roof was important to Janis because it helped to save the environment and save money at the same time. Janis almost never talked about environmental issues without also mentioning the financial impact of such issues. Her attention to the socioeconomic positioning of her family and her peers was central to the discourse she brought to the roof. While it has become acceptable in left-leaning white middle class culture for individuals to shoulder the burden of greening our world (i.e., by paying more for green electricity, etc.), Janis was adamant that environmental sustainability had to be affordable. In fact, Janis described the green roof as a roof that is "healthier" both financially and environmentally:

> It's basically like a roof that is healthier, like for finances and stuff like that. A word to describe it is efficient. It's helping you save the things you need, like money and electricity.

As indicated by the quote above, she suggested that the green roof was the best example of what it meant to be a make-a-difference expert because the youth know what they are doing, they have done all of the hard work, and they helped save the club money. According to Janis, it would be hypocritical for the club *not* to get a green roof after the youth worked so very hard to learn about green energy.

Janis' narrative focused in part on getting the right story to the right audience. Whereas Cathy (and to a lesser extent Jana) drew upon scientific research to build a coherent and convincing story that provided evidence on the relationship between urban design and climate change, Janis used science to link greening practices with economic and personal concerns. Of the activities that Janis pointed out as the most crucial, she highlighted the two audits that youth did of the club to determine how to save money and to make youth more comfortable. The first audit took place in Fall 2007 and focused on the electrical usage in each room in the club (appliances, wattage, and hours in use). The second audit took place 3 months before the roof installation in Spring 2009 and focused on determining potential locations for skylights (in the new roof).

Janis further stated that both of these activities were important in convincing others that the green roof was important because they showed how much money and electricity was used:

> We went around the club and saw how much computers were left on and how many watts that was. How many lights, and what kind of lights. We also calculated how much money the electricity uses, changing watts to dollars. We showed them how much money they would save and how much money each room used, and how many light bulbs in each room that could be changed for natural light. So now there are like 3 skylights, in the lobby, clubroom, and office. It helps the light because it's natural light. It's better than fluorescent bulbs, but what it means is that we save energy, and so we save money, and that is what we need to do.

Janis was not at all surprised that "the kids were the main reason the club got a new roof." Agreeing with her peer Kayla, she stated, "I didn't know that the roof cost $65,000 but I'm not surprised. Our work made the difference because they showed what green roofs were and how they would save money and the environment."

Discussion: Activism and Becoming a "Make-a-Difference Expert"

The roof signified scientific, social and personal accomplishments for the youth in ways that reify what it means to be a make-a-difference expert. The roof reflected the hard work and dedication of the youth to develop understandings of their own roles in global sustainability, and in their ability to take this message to those in power in relevant, fun and justifiable ways. The roof represented "many hours" and "extra days" of researching, getting evidence, and educating others about their own contribution to the larger global carbon footprint. The roof also reflected a "big step" ahead for the youth in GET City, the club and the community – a repositioning not only of the club as a leader within the community, but of tacit understandings of who is allowed to participate in the discourses of environmental stewardship. Yet, the differing emphases on the range of social, personal and scientific accomplishments also suggest that what it means to be a make-a-difference expert is dynamic, and that activism is a generative process deeply linked to knowing and being in science (see Fig. 28.7).

The youth took up many subject positions in their stories of the roof, which framed how and what it meant to be a make-a-difference science expert. They reflected a growing self-awareness of how their own practices (and by extension those of the club) contributed to a larger carbon footprint and to wasted money. Citing evidence and proof from audits they conducted, Cathy and Jana traced out carbon cycles and money cycles in relation to the roof. While the connection between socioeconomic positioning and environmental concerns were strongest with Janis, how and why understanding the carbon cycle mattered was coupled with both environmental concerns and financial well-being for all the youth.

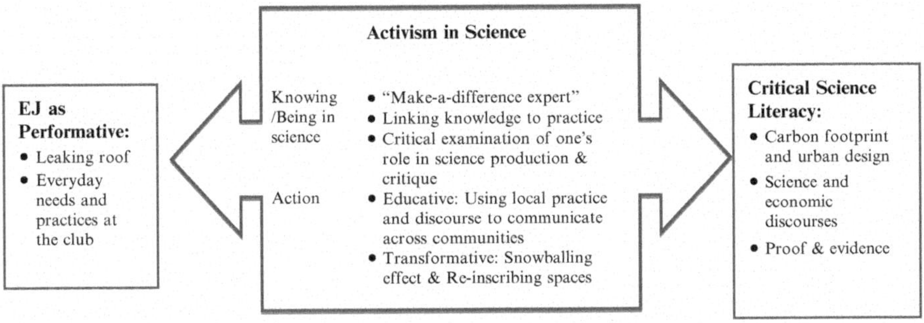

Fig. 28.7 Activism in science

Thus, becoming a make-a-difference expert required a localized knowledge of the scientific phenomenon at hand. Carbon cycling is a big idea (and an abstracted idea) in science, and yet, to be a make-a-difference expert meant that the youth could explain its value in terms that made sense scientifically as well as to their neighbors, in both cases with appropriate forms of evidence – as Janis reminded us, by "changing watts to dollars."

The youth also leveraged their growing expertise to break down power differentials around broader environmental issues in their community and city. Janis repeatedly referenced their "big break" when they got those in power to listen, and how important it was to show others. While youth used their expert position to break down power, they did not reposition themselves with power over others. Instead of activistic protesting, they took more of shared stance with their strong commitment to educate others. Cathy talked a lot about how "others didn't know" and they "had to make movies to show the evidence" and that movies were important because people would not read school-like papers.

What is interesting about taking on the responsibility for educating others and leading change was the merging of what is often viewed as contradictory subject positions in the context of science education, even in informal spaces. That is, by enacting science expertise that reflected both traditional scientific practice while also leveraging hip-hop, youth-speak, art and creativity the youth co-opted (un) desirable meanings of being a "club kid" with an urgency to build a more just world, fashioning a practice that was respected across two different worlds: local peer culture and white corporate/governmental America. Indeed, the youth's actions were legitimized by peers *and* authority figures (e.g., the roof funders, club leaders, mayor's office), and such maneuvering inscribed urban youth culture into doing science with a purpose. Simultaneously, it justified their growing power and leverage as green energy experts. Such an identity emphasized (rather than *averted*) how race/culture *and* science merged to transform being a "club kid" into something desirable: becoming a science expert to the local and global communities.

This latter point is fairly significant. As Cathy indicated in her interview, the issue of race was central to what it means to be a make-a-difference expert. Painfully aware of how her peer group is portrayed in the media, the roof signified a way to speak back to these deficit images. Both Jana and Janis further recognized how youth voices have been left out of environmental discourses. These youth, and others in GET City, spent hours scouring the internet looking for images of Black families recycling or engaging in environmentally friendly practices only to be frustrated by their inability to find them. That their work at the club led to the green roof destabilized stereotypical understandings of what it means to be a club kid or an environmental steward.

Finally, the green roof reflected just how much the club itself has been re-inscribed with new meaning for youth. It is not just a place for kids to hang out or to engage in after school programs. The club itself is a leader in a global movement towards sustainability and a local icon for making a difference. More importantly it is home to youth from lower-income and African American backgrounds who have created the power to enact change in a domain that has largely ignored their voices.

The power here is not in the roof itself, but in the everyday actions that made the roof possible. All of the youth pointed towards a "snowballing" effect of their moment-to-moment actions and the deeper understanding on the carbon footprint concept (knowledge-in-the-making) they gained as they explored different aspects of becoming more green in their local environment. While Cathy seemed most certain of the long term nature of their efforts, each of the youth pointed toward critical moments where their work took new form because they reached a new audience, which in all cases was an audience with more power to make decisions and with more money. Youth's critical engagement with the broader ideas of carbon cycling and climate change – and their relationship to broader political and economic themes – along with an examination of their own practices paved the way for the club leadership to imagine replacing their leaking roof with a green roof despite their limited budget. The green roof itself was not initially a target of the youth's efforts. Their investigation of the club's carbon footprint resulted in recommendations made regarding the lunch program, recycling, composting, and energy usage. The youth's ability to make these ideas a part of the way of thinking at the club opened up the possibility for the club leaders to ask "what they could do" when it came time to replace their roof.

Looking Ahead

We have posited a framework for activism in science education grounded largely in critical studies of environmental justice and science literacy. The youth's narratives on the new green roof offer a compelling way to make concrete and nuance this framework. Activism requires locally situated way of knowing the world that juxtaposes local culture and values on both how and why one engages in scientific practice. It also suggests that part of "knowing" is in how one takes actions in ways that are educative across a range of communities of practice and transformative to the spaces they claim. In the case of the green roof, such activism was made possible by youth's recognition as "community science experts", but at the same time their activism shaped that identity in ways that carried a range of meanings across the youth in the program.

Acknowledgment This material is based upon work supported by the National Science Foundation under Grant No. [Blind].

Any opinions, findings, and conclusions or recommendations expressed in this material are those of the author(s) and do not necessarily reflect the views of the National Science Foundation.

References

Bowers, C.A. (2002). Toward an eco-justice pedagogy. *Environmental Education Research, 8*(1), 21–34.

Brodkin, K. (2009). *Power politics: Environmental activism in South Los Angeles*. New Brunswick: Rutgers University Press.

Calabrese Barton, A., & Tan, E. (2010). We be burnin: Agency, identity and learning in a Green Energy Program. *Journal of the Learning Sciences, 19*(2), 187–229.

Cohen, C. (2006). *African American youth: Broadening our understanding of politics, civic engagement and activism. Youth activism: A web forum organized by the Social Science Research Council*. Retrieved May 11, 2010, from http://ya.ssrc.org/african/Cohen/

Hobson, K. (2006). Enacting environmental justice in Singapore: Performative justice and the Green Volunteer Network. *Geoforum, 37*(5), 671–681.

Kissing, M. (2010). *Step in the OK direction: Youth caring for the(ir) environment(s)*. Paper to be presented at the American Education Research Association Annual Meeting, Denver, CO.

Molina, G. (2000). *Supervisor Gloria Molina—List of accomplishments*. Retrieved May 11, 2010, from http://molina.lacounty.gov/PDFs/GM%20list.pdf

Nasir, N. S., & Hand, V. (2008). From the court to the classroom: Opportunities for engagement, learning, and identity in basketball and classroom mathematics. *Journal of the Learning Sciences, 17*(2), 143–179.

National Research Council. (1996). *National science education standards*. Washington, DC: National Academy Press.

Robins, S. (2006). From "rights" to "ritual": AIDS activism in South Africa. *American Anthropologist, 108*(2), 312.

Ryder, J. (2001). Identifying science understanding for functional scientific literacy. *Studies in Science Education, 36*, 1–44.

Stephens, S. (1996). Reflections on environmental justice: Children as victims and actors. *Social Justice : A Journal of Crime, Conflict & World Order, 23*(4), 62.

Subramaniam, M. (2000). Whose interest? Gender issues and wood-fire heating stoves. *American Behavioral Scientist, 43*(4), 707–278.

Teelucksingh, C. (2002). Spatiality and environmental justice in Parkdale (Toronto). *Ethnologies, 24*(1), 1–16.

Urrieta, L., Jr. (2004). Chicana/o activism and education: An introduction to the special issue. *The High School Journal, 87*(4), 1–9.

Weinstein, M. (2006). Slash writers and guinea pigs as models for a scientific multiliteracy. *Educational Theory and Philosophy, 38*(5), 583–599.

Weinstein, M. (2008a). Captain America, Tuskegee, Belmont, and righteous guinea pigs: Considering scientific ethics through official and subaltern perspectives. *Science & Education, 17*(8–9), 961–975.

Weinstein, M. (2008b). Finding science in the school body: Reflections on transgressing the boundaries of science education and the social studies of science. *Science Education, 92*(3), 389–403.

Part IV
Post-secondary Education

Larry Bencze and Steve Alsop

Preamble

In this final part of the book, four chapters are included that provide theory and practice dealing with socioscientific issues and actions to address them by students in a range of post-secondary contexts. Chantal Pouliot from Quebec City in Canada provides a rich, theory-informed, description of post-secondary/pre-university students' perspectives and priorities regarding biotechnological issues. Two other chapters, meanwhile, explore similar work with student-teachers. Lyn Carter, Carolina Castano and Meillita Jones from Melbourne, Australia, describe their collaborative efforts in use of transformative learning principles in encouraging student-teachers to address personally-meaningful socioscientific issues. Darren Hoeg and Larry Bencze from Toronto, Canada, describe a similar teacher education programme that also prioritizes student-led primary research as a basis for actions on issues and how such approaches were received by practice teaching supervisors. Finally, Jose Etcheverry, from Toronto, Canada, describes a university-based programme that attempts to educate a range of students (e.g., in schools and universities) about sustainability issues and actions and to engage them in community projects that may lead to tangible changes in energy use priorities and practices.

The Chapters

29. **Citizens as Concerned but Knowledge-Poor Watchdogs: Attributions of Legitimacy to Social Actors in the Management of Biotechnology Issues**
    Chantal Pouliot
30. **Transformative Learning in Science Education: Investigating Pedagogy for Action**
    Lyn Carter, Carolina Castano Rodriguez, and Mellita Jones
31. **Promoting Students' Collective Socio-scientific Activism: Teachers' Perspectives**
    Pedro Reis
32. **Counter Cultural Hegemony: Student Teachers' Experiences Implementing STSE-Activism**
    Darren Hoeg and Larry Bencze
33. **Implementing Practical Pedagogical Strategies for the Widespread Adoption of Renewable Energy**
    Jose Etcheverry

Chapter 29
Citizens as Concerned but Knowledge-Poor Watchdogs: Attributions of Legitimacy to Social Actors in the Management of Biotechnology Issues

Chantal Pouliot

Abstract The concepts of participation and deliberation have been invested with strong symbolic weight in the field of science education and, more specifically, in the teaching of socio-scientific issues (SSIs). However, the teaching of socio-scientific issues has not yet emerged as the "natural" or "self-obvious" place for focusing attention on the socio-political management of socio-scientific issues. In the first section of this chapter, I outline a number of conceptual contributions originating in political philosophy, a field that has engaged in sustained reflection concerning the participation of ordinary citizens in the deliberations surrounding socio-political decision making. In the second section, I present the viewpoints of post-secondary/pre-university students (who are also training to become primary or secondary school teachers) concerning the management of socio-scientific issues. I also provide illustrations of how these students describe the roles played by various actors – citizens, industry, government, and members of the scientific community. In the third section, I identify the opportunities offered by these descriptions for redistributing legitimacy and re-examining the modalities of citizen participation in the management of socio-scientific issues.

Keywords Socio-political action • Management of socioscientific issues • Citizens participation • Preservice teachers • Science studies • Political philosophy • Roles and capacities of social actors

---

C. Pouliot (✉)
Département d'études sur l'enseignement et l'apprentissage, Université Laval, 2320 rue des Bibliothèques, Québec, QC G1V 0A6, Canada
e-mail: chantal.pouliot@fse.ulaval.ca

L. Bencze and S. Alsop (eds.), *Activist Science and Technology Education*, Cultural Studies of Science Education 9, DOI 10.1007/978-94-007-4360-1_29,

Prologue (Quebec's Student Spring)

I wrote the first lines of this chapter as a crisis in democracy was unfolding. On the evening of May 22, 2012, an estimated crowd of 150,000 was moving through the streets of Montreal, Quebec. The marchers had gathered for two specific reasons: to protest against the 75 % hike in university tuition over a 5-year period (i.e., an increase of $1,625) and to denounce the Quebec government's Bill 78, which was designed to force the return to class of 175,000 postsecondary students, who had been on strike for 100 days, and which, in particular, made it mandatory to inform the police in writing of a demonstration at least 8 h before its scheduled start time. At 10 pm, the demonstration was declared illegal by the Montreal police department because it had not received the proposed march route and because some of the protestors were wearing masks.

On the evening of the 100th day of the strike, citizens were able to express the strength of convictions regarding the asymmetric exercise of power. Not only students (including high school students) but also union members, professors, parents and other people sympathetic to the cause all voiced their dissatisfaction with the government which, in their view, had failed to represent their interests and, worse yet, had made short shrift of their participation in the deliberations occurring upstream from the decision to increase university tuition.

Over a 100-day period, a series of demonstrations occurred in daytime and in evenings, producing scenes of violence to which Quebecers were little accustomed. The media broadcasted – repeatedly, on occasion – images of peaceful demonstrators who were clubbed, pepper-sprayed and arrested by the dozen. On May 23, the 101st day of the strike, 694 people were arrested in Montreal and Québec City. On the 102nd day, thousands of people defied the ban and took to the streets without notifying the police authorities of their march route (a situation that would occur repeatedly thereafter). Numerous arrests were made both that day and on the days following (making for a total of 3,400 arrests during the student protests; Dupuis-Déri 2012). In contrast with the agitation characterizing the spring protests, summer turned out to be relatively calm. And, just prior to fall, a new provincial government took power (September 4, 2012). The new premier quickly moved to rescind the tuition increase.

Science Education for Socio-political Action

The concepts of participation and deliberation have been invested with strong symbolic weight in the field of science education and, more specifically, in the teaching of socio-scientific issues (SSIs). Customarily, few details have been provided as to what is meant exactly by participation and deliberation, which are also used in the fields of political philosophy and science studies. More recently, however, Christensen and Fensham (2012) have strived to clarify the meaning and use of such key ideas as risk, uncertainty and complexity. The emancipative goals associated with teaching of socio-scientific issues have been explicitly discussed by numerous international authors. According to Robottom and Simmoneaux (2012), such ambitions flow from the observation that "scientific expertise continues to be highly regarded in society" (p. 3). From a similar perspective, Albe (2009) has stressed that the teaching of socio-scientific issues is aimed at fostering "a critical, reasoned use of expertise and democratic participation in public debates, expertise procedures [i.e., assessment processes] and decision making in techno-scientific affairs" (p. 16, our translation). Then, from a perspective centering on an activist approach to socio-scientific issues, Bencze et al. (2012) have pointed out that: "Potential personal, social and environmental problems associated with SSI [...] appear to be so severe that it is apparent we need activist-oriented societies to address them as soon as possible" (p. 132). Thus, it can be affirmed without too much risk of error that both the teaching of socio-scientific issues and the research dedicated to this subject are, as a rule, framed according to the idea of the participation of 'ordinary' citizens in democratic activity and informed decision making (Roth and Désautels 2002).

That being said, the teaching of socio-scientific issues has not yet emerged as the "natural" or "self-obvious" place for focusing attention on the socio-political management of socio-scientific issues. In effect, the existing research provides information about the values being mobilized, the arguments being advanced, the discipline-centred notions being used and the views of the nature of science being adduced. At the same time, it has, only to a very limited extent, documented students' points of view concerning roles that various social actors may assume as part of managing such issues. I believe, however, that the democratization of socio-scientific issues (or the absence thereof) inevitably brings into play not only the granting of the "right of voice" (and the right to be involved in the production of legitimate knowledge) but also the adoption of a critical view of dominant discourse concerning, particularly, the roles and capacities of citizens. With these considerations in mind, I believe that the teaching of socio-scientific issues should draw on models and theoretical notions that offer the opportunity to: (1) "problematize" the management of socio-scientific issues through theoretical frameworks that are sensitive to the (asymmetrical) granting of the right of voice and the right of knowledge production; and (2) where possible, break with the dominant "deficit model" of citizens' knowledge (Callon et al. 2001). From that point of view, I agree with Sadler et al. (2011) that the community needs additional conceptual tools for supporting SSI-based education (p. 80).

This chapter is divided into three sections. In the first section, I outline a number of conceptual contributions originating in political philosophy, a field that has engaged in sustained reflection concerning the participation of ordinary citizens in the deliberations surrounding socio-political decision making. Obviously, it will not be a question of delving into linkages between the ideas of participation and deliberation, or of going into detail about the tensions arising between them. The point, instead, will be to bring out how most of the new forms of democracy partake of both participatory democracy and deliberative democracy. I also highlight the contribution of science studies to reflection on citizen participation in the management of socio-scientific issues. Research in this field offers the opportunity to conceive of procedures for managing issues according to a context in which citizens appear as legitimate interlocutors and co-producers of knowledge. In the second section, I present the viewpoints of post-secondary/pre-university students[1] (who are also training to become primary or secondary school teachers) concerning the management of socio-scientific issues. I also provide illustrations of how these students describe the roles played by various actors – citizens, industry, government, and members of the scientific community. In the third section, I identify the opportunities offered by these descriptions for redistributing legitimacy and re-examining the modalities of citizen participation in the management of socio-scientific issues.

Political Philosophy: Participatory Democracy and Deliberative Democracy

In the field of political philosophy, the democratization of the public sphere is a weighty notion. For Cohen and Fung (2004), "[E]xpanding and deepening citizen participation may be the most promising strategy for challenging the inequalities that stem from asymmetric concentration of interests and from traditional social and political hierarchies" (p. 25). Researchers in this field have dedicated reflection to such subjects as the genealogy of participatory mechanisms and forums, as well as the interconnections between the notions of participatory and deliberative democracy. As a look at the contemporary literature in political philosophy serves to show, participatory democracy and deliberative democracy are descended from distinct traditions.

Participatory democracy came to the fore following publication of Pateman's *Participation and Democratic Theory* (1970). Out of a concern for raising questions over the limitations of representative democracy, research and discussion about participatory democracy have focused on the plurality of actors and the range of modalities and forums through which their commitment is expressed. The deliberative democracy current, which began to gain influence in the 1990s, draws on the theories of Rawls (1993, 2001) and Habermas (1997). Deliberative democracy does not run counter to the idea of representation, per se. It maintains a pronounced interest for the moments and processes of discursive interaction. However, according to the main premise, the quality of the back-and-forth of arguments is what serves to convince interlocutors and to facilitate shared decision making.

[1] These students are enrolled in a CÉGEP institution. 'CÉGEP' is the acronym for *Collège d'enseignement général et professionnel* (in English, this means 'General and Vocational College'). These institutions are unique to Québec, and is a mandatory step for secondary students pursuing university in that province.

Thus, reasoned argumentation has an important role to play, having the capacity to convince and win over interlocutors to a decision deemed as being in the interest of the common good.

For some philosophers, deliberative democracy represents an unattainable ideal. Mouffe (2005), among others, has criticized the notion of deliberative democracy, arguing that many socio-political conflicts are structured by irreconcilable points of view. These conflicts, for which there is no rational (and discursive) solution, unfold in a space of social inequalities and are necessarily settled through the exclusion of people concerned (she refers to this irreconcilable dimension as 'the political,' in contrast with 'politics,' a term she uses to refer to the political activity associated with living together). From that perspective, what appears to me as being of particular relevance to the teaching of socio-scientific issues are the notions that: (1) the existing socio-political order results from a particular expression of power relations; and (2) there is always a possibility of bringing about world situations different from the currently existing one.

Upon examining aims and theoretical tenets underlying notions of participatory and deliberative democracy, it becomes apparent that socio-political mechanisms and forums established in the last several years are both participatory and deliberative. For several authors in the field of political philosophy, such as Blondiaux (2004, 2005) and Cohen and Fung (2004), the end result has been to lessen the need for (and indeed to render obsolete, according to Blondiaux) contrasting analytical categories (e.g., participation/deliberation) when discussing the democratization of the management of socio-political issues.

Science Studies and Experiences in Citizen Participation

Science studies are driven by a concern for understanding the origins, dynamics and consequences of the connections between science, technology and public interests (Sismondo 2008). Research in this field has brought out a certain number of elements that are relevant to the thrust of this chapter.

Citizens Who Are Interested in Becoming Involved – And Capable of Doing So

A diverse body of research has helped to portray the sciences as social and political activities (Haraway 1991). In the second wave of science studies (Collins and Evans 2002), considerable effort was dedicated to documenting the ins and outs of citizen participation in management of local and national socio-scientific issues. The pioneering articles of Epstein (1995) and Wynne (1996) have helped to enrich our understanding of the ways citizens grapple with complex socio-scientific problems, configure the options available to them, and interact with people mandated to provide expert opinions. This widely-recognized research was followed by other projects which demonstrated that citizens are capable of successfully and usefully engaging with defining what constitutes a problem, establishing research protocols, producing legitimate knowledge, and critically comparing uncertainties and risks accompanying socio-scientific issues.

Among other significant contributions, this research has shed light on a twofold demand on the part of citizens. First, they are calling for additional opportunities to be included in the decision-making procedures surrounding socio-scientific issues. Secondly, and primarily in response to the unpredictable nature of impacts of techno-scientific developments, citizens are seeking to enhance the legitimacy accruing to the knowledge they develop and master.

Uncertainties at the Heart of Socio-scientific Issues

For many researchers, the discourse of science tends to disregard uncertainties – in a certain way to the benefit of risk:

> [Furthermore,] scientific and policy institutions often frame as 'risk' – implying calculable probabilities of known outcomes – what is actually uncertainty or even ignorance about the possible consequences of a given form of technological development, and ambiguity as to the proper meanings of the issue(s) at stake. These more challenging dimensions are thus concealed from formal public treatment and negotiation. (Leach et al. 2005, p. 10)

Discussions of the management of socio-scientific issues from the perspective of risk imply that there is some control over the outcomes of scientific practices. And yet, at the heart of socio-scientific issues are uncertainties whose effects, as Beck (1992) has pointed out, concern all citizens whenever there is no longer any externality. It is from this perspective that Callon et al. (2001) and Wynne (2005) stress the need not only to bring citizens into deliberations within the framework of 'hybrid forums,' but to also invite them to take part in producing knowledge that is relevant for reflecting on and prioritizing the available alternatives.

The 'Double Delegation': The Asymmetric Distribution of Legitimacy Regarding the Right of Voice and the Right of Knowledge Production

Empirical and analytical research on experiences of citizen participation began to mushroom in the 1990s. This work has brought out how the management of socio-scientific issues proceeds through power relations and the asymmetrical attribution of legitimacy to the actors and forms of knowledge thus mobilized (Leach et al. 2005, p. 9).

In a recent work (published in French in 2001 and in English in 2009), Callon, Lascoumes and Barthe developed a conceptual framework for interpreting various types of management of socio-scientific issues, ranging from a traditional type of management (termed 'representative') to a kind of management in which citizens are actively involved (also referred to as 'participatory' or 'dialogic'). According to their view, the 'representative' management of current socio-scientific issues gives birth to two couples: the layperson/scientist couple and the ordinary citizen/elected representative couple. Following Latour (1999), they argue that within each couple, the roles of political representation and of the production of legitimate knowledge (i.e., knowledge that influences decision making) are distributed asymmetrically. On the one hand, scientists are attributed the roles of producing knowledge, informing citizens and speaking for themselves. And on the other hand, citizens (assumed to be uninitiated in such matters) are associated with deficits of information and understanding; thus to them falls the role of being informed by scientists. In other words, in the context of the 'delegative' management of socio-scientific issues, citizens are, as a rule, stripped of both their right of voice and any role in producing legitimate knowledge. That is what these authors mean by 'double delegation.'

A Critical Look at Experiences of Citizen Participation

The analysis of new forms of citizen participation serves to bring out the diversity of both the modalities of participation as well as ways that authorities take into consideration viewpoints of citizens. To begin with, citizen participation generally fosters a fuller understanding of issues plus greater accountability. In contrast, hybrid sociopolitical forums are often consultative in nature and serve to legitimate choices previously decided on by elected representatives (Blondiaux 2004; Callon et al. 2001; Jasanoff 2012; Pestre 2006). In point of fact, although participatory and deliberative forums and mechanisms are highly valued in everyday political discourse, they are not always taken seriously in government action, as has been shown in a recent publication of the Institut du Nouveau Monde (Pion et al. 2009), which reviewed the experiences and methods of consultation regarding socio-scientific issues locally (i.e., Québec) and internationally.

Some Insightful Models

For anyone interested in conceptualizing the way socio-scientific issues are articulated and managed, the models of citizen/scientist interaction proposed by Callon (1999) (and borrowed from by different authors in varying degrees, depending on their research focuses) constitute particularly insightful heuristic tools (Rask 2003). The trio of deficit model, public debate model and co-production of knowledge model, along with the Translation model, have all helped to counter the still widely-held notion that ordinary citizens are unable to grasp the complexity of techno-scientific issues involved in, for example, the production and consumption of GMOs, the development of nanotechnologies, biotechnologies and food safety, or medical genetics (Jasanoff 2012; Leach et al. 2005). Accordingly, these models also appear to be of value for a kind of science education aimed at laying the foundations for an emancipative citizen identity (Pouliot 2008, 2009, 2011). The next few paragraphs are designed to sketch out the implications of this view.

The underlying premise of the deficit model is that scientists are the social actors best equipped to grasp the complexity of socio-scientific issues. Corresponding to the most widespread type of management of these issues, the deficit model has come in for severe criticism because it gives rise to an asymmetric distribution (between citizens and scientists) of the right of voice and the right to produce legitimate knowledge – i.e., knowledge that is taken into consideration by the actors engaged in the political processes surrounding decision making. In effect, in the deficit model, scientists are the ones who are attributed the roles of defining what counts as a problem, determining the make-up of research collectives, and producing and disseminating scientific knowledge. Furthermore, under this model, exchange between scientists and citizens is predominantly unidirectional and informational. Thus, from a certain point of view, the main consideration is to educate a public considered as being homogeneous with regard to its interests and misgivings, and as having a deficit of the knowledge relevant to understanding the issues at hand.

Under the public debate model, the right of voice is redistributed: scientists and citizens interact in spaces of public discussion (such as referendums or citizens' conferences). Citizens, who form sub-groups holding divergent and, indeed, antagonistic views do not necessarily speak with one voice (these sub-groups are referred to as 'concerned groups' in the formulation of Callon et al. 2001). Under this model, the knowledge produced and held by citizens, though different from that of scientists, is viewed as being likely to enrich efforts to define what counts as a problem and to examine potential avenues for action. One of the criticisms levelled at the public debate model concerns the inherent way it asymmetrically assigns roles in the production of scientific knowledge. Indeed, though communication has now become bi-directional, the production of legitimate knowledge remains, as in the deficit model, a role exclusively reserved for scientists.

The co-production of knowledge model is characterized by a redistribution of roles pertaining to participation in the production of legitimate knowledge. According to the main idea underlying redistribution, citizens may have experience that is relevant to the socio-scientific issues at stake and have the cognitive and discursive competencies required for taking part in defining problems, constituting research collectives and producing legitimate knowledge. This model presents a relationship between knowledge and power that is different from the preceding models. Specifically, it attributes greater legitimacy to citizens' knowledge, which is discredited under the deficit model.

The Translation model also constitutes an alternative way of conceiving of the management of socio-scientific issues. A threefold model, it aims to identify the moments or phases during which citizens are brought into the process of managing issues – in particular, when defining problems ("translation 1"), producing data and knowledge ("translation 2"), and disseminating the resulting knowledge ("translation 3").[2] The thrust of this approach is to redistribute the roles of representation and knowledge production between scientist and non-scientist actors.

The Viewpoints of Prospective Teachers Concerning the Management of Socio-scientific Issues

Methodological Considerations

The following viewpoints have been excerpted from a study conducted among post-secondary/pre-university students who are also prospective primary and secondary schoolteachers. This research project concerned students' relationship toward people considered to be scientific experts (article submitted). In this context, during an activity at the beginning of their course, students had to give their opinion in writing concerning a statement borrowed from the VOSTS questionnaire (Aikenhead and Ryan 1992), which was developed on the basis of the written views of 10,000 young Canadians concerning the social nature of science (Lederman et al. 2002). This questionnaire contains an inventory of statements to which survey participants are asked to reply using a pool of multiple-choice items; they are also requested to support their views in writing. The VOSTS questionnaire has been used in several studies (Fleming 1992; Zoller et al. 1991) to characterize teachers' viewpoints. I gathered 120 written responses to a VOSTS statement dealing specifically with the roles of scientific experts and citizens regarding the development of biotechnologies, worded as follows: "Scientists and engineers should be the ones to decide on future biotechnology in Canada (for example, recombinant DNA, gene splicing, developing ore-digging bacteria or snow-making bacteria, etc.) because scientists and engineers are the people who know the facts best." Students had to select from a set of responses the statement that best expressed their position for each item and justify their choice in writing (about one page).

[2] It is important to note that, in this model, citizens do not constitute a monolithic group in terms of the power they exert whenever they participate in any phase of the management of socioscientific issues. A number of authors have argued (and, in some cases, illustrated) that social actors concerned with socioscientific issues move and act in a context of structural social inequalities. Money, social status and professional relations are all examples of elements that give them greater leverage in decision-making processes (Alsop and Bencze 2010; Kerr et al. 2007; Kleinman 2000).

Descriptions of Actors

In this section, I illustrate ways that students, on the whole, describe various actors – citizens, industry, government, and members of the scientific community – who are concerned with management of biotechnologies (but also of other issues, such as shale gas, that were spontaneously identified by students). Hopefully, this will help the reader to develop a rich, in-depth, picture of student viewpoints. With this goal in mind, I set out several excerpts, some of which (pertaining to the roles and capacities of citizens) are at first glance surprisingly lengthy.

Citizens

The great majority of the viewpoints articulated by prospective teachers are framed in relation to the notion according to which citizens should be informed by scientists of the research performed to date and of the knowledge necessary for understanding the development of techno-sciences. Such is the opinion of Annie (all student names are pseudonyms).

> Annie: People should be informed clearly and objectively (without manipulation) of the risks and benefits, in order to make informed decisions. (PT 44)

For Julia, whose viewpoint appears below, the information provided to citizens regarding research practices would bring about a form of transparency concerning scientists' research intents.

> Julia: Everyone is entitled to know what is happening where the future of society is concerned. That is why scientists, engineers and other specialists should inform the public about their research intents. (PT19)

Concerning citizen participation in decision-making processes, some students mentioned that citizens are unable to take part on account of how little scientific knowledge they master. As is noted in the following description, formulated by Myriam, the attributing of legitimacy to citizens' viewpoints rests on the mastery of techno-scientific knowledge.

> Myriam: The public would obviously have its say, and its opinion would be as important as that of scientists *if* it were as well informed and better educated about the subject [...] Obviously, the public should be informed and consulted, but given its lack of knowledge about the subject, it would be unable to make decisions relating to science. That isn't to say that it shouldn't – quite to the contrary. With more scientific knowledge, information and access to real data, the public should in fact be consulted more and its opinion should be respected more. (Participant's emphasis, PT15)

The description articulated by the next participant sketches out a portrait of a ranking of statuses in which scientists occupy a leading position. Maïka argued that owing to their academic training, scientists and engineers were better placed to understand the issues, noting also that the public lacked legitimate knowledge.

For this prospective teacher, it would appear that citizens' viewpoints should be solicited but not necessarily taken into consideration in the decision-making process.

> Maïka: Scientists and engineers have the appropriate, necessary training for understanding this phenomenon. Indeed, I believe they are the ones who should have access to important data since they will be able to view the situation from a fuller perspective. Secondly, I think that the public in general does not have any training or credible information that would make them better placed to make decisions regarding the issues related to the problem. [..] the public can have its say and be informed about the situation all the same. Thus, they should also be consulted in important decisions, but the opinion and positions of scientists should nevertheless be given priority, as they are the ones who are most competent to improve things and to gain a clearer understanding of the phenomenon than the public in general would be able to. (PT16)

Rosie, the next participant, cited the potential effects of the participation of citizens who are little informed about the issue. Using a model of iterative relations between two distinct spheres – i.e., science and society – she suggested that citizens' lack of understanding of issues could result in social, political and economic conflicts.

> Rosie: Since science influences society, and society influences science, it's important to have the public's opinion, too. In this case, there is a relationship of reciprocity, and the ignorance [of public opinion] about such issues could give rise to several social, political, economic and other conflicts. (PT13)

Thomas held to a viewpoint that was even more radical than that of Myriam (i.e., the public lacks knowledge) and of Maïka (the public lacks any credible training and information). In his opinion, there is an apparently unbridgeable gap between citizens and the scientific community. By asserting that inviting citizens to take part in decision making would constitute an arbitrary, unjustified attribution of power to citizens, Thomas clearly showed he was aware that the management of biotechnological issues went hand in hand with the distribution and exercise of power.

> Thomas: Mixing up scientific (and economic!) decisions with the populace amounts to placing power in the hands of a people that is disconnected from the reality of a country and in large part disconnected from the world of science and true facts. (PT7)

Several students mentioned their hesitation – or doubts – about the need for citizen participation in decision making or about the weight to be accorded to citizens' viewpoints. Such is the case of Marie, who, all the same, proposed that citizens be given the opportunity to express themselves in socio-political forums for decision making. In her opinion, there was a possibility that decision-makers would be receptive to some of the ideas thus voiced.

> Marie: Society should be informed a bit and kept abreast of various project and future decisions. I don't know whether it should have an impact on the final decision. It would be worthwhile to perhaps hold a conference at which members of society gave their opinions and raised questions. Even though it wouldn't have a direct impact on the final question, between the lines, if a good idea is put forward, there will probably be a portion of it in the decision-making path. (PT 17)

According to several students, it is absolutely vital to consult citizens. This viewpoint, as shown in the comments of Andrée and Mathieu, is premised on the notion that decisions have an impact on citizens.

Andrée: The public should also be informed, since shale gas, for example, has *impacts on their lives and indeed on their health.* [...] The public *must* be consulted on this subject, since the homes of many people are at risk if the decision is made to go ahead with shale gas. (My emphasis, PT3)

Mathieu: The decisions made affect the public (100 times greater than the population of scientists). Each citizen is entitled to give *his or her opinion* in the decisions to be made, which are certain to have *major repercussions* on the future. (My emphasis, PT20)

In the following excerpt, Nika describes the possibility of a citizen response or reaction. As will be noted, this form of political action is made possible by the education of citizens by scientists.

Nika: I fully believe that the public should be informed by scientists regarding decisions, since it is something that directly concerns us. In addition, by being informed, we will be able to react if we are in disagreement with them [the scientists] and their decisions. It might happen that scientists do not consider the potential consequences of their decisions as being important; that is why, if we are informed, we will be able to slow down or act on the decisions we disagree with. (PT41)

For some participants, citizens have the duty to contribute to the management of socio-scientific issues. Camille sketched out a kind of active citizenship in which one of the duties consisted in contributing to the formulation of alternatives.

Camille: [...] Obviously, these are problems that affect the entire public. Therefore, we should be informed and aware, but we should also offer alternatives in order to aid experts in their research and work. It is our duty as active members of society. (PT1)

Several students brought up the possibility of a leading role in the management of the issue of biotechnology development. In Théo's view, the public would act in the capacity of third party in the event of disagreement between government representatives and scientists. In his description, citizens are depicted as actors having indubitable capacities for understanding but whose knowledge on the subject have in part been provided by scientists.

Théo: Furthermore, the general public would serve as a kind of moderator between the two. In a conflict between the ideas of scientists and those of government representatives, the public would be the one that tipped the scales one way or the other. However, the public would have to base its position on solid arguments, probably in relation to the briefs presented by the other two parties. From this perspective, I believe it is necessary to better inform the general public about such subjects, since it first will be affected and yet, despite this, it is quite often left in the dark. It isn't absolutely necessary to consult it for every decision (that is unimaginable, even) but it would be worthwhile keeping it informed and offering it more often the possibility of giving its opinion about the developments concerning it. (PT 10)

Government

On the whole, students discussed the role of government in the management of socio-scientific issues in terms of the model of representative democracy. Many of them stated that the government should act as a representative of citizens. The viewpoint of Catherine was widely shared:

Catherine: It must be remembered that all society is concerned by this subject and that government should, normally, be a kind of "representative" vis-à-vis society. (PT 22)

In its capacity of citizens' representative, the government should be transparent in its interactions with citizens and scientists. Thus, Théo noted that elected officials sometimes skew the presentation of the pros and cons associated with the courses of action opted for.

Théo: More than any other [actor], the government must make an informed choice and not merely focus on the political or economic advantage that it stands to reap. Because, quite often, we are witness to governments concealing the problems related to their decisions and showing only the advantages (for instance, the various bridges across Quebec that the government contents itself with leaving in their current condition while being fully aware that they are extremely dangerous). It should also strive to convey clearly to the other two parties the (political and economic) advantages and disadvantages that are likely to stem from a decision. (PT10)

Several students advanced descriptions that evoke a certain perplexity regarding the transparency of elected officials in relation to decision making. In the following excerpt, Thomas uses expressions such as "theoretically," "apolitical," "again, theoretically," "the situation is different" – all of which highlight the limitations of representative democracy.

Thomas: The government represents – "theoretically" – the entire population. In an increasingly apolitical country, this reality is slowly fading away but it nevertheless remains true – again in theory. [...] The entity best placed to manage this type of decision thus continues to be the government. Of course, there is corruption, but again, theoretically, the government represents the public. In actual practice, the situation is different... (PT7)

Some participants ascribed to elected representatives self-interested behaviour driven by personal motives. From this perspective, Richard performed an analytical categorization that lumped together both the government and private companies.

Richard: In order to make a decision on a subject having an impact on all society, I believe it is important that the people making the decision not be motivated by their personal profit. Clearly, government officials and private companies are not well placed to make such decisions. (PT33)

Finally, student descriptions of the government portrayed this actor in reference to its mastery of knowledge considered relevant and necessary for shedding light on the issue. Generally speaking, it is proposed that the government be informed by scientists of the state of the issue since it is not itself a producer of knowledge. The viewpoint of Andrée stands out from the others in that it cites the Premier[3] of Québec (at the time this research was being conducted) to support the notion

[3] A 'Premier,' is the leader of the political party that dominates (leads) the government in any province, like Québec, in Canada.

according to which the government does not possess expertise in the area of techno-scientific development.

> Andrée: The government should not take charge of these technologies, since, as [Premier] Jean Charest asserted on the subject of shale gas on the television show "Tout le monde en parle," the government is not an expert in this area. (PT3)

Industry

Students' descriptions of private companies offer a portrait of self-interested behaviour. It is thus vital to contemplate protective measures and ensure that decision making is not entirely left up to these actors, as brought out in the comments of Isabelle and Maëlle.

> Isabelle: The government, public servants, companies, etc. should be as informed of the problem to the same extent as scientists and the public, but they should not be entirely responsible for decision making, as they often seek to protect their interests. (PT27)

> Maëlle: It's true that private companies have interests to protect, so it's possible their decisions would not be made on the basis of the good of society but solely on what works to their advantage. (PT18)

Raphaëlle, Sandrine and Richard articulated their viewpoints from the perspective of the influence of private companies on research activities. In particular, Richard's description emphasized the potential association of scientists with private companies and brought up the possibility of bribery.

> Raphaëlle: If a company that produces GMOs hires researchers for a project, the latter will grant greater importance to the benefits (i.e., the company wants to sell its product and provides funding for the project) than to the drawbacks. (PT25)

> Sandrine: For example, if it was entirely left up to scientists to make decisions, they might be *manipulated or paid off by the gas companies* in order to be able to drill for shale gas. (My emphasis, PT8)

> Richard: In order to prevent research from being biased by the bribing of scientists, the latter should not have the sensitive task of making the final decision. [. . .] Scientific experts and engineers have the power to inform us, but because of the research funding provided by a private company for its own ends, experts should not make the final decision. (PT 33)

Scientists

In most cases, students mentioned that scientists are the people best placed for understanding the issue. Their academic training and their participation in the production of scientific knowledge make them particularly competent for grasping the ins and outs of controversial techno-scientific developments.

Rosie: Scientists are best placed to contemplate the components of a problem, and they are the ones who are certainly the most aware of the components of this problem. (PT13)

As might well be imagined, one of the roles attributed to scientists is the production of knowledge. That is the view of Florence, who in passing notes the difficulty that citizens have producing empirical knowledge.

Florence: Obviously, people do not have the resources required to conduct experiments; that is why experienced people such as scientists and engineers are around. (PT24)

As is clear from an analysis of the descriptions of the management of biotechnology developments, students view scientists as having the role of informing citizens (and occasionally the government). According to Théo, scientists are not only responsible for providing information but also of defining what counts as a problem.

Théo: There's no doubt about it: scientists and other engineers are masters in their field. Thus, it is up to them to grasp, picture and explain (through popularization) all the ins and outs of the issue – all with a view to enabling the other parties to make an informed choice. (PT10)

The following two excerpts serve to show that the moments when citizens are informed by scientists are the subject of diverging views. For Mathieu, scientists should inform citizens once the results and findings have been produced. For Mathilde, communication between scientists and citizens (or, to more precisely, *from* scientists *to* citizens) should occur at several different times and should not be limited to the start and end of the process.

Mathieu Once they have finished conducting their research, scientists and engineers should convey the findings to society. (PT20)

Mathilde: Engineers and scientists should inform the public of the steps and procedures they are undertaking throughout the entire process. Which means not merely making do with informing the public at the beginning or the conclusion of the project, but providing information and results as the study progresses. (PT29)

Generally speaking, in the view of students, scientists are well placed to grapple with the scientific issues involved because they have scientific knowledge Some students attributed to scientists a considerable role in the management of socio-scientific issues, as is illustrated by a few of the following excerpts. For Julien and Maëlle, decision making should be left up to scientists.

Julien: Who is in a better position than scientists to more fully inform people about this subject? No one. Obviously, we need the opinion of these specialists – otherwise, we are likely to make the wrong decisions. They are fully equipped to guide us in these decisions. They deserve our fullest confidence and, by the same token, they should make decisions regarding the future of biotechnologies in Canada. (PT 45)

Maëlle: It would be better to have scientists make decisions about the future of biotechnology in Canada or shale gas, as they have the requisite education as well as access to essential data. (PT18)

Other students took a different view of the participation of scientists in decision making, holding to the idea that the responsibility for making decisions could not be left up to scientists. That said, these students generally accorded considerable legitimacy to the knowledge held and produced by scientists. Such is the case of Mathieu.

> Mathieu: I think it is out of the question to leave it up to scientists to make decisions. It is true they are more familiar with their subject and have findings. However, I don't see why we shouldn't have our say on the matter. (PT20)

Students who did not think it was desirable to leave it up to scientists to make decisions justified their viewpoint in two main ways. Some of them emphasized the effects of decisions on the lives of citizens. This was the line of argument adopted by Manon, who noted that decisions were accompanied by unexpected consequences.

> Manon: It cannot be asserted with certainty that scientists have the competencies required for making decisions, since *many consequences* of the development of science and technologies *cannot be predicted*. (My emphasis, PT38)

Other students pointed out that scientists did not make a priority of examining any issues other than socio-scientific issues. That was the point of view of Thomas.

> Thomas: Asking experts to make decisions would amount to throwing yourself off the cliff, since decisions would often be made without any economic, international or other kind of awareness. (PT7)

The Hybrid Management of Issues: A Second Look at Descriptions

In the preceding section, I illustrated ways in which students who are also prospective primary and secondary teachers described actors concerned with controversial development of biotechnologies. In particular, it should be clear that students attributed to citizens both the role of being informed and the role of contributing, in varying degrees, to discussions. They attributed to government the roles of representing citizens and informing them about the ins and outs of socio-scientific issues as well the decisions to be made. Finally, they attributed to scientists the dual role of conducting research and communicating the most relevant knowledge to citizens and government. That said, it is readily apparent that the deficit model forms a backdrop to the formulation of descriptions.

The deficit model packs considerable weight. Given its omnipresence in students' descriptions, the task of thinking outside the box becomes all the more difficult. As proof, I note that this model structures even the viewpoints of those students who are persuaded of the need for a hybrid management of socio-scientific issues. Indeed, their descriptions of forms of hybrid management of issues are laced with asymmetrical attributions of legitimacy, the most striking one of which concerns scientific-versus-lay knowledge and practices. In the opinion of students, there are kinds of knowledge that make it possible to gain a more sophisticated understanding of all the ins and outs of issues and to better position oneself. It is on the basis of the mastery of this knowledge that students propose, on the one hand, having scientists participate in decision making and, on the other hand, ensuring that citizens are educated (i.e., informed) by scientists.

In my view, the crux of the 'problematization' of teaching socio-scientific issues should be framed, in part, in terms of the legitimacy of the knowledge and practices of the various actors concerned – chief of which being the knowledge and practices of citizens and scientists. Upon closer examination of students' descriptions in light of the first section of this chapter on the subject of the contributions of political philosophy and science studies, it becomes clear that students' viewpoints offer opportunities for redistributing legitimacy and reconsidering more substantial modalities of citizen participation in the management of socio-scientific issues.

The comments of prospective teachers persuaded of the need for citizen participation in the management of socio-scientific issues are premised on the notion that the decisions to be made will have impacts on the lives of citizens. In the field of science studies, it is precisely the fact that socio-scientific issues come accompanied by numerous uncertainties that relevance accrues to the participation of citizens in defining what counts as a problem (Callon et al. 2001; Irwin 2001; Jasanoff 2003; Wynne 2003). From this perspective, it would be worthwhile raising questions in the classroom about risk discourse (which gives short shrift to uncertainties) and strengthening the legitimacy of citizen participation in debates and the co-production of knowledge by configuring the first moments or phases of the management of issues from the perspective of a hybrid definition of what counts as a problem. To this end, distinguishing between the theoretical notions of risks and uncertainties could prove valuable, as could using the Translation model described by Callon et al. (2001), which subdivides the procedures for producing scientific knowledge into three distinct moments (i.e., "translation 1," "translation 2" and "translation 3").

Furthermore, comments of students persuaded of the need for citizen participation in the management of socio-scientific issues stand out on account of how they show that in the eyes of students, such participation constitutes a way of managing the risks and uncertainties of political and techno-scientific misadventures. From the perspective of political philosophy and science studies, students 'problematized' the limitations of representative democracy. Indeed, many students brought up potential situations of corruption and the biased representation of issues and avenues for action. Students also mentioned that even if scientists were "best placed" for understanding the issue at hand, they were also capable of deliberately concealing certain research findings or downplaying issues of a non-purely scientific nature. According to their descriptions, students attributed to citizens the role of concerned watchdogs who nevertheless present a deficit in terms of the capacity for taking action and producing legitimate knowledge. In light of the preceding considerations, it would thus be worthwhile using models such as the public debate model or the co-production of knowledge model to: (1) show that socio-political management forums and mechanisms come accompanied by widely varying levels of legitimacy in regard to the right of voice and the capacity for knowledge production; and (2) convey the idea that it is possible to conceive of management procedures in which citizens are considered as legitimate interlocutors able to contribute their knowledge – on equal footing with scientists, moreover.

Concluding Remarks and Epilogue

As I was wrapping up this chapter, a Montreal police officer, well known among the public for having gratuitously pepper-sprayed demonstrators back in spring, was using excessive force against four citizens in an apartment in the Plateau Mont-Royal neighbourhood (October 2, 2012). A series of gripping video images, coupled with a phone conversation with fellow officers about the arrests that was captured unbeknownst to her, led to a full-blown media uproar. This officer not only intervened physically, she also vented a series of derogatory opinions concerning the social ranking (i.e., "artists") and presumed political identity (i.e., "red squares" – i.e., participants in the 2012 Quebec student protests and their sympathizers) of the people arrested. She could be heard saying: "All the rats that were upstairs . . . these guitar-playing [expletive] . . . all a bunch of red square types, all artists, basically a bunch of a–holes, and they all started coming out of the apartment."

On November 1, 2012, a judge of the Superior Court of Québec handed down a landmark decision. For the first time during the protests, a student leader was found guilty of contempt of court. Gabriel Nadeau-Dubois was, thus, confronted with the possibility of doing jail time. Later in the evening, approximately 100 people demonstrated in Montréal in support of this figure, who many considered to be the spokesman of a generation.

There is room for establishing a variety of connections between science education and social action in relation to what turned into the longest student crisis in Québec's history. In particular, this movement laid to rest the dominant representation according to which ordinary citizens (to use the term of Callon et al. 2001) and students lack the discursive and cognitive capacities required to understand complex situations. For anyone living in Québec between March 22 and September 4, 2012 or who had access to national and international media, the students and numerous concerned groups offered, for more than 100 days, a living demonstration not only of courage but also of a capacity for articulating arguments and concepts, thereby enabling them to hold their ground vis-à-vis the reigning Liberal (political party) government's discourse about paying one's "fair share." These students showed that they were able to produce legitimate knowledge – i.e., knowledge useful for understanding the issue and participating in its management (thereby becoming co-producers of knowledge) – while, throughout this time, they were viewed as enemies rather than as adversaries (Mouffe 2005) by the government of the day.

As brought out in this article, the politicization of science in the classroom calls for identifying the social actors concerned and the arguments invoked, legitimizing a plurality of knowledges, and configuring the management of socio-scientific issues according to a model at considerable remove from a deficit-centred interpretive framework. That being said, I do not claim to have identified *the most worthwhile* theoretical tools but rather *some tools among others* that, in addition to being relevant and useful in science teaching contexts, offer the requisite flexibility and, to my way of thinking, a stupendous opportunity for citizen emancipation and the 'democratization of democracy' (Callon et al. 2001; Désautels 2002). In my view, the emphasis on introducing political notions and considerations into science in the classroom squares with Hodson's recent acute observation: "Those teachers who promote involvement and develop action skills are 'riding a tiger,' but it is a tiger that may well have to be ridden if we really mean what we say about education for civic participation" (2010, p. 205).

References

Aikenhead, G. S., & Ryan, A. (1992). The development of a multiple choice instrument: "Views on Science-Technology-Society" VOSTS. *Science Education, 76*(5), 477–491.
Albe, V. (2009). *Enseigner des controverses*. Rennes: Presses Universitaires de Rennes.
Alsop, S., & Bencze, L. (2010). Introduction to the special issue on activism: SMT education in the claws of the hegemon. *Canadian Journal of Science, Mathematics, and Technology Education, 10*(3), 177–196.
Beck, U. (1992). *Risk society: Towards a new modernity*. London: Sage.
Bencze, L., Sperling, E., & Carter, L. (2012). Students' research-informed socioscientific activism: Re/visions for a sustainable future. *Research in Science Education, 42*(1), 129–148.
Blondiaux, L. (2004). Prendre au sérieux l'idéal délibératif: un programme de recherche. *Revue Suisse de Science Politique, 10*(4), 158–168.
Blondiaux, L. (2005). Démocratie délibérative et démocratie participative: une lecture critique. In M. H. Bacque, H. Rey, & Y. Sintomer (Eds.), *Démocratie participative et gestion de proximité*. Paris: La découverte. November 11, 2004.
Callon, M. (1999). The role of lay people in the production and dissemination of scientific knowledge. *Science, Technology, & Society, 4*(1), 81–94.
Callon, M., Lascoumes, P., & Barthe, Y. (2001). *Agir dans un monde incertain. Essai sur la dé mocratie technique*. Paris: Seuil [published in English (2009) as *Acting in an Uncertain World; An Essay on Technical Democracy* (trans: Burchell, G.). Cambridge, MA: The MIT Press].
Christensen, C., & Fensham, P. J. (2012). Risk, uncertainty and complexity in science education. In B. J. Fraser, K. Tobin, & C. J. McRobbie (Eds.), *Second international handbook of science education* (pp. 751–769). Dordrecht: Springer.
Cohen, J., & Fung, A. (2004). Radical democracy. *Swiss Political Science Review, 10*(4), 23–34.
Collins, H. M., & Evans, R. (2002). The third wave of science studies: Studies of expertise and experience. *Social Studies of Science, 32*(2), 235–296.
Désautels, J. (2002). L'alphabétisation technoscientifique et la démocratisation de la démocratie. *Canadian Journal of Science, Mathematics and Technology Education, 2*(2), 189–195.
Dupuis-Déri, F. (2012, October 12). Matricule 728 – Un cas isolé ou un cas qu'on tente d'isoler? *Le Devoir*.
Epstein, S. (1995). The construction of lay expertise: AIDS activism and the forging of credibility in the reform of clinical trials. *Science, Technology & Human Values, 20*(4), 408–437.
Fleming, R. (1992). Teachers' views of technology. *The Alberta Journal of Educational Research, 38*(2), 141–153.
Habermas, J. (1997 première parution en 1992). *Droit et démocratie. Entre faits et normes*, Paris: Gallimard.
Haraway, D. (1991). *Simians, cyborgs and women: The reinvention of nature*. New York: Routledge.
Hodson, D. (2010). Science education as a call to action. *Canadian Journal of Science, Mathematics, and Technology Education, 10*(3), 197–206.
Irwin, A. (2001). Constructing the scientific citizen: Science and democracy in the biosciences. *Public Understanding of Science, 10*(1), 1–18.
Jasanoff, S. (2003). Breaking the waves in science studies: Comment on H. M Collins and Robert Evans, 'The third wave of science studies'. *Social Studies of Science, 33*(3), 389–400.
Jasanoff, S. (2012). *Science and public reason*. Abingdon: Routledge.

Kerr, A., Cunningham-Burley, S., & Tutton, R. (2007). Shifting subject positions: Experts and lay people in public dialogue. *Social Studies of Science, 37*(3), 385–411.
Kleinman, D. L. (2000). Introduction. In D. L. Kleinman (Ed.), *Science, technology, and democracy* (pp. 1–12). Albany: State University of New York Press.
Latour, B. (1999). *Politiques de la nature*. Paris: Éditions La Découverte [also published in English (2004) as *Politics of Nature – How to Bring the Sciences into Democracy* (trans: Porter, C.). Cambridge, MA: Harvard University Press].
Leach, M., Scoones, I., & Wynne, B. (2005). Introduction: Science, citizenship and globalization. In M. Leach, I. Scoones, & B. Wynne (Eds.), *Science and citizens* (pp. 3–14). New York: Zed Books.
Lederman, N. G., Abd-El-Khalick, F., Bell, R. L., & Schwartz, R. S. (2002). Views of nature of science questionnaire (VNOS): Toward valid and meaningful assessment of learners' conceptions of nature of science. *Journal of Research in Science Teaching, 39*(6), 497–521.
Mouffe, C. (2005). *The democratic paradox*. London: Verso.
Pateman, C. (1970). *Participation and democratic theory*. Cambridge, UK: Cambridge University Press.
Pestre, D. (2006). *Introduction aux science studies*. Paris: La Découverte.
Pion, L., Piron, F., & Duranceau, M. (2009). *Aux sciences citoyens!: Expériences et méthodes de consultation sur les enjeux scientifiques de notre temps*. Montréal: Les Presses de l'Université de Montréal.
Pouliot, C. (2008). Students' inventory of social actors concerned by the controversy surrounding cellular telephones: A case study. *Science Education, 92* (3), 543–559. (Special issue of *Science Education and Science Studies*).
Pouliot, C. (2009). Using the deficit model, public debate model and co-production of knowledge models to interpret points of view of students concerning citizens' participation in socioscientific issues. *International Journal of Environmental and Science Education, 4*(1), 49–73.
Pouliot, C. (2011). Post-secondary students' relationship to people they consider to be scientific experts. *Research in Science Education, 41*(2), 225–243.
Rask, M. (2003). The problem of citizens' participation in Finnish biotechnology policy. *Science and Public Policy, 30*(6), 441–454.
Rawls, J. (1993). *Political liberalism*. New York: Columbia University Press.
Rawls, J. (2001). *Justice as fairness: A restatement*. Cambridge, MA: Belknap.
Robottom, I., & Simonneaux, L. (2012). Editorial: Socio-scientific issues and education for sustainability in contemporary education. *Research in Science Education, 42*(1), 1–4.
Roth, W.-M., & Désautels, J. (2002). Science education as/for sociopolitical action: Charting the landscape. In W.-M. Roth & J. Désautels (Eds.), *Science education as/for sociopolitical action* (pp. 1–16). New York: Peter Lang.
Sadler, T. D., Dawson, V. M., Klosterman, M. L., Eastwood, J. L., & Zeidler, D. L. (2011). Metalogue: Issues in the conceptualization of research constructs and design for SSI-related work. In T. D. Sadler (Ed.), *Socio-scientific issues in the classroom: Teaching, learning and research* (pp. 79–87). Dordrecht: Springer.
Sismondo, S. (2008). Science and technology studies and an engaged program. In E. J. Hackett, O. Amsterdamska, M. Lynch, & J. Wajcman (Eds.), *The handbook of science and technologies studies* (pp. 13–31). Cambridge, MA: MIT Press.
Wynne, B. (1996). May the sheep safely graze? A reflexive view of the expert-lay knowledge divide. In S. Lash, B. Szerszynski, & B. Wynne (Eds.), *Risk, environment and modernity: Towards a new ecology* (pp. 44–85). London: Sage.
Wynne, B. (2003). Seasick on the third wave? Subverting the hegemony of propositionalism: Response to Collins and Evans. *Social Studies of Science, 33*(3), 401–417.
Wynne, B. (2005). Risk as globalizing "democratic" discourse? Framing subjects and citizens. In M. Leach, I. Scoones, & B. Wynne (Eds.), *Science and citizens* (pp. 66–82). New York: Zed Books.
Zoller, U., Donn, S., & Beckett, P. (1991). Students' versus their teachers' beliefs and positions on science/technology/society oriented issues. *International Journal of Science Education, 13*(1), 25–36.

Chapter 30
Transformative Learning in Science Education: Investigating Pedagogy for Action

Lyn Carter, Carolina Castano Rodriguez, and Mellita Jones

Abstract Sociocultural approaches to science education that aim towards a kind of scientific literacy for active citizenship have been increasing in recent years and are now firmly entrenched. Areas of interest are broad and include calls to socio-political action like those from Derek Hodson (Int J Sci Educ 25(6):645–670, 2003) and Larry Bencze (Can J Sci Math Tech Educ 8(4):297–312, 2008). Hodson argues that if contemporary social and environmental problems are to be solved, we need to orient science education strongly towards action. Included in much of this literature, either explicitly or implicitly, is the notion of a *transformation* in attitudes, behaviours, values, beliefs, and actions. While more often than not used colloquially, transformation is also a *term d'art* within field of transformative learning (TL). TL in not well known within science education, and we believe it may offer new insights into ways of progressing some of our sociocultural agendas. This chapter outlines research that uses the key precepts and processes of TL and presents findings from a pre-service teaching unit that was framed within TL theory. We ask what real TL may look like in a science classroom and what challenges and issues accompany its implementation.

Keywords Transformative learning • Activist science education • Mezirow • Disorientating dilemmas

Introduction

Sociocultural approaches to science education that aim towards a kind of scientific literacy for active citizenship have been increasing in recent years and are now firmly entrenched. One well-regarded perspective comes from Derek Hodson (2003) who argues that "if current social and environmental problems are to be

---

L. Carter (✉) • C. Castano Rodriguez • M. Jones
School of Education, Australian Catholic University, 115 Victoria Parade Fitzroy, Melbourne, VIC 3065, Australia
e-mail: Lyn.Carter@acu.edu.au

L. Bencze and S. Alsop (eds.), *Activist Science and Technology Education*, Cultural Studies of Science Education 9, DOI 10.1007/978-94-007-4360-1_30,

solved" (p. 645) we need to rethink science education. What is required he suggests, is science curricula oriented toward socio-political action which would "produce activists: people who will fight for what is right, good and just; people who will work to re-fashion society along more socially just lines; (and) people who will work vigorously in the best interests of the biosphere" (p. 645).

A decade after Hodson's (2003) call, we have made some advances towards this purpose. Erminia Pedretti and Joanne Nazir's (2011) typology of sociocultural currents within science education identify Hodson's work, and others like that of Larry Bencze (2008), Wolff-Michael Roth and Angela Calabrese-Barton (2004) and Deborah Tippins, Michael Mueller, Michiel van Eijck and Jennifer Adams (2010), within what they call the *socio-ecojustice current.* One of six strands or currents that Pedretti and Nazir (2011) construct, the socio-ecojustice approach believes that traditional science education does not go "far enough in educating students about the political, social and economic factors influencing science and science education, nor does it provide them with the tools necessary to actively transform society" (p. 617). The socio-ecojustice approach, or in Hodson's terminology *socio-political action*, can be distinguished from the currently popular socioscientific issues (SSI) approaches of the type developed by Dana Zeidler and his colleagues (Zeidler et al. 2009). These SSI approaches largely privilege cognitive moral development, suggest Pedretti and Nazir (2011), through the use of carefully selected science moral problems and values-centred discussion. In practice though, they are often vehicles or contexts to teach science content identified elsewhere (usually National/State standards) as essential learning (Zeidler 2010). Pedretti and Nazir's (2011) other sociocultural currents include *application/design,* which promotes the application of technology to utilitarian problems, the *historical* as extending student's understanding of the historical and sociocultural embeddedness of science, and the *logical reasoning* as the application of reason to socioscientific issues. Clearly, there are elements of overlap across all their six strands.

Like Hodson (2003), we believe that the "aim of science education should be the promotion of a certain type of citizenship and civic responsibility of which transformation, agency, and emancipation are key features" (Pedretti and Nazir 2011, p. 617); in short, science education for socio-political action. This view is consistent with those like Ira Shor (1992) who argue for an empowering education in which students take an active role through learning *about* action as well as in *taking* action to address societal problems. Having a clearly articulated intention or goal however, is one thing – it is quite another to conceptualise the *how* to teach science for socio-political action. We mean here the development and implementation of appropriate pedagogies and curricula. Nicholas Fox (2012, p. 15) is one scholar who has neatly encapsulated some of the difficulties of such teaching:

> As much as we talk politics with our students, read political novels, and highlight the activism of the past, the walls of the classroom present a problem for radical teachers. Our meetings host passionate discussions where students begin to tackle assumptions, dismantle ideas of privilege, even critique capitalism. But when class ends, what happens to the political fervor? Where does that revolutionary spark go? Does it spread out into the streets? Or does it end up at the bottom of backpacks, forgotten like last week's homework?

While activist pedagogies such as critical pedagogy or pedagogies of resistance from Paolo Friere (1986) and Henry Giroux (2011) for example, are well known within the general education literature, science education has been slow to explore these areas. Notable exceptions include the Special Issue of the *Canadian Journal of Science, Mathematics and Technology Education* (Alsop and Bencze 2009) on the role of activism in science, mathematics, and technology education, the subsequent response from David Burns and Stephen Norris (2012) and Alsop and Bencze's (2009) rejoinder. Hodson (2010) has also made a strong contribution with his four levels of curriculum starting with appreciating the "societal impact of scientific and technological change and recognizing that science and technology are, in substantial measure, culturally determined" (p. 199), and moving to acknowledging the coercive and legitimating power relations of science and technology. Hodson's (2010) third level argues the need to develop one's own views and query underlying value positions so as, fourthly, one can prepare for and take action. Bencze's STEPWISE project (Science and Technology Education Promoting Well-being for Individuals, Societies and Environments) project extends Hodson (2010, 2003) and others into the pedagogical space with a conceptual model that includes student-led inquiry based on their own primary and secondary research of SSIs. Taking actions is integral for Bencze which "could begin with making changes to their own lives …(but) … a key aspect of STEPWISE is to encourage and enable students to use their education in science and technology for helping others and the planet."

Implicit in all of these pedagogical and curricula stances is the *transformation* of attitudes, values, behaviours and beliefs leading to a willingness, desire and ability to take action. *Transformation* is the essential term here if the purpose of learning as Dewey extols us to believe, is to effect change in the learner (Ukpokodu 2009). While more often than not used within science education and indeed within most other educational literatures in its common parlance meaning, transformation is also a *term d'art* that appellates the field of transformative learning. First applied by Jack Mezirow in the late 1970s, transformative learning refers to adult learning that merges "critical thought with critical action to effect change" (Kitchenham 2008, p. 108). Mezirow (2000) defines transformative learning as a process whereby "we transform our taken-for-granted frames of reference to make them more inclusive, discriminating, open, [changeable], and reflective so that they may generate beliefs and opinions that will prove more true or justified to guide action" (p. 214). Transformative learning theory is in not well known within science education apart from some work by Kevin Pugh and his colleagues (Pugh et al. 2010) who described transformative experiences adapted from Dewey. We believe that transformative learning theory may offer unique pedagogical approaches to further the socio-political agenda within science education.

This chapter then, outlines our research that uses some key precepts and processes of transformative learning theory to move beyond the usual rhetorical devices surrounding learning to truly 'transform' our classroom practice towards action. Transformative learning's focus on action makes it a natural ally of social-political science education. We ask what transformational pedagogy may look like in a science classroom and what issues accompany its implementation. We begin here with a brief overview of transformation learning theory and move onto a view of transformative learning we believe appropriate for socio-political science education. We then describe the research setting where we implemented our transformative approach and, after reporting our data, we finish with a brief discussion on transformative learning pedagogy contribution to the socio-political science education agenda.

Transformative Learning

Mezirow's early theory of transformative learning drew from Jürgen Habermas and proposed three types of learning identified as *instrumental*, *dialogic* and *self-reflective*. Andrew Kitchenham's (2008) analysis of the evolution of Mezirow's theory describes these as how we best learn information (instrumental), where and when learning could best take place (dialogical), and why we should learn (self-reflective). Mezirow later added the Habermasian concept of emancipatory knowledge with its association to learning through critical reflection. For Mezirow (1991), emancipatory knowledge enabled learners to "identify and challenge distorted meaning perspectives" (p. 87) that constrain the way we view the world allowing previously unimagined alternatives to emerge. Meaning schemes can then be "either replaced or reorganised to incorporate new insights" (p. 88). Mezirow (2003) went on to describe transformative learning as a rational process involving a "metacognitive application of critical thinking that transforms an acquired frame of reference by assessing its epistemic assumptions" (p. 124). This process is described within an awareness framework that involves:

1. Recognition that an alternative way of understanding may provide new insights into a problem;
2. Context awareness of the sources, nature, and consequences of an established belief;
3. Critical reflection of the established belief's supporting epistemic assumptions;
4. Validating a new belief by an empirical test of the truth of its claims, when feasible, or by a broad-based, continuing, discursive assessment of its justification to arrive at a tentative best judgment;
5. Coping with anxiety over the consequences of taking action; and
6. Taking reflective action on the validated belief (p. 124).

In sum then, Mezirow's theory of transformative learning and its various iterations over several decades can be characterised by critical reflection that transforms a person's perspective through identifying, questioning, and revising assumptions and prevailing knowledge at both personal and socially embedded levels (Cranton 2006, 2011; Kitchenham 2008; Mezirow 1991, 2003). It is not a stretch to see that Mezirow's approach has much in common with what Hodson (2003) calls his four levels of curriculum.

Since its early beginnings with Mezirow, transformative learning has become an established paradigm within adult learning occupying diverse positionalities, sometimes complimentary and sometimes contradictory (Kitchenham 2008). Patricia Cranton (2011), for instance, argues that the acquisition of emancipatory knowledge is not enough on its own to indicate that transformative learning has occurred. Instead, transformation requires "a deep shift in perspective and noticeable changes in actions as a result of the shift" (p. 76). This means that critical reflection and emancipatory education practices are necessary but not sufficient conditions for transformative learning. In addition, each individual must experience their own *triggering event* to make the learning transform into action before transformative learning can be said to have taken place. The conditions for transformative learning occur when an individual on their own or as a part of a group, is triggered to challenge the prevailing view on a topic or issue (Cranton 2011).

Though Mezirow's work is mostly known for the rational and cognitive aspects of the transformational shift, he has lately recognized that the process of transforming what was once believed and is no longer valid, could be a painful realisation (Dirkx et al. 2006). For John Dirkx (2006), this means that transformative learning is a "holistic process (also) involving spiritual, intellectual, emotional and moral dimensions of being in the world" (p. 125). His view integrates experiences of the *outer world* that includes cognitive, epistemic and socio-cultural dimensions of the process of transformative learning, with the *inner world* involving emotional, moral and spiritual dimensions.

Other scholars have described further elements of transformative learning based around these subjective and inner processes. For instance, Knud Illeris (2009) likens transformation to "a break of orientation that typically occurs as the result of a crisis-like situation caused by challenges experienced as urgent and unavoidable, making it necessary to change oneself in order to get any further" (p. 14). Similarly, Cranton (2006) explains that the deep shifts and challenges of mindset associated with transformative learning can lead to feelings of discomfort or even distress in learners. Kaisu Malkki (2010) is another who uses instead, the term *comfort zone* to describe the interconnectedness of emotions and feelings:

> On one hand, when continuous and coherent interpretation (cognitive component) is not possible, we feel unpleasant feelings (emotional component). On the other hand, when we are able to interpret situations coherently and continuously (cognitive component) within the light of our meaning perspective, we may experience comfortable feelings (emotional component), which often go unnoticed, as there are no unpleasant emotions present. (p. 49)

The tensions between the cognitive and emotional states can lead to edge-emotions which are for Malkki (2010) "the unpleasant emotions which arise at the edges of the comfort zone, that is, when the meaning perspective becomes challenged" (p. 49).

This level of discomfort or crisis can, naturally, lead to some resistance from learners as they confront their existing beliefs and assumptions. It underscores the difficulty learners' encounter when experiencing transformative learning (also see Sterling 2010). The process is not one all would wish or, indeed, be able to undertake. This confrontation aspect of transformation presents perhaps the greatest challenge for educators wanting to encourage transformative learning. While well used to providing students with cognitive or intellectual challenges, educators are less skilled at, and secure about, offering emotional challenges. For example, how much of an emotional challenge can be viewed as appropriate before the political correctness and tacit rules of the formal educative relationships prevail? What are the ethical issues at large of structuring the discomfort essential to transformation?

Perhaps an indication of the fraught nature of this component of transformative learning is that scholars variously refer to it in relatively benign terms like Cranton's (2011) triggering events, to Ted Fleming's (2011) use of the more potent *disorienting dilemmas* and *disjunctures,* through to Illeris' (2009) brazen *crisis-like situation.* Whatever the terminology, educators need to understand how to constructively use such confrontation to transform learners' acquired frame of reference and action. This holistic view of transformative learning also engenders a moral and ethical responsibility from transformative educators to ensure that adequate support structures are available to assist learners to manage the potential distress and challenge they face in the transformative learning process (Cranton 2006).

From all these authors, it is clear that transformative learning not only involves a cognitive and rational process, but also reflects and evokes emotions and feelings that are integral to the process of transformation.

Transformative learning is not without its critics, though. Michael Newman (2010) argues for example, that there can be no such thing as transformative learning as "perhaps there is just good learning" (p. 37). He suggests that any significant change that occurs through the learning process is how learning is defined, hence there is nothing exceptional about the tenets of transformative learning. As a counterpoint, Illeris (2009) argues transformative learning theory,

> . . . implies what could be termed personality changes, or changes in the organisation of the self, and is characterised by simultaneous restructuring of a whole cluster of schemes and patterns in all of the three learning dimensions (content, incentive and environment). (p. 14)

There are also broader critiques from within the transformative community themselves. Edward Taylor (2008) outlines some of the newer and more ideologically driven analyses from feminist, race and postmodern positions. These critiques seek to identify and address inscribed political and cultural assumptions like patriarchy and Eurocentrism, in particular, in Mezirow's rational view of transformative learning.

Our Reading of the Transformative Learning Processes

We identify four characteristics within the holistic rendering of transformative learning theory that we believe make it different from other types of critical reflective pedagogies. These are, firstly, critical reflection and reflexivity, secondly, the mobilization of disorientation and conflict (or triggering events), thirdly, the embeddedness of the individual as the social emotional learning self and, finally, the focus on action.

Critical reflection, self-reflection and reflexivity are all well-rehearsed ideas within the education literature and do not require any elaboration here. In transformative learning processes, the reflection is initiated when a confrontation or disorienting dilemma encourages the learner to critically identify and revise their previously held assumptions. Prevailing knowledge, understanding, emotions and beliefs about an experience, an issue or an event are called into question. This process brings with it a degree of anxiety and discomfort. Consequently, transformative learning does not offer a benign pedagogy. Reflection and the concomitant development of new knowledge and understanding enable the possibility of alternative views that can ultimately lead to a reauthorizing of personal frameworks or biographies. While this is the desirable outcome, a re-accommodating of the learner's comfort zone with no discernable change is also a feasible result. Where personal framework, beliefs and values are changed, action and activism is much more likely.

Not surprisingly, transformative learning is very difficult to implement at the upstream end (confrontation in the education setting, more often than not – the classroom), and recognise its attainment downstream (transformed beliefs and actions which may occur somewhat later). It can be a highly unique process for each learner requiring just the 'right' triggers, which obviously extends well beyond the usual allotted class learning time. Karen Swanson (2010) is one practitioner who has tackled some of the thorny praxis issues of transformative learning. She encouraged the use of transformative learning to engender reflective practice in a teacher education context so as teachers could "consider and challenge their personal assumptions about teaching and learning" (p. 260). The aim was to promote the development of questions rather than answers to their issues. In another study, Omiunota Nelly Ukpokodu (2009) investigated pedagogies that fostered transformative learning in a multicultural education course. She found that experiencing a humanizing pedagogy including dialogic relationships in the learning community, various experiential activities and writing pre-post narrative inquiries "moved (students) from color-blindness to color-vision" (p. 7). Clearly, more studies like Swanson's (2010) and Ukpokodu's (2009) are needed and we hope our study adds insight to the implementation literature.

Implementing Transformative Science Education

Reported in part here, our research investigated whether transformative learning theory could become pedagogical for socio-political activism within science education. Our aim was to apply the principles and processes of transformative learning to a pre-service teacher's science class to explore the issues of its implementation, with a particular focus on the suitability and effects of 'triggering events' or 'disorientating dilemmas.' The research was undertaken within *EDST429 Education for the 21st Century,* a Bachelor of Education and Early Childhood elective unit (also known as a 'course' in Northern American and other settings) at the Australian Catholic University, Melbourne Campus. The unit was implemented as a 4-day intensive during the second semester (July–October) of 2012 for 23 students. Our hope was to truly transform students' values, beliefs, concepts, attitudes and behaviour towards scientific socio-political issues so their science learning leads to more ethical decision-making. In turn, this would influence what they believe is important for science education in their own teaching careers.

The Research Context

The first day of the four opened with us explicitly communicating our intention to use transformative learning theory as a basis for the unit. Being explicit is necessary argues Swanson (2010), to foster connections between students and teachers required to create the necessary openness of a transformative learning space. We then presented a 'disorienting dilemma' to the students, challenging them with the statement that any sort of egg/chicken consumption contributes to animal cruelty. This topic selection was based on our underlying assumption that the majority, if not all, students were consumers of eggs and/or chickens, and that students had limited knowledge of how agribusiness meets consumer demand for 'affordable' meat and eggs. Explicit teaching, together with personal student research about chicken and egg farming in cage, barn and free range environments, was used to engage students in Mezirow's (2003) steps of transformative learning; that is, becoming aware of the source, nature and consequences of beliefs, identifying and questioning their assumptions, and critically reflecting on and validating their beliefs through empirical evidence. Students were also encouraged to brainstorm a range of potential actions that could resolve any disorientation they or others experienced consequent to being more informed about the topic. They were then asked to commit to one of the levels of action, such as giving up eating eggs/chicken or buying only certified free range or the duration of the unit.

To expand students' knowledge and awareness of activism and what was required, a panel of special guests was invited to the first part of the second day of the unit. Pam Ahern, Director and creator of Edgar's Mission, a not-for-profit sanctuary for rescued animals; Helen Marston, Chief Executive Officer of Humane Research Australia; and Tony Gleeson, a climate activist, were the panel participants. Each speaker provided some background about the causes they represented and their personal journey towards becoming an activist.

The second part of the second day and the third and fourth days involved students working with other socio-political issues they had identified as personally important to them. Following Mezirow's steps of interrogating their assumptions using primary and secondary information sources, as well as brainstorming potential actions, students presented an activist campaign to their peers on their chosen issue. Students engaged in debate and discussion about the levels of action they would or would not be willing to undertake, and identified reasons for these decisions. These strategies enabled processes of discourse that Mezirow (2003) has identified as essential for transformative learning.

The primarily qualitative data from 21 students included their work artefacts, open responses to key questions at the end of each day, researcher observation notes, and semi-structured interviews conducted at the end of the unit. Some numeric data were derived from emotional graphics students completed periodically, indicating on a scale of 1–5 their levels of confrontation experienced during selected unit activities. A random sample of 11 of the 21 students participated in the final interviews. Data analysis was approached in two key ways. All numeric data was collated for the different activities/topics and an average taken. This form of descriptive analysis is appropriate given that the data is being used to support the qualitative data and that the sample size was not sufficient for rigorous statistical analysis. Qualitative data from the open responses and interview data were analysed for themes using a process of analytical induction (Bernard and Ryan 2010). This involves multiple passes of the data to identify and collate key responses into thematic categories. Finally, to enhance the trustworthiness of findings, the multiple sources of data were compared and contrasted until themes supported across the different forms of data emerged.

Findings and Discussion

The findings from this study suggest that transformative learning theory offers a relevant and powerful framework for students to increase their awareness of contemporary science issues and contribute to making informed decisions leading to action; in other words, socio-political science education. Most students identified it as a powerful pedagogy for challenging current beliefs of which the following comment is typical:

> This unit was so interesting and engaging! Each session I came away with new information and ideas. It challenged my learning and beliefs in a safe and open environment which was refreshing. . . .This was different to anything I had done before and I was grateful for the opportunity to challenge myself.

Here, we report findings particularly relevant to the *disorienting dilemmas* and critical reflection essential to question prevailing *assumptions* in terms of knowledge, understanding, views, emotions and beliefs about an experience or issue. Such confrontation is central within transformative learning theory for mobilizing people to take action. Overall, we found that the level of confrontation involved in disorienting dilemmas and the reflexive nature of the transformative learning process, as well as the emotional impact and the familiarity with the issue, influenced students' considerations to take action.

Level of Confrontation, Emotional Reactions and Familiarity with Issue

As Cranton (2006), Malkki (2010) and others suggest, confrontational aspects of the transformative experience generate strong emotional reactions including feelings of discomfort and distress. In our study, the emotional graphics showed that more than two thirds of the students (16 out of 22) expressed negative emotions following our introduction of the disorientating dilemma claiming that egg/chicken consumption contributes to animal cruelty. They described emotions such as 'sadness,' 'feeling upset,' 'uneasy,' 'uncomfortable,' 'overwhelmed,' 'confronted' and 'shocked,' with most students make comments such as 'this activity is confronting and stirs up emotions about egg production and the process in this country.' Students identified a range of possible actions that could be taken to address this issue including not eating eggs at all, only eating free range eggs if there were sufficient research demonstrating that the chickens were treated humanely, raising your own chickens, and lobbying government to better regulate chicken farming and increase consumer awareness about the ethical treatment of chickens. Most of the students were willing to commit to one of these forms of action, particularly if it was not too 'extreme' like stopping eating eggs altogether.

Similarly, students were confronted by some of the socio-political issues they or their peers investigated and presented on Day 3. Data indicates 'genetically modified (GM) foods' and 'puppy mills' were the most confronting and about which the students expressed a desire to act. Their suggested actions ranged from the social such as signing petitions and closing down puppy mills, to personal choices like checking labels in supermarket products to be more conscious consumers and avoid GM foods. The students' emotional connection to the issue, and that there were various levels of actions possible from the more to less 'extreme' seemed to be important in their willingness to consider taking action.

Where issues generated a higher level of discomfort and confrontation precipitating unpleasant emotions like anger, students displayed a greater range of reactions – from feeling empowered to being overwhelmed and without clarity on how to act. This was the case on Day 2, when the panel of three activist leaders talked about their causes. The emotional graphics indicated that the majority (14 out of 17) of the students felt *highly* confronted with some expressing 'anger of how things are, that we accept things to be as they are, a lifestyle fuelled by suffering,' while some others (5 out the 14) felt empowered: 'I feel inspired to change my attitudes to educate others and change the way the world justifies actions which are not ethical,' (Interview Student 7). The high levels of confrontation suggest that the 'dilemmas' presented by the panellists disoriented most students, but only a few expressed the desire to take action. It is not clear whether it was the high level of confrontation, the complexity of the issues involved or the significant changes in lifestyles represented by the activists or a combination of all three that discouraged the majority of students to consider action.

The familiarity and experience the students had with an issue or disorienting dilemma also seemed to influence their level of confrontation and commitment. Where students believed they were very familiar with an issue and/or it did not connect with their lives, their emotional graphics were neutral and they were reluctant to consider it further. This occurred for four of the students during the chicken/egg production industry dilemma: 'I was already aware of the debate about caged and free range hens – but it was interesting to hear other opinions'

(Interview Student 4). If however, students were familiar with the issue but the evidence presented was new and confronting, it seemed to generate a will to act. For example, students researching GM foods presumed they were not too bad as many of us probably consumed GM on a regular base. The information they found though was unexpected and confronting, encouraging action that Jim notes during the interviews:

> When you come across stuff like big companies like Monsanto when they develop a corn that is tolerant to everything except weeds and they also sell weed killer that sort of stuff just doesn't leave you. So you've got to tell people about that (Jim).

Similarly, some of the 16 students who felt confronted by the egg/chicken dilemma explained that they were not aware of the full story about the egg production industry: '[I feel] disappointed that I was so unaware about how chickens are treated. Finding out about how they are treated makes me sad' (Interview Student 3). When one student came across a video about the fate of male chicks, she described being highly confronted, as she did not know that these activities occurred. She explained that in documentaries exposing animal cruelty such as *Earthlings,* she would expect to see strong footage and would be less shocked than she was with the unexpected footage of killing male chicks. In these examples, students had some level of experience with the issues, but were not fully aware of their complexity and diversity.

When read together, these findings suggest that the degree of confrontation, relevance and familiarity of the issue are all important when considering the uses of disorienting dilemmas as pedagogical devices. Based on these results, it could be argued that students need stronger or more confronting situations for real transformation to occur. However, caution is necessary as confrontation generates anxiety and discomfort that might encourage students to accommodate the new information within their previous beliefs thereby avoiding unpleasant feelings. Students may feel disempowered to act and even disengage from the class as a strategy to deal with their discomfort and distress. This clearly occurred for some of the students as a consequence of the Day 2 panel discussion when the life choices of the activists appeared too confronting. Consequently, it remains a challenge when implementing disorientating dilemmas as an essential process in transformative learning to pitch just the right amount of confrontation, and clearly that pitch is different for everybody. It is an area that requires more research.

Critical Reflection: Unpacking Assumptions

Mezirow (2003) as we noted above, believes that real transformative learning requires the analysis of our assumptions and their underlying structures. As one of the key precepts of transformative learning theory, we asked students to identify the assumptions they held regarding the socio-political issue they investigated for Day 3. This involved researching the relevant topic information, evidence and supporting social structures, as well as reflective dialogue and critical thinking about their awareness of their own choices and possible actions. This is apparent in the following quote, in which a student reflects on her experiences and assumptions involved with deciding to investigate GM food:

> I've seen a really unusually sized fruit and so I'm thinking "Well hang on" and then we're talking about labels and I'm thinking "How do you know that what you're eating now isn't a modified organism?" Because when you buy fruit it doesn't say. It just says where it's from and it's a product of Australia or a product of so and so and that's it. [...] So then we're like "Yeah let's do that because we don't know much about it". And I assume that it's okay. My assumption is that it's okay because we probably all eat it. So we thought "How about the assumption be genetically modified foods are safe and beneficial". Well that was my assumption. (Mary, Final Interview p. 1)

Similarly, all interviewed students suggested that recognising their assumptions and reflecting on what supported those assumptions were critical to generating awareness of their own choices and actions. The interviews with Nicola and Jane about the fair trade issue Jane studied exemplify this:

> Jane: If I hadn't done this project, I wouldn't just randomly think, "let's see if there's any anti Nestle things"...You'd just be thinking, "chocolate, cool, I'm going to eat this".
>
> Nicola: When you buy something, you don't think, "where did this come from?" You just buy it. [...] If I didn't do this unit I would never sit there and think "why did I think this way? Why are my beliefs like that? What's made me believe that?"... I would have just gone on with my life". (p. 17)

Students were also asked to investigate contrary information that confronted their assumptions about their issue. This gave them the opportunity to recognize alternative ways of understanding and how they provide new insights into a problem (Mezirow 2003). The students valued this approach as a different way of learning. Cassie describes it thus:

> It's just given me a different way to learn, for me personally, because I've never done an assignment like that. So basically every time we've ever done an assignment or anything, we pick a topic or an issue and you research that and give your opinion on it. So I've enjoyed this one because we did research our opinion, but then we had to look at the other side. So it's basically like you're debating yourself and I've never done that before in an assignment. And that's why I think I liked it, because I've seen a different point to what my original opinion was, and it's actually changed my opinion now. (Final Interview)

Other students also noted that they did not attempt to identify or question assumptions, nor did they challenge their beliefs when they usually investigated topics within other units of study. They recognized that this transformative learning process was unique, and that they could identify views and aspects of contemporary issues that they may not have explored otherwise.

Moreover in their presentations for Day 3, where students were less able to identify and articulate the underlying assumptions for their issues, and did not provide strong supporting and contrary evidence, the emotional graphics data showed little confrontation and almost no emotional reactions by the rest of the students. The possible link between mobile phones and cancer, and parenting styles and influence in children's development, were too such examples. The ambivalence the students displayed regarding their desire to continue using mobile phones, not probing their assumptions too far, saw them provide little evidence confronting this assertion and only weak supporting evidence. Consequently, the other students didn't engage with the issue as a disorientating dilemma or question their personal assumptions.

Taken together, these findings suggest that guiding students in choosing socio-political issues on the bases of their assumptions, and understanding that assumptions are connected with our beliefs, choices and daily actions, is critical to generating disorienting dilemmas that encourage transformation. They also support authors like Dirkx and colleagues (2006), who regard internal dimensions of emotionality and external dimensions such as socio-cultural experiences as influencing individual student confrontation and their place in the transformative learning process.

Conclusion

This study was our first attempt to use transformative learning theory within science education. It is ongoing, and we plan to complete follow-up interviews with the students 9 months after completion of the unit. Our aim here is to see if any of the actions undertaken during the unit have persisted or if they have adopted new any actions relevant to the socio-political issues explored within the unit. We would also like to know whether the students are more likely to reflect on their own assumptions about various issues. Finally, for those students who are now teaching, we would like to know if they have incorporated any of the precepts of transformative learning into their own pedagogies and if they include socio-political science issues in their content. Moreover, in Semester 2, 2013, we plan to teach the unit again, better informed about some of our strategies based on our experiences in 2012.

With our results to date though, we belief that transformative learning theory furthers the socio-political agenda of those like Bencze and Hodson by focusing on empowering people to identify the underlying assumptions and structures supporting their beliefs and actions. Typically, socio-cultural approaches to science education such as SSI or STS have emphasized issues-based teaching for generating awareness and participation. The findings presented here suggest that it is not only the exposure to issues that matters, but also the requirement to unpack assumptions underlining the issues that is critical. It exposes learners to confronting information that is more likely to lead to action.

To finish, we can do no better than agree with Ukpokodu (2009) when she points out that:

> [m]ost people would argue that it is difficult, if not impossible, for students to experience transformation in learning based on one university course. But scholars of transformative learning believe that it is possible, and remind us that transformation can occur in discrete classes because transformation comes in different sizes that may entail "a moment of transition from passivity to naiveté to some animation and critical awareness" (Shor 1987, p. 34). Further, Shor (1987) warns us that in "looking only for big changes, [we] lose touch with the transformative potential in any activity" (p. 35). The process of

transformation may occur in stages. University courses can at least provide an avenue for understanding the nature of society as a contested terrain (Shor and Freire 1987). As Freire (1986) also suggests, "for transformation, we need first of all to understand the social context of teaching, and then ask how this context distinguishes liberating education from traditional methods" (p. 33). That is, the transformative focus may become one of first, developing critical knowledge and a lens for reflecting and then making plans for action. Thus, for me, the students' development of self, a new awareness of societal and educational reality, and praxis is a form of learning transformation. Freire (1986) suggests that, in the final analysis, liberatory education must be understood as a moment or process or practice where we challenge people to mobilize or organize themselves to get power. (p. 6)

References

Bencze, J. L. (2008). Private profit, science and science education: Critical problems and possibilities for action. *Canadian Journal of Science, Mathematics & Technology Education, 8*(4), 297–312.

Bencze, J. L., & Alsop, S. (2009). A critical and creative inquiry into school science inquiry. In W.-M. Roth & K. Tobin (Eds.), *World of science education: North America* (pp. 27–47). Rotterdam: Sense.

Bernard, H. R., & Ryan, G. W. (2010). *Analyzing qualitative data: Systematic approaches.* Thousand Oaks: Sage.

Burns, D., & Norris, S. P. (2012). Activist environmental education and moral philosophy. *Canadian Journal of Science, Mathematics, and Technology Education, 12*(4), 380–393.

Cranton, P. (2006). *Understanding and promoting transformative learning: A guide for educators of adults* (2nd ed.). San Francisco: Jossey-Bass.

Cranton, P. (2011). A transformative perspective on the scholarship of teaching and learning. *Higher Education Research and Development, 30*(1), 25–86.

Dirkx, J. M. (2006). Engaging emotions in adult learning: A Jungian perspective on emotion and transformative learning. *New Directions for Adult and Continuing Education, 109*, 15–26.

Dirkx, J. M., Mezirow, J., & Cranton, P. (2006). Musings and reflections on the meaning, context, and process of transformative learning. *Journal of Transformative Education, 4*(2), 123–139.

Fleming, T. (2011, May 28–29). *Recognition in the work of Axel Honneth: Implications for transformative learning theory*. Paper presented at the International Transformative Learning Conference, Athens.

Fox, N. (2012). Teaching (is not) activism. *The Radical Teacher Issue, 94*, 14–23.

Freire, P. (1986). *Pedagogy of the oppressed*. New York: Continuum.

Giroux, H. (2011). *On critical pedagogy*. New York: Continuum.

Hodson, D. (2003). Time for action: Science education for an alternative future. *International Journal of Science Education, 25*(6), 645–670.

Hodson, D. (2010). Science education as a call to action. *Canadian Journal of Science, Mathematics, and Technology Education, 10*(3), 197–206.

Illeris, K. (2009). A comprehensive understanding of human learning. In K. Illeris (Ed.), *Contemporary theories of learning: Learning theorists. . .in their own words* (pp. 7–20). Milton Park, Abington: Routledge.

Kitchenham, A. (2008). The evolution of Jon Mexirow's transformative learning theory. *Journal of Transformative Education, 6*(2), 104–123.

Malkki, K. (2010). Building on Mezirow's theory of transformative learning: Theorizing the challenges to reflection. *Journal of Transformative Education, 8*(1), 42–62.

Mezirow, J. (1991). *Transformative dimensions of adult learning*. San Francisco: Jossey-Bass.

Mezirow, J. (2000). *Learning as transformation: Critical perspectives on a theory in progress*. San Francisco: Jossey Bass.

Mezirow, J. (2003). Transformative learning as discourse. *Journal of Transformative Education, 1*(1), 58–63.
Newman, M. (2010). Calling transformative learning into question: Some mutinous thoughts. *Adult Education Quarterly, 62*(1), 36–55.
Pedretti, E., & Nazir, J. (2011). Currents in STSE education: Mapping a complex field, 40 years on. *Science Education, 95*(4), 601–626.
Pugh, K., Linnenbrink-Garcia, L., Koskey, K., Stewart, V., & Manzey, C. (2010). Motivation, learning, and transformative experience: A study of deep engagement in science. *Science Education, 94*(1), 1–28.
Roth, W. M., & Calabrese-Barton, A. (2004). *Rethinking scientific literacy*. New York: RoutledgeFalmer.
Shor, I. (1987). *Critical teaching and everyday life*. Chicago: University of Chicago Press.
Shor, I. (1992). *Empowering education: Critical teaching for social change*. London/Chicago: University of Chicago Press.
Shor, I., & Freire, P. (1987). *A pedagogy of liberation: Dialogues on transforming education*. New York: Bergin and Garvey.
Sterling, S. (2010). Transformative learning and sustainability: Sketching the conceptual ground. *Learning and Teaching in Higher Education, 5*, 17–33.
Swanson, K. (2010). Constructing a learning partnership in transformative teacher development. *Reflective Practice: International and Multidisciplinary Perspectives, 11*(2), 259–269.
Taylor, E. W. (2008). Transformative learning theory. *New Directions for Adult and Continuing Education, 119*, 5–15.
Tippins, D., Mueller, M., van Eijck, M., & Adams, J. (Eds.). (2010). *Cultural studies and environmentalism: The confluence of ecojustice, place-based (science) education, and indigenous knowledge*. Dordrecht: Springer.
Ukpokodu, N. (2009). Pedagogies that foster transformative learning in a multicultural education course: A reflection. *Journal of Praxis in Multicultural Education, 4*(10), 1–9.
Zeidler, D. L. (2010). Developing sociocultural and human values in science and technology through socio-scientific Issues. In S. Dolinšek & T. Lyons (Eds.), *Socio-cultural and human values in science and technology education* (pp. 33–39). Bled: IRI UL, Institute for Innovation and Development of University of Ljubljana.
Zeidler, D. L., Sadler, T. D., Applebaum, T., & Callahan, B. E. (2009). Advancing reflective judgment through socioscientific issues. *Journal of Research in Science Teaching, 46*(1), 74–101.

Chapter 31
Promoting Students' Collective Socio-scientific Activism: Teachers' Perspectives

Pedro Reis

Abstract This chapter describes analyses for the 'We Act' project, which is intended to foster development, implementation and study of materials and methodologies aimed at supporting teachers and students in taking informed and negotiated actions to address social and environmental issues associated with the fields of science and technology. This project combines development, action and research components and crosses three different areas: (1) the promotion of an active inquiry-based learning regarding real-life controversies associated with science and technology; (2) the stimulation of students' participation in collective democratic problem-solving action; and (3) the support of the first two areas with art initiatives and uses of Web 2.0 tools. Through a qualitative approach, this chapter discusses teachers' motivations for participating in such a project and also difficulties they identify in their students' enrolment in activism on social and environmental issues associated with science and technology. The data obtained through an online questionnaire and semi-structured interviews involving all the teachers in the project were submitted to content analysis. Teachers' participation in the project is mainly motivated by a strong willingness to find ways to: (a) change what they consider to be students' apathy concerning school science activities; and (b) empower them for democratic problem-solving action regarding social and environmental problems affecting society. Through the involvement in the project and actions implemented by their students, they began: (a) considering research-informed action as a major aspect of scientific literacy; and (b) recognizing students as important agents of change (capable of implementing, with success, impacting actions on their families and groups of friends) and, consequently, as 'citizens' (as opposed to 'future citizens'). However, they face important obstacles in the

Coordinator of the 'We Act' Project, which was funded with National Funds by the Foundation for Science and Technology (FCT) under the grant PEst-OE/CED/UI4107/2011

P. Reis (✉)
Instituto de Educação, Universidade de Lisboa,
Alameda da Universidade, Lisbon 1649-013, Portugal
e-mail: preis@ie.ul.pt

L. Bencze and S. Alsop (eds.), *Activist Science and Technology Education*, Cultural Studies of Science Education 9, DOI 10.1007/978-94-007-4360-1_31,

implementation of these collective actions; namely, time restrictions imposed by overcrowded curricula and difficulties in finding the 'proper' controversial topics capable of overcoming student apathy.

Keywords Activism • Science education • Socio-scientific issues • Community of practice • Art • Web 2.0 tools

The Project 'We Act' – Promoting Collective Activism on Socio-scientific Issues

Contemporary society is marked by controversial scientific and technological proposals and by social tensions between individual rights and social aims, political priorities and environmental values, economic interests and health concerns (Nelkin 1992). The wellbeing of individuals, societies and environments are threatened by complex problems, some of them caused by the controversial relation of business with science and technology: the pressure for profit can compromise the quality of researchers' practices and products, raising personal, social and environmental issues (Bencze 2008; Ziman 2000).

The seriousness of the socio-scientific controversies affecting our society requires a citizenry that is well informed and empowered to take appropriate actions about such issues (Gray et al. 2009). Community research-informed action is frequently considered a major aspect of scientific literacy (Hodson 1998) and a way to empower students as critics and creators of knowledge, instead of placing them in the role of consumers of knowledge as school science systems often appear to support (Bencze and Sperling 2012).

In a context like this, school science practices must be transformed and the concept of scientific literacy must be broadened. In many science classrooms, the emphasis is on the products of professional science and technology, through teaching modes that suppress students' desires to ask questions, pursue their own inquiry paths, discuss/critique different perspectives and develop their own conclusions (Bencze and Carter 2011). The focus of school science on consensual, well-established knowledge promotes a simplistic positivist conception of the practice of science and the notion that its findings are absolute and unequivocal (Driver et al. 1996; Levinson 2008). However, science-in-the-making is often uncertain, tentative and controversial (Ziman 2000). According to Derek Hodson (2003), instruction must be broadened in order to promote knowledge about the nature of science and technology, science inquiry skills, and socio-political activism on socio-scientific issues (SSI). In a society threatened by complex SSI, an explicit analysis and recognition of social injustices and the resultant importance of socio-political action becomes critical. Therefore, the concept of scientific literacy must include the development of students' "capacity and commitment to take appropriate, responsible and effective action on matters of social, economic, environmental and moral-ethical concern" (Hodson 2003, p. 658). Some authors suggest that students' activism on SSI have the power of improving: (a) their knowledge of these issues; (b) their inquiry and citizenship competences; and, eventually, (c) the wellbeing of individuals, societies, and environments (Bencze and Carter 2011; Roth and Désautels 2002).

The project 'We Act – Promoting Collective Activism on Socio-Scientific Issues' (initiated on September of 2012) represents the most recent step in a line of research and intervention aimed at supporting the discussion of SSI in Portuguese schools as a way of preparing students for an active, informed participation in society. Former initiatives involved the development of materials and approaches to support the discussion of SSI (face-to-face and through online interfaces) in science and philosophy classes and the evaluation of their impact on students' competences (Hilário and Reis 2009; Reis 1997, 2003). Other studies identified factors that positively influence the classroom discussion of SSI and built knowledge in supporting teachers with confidence, motivation and knowledge required for the implementation of such activities (Freire et al. 2012; Galvão et al. 2011; Reis 2003, 2004, 2008, 2013; Reis and Galvão 2004a, b, 2009).

Following a critical pedagogy, this project assumes education as a democratizing force and a catalyst for individual development and social transformation (Dewey 1916; Freire 1970/1987). It assumes school as a live forum for liberating dialogue instead of an institution aimed at teaching for testing, social conformity and competition between individuals and societies (Kellner and Kim 2010).

The main objective of the 'We Act' project is the development, implementation and study of materials and methodologies aimed at supporting/coaching teachers and students (from primary school to the university) in taking informed and negotiated actions to address social and environmental issues associated with the fields of science and technology (also denominated as Science, Technology, Society and Environment (STSE) controversies, or SSI). It intends to identify factors that positively and negatively influence involvement in this type of action and to build knowledge on appropriate intervention processes that can support teachers with the confidence, motivation and knowledge required for stimulation of such research-informed actions. The project combines development, action and research components and crosses three different areas: (a) the promotion of an active inquiry-based learning regarding real-life controversies associated with science and technology; (b) the stimulation of students' participation in collective democratic problem-solving action; and (c) the support of the first two areas with art initiatives (e.g. drama with a role play component, cartoons, comic strips and posters) and the use of Web 2.0 tools (e.g. for the production and dissemination of podcasts, vodcasts, discussion forums, blogs, comic strips, posters and brochures). While there is a significant body of literature in the area of the discussion of SSI, there is significantly less literature related to the synergic combination of this area with the use of arts-based approaches and Web 2.0 tools mobilized towards activism and social intervention on such controversies/issues. This synergic combination with the aim of promoting collective democratic problem-solving action on SSI is the major novelty of this project.

The 'We Act' project is aimed at stimulating the reconstruction of scientific literacy in schools as collective practices, fostering community action(s) on SSI and recognizing students and teachers as agents of change "using science to address their own problems and, as a result of trying to find solutions, produce new knowledge" (Levinson 2008, p. 144). In this context, students from all ages are considered as a 'citizens,' as opposed to a 'future citizens' and "science is a means of promoting a democracy where citizens act in socially responsible ways" (Levinson 2008, p. 145).

The project team integrates participants (teachers) from different levels of education and institutions (basic schools, secondary schools, polytechnic institutes, universities, research centres in education, research centres in science) with common interests in socio-scientific and socio-environmental issues (SEI). Many teachers are, or were, involved in Master and Doctoral programs on education at the Institute of Education – University of Lisbon. So, they have some level of experience with educational research. During the school year of 2011–2012 contacts were established in order to invite teachers for this community. All the invited teachers share a strong interest in the classroom discussion of SSI (and SEI) and a strong belief in school as a major force for individual development and social transformation. Some of them are deeply involved in environmental education. During the last years, six of these teachers developed academic researches (four master dissertations and two doctoral thesis) about the educational potentialities of SSI's classroom discussion, under the supervision of the 'We Act' coordinator. Other seven teachers wanted to develop their academic researches (five master dissertations, one doctoral thesis, and one post-doctoral research) centred in collective activism on SSI and SEI. So, they considered the 'We Act' project to be an important support for their action-research initiatives. During their master course, several teachers developed knowledge about collective activism and Web 2.0 tools in two subjects (one semester long): 'Environmental Education' (two teachers) and 'Conception of Digital Educational Resources for the WWW' (seven teachers). All this common interests, practices and paths represent evidences of a shared history of learning, that teachers want to proceed and deepen through their involvement in the 'We Act' community.

Despite the existence of a common learning history, at the beginning of the community, several meetings took place at the university in order to develop a shared vision for the project, through an active and dynamic negotiation of meaning and future actions' planning. Several themes were discussed: (a) possible relations between Portuguese curricula and different SSI or SEI; (b) community research-informed action as a major aspect of scientific literacy and students' empowerment as citizens; (c) the mobilization of arts-based approaches and Web 2.0 tools towards activism and social intervention; (d) possible techniques to assess students' knowledge and competences during activism projects. The more experienced members shared their experiences of school activism in order to pull the community's competence and stimulate the less experienced into the planning and implementation of collective activism initiatives about SSI and SEI. A special focus on pedagogical knowledge intended to empower the less skilled, helping them to avoid or surpass some common obstacles associated with this sort of project.

The 'We Act' community had a supportive/coaching role during the entire process of actions planning and implementation. The different competences of the team were mobilized in order to help teachers develop the knowledge and the confidence for supporting students' involvement in collective democratic problem-solving action.

All the action(s) supported in the context of this project were negotiated collectively between each team participant and her/his students to address SSI or SEI (from local, national and/or international level) that students consider socially relevant (allowing the development of inquiry-based activities in real life situations).

During the school year of 2012–2013, coaching meetings took place at specific teachers' requests (involving between two and four community members, according to their competences and schedules) in order to discuss their projects' ideas, aims, materials, difficulties and obstacles. These moments were also a context for coordinating perspectives, actions and contexts according to the project's aims and the expected effects. Occasionally, some of these members were posteriorly involved in their colleagues' classroom activities, exemplifying practices or just observing/discussing them in order to support the development of their colleagues' competence and confidence.

Based on their expertise, some community members collaborated in the development of assessment materials for activism activities (e.g., observation rubrics) and pre and post on-line questionnaires aimed at evaluating the project's impact on students' conceptions about: (a) science education; (b) collective activism; and (c) the nature of science. These materials, developed according to project aims, were made available to all community members, which applied, at least, the pre and post questionnaire centred on students' conceptions about activism. Right now, at the end of the school year, all collected data are being analysed in order to determine statistically significant impacts of the project on each group and on the total group of students. For the specific project implemented by each community member, qualitative data were collected through participatory observation and interviews.

Both in the middle and at the end of the school year, two general meetings took place (with all the community) in order to share and discuss the members' experiences (and the results attained), coordinate actions according to the project's aims and evaluate the project implementation.

Over the school year, the different meetings strengthened the bounds between members and the community's identity through the accumulation of experiences (both successes and failures), stories, classroom materials and competencies.

Given the characteristics and aims of our project, an action-research methodology was chosen (Carr and Kemmis 1986). Among other aspects, the process of action research will focus on: (a) stimulating the willingness to act, innovate and change; (b) problem-solving in school contexts; and (c) developing knowledge/expertise. The implementation of an action research project in schools and classrooms from all levels of education intends to improve the interaction (and to bridge the gap) between different communities: (a) teachers and researchers; (b) science and educational researchers; (c) basic, secondary and higher education teachers. Simultaneously, it allows a wider and collective dissemination of the materials, methodologies and approaches developed during the project in those different communities.

Activism on SSI Through Arts-Based Approaches and Web 2.0 Tools

Project We Act's major novelty is the synergic combination of SSI discussion with the use of arts-based approaches and Web 2.0 tools towards the promotion of collective activism on those issues.

Accordingly to Maria Varelas and colleagues (2010), drama activities foster the development of conversational spaces, allowing the expression of multiple discourses through the combination of everyday and scientific ways of thinking, communicating and acting. Students' imagined roles allow the exploration of issues, events and relationships, contributing to infusion of "science with emotions, excitement, fun, interaction, and shorten[ing] the distance between object of study and subject" (Varelas et al. 2010, p. 322). The imaginary worlds created during drama allow students to establish connections between their own experiences and the socio-environmental and socio-scientific issues with which they are confronted. Science theatre at school can be a useful method to contextualise SSI, to foster students' interest for these issues and to counterbalance the more theoretical side of school science (Wieringa et al. 2012). Drama performances have the potential to stimulate thought and reflection among both the performers and the target audience (Wieringa et al. 2012). These situations may even function as a discussion catalyst between the students (involved in the dramatization) and the audience, increasing their understanding and creating a good opportunity for socio-political action regarding the socio-scientific issues at stake.

Drama activities about SSI, with a component of role-play, can be used to promote the appreciation of science and the understanding about scientific concepts, the processes and nature of science, the interactions between science, technology, society and environment (Simonneaux 2001). When curricula are framed on mastery of content, the implementation of these activities can be considered a distraction. However, role-play activities can promote the development of scientific concepts, when students have to do inquiry about their role and the facts involved in the drama (Hilário and Reis 2009). Students experiment with the complexity of scientific and ethical decision-making when they produce and develop their own characters. This kind of experiential drama invites students and the audience to live an experience and to adopt motivations, opinions or attitudes (Levinson et al. 2008; Ødegaard 2003). The use of role-play activities allow students to explore different perspectives in socio-scientific issues, fostering their reflection and understanding about such controversies and its complexities (Colucci-Gray et al. 2006).

The production and presentation of cartoons and comic strips on SSI is another educational approach explored in this research project. This methodology can involve students in inquiry and discussion, and instigate their involvement in community action on SSI. When students draw cartoons on specific issues, they have to: (a) apply their knowledge and understanding in ways that demonstrate and enhance their critical thinking skills; and (b) explore and clarify their value systems (Kleeman 2006). Cartoons and comic strips allow expression of feelings, anxieties and other emotions that may not emerge through more traditional techniques, facilitating the presentation of sensitive opinions (Beard and Rhodes 2002). Since the beginning of the twentieth century, cartoons and comic strips have been used to challenge values of dominant social and political movements while envisioning new ideologies and new concepts of society that may replace the old. Some authors have transformed what seemed to be a harmless source of entertainment into a major vehicle of radical mass communication, illustrating and exposing through abstraction the sources of social conflicts and the roots of inequalities (Cohen 2007). Current comic-authoring Web 2.0 tools, moreover, allow students to avoid the time-consuming and often frustrating activity of creating traditional comics by hand.

The discussion inherent to the preparation of drama activities and exhibits on SSI can be particularly useful, both in terms of learning about the contents, the processes and the nature of science and technology, and in terms of the students' cognitive, social, political, moral and ethical development (Kolstø 2001). Drama activities and exhibitions (namely, of cartoons and/or comic strips) about socio-scientific issues, as a socio-cultural context, can raise questions, elicit personal reflection and stimulate conversations between students and visitors, transforming both of them into learners and political activists (Braund and Reiss 2004; Levinson et al. 2008).

The development of discussion forums, blogs, vodcasts, podcasts and posters by students (using Web 2.0 tools) is based on a socio-constructivist learning perspective: knowledge is actively built through constant interaction with the worlds inside and outside the classroom. According to this perspective, it is not enough that students listen to scientific explanations; students must be given the opportunity of develop ideas, use and defend them. Online tools have great potential in developing communication and argumentation skills and can be very useful for activist initiatives (Stegmann et al. 2007). Web 2.0 tools, especially those allowing collective communication (namely through social networks), provide students (and all citizens) with powerful means to express their voices and visions, fostering interactive and decentralized forms of communication/intervention and a participatory model of democracy. Through these forms of communication and intervention, students assume a role of active problem solvers (and socio-political activists) and not just a role of spectators relying on experts to point out directions. Teachers have an important role in developing students' (and all citizens in general) awareness of "the vast potential of Internet media for their cultural/social/political empowerment" (Kellner and Kim 2010, p. 20).

Methodology

This paper presents a qualitative study centred on the teachers' motivations for participating in the project 'We Act' and also the difficulties they identify in their students' enrolment in activism on social and environmental issues associated with science and technology. Data were collected during the first year of the project's implementation through an online questionnaire (with open-ended items) and semi-structured interviews conducted with the 25 participating teachers. The questionnaire intended to gather evidence about the teachers' opinions regarding the three aspects under study: (1) personal motivations for participating in the project; and (2) the difficulties they are facing stimulating and supporting that process of collective democratic activism. The interviews were aimed at clarifying and deepening information gathered with the questionnaire.

Both the text with the answers to the questionnaire and the audio files from the interviews were submitted to content analysis (with the help of NVivo10 software) which sought to extract the implicit conceptions about several aspects under study. This kind of analysis involved the classification of meaningful elements, according to certain categories that could bring order to the apparent disorder of the raw data. The category construction process was influenced by the aims and theoretical background of the study. This analysis was discussed between three researchers from the project: all discrepancies were resolved by agreement among them.[1]

Results

This section presents the results concerning teachers' motivations for participating in the 'We Act' project and also the difficulties identified in their students' enrolment in activism on socio-scientific and socio-environmental issues. The data regarding each teacher is condensed and displayed in a meta-matrix by fields of interest (the aspects in study) in a form that allows a systematic visualization and comparison (Table 31.1) (Miles and Huberman 1994).

Teachers' participation in the project has been mainly motivated by a strong willingness to develop professionally in order to: (a) implement more dynamic and creative classroom practices capable of fostering students' motivation and changing students' apathy concerning school science activities (specifically through focus in local real-life and socially-relevant situations); and (b) empower both themselves and students as active citizens and agents of change in society.

Another important motivation to engage in the 'We Act' project has been the possibility of building knowledge and competence through the interaction with other colleagues with different backgrounds and experiences about how to promote students' scientific literacy (specially in what concerns collective and informed activism regarding socio-scientific issues). The meta-matrix shows the participants' diversity of experiences, backgrounds and working contexts. This diversity of knowledge and skills is considered a major reason for the participants' involvement in this project. They particularly value the support and the collective construction of theoretical, pedagogical and technical knowledge (regarding collective democratic problem-solving action) through the interaction in the context of this community.

[1] The author expresses his gratitude to all the members of the 'We Act' project that collaborated in this research.

Table 31.1 Teachers' data meta-matrix

Teachers' Name (and highest school degree)	Teaching experience (years)	Level of education (and subject) where they work	Students' age	Brief description of the activities being implemented within "We Act" project	Motivations for participating in the project "We Act"	Difficulties identified in their students' enrolment in activism (and possible actions to overcome them)
1. Maria (PhD Geology)	≥ 21	University (Geology)	≥ 18	Teacher participating in the project but not yet supporting her students in activism. Still developing the technical and pedagogical knowledge necessary to support activism through Web 2.0 tools	To learn new educational strategies (professional development) To learn the technical and the pedagogical knowledge necessary to Web 2.0 tools classroom integration (allowing the implementation of learning methodologies more in line with students' interests) To learn ways of empowering students with the necessary competences for democratic problem-solving action regarding social and environmental problems affecting society	Students' difficulties not yet identified Teacher difficulties in integrating Web 2.0 tools in her classes: lack of technical and pedagogical knowledge Teacher difficulties in planning educational activities and scenarios for stimulating students' activism Teacher difficulties in finding the time and the opportunity for activism in a curriculum overcrowded with substantive scientific knowledge (Maria recognises the important role of the community in the development of knowledge about how to surpass these difficulties)

(continued)

Table 31.1 (continued)

Teachers' Name (and highest school degree)	Teaching experience (years)	Level of education (and subject) where they work	Students' age	Brief description of the activities being implemented within "We Act" project	Motivations for participating in the project "We Act"	Difficulties identified in their students' enrolment in activism (and possible actions to overcome them)
2. Celia (Grad)	≥21	1st Cycle of Basic Education (all subjects)	6–10	Teacher working in a rural area and supporting her students in research and activism (mainly through drama, exhibitions and school newspaper articles open to the local community) about the problems associated with the use of chemicals in agriculture	To build knowledge (through the interaction with other colleagues with different backgrounds and experiences) about how to promote students' scientific literacy (specially in what concerns collective and informed activism regarding socio-scientific issues) To empower students as active citizens and agents of change	Students presented some difficulties in identifying problems associated with the use of chemicals in agriculture: their parents' agricultural practices are assumed as "normal" and (many times) not questioned Teacher difficulty in assessing competences associated with activism (The problems with assessment have been solved with the support from the community)
3. Amélia (Ms Ed)	11–20	3rd Cycle of Basic Education (Physical and Chemical Sciences)	12–15	Teacher supporting her students in the development of podcasts centred on socio-scientific or socio-environmental issues (related with the Physical and Chemical Sciences curriculum) they consider relevant. The podcast scripts are based on students' research and discussed with the teacher	To build knowledge (through the interaction with other colleagues with different backgrounds and experiences)	Students' lack of technical knowledge about what a podcast is, its utility and how to produce it Students' difficulties with the issues' identification and the production of podcasts' scripts Students' difficulties with collaborative/team work (These problems were overcome by Amélia's support in the

						identification of SSI, the exploration of audio recording computer tools and the attribution of specific roles to each team member)
4. João (Ms Ed)	≥21	3rd Cycle of Basic Education and Secondary Education (Physical and Chemical Sciences; Physics)	12–18	Teacher supporting his students in: (a) the development of an exhibition about socio-scientific issues related with different sources of energy (3rd Cycle students); and (b) the organization of a "Science Café" where socio-scientific or socio-environmental issues selected by the students will be discussed (12th grade students). Both these initiatives area based on students' research and will be open to the community	To develop activities and initiatives aimed at fostering students' competences of reflexion, critical analysis and activism regarding socio-scientific issues To empower students as active citizens and agents of change	The 3rd Cycle students (the youngest) had difficulties in identifying and selecting the SSI (lack of autonomy). The 12th grade students were much more autonomous (So, João had a more directive actuation with the 3rd Cycle students, proposing SSI and defining the working methodology)
5. Heloísa (PhD Ed)	≥21	University (Teacher education for Kindergarten and 1st and 2nd Cycles – Grad and Ms)	≥18	Teachers working together in supporting their student teachers through the development of activism initiatives (related with socio-scientific and/or socio-environmental issues) both at the university and at the schools where they are doing the practicum	To stimulate student teachers into activism about socio-scientific and socio-environmental issues	Some student teachers are having difficulties in motivating both their students and the local practicum supervisors into activism: it's always easier to work the way they are used to and activism requires a strong shift in classroom management (demanding more student centred practices than those they are used to)
6. Horácio (Ms Ed)	≤10	University (Teacher education for Kindergarten and 1st and 2nd Cycles – Grad)	≥18		To develop activities and initiatives aimed at fostering student teachers' scientific and pedagogical knowledge regarding activism about socio-scientific and socio-environmental issues	

(continued)

Table 31.1 (continued)

Teachers' Name (and highest school degree)	Teaching experience (years)	Level of education (and subject) where they work	Students' age	Brief description of the activities being implemented within "We Act" project	Motivations for participating in the project "We Act"	Difficulties identified in their students' enrolment in activism (and possible actions to overcome them)
					To develop knowledge about new strategies with potentialities for citizenship education To develop as a professional	(Heloísa and Horácio, as university supervisors – and during the meetings they have with school supervisors – try to suggest ways of solving problems and dealing with obstacles)
7. Marta (Grad)[a]	11–20	3rd Cycle of Basic Education (Physical and Chemical Sciences)	12–15	Teacher supporting her students (from a rural community) with the preparation (researching and writing the scripts) of dramatizations and radio broadcasts (both in a school and a local radio) about socio-scientific and/or socio-environmental issues they considered relevant (e.g. water treatment, genetic modified seeds, and agro toxics)	To find ways of promoting students' involvement with socio-scientific issues (affecting their community) they consider important and relevant To develop activities and initiatives aimed at fostering students' competences of decision making and activism regarding socio-scientific issues	Students are very distant from their community problems and some of them have no interest in school activities (Trying to solve these problems, Marta had a very positive actuation identifying local problems with strong impact on students' and their families' life and making them conscious about the important role they can have as spokesmen of the marginalized members of their community)
8. André (Ms Ed)	11–20	Secondary Education (Biology)	17–18	Teachers from the same school, supporting each group of	To develop as persons and professionals	Teachers' difficulties in finding the time and the opportunity

9. Tânia (Ms Ed)	≥21	Secondary Education (Biology)	17-18	students (from their classes) through: (a) the identification of a socio-scientific issue; (b) the research of the different dimensions/aspects of that issue; and (c) the organization of different forms of activism initiatives (chosen by the students)	To empower students as active citizens and agents of change	for activism in a curriculum overcrowded with substantive scientific knowledge (These three teachers tried to solve this problem using extra hours to support students, either in person or through Internet – e-mail and videoconference)
10. Carla (PhD Ed)	≥21	Secondary Education (Biology)	17–18			
11. Iris (PhD Ed)	≥21	University (Education – Ms and PhD)	≥21	This teacher will propose to her Master students the development of dissertations about informed and negotiated actions to address social and environmental issues associated with the fields of science and technology	The relevance of the project's theme; To learn new educational strategies (professional development) To build knowledge through the interaction with other colleagues with different backgrounds and experiences	Not applicable
12. Catarina (Grad)[a]	11–20	2rd Cycle of Basic Education (Natural Sciences)	10–12	Students are researching ways to save water, reduce energy consumption and mitigate the environmental impact of garbage. Their conclusions will be presented to their parents through an open session	To develop research competences about her own teaching practice To develop knowledge about new educational strategies (professional development) To empower students as active citizens and agents of change	Because some students don't have access to the Internet at their home (4–5 in each class), Catarina had to change her initial plans involving the use of Web 2.0 tools for activism purposes Students had difficulties in identifying and selecting the SSI (due to a lack of autonomy) Students' apathy concerning school activities (they seem to prefer to listen rather than to act)

(continued)

Table 31.1 (continued)

Teachers' Name (and highest school degree)	Teaching experience (years)	Level of education (and subject) where they work	Students' age	Brief description of the activities being implemented within "We Act" project	Motivations for participating in the project "We Act"	Difficulties identified in their students' enrolment in activism (and possible actions to overcome them)
						(So, Catarina had a more directive actuation, proposing the SSI and defining the working methodology)
13. Vanda (Ms Sc)[b]	$\leq$10	University (Teacher education for Kindergarten and 1st and 2nd Cycles – Grad)	$\geq$18	In the subject "Ecology and Environment", the student teachers are conducting researches and preparing posters and seminars about socio-environmental issues	To implement more dynamic and creative classroom practices capable of fostering students' motivation To create new curricular materials that make science education relevant and popular in the eyes of the students	Teacher's difficulties assessing the impact of the activism initiatives on students' competences (This problem has been solved with the support of the community through the collective development of assessment tools – questionnaires and observation schedules)
14. Miriam (PhD Ed)	$\leq$10	University (Science Education – Ms and PhD)	$\geq$21	This teacher is supervising Master students' dissertations about informed and negotiated actions to address social and environmental issues associated with the fields of science and technology	The relevance and interest of the project's theme To collaborate in the development of activities capable of promoting students' scientific literacy competences, especially in what concerns their ability to act in order to foster social change	Not applicable

15. Joaquim (Ms Ed)	$\leq$10	3rd Cycle of Basic Education (Physical and Chemical Sciences)	12–15	Students are preparing a drama activity about "Genetic Modified Foods", with a component of role-play, which will be presented to the school community	To develop as a person and as a professional To contribute to the improvement of science education with new and more powerful educational strategies To empower students as active citizens and agents of change	Teacher's difficulty in motivating students for the activities – they seem to prefer to listen rather than to act Students had some difficulties in adapting to the demands of more student centred activities (Joaquim had an important role in supporting the students during the entire process, suggesting methodologies and resources and helping them to surpass difficulties)
16. Nicolau (Ms Ed)	11–20	Vocational Secondary Education (Multimedia)	15–18	Multimedia students are identifying Non-Governmental Organizations responsible for relevant activism initiatives, and establishing partnerships to develop marketing/communication materials (e.g. webpages, films, posters and brochures) for those organizations	To motivate students through the implementation of projects centred in real situations To develop knowledge about new educational strategies (professional development) To promote a direct intervention in society To empower students as active citizens and agents of change To collaborate in the development of initiatives capable of developing students' civic conscience	Nicolau had some hard work combining the activism activities chosen by the students with the topics from the multimedia curriculum: but he was very successful in doing so

(continued)

Table 31.1 (continued)

Teachers' Name (and highest school degree)	Teaching experience (years)	Level of education (and subject) where they work	Students' age	Brief description of the activities being implemented within "We Act" project	Motivations for participating in the project "We Act"	Difficulties identified in their students' enrolment in activism (and possible actions to overcome them)
17. Rosa (Grad)[a]	≤ 10	2rd Cycle of Basic Education (Civic Education)	10–12	Students are studying their homes water consumption in order to find ways to save water. Their conclusions are published in a blog open to their families. Students are stimulated to participate in campaigns aimed at collecting materials (e.g. radiographs, batteries and empty ink cartridges) to raise money for humanitarian purposes	To allow and stimulate students to act on socio-scientific and socio-environmental issues they consider important and relevant To empower students as active citizens and agents of change To share educational resources, ideas and results with other colleagues in order to support each other in the development of activism initiatives	Students' difficulties in searching, selecting and combining information: usually they only copy and paste information they find in the Internet without referring the sources Students' apathy concerning school activities (they seem to prefer to listen rather than to act) Students' difficulties with team/collaborative work and to meet deadlines Students' lack of commitment during the activities because they don't value the subject of Civic Education like they do with Mathematics, Natural Sciences or Portuguese and they don't identify these "different" activities as "proper" school activities Students' lack of participation in the blog (Rosa, with the support from the community, tried to solve these problems:

						(a) supporting her students during the process of searching, selecting, combining and re-writing information; (b) assigning specific roles to each group member and defining deadlines for each small task; (c) assessing each task; (d) working in collaboration with the Natural Sciences subject)
18. Alberto (PhD Ed)	11–20	University (Teacher education for Kindergarten and 1st and 2nd Cycles – Grad and Ms)	≥18	In the subject "Investigations in the Environment", the student teachers are conducting researches and preparing an interactive multimedia exhibition about socio-environmental issues. This exhibition will be open to all the community of the School of Education	To stimulate student teachers into activism about socio-scientific and socio-environmental issues To contribute to Portuguese students' involvement in decision making processes related with science and technology To learn ways of empowering students with the necessary competences for democratic problem-solving action regarding social and environmental problems affecting society To build knowledge about how to promote students' active and informed citizenship	Teacher's difficulty in motivating some student teachers for the activities Teacher's difficulty in supporting all the projects developed by different groups – a time consuming task Student teachers had some difficulties in adapting to the demands of more student centred activities (they are more used to listen than to act) (Alberto, with the help from the community, had an important role in defining objectives, supporting the student teachers during the entire process, suggesting methodologies and resources and helping them to surpass

(continued)

Table 31.1 (continued)

Teachers' Name (and highest school degree)	Teaching experience (years)	Level of education (and subject) where they work	Students' age	Brief description of the activities being implemented within "We Act" project	Motivations for participating in the project "We Act"	Difficulties identified in their students' enrolment in activism (and possible actions to overcome them)
						difficulties. Students' performance was assessed in the each phase of the process)
19. Marisa (Ms Ed)[b]	≥21	3rd Cycle of Basic Education (Natural Sciences)	12–15	Teacher supporting students' activism initiatives about the causes and ways to prevent cardiovascular diseases	Bring to the classroom strategies capable of promoting students' personal and social development To prepare students to assume leadership (and stop just being led) To empower students as active citizens and agents of change	Teacher's difficulty in finding the necessary time (during classroom time) to support students' initiatives – a time consuming task (Marisa solved this problem proposing autonomous work – outside the classroom – to the students and supporting the initiatives during one class a week and through an online forum developed in the school Moodle learning platform)
20. Elis (PhD Ed)	11–20	University (Teacher education for Kindergarten and 1st and 2nd Cycles – Grad and Ms)	≥18	In the subject "Environmental Education", the student teachers are conducting researches and preparing an interactive multimedia exhibition about socio-environmental issues. This exhibition will be open to all the community of the School of Education	To stimulate students into informed and responsible activism about socio-scientific and socio-environmental issues To develop activities and initiatives aimed at fostering students' competences: e.g. knowledge, thinking, communication and attitudes	Teacher's difficulty in supporting all the projects developed by different groups – a time consuming task Student teachers had some difficulties in adapting to the demands of more student centred activities (they are more used to listen than to act)

						(Elis had an important role in defining objectives and deadlines for each phase of the project, and assessing students' performance during each one of those phases)
21. Fausto (PhD Ed)	≥21	University (ICT and Education – Ms and PhD)	≥21	Teacher involved in the development of an online multimedia platform aimed at disseminating the initiatives and artefacts produced in the context of collective actions on SSI – texts, podcasts, vodcasts, personal comments and descriptions, posters, cartoons, comic strips, etc	To study ways of using Web 2.0 tools in the support of activism initiatives	Not applicable
22. Mariana (Ms Ed)	≥21	3rd Cycle of Basic Education (Natural Sciences)	12–15	Teacher supporting students in the identification of a local environmental problem, the research of possible solutions and the presentation of those solutions in a special session at the city council. In this session several students' proposals will be presented and voted: the best will be implemented by the city council	To build knowledge (through the interaction with other colleagues with different backgrounds and experiences) To involve students in research projects To learn new educational strategies (professional development) To motivate students through the implementation of projects centred in real life situations	Teacher difficulties in finding the necessary time to support activism initiatives – a time demanding activity – in a curriculum overcrowded with topics (Mariana managed to solve this problem, motivating her students to some homework – with the perspective of winning the city council contest)

(continued)

Table 31.1 (continued)

Teachers' Name (and highest school degree)	Teaching experience (years)	Level of education (and subject) where they work	Students' age	Brief description of the activities being implemented within "We Act" project	Motivations for participating in the project "We Act"	Difficulties identified in their students' enrolment in activism (and possible actions to overcome them)
23. Paula (Grad)[a]	11–20	3rd Cycle of Basic Education (Civic Education)	12–15	Teacher supervising students during the development of activism initiatives related with traffic education (resorting to comic strips, video animations and posters built with Web 2.0 tools)	To motivate students through the implementation of projects centred in real life situations To stimulate students into informed and responsible activism about socio-scientific and socio-environmental issues To learn new educational strategies (professional development)	Paula felt some difficulties in finding the necessary time to support activism initiatives – a time demanding activity Students had some difficulties in adapting to the demands of more student centred activities (they are more used to listen than to act) (Paula, with the support from the community, managed to solve students' difficulties)
24. Artur (Ms Sc)[b]	11–20	University (Teacher education for Secondary – Grad)	17–18	Teacher supporting other science teachers and their students in the research about the causes of local environmental problems (in Cape Verde – Africa) and in the implementation of activism initiatives with the aim of promoting possible solutions	To empower students with the necessary competences for democratic problem-solving action regarding social and environmental problems affecting society To empower students as active citizens and agents of change To learn new educational strategies (professional development)	Students had some difficulties in adapting to the demands of more student centred activities (they are more used to listen than to act) (Artur and the teachers working with him, with the support from the community, had a very positive actuation supporting the students through the entire process)

25. Rita (Grad)[a]	≤10	3rd Cycle of Basic Education (Natural Sciences)	12–15	Science teacher working together with an ICT teacher in order to support her students in the production (research, script writing and technical development) of vodcasts about possible solutions for different environmental problems (local and global). This vodcasts will be broadcast through social networks and the school website	To motivate students through the implementation of projects centred in real life situations To stimulate students into informed and responsible activism about socio-scientific and socio-environmental issues To develop as a professional To study ways of using Web 2.0 tools in the support of activism initiatives	Rita felt some difficulties in finding the necessary time (exclusively in the Natural Sciences classes) to support her students in the production of vodcasts (Rita, with the support from the community, managed to solve this problem involving the ICT teacher in the project and planning with precision the tasks for each class)

[a]Teacher doing Master on Education
[b]Teacher doing PhD on Education

Despite their different competences, all the participants share an interest in socio-scientific and socio-environmental issues and the sociological, political, ethical and economic aspects of these issues.

The support from the community has been considered particularly positive and effective in the development of: (a) pedagogical knowledge regarding the organization, implementation and assessment of the initiatives (involving Web 2.0 tools and art based approaches) according to the different realities and educational objectives defined by the teachers; and (b) assessment instruments centred on the different competences on which teachers considered important to work.

> I had no idea about how to assess students' competences during a discussion activity or a project involving research and informed action about environmental problems. This is one of the aspects I have learned a lot about. (Catarina, interview)
>
> Contacting with other colleagues' experiences and having the opportunity to discuss with them the strategies they have developed in order to solve the problems they faced was very formative. Despite the different contexts where we work, our problems are very similar. So, I learned a lot with the colleagues during the meetings. (Horácio, interview)
>
> The fact of belonging to a community gave me the necessary support and strength to experiment new approaches and activities. It was a constant opportunity to contact with different projects, ideas, methods, materials, ways of overcoming difficulties... an opportunity to surpass myself... to surpass my fears and uncertainties... to gain courage to risk... It's easier when we 'walk' side-by-side with other colleagues: we have fun, we share our successes, and we reflect together on our failures trying to find the causes and develop possible solutions. Some times we feel isolated at school and then it's very difficult to gain the courage to experiment new methods. (Elis, interview)

Through the involvement in the project and the interesting actions implemented by the community, teachers reinforced some conceptions: (a) considering research-informed action as a major aspect of scientific literacy; and (b) recognizing students as important agents of change (capable of implementing, with success, impacting actions at their families and groups of friends) and, consequently, as "citizens" (as opposed to "future citizens"). Children and young people are seen as social actors in their own right ("citizen now"), and not merely objects of socialisation ('citizen becoming') (Invernizzi and Williams 2009).

> For me it was a complete new perspective: to see young children as citizens. All the examples of students' actions that we have discussed [in the community meetings] ... And all the experiences implemented by our colleagues' students... Reflecting on all this made me understand the necessity of empowering children for informed action. They must feel that both their opinions and actions are valued and stimulated. And this requires a shift in my classroom practice: a shift into problem centred action. (Rosa, interview)
>
> Students can be extremely insistent and persuasive when they believe in something, when they are convinced of something. During this year we saw some examples of students' dedication to causes they believe in. It would be a crime to suppress (or even to ignore) this impetus. (...) These situations were great examples of active citizenry. (Nicolau, interview)

Despite their strong motivation and the positive impacts associated with the involvement in the 'We Act' project, participating teachers have been facing important obstacles in the implementation of collective actions, namely: (a) time restrictions imposed by overcrowded curriculums (Maria, André, Tânia and Carla);

(b) difficulty in finding the necessary time (during classroom time) to support students' initiatives – a time consuming task (Elis and Marisa); (c) difficulty in finding the 'proper' controversial topics capable of breaking out students' apathy (Marta and Catarina); (d) students' lack of commitment when the activities are implemented in less "valued" subjects (e.g. Civic Education) or because they don't identify this "different and strange" activities as "proper" school activities (Rosa); and (e) some students negative reactions to classroom practices more centred on "doing and collaborating" and less centred on "listening" to the teacher (habits are difficult to break) (Amélia, Heloísa, Horácio, Joaquim and Rosa).

> I felt a lot of difficulties [in implementing at activism project with the students]. It was very difficult to motivate them for the project. They were unable to identify any environmental problem they would to act on. Of course they are very young [10 and 11 years old]. But I always felt their apathy concerning school activities: they seem to prefer to listen rather than to act. They are not used to active teaching methods or to collaborative work. Another important aspect: they didn't show commitment during the activities because they don't value the subject of Civic Education. It's not important for them. It's not like Maths or Portuguese Language.... (Rosa, interview)
>
> I felt difficulties in supporting at the same time all the projects developed by different groups. Students didn't have the necessary autonomy for this type of project. They had some difficulties in adapting to the demands of more student centred activities. So, it was quite a formative experience for them... and for me. (Elis, interview)

Teachers are also facing difficulties in motivating other colleagues into activism: it's always easier to work the way they are used to and activism requires a strong shift in classroom management, demanding more student centred practices than those they are used to (Heloísa and Horácio).

> Our students' supervisors [in a teacher education course] aren't used to student centred activities. So, they learned a lot with the projects implemented by our students. (Horácio, interview)

One teacher (Maria), despite her willingness to implement activism initiatives about socio-scientific issues (and her conviction about the educational potential of these activities), is still developing (through the involvement in the community) what she considers to be the necessary technical and pedagogical knowledge regarding classroom integration of Web 2.0 tools and activism initiatives planning. She also feels difficulties in finding the time and the opportunity for activism in what she considers to be "a curriculum overcrowded with substantive scientific knowledge" (Maria, interview). Maria seems to put too much emphasis on the scientific content, which is viewed as an aim in itself, leaving any consideration about the nature of science and the interrelations between science, technology and society as a mere footnote of her classes. She seems to perceive the curriculum as a list of topics that must be covered thoroughly.

> [In the Geology course], we only have this subject [General Geology] to address basic concepts of geology indispensable for understanding the contents of the next subjects. I don't get any time available to address controversies related to the interactions between science, technology, society and the environment. Indeed, the reaction of my colleagues would not be very positive. However, I think the discussion of socio-scientific controversies is extremely useful for the understanding of the scientific and technological endeavours and

> their interactions with society. Next year I will try to promote some session outside the classroom (eventually, a session organized by Students' Association) where students can discuss socio-scientific controversies with experts with opposing perspectives. Maybe, I will be able to propose (with the help of our community) some activity based on an on-line forum. But first I need to learn how to develop such an activity. (Maria, interview)

Other teachers, despite feeling the same lack of time, have the capacity to interpret the curriculum so as to address the topics and carry out the activities they consider important for their students' development (in terms of knowledge, skills and attitudes) and relevant for the specific communities they live in and the society in general. These teachers share a conception of curriculum allowing for levels of decision-making suited to the specific needs of society:

> I always manage to find a way to involve students in the discussion of some SSI or SEI that I consider particularly relevant for their personal competences or the community where they live. Of course, the issue also depends on the curricular topics: it must have some relation with them. But, in my opinion, teachers must be particularly skilled in showing the social relevance of the topics they teach. I consider that skill to be extremely important to raise students' interest in science education. (Mariana, interview)
>
> For me it was very easy. The multimedia curriculum doesn't have any topic about activism... and nothing about SSI or SEI. However, the activism about environmental or social problems is fantastic material for the development of students' multimedia projects. I have several friends involved in Non-Governmental Organizations. So, I challenged students to contact the NGO with whom they feel some personal connection and to develop a multimedia campaign to divulge their objectives. It was quite a success! (Nicolau, interview)

These teachers assume a role of active curriculum constructors (and not just consumers/executors) (Hargreaves 2000), managing content and choosing the educational experiences according to students' specific characteristics, the contexts in which they live and the needs of society. So, teachers' conceptions about the curriculum (and not the curriculum itself) represent an important inhibitor/stimulator of their support (and stimulation) of activism initiatives in their classes.

Another barrier to activism initiatives in school has been the final national examination at the end of the 11th grade. In spite of teaching both the 3rd Cycle of Basic Education (7th to 9th grade) and the Secondary Level (10th to 12th grade), many participating teachers decided to implement their activism initiatives in the former period rather in the latter. This selection was justified with teachers' difficulties in finding the time and the opportunity for activism in a curriculum overcrowded with substantive scientific knowledge (in 10th and 11th grade) and followed by a national exam with strong impact on students' access to the university.

> In the 10th and 11th grade science subjects it's almost impossible to spend time addressing SSI or SEI: there is a great quantity of curricular topics and a final exam that doesn't cover controversial issues related with science-technology-society-environment interactions. We feel a constant pressure from the students (and also their parents) to 'train' them for the sort of items normally included in the exams. (João, interview)
>
> There is a huge pressure from the families: students must have the highest marks in order to attend their favourite university courses. So, we teach for testing. Many active classroom practices (like classroom discussions or activism projects) are seen as 'distractions', 'folklore', a 'waste of time'. (Carla, interview)

> In these school years [10th and 11th], aims like education for active citizenship and social justice are sidelined by the priority of getting the highest grades in the final exam. Individualism dominates social interest. (Tânia 1st general meeting)

Formal evaluation (and the kind of national exam that is proposed) is crucial in establishing the importance and the priority given to specific topics and to certain educational practices (Galvão et al. 2007). The national exam at the end of the 11th grade has a harmful effect on the organisation and management of the 10th and 11th grade science classes. Despite teachers' total freedom regarding approaches, methodologies and activities implemented in her/his classes, students' external assessment on a national level plays a strong regulatory role. The absence of assessment items centred on the discussion of the social, economic, ethical and moral implications of science and technology (an important dimension of the secondary science education curricula) has an undesired effect on classroom practices, compromising the implementation of research, discussion and activism initiatives about socio-scientific and socio-environmental issues. This national exam, instead of inducing a positive change in classroom practices (accordingly with curricular orientations calling for argumentation, discussion and action regarding those issues), has a limitative effect on the implemented range of educational approaches, methodologies and activities. Too much emphasis is placed on memorisation (of a wide list of terms, concepts and facts) and on prescriptive laboratory work.

Conclusions

The results presented in this article show that teachers' support of informed and negotiated activism regarding socio-scientific and socio-environmental issues is not an easy task, requiring: (a) strong beliefs about the educational potentialities of such approach in students empowerment as citizens; (b) knowledge about the interactions between science, technology, society and environment; (c) pedagogic knowledge concerning the implementation of specific activism initiatives; (d) the empowerment of teachers and students as active citizens and agents of change; and (e) the willingness and the capacity to change the school, the community and/or the society. These factors are quite important for teachers' liberation of the oppression exercised by the curricula and national exams.

The "We Act" project has sparked teachers' interest in activism and paved the way for the development of a community of practice (Wenger 1998, 2010) formed by teachers with shared interests and engaged in a process of collective development, implementation and study of activities and methodologies aimed at supporting students (from all education levels) in taking informed and negotiated actions to address socio-scientific and socio-environmental issues. Through interaction and mutual support (learning with each other) they have been enrolled in a process of action-research about their own classroom practices in order to improve their competences as educators and their students' competences of active citizenship.

The ‘We Act’ community members are bound by the idea that through collaborative and continuous involvement in activism initiatives both students and teachers develop the necessary competences for a more active citizenry in a more democratic and socially just society. This involvement has been providing them with: (a) understandings about how science, technology, society and environment interact with each other; (b) ideas about how society should work; and (c) the power, the willingness and the opportunities for transforming classroom, school and society.

The dynamic learning environment developed by the ‘We Act’ members exhibit several characteristics of what Wenger (1998, 2010) labels as a ‘community of practice’: (1) a shared interest; (2) a sustained mutual relationship focused on learning in action (a learning partnership); (3) a dynamic negotiation of meaning; (4) shared ways of engaging in doing things together; (5) a rapid flow of information and innovation; (6) a shared knowledge of each member competences; (7) the development of an identity through the accumulation of experiences, stories, classroom materials, ways of addressing recurring problems, knowledge and competencies; (8) a shared discourse reflecting a certain worldview.

The ‘We Act’ community of practice has been providing teachers with a ‘security net,’ supporting them during their innovation efforts and allowing the sharing of successes and the dilution and attenuation of failures. This community has developed the power of fighting teachers’ feelings of isolation and discouragement that inhibit classroom innovation, encouraging individual development and social transformation at the same time.

References

Beard, C., & Rhodes, T. (2002). Experiential learning: Using comic strips as ‘reflective tools’ in adult learning. *Australian Journal of Outdoor Education, 6*(2), 58–65.

Bencze, J. L. (2008). Private profit, science and science education: Critical problems and possibilities for action. *Canadian Journal of Science, Mathematics & Technology Education, 8*(4), 297–312.

Bencze, L., & Carter, L. (2011). Globalizing students acting for the common good. *Journal of Research in Science Teaching, 48*(6), 648–669.

Bencze, J. L., & Sperling, E. R. (2012). Student-teachers as advocates for student-led research-informed socioscientific activism. *Canadian Journal of Science, Mathematics & Technology Education, 12*(1), 62–85.

Braund, M., & Reiss, M. (Eds.). (2004). *Learning science outside the classroom*. London: Routledge Falmer.

Carr, W., & Kemmis, S. (1986). *Becoming critical: Education, knowledge and action research*. Falmer: Lewes.

Cohen, M. (2007). “Cartooning capitalism”: Radical cartooning and the making of American popular radicalism in the early twentieth century. *International Review of Social History, 52*, 35–58.

Colucci-Gray, L., Camino, E., Barbiero, G., & Gray, D. (2006). From scientific literacy to sustainability literacy: An ecological framework for education. *Science Education, 90*, 227–252.

Dewey, J. (1916). *Democracy and education: An introduction to the philosophy of education.* New York: Free Press.

Driver, R., Leach, J., Millar, R., & Scott, P. (1996). *Young people's images of science.* Buckingham: Open University Press.

Freire, P. (1970/1987). *Pedagogia do oprimido* [Pedagogy of the oppressed], 17.ª edição. Rio de Janeiro: Paz e Terra.

Freire, S., Faria, C., Galvão, C., & Reis, P. (2012). New curricular material for science classes: How do students evaluate it? *Research in Science Education, 43*(1), 163–178.

Galvão, C., Reis, P., Freire, A. M., & Oliveira, M. T. (2007). Science curriculum in Portugal: From the development to the evaluation of students' competences. In D. Waddington, P. Nentwig, & S. Schanze (Eds.), *Making it comparable. Standards in science education* (pp. 237–253). Münster: Waxmann.

Galvão, C., Reis, P., Freire, S., & Almeida, P. (2011). Enhancing the popularity and the relevance of science teaching in Portuguese Science classes. *Research in Science Education, 41*(5), 651–666.

Gray, D., Colucci-Gray, L., & Camino, E. (Eds.). (2009). *Science, society and sustainability: Education and empowerment for an uncertain world.* London: Routledge Research.

Hargreaves, A. (2000). Four ages of professionalism and professional learning. *Teachers and Teaching: History and Practice, 6*(2), 151–182.

Hilário, T., & Reis, P. (2009). Potencialidades e limitações de sessões de discussão de controvérsias sociocientíficas como contributos para a literacia científica [Potentialities and limitations of socio-scientific controversies' discussion sessions as contributors for scientific literacy]. *Revista de Estudos Universitários, 35*(2), 167–183. (Brazil).

Hodson, D. (1998). *Teaching and learning science: Towards a personalized approach.* Buckingham: Open University Press.

Hodson, D. (2003). Time for action: Science education for an alternative future. *International Journal of Science Education, 25*(6), 645–670.

Invernizzi, A., & Williams, J. (2009). *Children and citizenship.* London: Sage.

Kellner, D., & Kim, G. (2010). YouTube, critical pedagogy, and media activism. *The Review of Education, Pedagogy, and Cultural Studies, 32*(1), 3–36.

Kleeman, G. (2006). Not just for fun: Using cartoons to investigate geographical issues. *New Zealand Geographer, 62*, 144–151.

Kolstø, S. (2001). Scientific literacy for citizenship: Tools for dealing with the science dimension of controversial socioscientific issues. *Science Education, 85*(3), 291–310.

Levinson, R. (2008). A theory of curricular approaches to the teaching of socio-scientific issues. *Alexandria, 1*(1), 133–151.

Levinson, R., Nicholson, H., & Parry, S. (Eds.). (2008). *Creative encounters: New conversations in science, education and the arts.* London: The Wellcome Trust.

Miles, M. B., & Huberman, A. M. (1994). *Qualitative data analysis.* Thousand Oaks: Sage.

Nelkin, D. (Ed.). (1992). *Controversy: Politics of technical decisions.* London: Sage.

Ødegaard, M. (2003). Dramatic science. A critical review of drama in science education. *Studies in Science Education, 39*, 75–101.

Reis, P. (1997). A Promoção do Pensamento através da Discussão dos Novos Avanços na Área da Biotecnologia e da Genética [Promoting thinking through the discussion of new developments in biotechnology and genetics]. Master dissertation, Departamento de Educação da Faculdade de Ciências da Universidade de Lisboa.

Reis, P. (2003). *O "admirável mundo novo" em discussão [Discussing the "Brave New World"].* Lisboa: Ministério da Educação, Instituto de Inovação Educacional.

Reis, P. (2004). *Controvérsias sócio-científicas: Discutir ou não discutir? Percursos de aprendizagem na disciplina de Ciências da Terra e da Vida* [Socio-scientific controversies: To discuss or not to discuss? Ways of learning in earth and life sciences]. Doctoral thesis, Departamento de Educação da Faculdade de Ciências da Universidade de Lisboa.

Reis, P. (2008). *A escola e as controvérsias sociocientíficas: Perspectivas de alunos e professores [School and socio-scientific controversies: Students' and teachers' perspectives]*. Lisboa: Escolar Editora.

Reis, P. (2013). Factors of success regarding classroom discussions of SSI: A cross-case study. *Les Dossiers des sciences de l'éducation, 29*, 67–80.

Reis, P., & Galvão, C. (2004a). Socio-scientific controversies and students' conceptions about scientists. *International Journal of Science Education, 26*(13), 1621–1633.

Reis, P., & Galvão, C. (2004b). The impact of socio-scientific controversies in Portuguese natural science teachers' conceptions and practices. *Research in Science Education, 34*(2), 153–171.

Reis, P., & Galvão, C. (2009). Teaching controversial socio-scientific issues in biology and geology classes: A case study. *Electronic Journal of Science Education, 13*(1), 162–185. Available at: http://ejse.southwestern.edu/volumes/v13n1/articles/art7-reis.pdf

Roth, W.-M., & Désautels, J. (Eds.). (2002). *Science education as/for sociopolitical action*. New York: Peter Lang.

Simonneaux, L. (2001). Role play or debate to promote students' argumentation and justification on an issue in animal transgenesis. *International Journal of Science Education, 23*(9), 903–928.

Stegmann, K., Weinberg, A., & Fischer, F. (2007). Facilitating argumentative knowledge construction with computer-supported collaboration scripts. *Computer-Supported Collaborative Learning, 2*, 421–447.

Varelas, M., Pappas, C. C., Tucker-Raymond, E., Kane, J., Hankes, J., Ortiz, I., & Keblawe-Shamah, N. (2010). Drama activities as ideational resources for primary-grade children in urban science classrooms. *Journal of Research in Science Teaching, 47*(3), 302–325.

Wenger, E. (1998). *Communities of practice: Learning, meaning, and identity*. Cambridge: Cambridge University Press.

Wenger, E. (2010). Communities of practice and social learning systems: The career of a concept. In C. Blackmore (Ed.), *Social learning systems and communities of practice* (pp. 179–198). London: Springer.

Wieringa, N. F., Swart, J. A. A., Maples, T., Witmondt, L., Tobi, H., & van der Windt, H. J. (2012). Science theatre at school: Providing a context to learn about socio-scientific issues. *International Journal of Science Education, Part B, 1*(1), 71–96.

Ziman, J. (2000). *Real science: What it is, and what it means*. Cambridge: Cambridge University Press.

Chapter 32
Counter Cultural Hegemony: Student Teachers' Experiences Implementing STSE-Activism

Darren Hoeg and Larry Bencze

Abstract In this chapter we outline a study on student teachers' experiences attempting to implement STSE-activism during a required 1-month practice teaching session at local public secondary schools. We utilized semi-structured interviews and several heuristic survey instruments to collect participants' beliefs and orientations about: the relationship between science and society; school science; activism; their experiences teaching science during their practicum, including their ability to implement STSE-activism; and what resisted its implementation. Participants demonstrated considerable similarity in their beliefs and orientations to STSE-activism both before and after their practicum, and they described very similar experiences during the practicum. Pre-practicum beliefs and orientations appeared to be highly amenable to teaching STSE-activism. Post-practicum, however, participants expressed considerable skepticism about their ability and desire to teach STSE-activism and demonstrated a growing prioritisation for didactic content-teaching. We suggest these changes may stem from the influence of a hegemonic school science culture participants experienced during their practicum that is resistant to STSE-activism. This culture appears to prioritise knowledge consumption, rather than knowledge production, a characteristic feature of school education influenced by neoliberal values. This culture was described by participants as one that; prioritised teacher-directed content learning and other authoritarian classroom approaches; is seen by the school as necessary to prepare students for university; provides inadequate support for student-led STSE-activism; and includes student who are largely resistant to the open-ended and action oriented educational experiences typically enacted in STSE-activism. We suggest their experiences during practicum aligned student teachers' practice towards conservative, traditional, neoliberal structures and culture that restricted their agency to implement STSE-activism. If we are to enable school science education that strives to affect sociocultural criticism and change, we

D. Hoeg (✉) • L. Bencze
Ontario Institute for Studies in Education, University of Toronto, Bloor St W 252, Toronto, ON M5S 1V6, Canada
e-mail: darren.hoeg@utoronto.ca; larry.bencze@utoronto.ca

L. Bencze and S. Alsop (eds.), *Activist Science and Technology Education*, Cultural Studies of Science Education 9, DOI 10.1007/978-94-007-4360-1_32,

believe it is imperative that student teacher education and school experiences are supportive of their attempts to implement STSE-activism.

Keywords STSE-activism • Student-teachers • Practicum • Cultural hegemony • Knowledge-consumers • School structures • Neoliberalism

Introduction

The challenges present for new teachers implementing an activism oriented science education were made prescient during a recent symposium on activism in school science at our university. Much of the discussion in the symposium was grounded in a general agreement on the merits of activism in science, validated by recognition of its appropriateness within the Science Technology Society and Environment (STSE) component of the local provincial curriculum. It was ironic, therefore, that much of the discussion articulated caution about including activism in school science. Great effort was made by several panelists, for example, to problematize activism on the grounds of it being unrealistic given the contemporary knowledge expectations in school science. Detailed argument ensued about the need to instill in students the 'canons of science' in order to provide them with the 'cultural capital' (Bourdieu 1990) they need to be engaged citizens and to be able to do well on formal assessments. Another branch of discussion questioned the political orientation of activism, and whether it is ethical for teachers to influence students to take certain views or actions (as if this is not already occurring, subversively). The institutional and disciplinary structures constraining teachers' abilities to implement STSE-activism in the classroom were identified and questioned: Will students be able to learn what they need to know to succeed in university science? Will there be time for STSE-activism based projects? Will they adequately learn 'the scientific method?' Some misunderstandings of the scope and potential of activism in science is clear in these debates. Also clear is the failure of many contributors to see that these very resistances to activism are symptomatic of the insidious and hegemonic social power structures this type of education aims to change. We use this commentary as an opening to this chapter because the discussions initiated there accurately depict the actualities of the resistances and challenges new teachers face attempting to implement activist-based education in their classroom. This chapter will broadly outline some of these realities by discussing the experiences of student teachers developing activist education in school science.

Socioscientific Issues and the Need for Activism

It can be argued that social issues related to the co-dependence of science, technology, society and environments are the most pressing issues of contemporary human populations. A sense of urgency on issues such as global climate change, environmental degradation, and related human health concerns are shared by many scholars who recommend that educational systems need to prepare societies to actively address these issues. We likely need scientific understanding to solve many of these problems, and science has inarguably contributed to immeasurable improvements to human-well being and standards of living. However, the processes and products of science have also contributed to, for example, the enormous amounts of carbon dioxide gas released into the atmosphere from the burning of fossil fuels that contribute to climate change (Ehrlich and Ehrlich 1972; Stobaugh and Yergin 1979; Likens et al. 1979), making blind adherence to science for solutions to environmental and social problems perhaps unwise.

The consequences of the misapplication and misuse of the products and processes of science, technology and engineering has lead to disagreement about their use; these disagreements have been termed socioscientific issues (SSIs) (e.g., Zeidler et al. 2005). Disagreement also focuses on how socio-economic factors are related to science and SSI's; for example, it is apparent that science and engineering enable the rapid cycles of production and consumption of for-profit products and services that sustain the economy (Kleinman 2003; Krimsky 2003; Ziman 2000). The manipulation of science by corporations to support the intensification of this global consumerism potentially compromises the integrity of science knowledge gains and dissemination. This intensification appears to be an aggressive form of capitalism – known as *neoliberalism* – that has come to dominate the *Zeitgeist* (Hegel 1837/1975) of many societies (Gabbard 2000; McMurtry 1999).

The term neoliberal, in its most common usage, refers to the social and economic values underlying reform policies, such as eliminating price controls, deregulating capital markets, lowering trade barriers, and reducing state influence on the economy especially by privatization and fiscal austerity (Boas and Gans-Morse 2009). The consequences of neoliberalism for education are widespread, but likely include the high prioritisation of content teaching and assessment that positions science knowledge as a commodity that is to be consumed by students (Eisenhart et al. 1996). Competition to acquire this commodity potentially inculcates values in students, associated with liberalism, conducive to gaining more knowledge than peers, such as competition, consumption and individualism, (Bencze and Carter 2011). The economic 'value' of the student is then determined through examinations that assess the amount of knowledge acquired, as evidence of their potential future contribution to society (e.g. the economy). This type of learning may also be antithetical to developing activist oriented skills and knowledge students may need as future citizens (Stetsenko 2012; Wood 1998). Instead, school science students might be engaged in individual and community action required to address SSI's.

SSIs Education and Activism in Western School Science

The potential inclusion of activist oriented science education lies in the call for students' to acquire *scientific literacy*. Many national curricula "recognise the importance of broadly conceptualising scientific literacy to include informed decision making; the ability to analyse, synthesise, and evaluate information; deal sensibly with moral reasoning and ethical issues; and understand connections inherent among SSIs" (Zeidler et al. 2005, pp. 357–358). Although scientific literacy is a term that remains contested, Derek Hodson (2003) describes it in the following four domains: (1) learning science and technology; (2) learning about science and technology; (3) doing science and technology; (4) engaging in sociopolitical action. Of engaging in sociopolitical action, he writes "acquiring the capacity and commitment to take appropriate, responsible, and effective action on matters of social, economic, environmental and moral-ethical concern" (p. 658).

The first of these domains appears to be a priority for school science, compromising teachers' implementation of the other three domains (Bencze and Sperling 2012). Learning canonical and commodified scientific knowledge alone, stripped of the sociocultural and socioscientific realities that went into its development, has the potential to lead to unrealistically positive, even *mythical* (Barthes 1972) conceptions about the nature of science, engineering, and their products and services (Bencze and Sperling 2012). These conceptions appear to be based on positivistic epistemological and realist ontological beliefs which were first identified 30 years ago by Robert Nadeau and Jacques Désautels (1984) including: (i) Naïve realism – that scientific knowledge is a reflection of things as they actually are; (ii) Blissful empiricism – that all scientific knowledge derives directly and exclusively from observation of phenomena; (iii) Credulous experimentalism – that experimentation makes possible conclusive verification of hypothesis; (iv) Blind idealism – that the scientist is a completely disinterested, objective being; (v) Excessive rationalism – that science brings us gradually closer to the truth. These beliefs have been affirmed and reaffirmed as extant in the science education research literature, even up to contemporary times (e.g., Hodson 2009), suggesting a dominant culture.

Although notions of culture can and have been contested in, for example, anthropology (Sewell 1999), here we mean by culture "the unique qualities represented in patterns of traditions, symbols, rituals and artifacts" (Wax 1993, p. 100). Although Wax goes on to describe culture as "a thing of shreds and patches" (p. 101), there does appear to be consistent conservative principles in Western school science, indicative of culture, that are greatly influenced by neoliberal capitalism (Bencze and Carter 2011). A form of neoliberal consumerist culture characteristic of neoliberal discourse is inscribed in teachers' and students' habits of ingesting a steady diet of conclusions (products) that can suppress perhaps innate inclinations to be curious, ask questions, and inquire. Instead, they develop tendencies of conformity that prevent them from drawing their own conclusions, and critiquing knowledge and those who control it (Wood 1998). Quick delivery of scientific content likely makes students (and teachers) become individually competitive (Eisenhart et al. 1996). These traits, if followed through to a logical progression, could lead to ambivalence towards those less successful than themselves (Beyer 1998).

A *cultural hegemony* (Gramsci 1982) appears to exist when these foundational beliefs become dominant. Gramsci described hegemonic culture as one in which values of the bourgeoisie become the 'common sense' values of everyone.

Thus, a consensus culture develops in which people in the working-class identify their own good with the good of the bourgeoisie, helping to maintain the *status quo*. We suggest the dominant culture of Western school science we described has the potential to reproduce the neoliberal *status quo*, maintaining a cultural hegemony that becomes recognised by students and others producing it, as Gramsci might say, as for their own good.

SSIs and Activism in North American Science Curricula

In North American school science, coverage of SSIs have become common in curriculum, particularly in curricular strands termed Science Technology and Society (STS) in the United States, and Science, Technology, Society and the Environment (STSE) in Canada. STS and STSE provide the space in science curriculum for teachers and students to explore issues connected to the well-being of individuals, societies, and environments (WISE), making necessary the preparation of teachers to teach STSE. Yet, divergent views on how to address SSIs has resulted in a multiplicity of approaches and interpretations. One such interpretation is that students will have opportunities to critically engage with SSIs, draw their own conclusions, and take actions. We suggest there are many reasons educators promote student/citizen actions on SSIs. From sociocultural perspectives, several prominent theories on social development, such as *activity theory* (Engestrom 1999) and *actor network theory* (Latour 2004) postulate that humans are innately programmed to purposively, actively transform our environment; thus, to be active and an activist is human-nature (Stetsenko 2012). However, school-based learning centered on the consumption of ready-made scientific fact may repress these natural learning and developmental tendencies (Wood 1998), forcing students to acquire skills of conformity rather than to develop natural abilities of agency and activism. Perhaps the most important argument for including activist-based science education is the growing need for a citizenry that has agency to change societies with ever-more serious personal, social and environmental problems.

Preparing New Science Teachers for School Science Activism

The landscape defined by the conservative beliefs we have described represents potentially rocky ground for new science teachers. A school culture based on these beliefs might be particularly resistant to considerations of the social-construction of scientific knowledge, the inherent bias in this knowledge, the political nature of scientific research, and students as knowledge creators rather than knowledge consumers. These considerations are important, however, to education involving SSIs and activism (Bencze 2008). A school-wide systemic preoccupation with providing students with knowledge to consume potentially reconstitutes SSIs as simply another type of content knowledge (Hodson 2003; Lester et al. 2006; dos Santos 2009). New science teachers entering schools, regardless of their personal subjective beliefs, are likely burdened with the task of 'fitting in' and 'learning the system,' requiring the acquisition of a large degree of cultural capital (Bourdieu 1990). However, many progressive forms of science education, such as teaching SSIs, ideally, may require *counter-cultural capital* consisting of new pedagogical values, traditions and practices that resist and are antithetical to the dominant ones. Yet, these may be difficult to develop in a conservative hegemonic culture.

Although various theoretical stances have been taken to explain the social relations reproducing culture in schools, we see great potential in using actor-network theory (ANT) (Latour 2004). ANT, developed within the fields of Science and Technology Studies (STS) by Bruno Latour, Michael Callon and John Law, is used to understand interactions in systems, or *networks*. The theory can be described as 'material-semiotic,' meaning it considers both the material and symbolic meaning of *actants* in a network, and how these come together to act as a whole. New teachers in schools can be said through ANT to be actants entering an existing network; they potentially both change and are changed by this contact. Yet, the conservative culture of school science may be a relatively stable network, and therefore not easily changed by new actants, such as student teachers. New actants can however 'mobilise' other actants, such as other teachers or material resources, which can become similarly aligned with the initial actant, potentially changing the network.

To become actants that can change the system, teachers may need skills and orientations representative of a form of counter-culture capital; that is, cultural skills and literacy that will allow individuals to go against "the very grain" (Foucault 1980, p. 39) of school that appears to be structured according to neoliberal ideology. Re-educating practising teachers may be difficult because, as noted previously, they may be constrained by a dominant school culture that is ill-suited to pedagogies characteristic of activism, and the uncertain nature of SSIs. *Student* teachers may be better candidates to promote activism in SSIs because new teacher education offers a chance to intervene in the reproduction of the conservative practices of schools. Preparation to teach STSE-activism can offer alternative views to new teachers who may have been conditioned through years of being a student in science education to expect to use didactic instruction to teach science.

Understanding how new teachers navigate the cultural landscape of school science and how this navigation influences their beliefs and practices is therefore of central concern to those of us involved in science teacher education. Our efforts to inculcate a socially critical, activist-oriented disposition and skill-set in new teachers is one way of introducing counter-cultural change to what appears to be a reproductive 'system' or culture.

Science and Technology Education Promoting Wellbeing for Individuals, Societies and Environments (STEPWISE)

Since 2006 we have been working with student teachers to promote student-led activism to address SSIs. This work has been organised around the curriculum and instructional framework known as 'STEPWISE,' which is the acronym for Science and Technology Education Promoting Wellbeing for Individuals, Societies and Environments. STEPWISE is a pedagogical framework that orients student learning and activity toward activism. The research and learning activities described in the framework include constructivist and student-directed activities, resulting in students constructing their own knowledge. A key feature of STEPWISE is to engage students in primary (e.g., their own studies) and secondary (e.g., internet searches) research to help inform their decisions about SSIs and taking action, culminating with students directing their own research-informed actions to address SSIs. In previous research with student teachers using STEPWISE (Bencze et al. 2012), it was concluded that activities involving them in secondary and primary research potentially sensitises them to teaching activism in school science. Figure 32.1 shows a linear version of STEPWISE, as opposed to the ideal, cyclical, tetrahedral model we typically promote. Although this model might be seen as a deficit learning approach, resembling the problematic 'banking' type of education described by Freire (1997), it seems to reflect many students' and teachers' existing conceptions about the progressive nature of gaining expertise and confidence, so we, therefore, view this compromise as a necessary start to introduce what is usually an unfamiliar and potentially uncomfortable pedagogy.

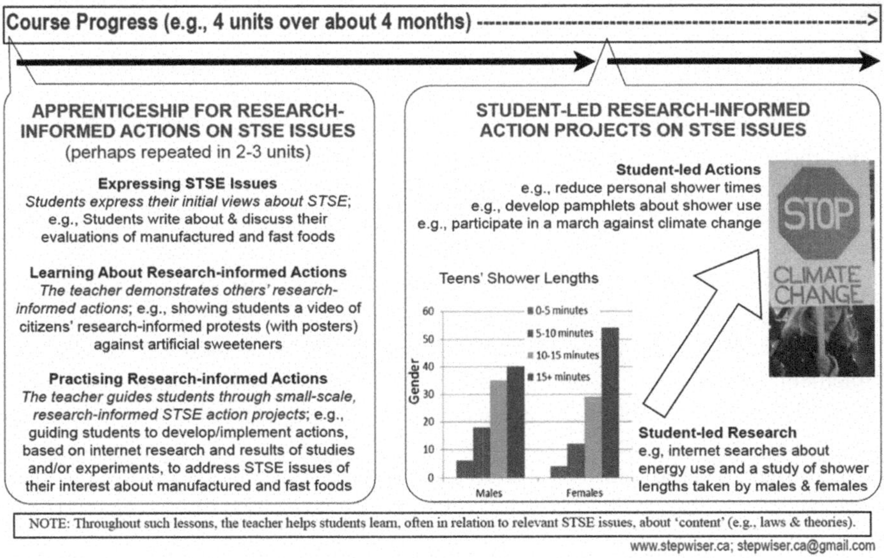

Fig. 32.1 The STEPWISE program and timeline

Methodology

In the study outlined below, we wanted to gain insight into student teachers' experiences attempting to implement STEPWISE during a required 1-month practice teaching session at local public secondary schools. Although the intention was that each teacher would implement as much of a STEPWISE research-informed action project as possible, we were aware of the potential obstacles to this and were therefore also interested in identifying these.

All of the participants in this study were enrolled in an elective course called *Science and Technology in Context*. Offered as a 36-h elective course from September to December 2010, this course provided 22 student teachers with opportunities to enhance their expertise in the two learning domains least likely to be addressed in schools; that is, skills education (e.g., experiment design) and STSE education. The other learning domain mandated by the local government (Ministry of Education [MoE] 2008), content education, is addressed in required science methods courses not taught by us. Student teachers were engaged in lectures, course activities, and homework assignments that addressed research-informed actions on STSE issues, the main pedagogical element in Fig. 32.1 (above).

Although the course has many components that can and have been studied (e.g., Bencze and Sperling 2012), our interests in this research were student teachers' experiences during their practicum. We collected data on 6 male and 1 female student teachers enrolled in the course, both before and after their practicum. All of the participants were between the ages of 25 and 40, had at least a bachelor degree in science, and had some experience teaching science prior to the teacher education program.

To obtain data on participants' orientations to science, research-informed activism and STSE, we utilized semi-structured interviews as a primary data source. Interviews were designed to collect participants' *espoused beliefs* about the relationship between science and society, school science, and activism, both broadly and in science. Beliefs are the theoretical frameworks through which one views the world and makes decisions (Kane et al. 2002). Pre-service teachers' espoused beliefs are of interest because they often do eventually translate into beliefs-in-use once attaining a full-time teaching position (Marbach-Ad and McGinness 2008).

In second and third interviews (when possible), we re-questioned participants about their general orientations to and beliefs about science and activism, as well as their experiences teaching science during their practicum, including their ability to implement components of STEPWISE, and what resisted its implementation. Table 32.1 provides a brief description of participants educational and research backgrounds, and their research participation.

In addition, several heuristic devices, as outlined below, were used to evaluate participants' orientations to SSI's and science education. These aided us in developing suitable interview questions, and in interpreting the data.

Table 32.1 Research participants

Name	Grades taught in practicum	Educational/research background	Interview and research participation
David	12 Chem.	PhD in chemistry, worked in industry	2+ P.C + STP + RepGrid + course work
Mark	9 Acad. Sci./11 Bio	Honors biology/psychology degree	3+ P.C + STP + RepGrid + course work
Stan	11/12 Chem.	Honors biochemistry degree	2+ P.C + STP + RepGrid + course work
Kirk	9 Acad. Sci./11 Bio	Bachelors degree information technology	2+ P.C + STP + RepGrid + course work
Jason	6 General science	Bachelors degree science/ environment	2+ P.C + STP + RepGrid + course work
Barb	11/12 Chem.	Honours chemistry degree	2+ course work
Larry	11/12 Biology	Honours biology degree	3+ course work

1. Repertory Grids (or RepGrids) were used to collect information about participants' preferred methods of teaching science. The RepGrid technique is a method for eliciting what people think about a given topic. Based on Personal Construct Theory (Kelly 1955), a RepGrid quantifies people's view of objects they interact with according to a collection of related similarity–difference dimensions, referred to as personal constructs. In this study, participants were asked to indicate a level of agreement (1–9), in the context of nine different learning scenarios, with the following five constructs (teaching approaches); (1) Teacher-directed/closed-ended (TD/CE) presentations; (2) Teacher-directed/closed-ended (TD/CE) practical activities; (3) Teacher-directed or student directed/open-ended (TD or SD/OE) practice inquiry and/or design activities; (4) Student-directed/open-ended (SD/OE) inquiry and/or design projects; and, (5) Student-led STSE activism projects. This quantitative data allows for numerous potential statistical treatments; however we were most interested in participants' general level of support for each of the five teaching approaches, therefore we only calculated average scores.
2. Cathleen Loving's (1991) *Scientific Theory Profile* (STP), designed for use with science teachers, was used to gain understanding of participants' epistemological and ontological orientations related to science. An individuals' STP is evaluated by their location on an x–y axis, with the x-axis representing methods for judging theories (rational vs. natural) and the y-axis representing their views on the 'truth' of scientific theories (realism) versus models of what works best (anti-realism). It is apparent that school science tends to portray professional science as a rationalist–realist endeavour – meaning that it is, for example, highly logical and successful in determining truths. A naturalist/anti-realist orientation is, on the other hand, likely more aligned with epistemology and ontology most conducive to activism in science education.
3. The *Political Compass* (www.politicalcompass.org) is a heuristic device we used to evaluate participants' general political orientation. Orientations are described on an X (right or left) and Y (authoritarian/libertarian) axis according

to participants' responses to a series of questions. Political orientations in the lower left, left/libertarian quadrant are presumably most aligned with activism.

A general inductive approach (Thomas 2006) was used to analyse the data. The analysis involved a repetitive and circular coding procedure – deriving, defining and modifying coding categories while reading, rereading and assigning responses to the categories. Codes, categories, and themes were then negotiated between us until consensus was reached (Wasser and Bresler 1996). This process was aided using NVivo (QSR International), a qualitative research software that allows the researcher to import data sources such as transcripts, and code them electronically, resulting in *nodes* (themes) composed of the various related codes. To help ensure the trustworthiness of claims, each was based on a minimum of three corroborating data sources.

Findings

Participants demonstrated considerable similarity in their beliefs and orientations to STSE-activism both before and after their practicum, and they described very similar experiences during the practicum. We use the data collected to illustrate how their beliefs and orientations changed during the course of the study, and how structures they perceived in school science networks influenced their ability to implement STEPWISE.

Pre-practicum Orientations to STSE and Activism

Using Repertory Grids, Cathleen Loving's (1991) Scientific Theory Profile and the Political Compass, each participant demonstrated pre-practicum orientations to school science that appeared to be amenable to teaching STSE-activism. In RepGrids completed before practicum, all seven participants indicated support for student-directed/open-ended inquiry and student-lead STSE-activism projects. Teacher directed/closed-ended strategies, such didactic presentations and lecture were seen as less valuable in all cases. All of the participants indicated they supported positions in the anti-realist-naturalist quadrant of the STP (except David, who was only just outside of this quadrant). Each student-teacher also was located in the left-libertarian quadrant of the political compass, indicating beliefs of economic cooperation and collectivism, and that personal freedom should be maximised. These preferences should be conducive to implementing educational experiences based on STEPWISE; their anti-realist/naturalist orientation suggests they view scientific fact and theory as sociocultural constructions rather than absolute truth and therefore value knowledge construction as opposed to knowledge consumption. A left-libertarian political position implies criticism of aggressive capitalist ideologies such as neoliberalism, and therefore may predispose teachers to engage students in action on SSIs.

Interviews conducted before their practicum focused on participants general beliefs and orientations to science, school science, STSE and activism/STEPWISE. In most cases these mirrored and were mirrored by their results in the RepGrids, STP and Political Compass. Each participant expressed during interviews personal experiences that caused disillusionment about science due to science-business partnerships. For example, David, who has a Ph.D. in organic chemistry, said he felt "like a butcher" (second interview) being forced by private pharmaceuticals firms to produce drugs that were not sufficiently tested for public consumption. Jason, working for an environmental assessment company, became skeptical of professional science once realising he was "working for companies that sort of told you what the end result will be and you are obliged to produce a report that fits that result" (second interview).

Participants also held views that were critical of many of the traditions of school science outlined elsewhere in this chapter. Socioculturally critical stances, believed to be important to activism (Watts et al. 2003), were apparent in participants' responses; for example, Kirk claimed, "I am against allowing well off people to become more well off, and your creating a beast that's only going to flow into a classed society, haves and haves not's, so that was something I struggled with quite a bit" (first interview). Participants were generally positive about teaching STSE-activism in school science, and described the benefits of this type of education; for example, Barb explained, "students are more personally involved (in STSE-activism), it's their data, rather than just going to the internet and reading out of a book, so they have ownership", and that this "knowledge will help them in their decisions and actions help them to achieve or do things" (first Interview). Endorsement of student-led STSE–activism was demonstrated through descriptions such as "it motivates students," "its engaging" and "it shows them (students) how to be active citizens, which they will need" (first Interviews).

Contrary to these liberal views, we also noted more traditional and conservative beliefs among the participants, although not as frequently. Lee summarised the conflict between an 'ideal' science education with the 'realities' of school in this statement:

> I like the idea of student directed open ended research projects too, but have to mix it with a little bit of reality, according to what universities do, with what you wish universities could be. (first interview)

Although pessimism about STSE-activism was less common than optimism, Lee's statement suggests an existing awareness of the traditional culture of school, and this reality was identified by some participants as problematic to conducting STSE-activism oriented activities.

Post-practicum Orientations to STSE and Activism

We administered pre and post-practicum RepGrids to three student teachers, and these demonstrate some changes in their teaching preferences (See Table 32.2). All three participants indicated they supported teacher-directed/closed-ended presentation more after the practicum than before. Support for STSE-activism generally decreased after the practicum. David had approximately equal support for activism both before and after the practicum, however, all of David's scores increased post-practicum except that for STSE-activism.

Although we do not claim statistical validity for the RepGrid data, and we cannot account for any potential *Hawthorne* effects (telling us what they think we want to hear) (Gall et al. 2007), these results suggest participants' attempts to implement STSE-activism were discouraging, and positive reinforcement was obtained through more traditional teaching practices. Although some participants still expressed interest in implementing STSE-activism after the practicum, unfortunately, more frequently, participants were skeptical about activism and doubted their ability to implement it in the future.

Table 32.2 Participant repertory grid scores before and after practicum

Participant	Construct	Before practicum	After practicum	Difference
David	TD/CE presentations	4.5	5	+0.5
	TD/CE practical activities	3.75	5.65	+1.9
	TD, SD/OE practical inquiry and/or design activities	5.3	6.3	+1.0
	SD/OE inquiry and/or design projects	6.0	6.25	+0.25
	Student-led STSE activism projects	6.0	6.05	+0.05
Mark	TD/CE presentations	4.15	4.7	+0.55
	TD/CE practical activities	7.55	5.65	−1.9
	TD, SD/OE practical inquiry and/or design activities	8.45	7.25	−1.2
	SD/OE inquiry and/or design projects	7.9	7.75	−0.15
	Student-led STSE activism projects	8.25	7.75	−0.5
Stan	TD/CE presentations	4.5	5.2	+0.7
	TD/CE practical activities	6.65	6.45	−0.2
	TD, SD/OE practical inquiry and/or design activities	6.85	5.85	−1.0
	SD/OE inquiry and/or design projects	6.45	6.45	0.0
	Student-led STSE activism projects	6.0	5.80	−0.2

Post-practicum Skepticism of Activism

Student teachers' expressed considerable skepticism about teaching STSE-activism after their practicum. For example, Stan commented that "after going through practicum, I don't know how practical STEPWISE would be in a normal school environment," (second interview) and Jason added that "there seemed to be a belief that, or value for, getting the information, knowledge students needed, and the whole school was geared for that, and there was no place for activism." (second interview). These experiences caused many participants concern that their inability to develop activist-oriented education might result in future ambivalence, as expressed by David, who said "my fear is that this type of thing will disappear from my conscience...I don't have a lot of passion for it right now, so I kind of wonder if this sort of thing might just disappear from my memory" (second interview).

Skepticism had progressed to near rejection for several participants; Stan was most outspoken about this, stating, "Asking the kids to be activists – like, in STEPWISE, asking the kids to be activists seems to be what ties the entire thing together, and I think that might be too much" (second interview). Many participants felt that activism is a personal decision that may not need to be part of school; for example, David claimed: "It [activism] doesn't have to end there, but I think it can and maybe should end there." (third interview). When asked how he felt about activism after practicum, Stan, who was very supportive of activism before the practicum, said:

> I don't like it. I mean people shouldn't....if people have a passion for something, I mean, sure, go march, or whatever...but if you don't have a passion, to require someone to do it? I think that might have the opposite effect. (second interview)

Post-practicum Prioritisation of Content Teaching

Another noteworthy change in participants' post-practicum beliefs was a growing prioritisation for content-teaching. For example, when Jason was asked about his AT's frequent use of teacher-directed lecture to teach content, he answered, "I sort of could understand where she came from, because most of them (students) would be going to university, and we are obligated to give them the knowledge they need to do well in university the first year" (second interview). Mark said that he felt he "couldn't allow them to completely skip over the many terms and processes and function in the textbook that you have to know if you are learning homeostasis." (second interview). Kirk, expressing a high priority for content similar to other participants, stated "But practically, the science curriculum has so much stuff to get through, if I spend time on other than just the content, I'm not going to be able to get through the content" (second interview).

These apparent changes in beliefs and priorities after the practicum are concerning to us. Regardless of whether students initial enthusiasm and commitment to STSE-activism was genuine or not, it was apparent that powerful structural pressures they encountered in schools oriented or re-oriented them towards the familiar hegemonic culture that is resistant to STSE-activism. With this in mind, the question then becomes: What are the sources of resistance to STSE-activism in schools?

Aspects of School Science Culture Resistant to STSE-Activism

Each participant planned on attempting a STEPWISE project (or components of one) during their practicum. However, they were all constrained by various resistances, or structures, they described during interviews. As a result, no participant was able to initiate a full project, and most were unable to teach even smaller components of STEPWISE. Although the reasons for this were varied and complex, Mark summarised the group well in saying "I wanted to do all of this, but I ended up just needing to get my feet wet, and get along with my AT" (second interview). Stan added to the growing consensus, explaining "I was just too busy getting down what I really needed to get down" (second interview); this suggests Stan, after experiencing the culture of school, realised STEPWISE and STSE-activism was not something that was really valued or wanted. Only three participants claimed doing any component of STEPWISE; Kirk did "a little bit of secondary research because I got them to look up a little bit about the Walkerton disasters[1] and stuff like that" (second interview), Mark "touched on aspects of these (STSE-actions) as far as how to elaborate on them in the context of certain topics" (second interview) and Barb did a project which was STEPWISE oriented, but found it too complicated to implement (third interview).

The AT as a Representative of School Culture

The associate teacher (AT) appeared to be a direct-line to, and act as a representative of, the wider school culture and was seen by each participant as an obstacle to their implementation of STEPWISE. Dominant cultural characteristics communicated through the student-teachers' interaction with the AT are presented below.

[1] In the small town of Walkerton, ON, seven people died as a result of drinking water that had been contaminated with high levels of E. coli bacteria – an apparent result of reductions in inspections implemented by the conservative government of the period.

Prioritising Content

Most of what Stan described earlier as what he "needed to get down" (second interview) was learning how to teach content, usually at the expense of student led activities that might satisfy the STSE component of the curriculum. Each participant identified teacher-directed content teaching as the top priority during their practicum. For example, David claimed "there is still so much of a focus on content that they still teach according to content and just throw in the STSE as an example of that content" (second interview). Stan explained: "In my school, teachers felt the science curriculum has so much stuff to get through, they had to spend most of their time on the content' (second interview). Each participant thought the AT inhibited them from implementing STSE-activism. For example, Kirk said "with the AT I had it (STSE-activism) wouldn't have gone down well" (second interview) and that "I pitched it to the AT, and the response I got was "well, they already covered that, and they can join the environmental club if they want to do that" (second interview). As a result, participants said that they felt they needed to cover the content that was identified by their AT.

AT Used Deficit-Type Teaching Approaches

In discussing the AT's teaching styles, participants frequently described very teacher-directed approaches; Mark, for example, said "He (the AT) was more about lecturing, and showing overheads and diagrams, he didn't have any student directed [activities] at all." (third interview). Kirk described his AT as a "a little bit of a 'chalk-and-talk merchant'" (second interview), indicating didactic instruction. When asked about his AT's teaching style, Mark said: "His attitude was, here is what they need to know, I will teach it, it is their responsibility to learn it" (second interview). Stan described his AT as "very Socratic, I guess you would call it, a lot of work sheets, a lot of him teaching, talking" (second interview). Jason explained his AT, "lets them sort of do group projects and stuff, but on a day to day basis, he puts stuff up on the board and they write stuff down" (second interview).

AT Prioritised University Preparation

Participants generally agreed the prioritisation of content coverage was due to the AT's perception that students need this knowledge and the memorisation skills for university. Stan, for example, stated "His main approach was content teaching and work habits, not research-based activities; more like developing the skills they need to get into university." (seconded interview). Kirk observed his AT's reasons for covering certain content was even more specific: "She really wanted to get to that [pertinent content] in because in her view, she felt some of them would be going on as nurses" (second interview). Mark also observed that his AT had very specific post-secondary targets for his students; "It seemed, it was grade 11 and 12 university academic courses, and if I was to sum it up, he was preparing them for 1st year Bio" (third interview). When asked about his AT's priorities, Stan said "teaching the skills and knowledge to the kids so they could find a job or go to university" (second interview). The university appears to be a powerful actant in the school science network, communicating expectations that appear to elicit the conservative priorities of the AT's resulting in the reproduction of knowledge consumers. University preparation, then, appears to be another school structure resistant to the implementation of STSE-activism in school science.

Lack of Community Support for Activism

Participants also felt there was a clear lack of support for student-led STSE-activism oriented education among science departments, administration, and parents. For example, Kirk had this observation about community expectations:

> In practice, I guess the question is how much can you cover in any given semester, and how do you weigh that with what the administration wants to see, and how do you reconcile that with what other science teachers want to see and what they do, the 3-4 other teachers who are teaching the same curriculum, who all want to be on the same level, have tests the same day, etc. So, there are restraints..... (second interview)

Many participants described potential resistance from the wider community; for example, David claimed "you run a danger of parents or administrators asking 'exactly what is this teacher trying to do here?' I'm not sure how involved parents are, but I'd worry that parents might be critical of these things" (second interview).

Students' Resistance to Activism

Students' responses to STSE-activism are perhaps of primary interest, since they are the target of this education. However, participants reported students were largely resistant to the open-ended and action oriented educational experiences in ideal forms of STSE-activism. In Marks's practicum, he observed that "students were not that engaged, they just didn't care, unless they were being marked or tested, they didn't care that much" (third interview). Kirk expressed similar concerns, describing a lesson in which he "asked students to bring in water from their area, and it was like asking for blood from a stone, not one of them wanted to bring in any water" (second interview). Jason commented that students were not interested in learning anything outside of the textbook content: "If they had more interest outside of the textbook, wonder about the environment and the world, like 'What's going on with this or what is the explanation for that?' he (the AT) would have been great to carry it on, but they didn't seem to" (second interview). These observations support the claims made by, for example, Wood (1998) that students become disinterested in developing their own knowledge when bombarded by a steady stream of facts and conclusions in didactic lectures, and that they may even loose the skills needed for activism (Stetsenko 2012). They also support claims of a neoliberal cultural hegemony.

Not Understanding How to Teach STSE-Activism

Participant's skepticism also appeared to stem from their lack of knowledge in developing an STSE-activism project. This was not apparent in students work (final project) in *Science and Technology in Context*, or during initial interviews. For example, when asked whether they would attempt to implement STSE-activism in the future, Stan commented: "I don't know. . ..I'm skeptical. . .unclear about how I would do it" (second interview) and Jason stated "I don't feel I have the confidence to do something like that." (second interview). Participants also appeared to have poor understanding of the scope of actions students might take, causing them to have conflicting feelings about asking students to be activists. For example, David stated: "I think I question if the student doesn't feel like acting on anything, is it our place to force them to act on anything?" (third interview); Jason expressed similar uncertainty about developing student activism: "I'm afraid that if you don't allow the students to pick a topic of interest, they might be disengaged in further activism" (second interview). Uncertainty about how to overcome students' apathy towards activism was also a common theme among participants. Mark, for example, "didn't see them as engaged with the issue itself, and I didn't know how to make them interested" (third interview). We speculate apparent lack of understanding and confidence about developing STSE-activism derives from their inability to surmount the barriers to activism they observed in schools.

Discussion

It is perhaps not surprising to anyone who has taught student teachers that school structures greatly influence, and often restrict the experiences available to them during their practicum and in their first years of teaching (Hodson 2009). These structures likely constitute a reality that must be identified, understood and changed if progressive approaches to science education are able to gain presence and momentum in school science. We take the view that the research presented here, although appearing to demonstrate somewhat pessimistic, if not surprising, findings, provides opportunity to better understand the structures among networks that new teachers entering school science classrooms must navigate so that teacher preparation might better prepare them to teach STSE-activism in school science.

Neoliberal Structures

Many of the structures apparent in the networks described by student teachers might be identified as originating from aggressive capitalist-based discourses that have been associated with science and school science, such as neoliberalism (Bencze 2010). The high prioritisation of content teaching and assessment of content learning is typically neoliberal in that it positions science knowledge as a commodity that is to be consumed by students (Bencze and Carter 2011). Observations made by participants identified this as the top priority of their AT, and the school in general, and evidence of a cultural zeitgeist based on neoliberalism that exists in their host schools. These are the hegemonic discourses and ideals that science education such STEPWISE attempts to challenge, yet these are deeply enmeshed in stable networks, resisting actants such as student teachers (or STEPWISE) that attempt to change the network. The student teachers did not have the power or knowledge to be counter-cultural actants in the schools science networks they entered and this appeared to affect their motivation and commitment to STSE-activism.

Science teacher educators may have minimal effect on changing ideology and structures, such as neoliberalism, at the level of society. Instead, we attempt to exert influence from the direction of school, changing school science networks by preparing teachers to develop educational experiences that challenge the hegemonic and socially unjust traditions of the discipline, which then can potentially inculcate socially critical and activist orientations in students (Watts et al. 2003). Yet, science teacher education does not appear to prepare teachers for this and it is unclear what constitutes adequate preparation. What follows are suggestions based on what we have learned in this research about preparing student teachers to teach STSE-activism.

The Need for Student Teachers with Pro-activist Orientations

Through our work with student teachers in the course *Science and Technology in Context*, we have observed that student teacher commitment to activism is likely a requirement for successful implementation. We found, for example, student teachers motivation to teach STSE-activism was positively influenced by opportunities to engage in research informed activism (Bencze and Sperling 2012). However, commitment to activism appears to be malleable and tenuous and many variables appear to be involved. It therefore seems critical that teachers with existing political, epistemological and ontological orientations that are beneficial to STSE-activism are involved in its implementation, and that science teacher preparation aims to develop these in student teachers. Recognising that new teachers are still developing philosophies and teaching approaches, we suggest that more concerted and explicit effort from all teacher education courses would be beneficial to this aim. These skills are needed to develop activist-oriented education not only in science but in other school subjects as well. *Science and Technology in Context* was the only experience participants of this research had in their program that explicitly developed activist orientations and knowledge needed to implement activist education in school science; the required science methods course prioritises teaching content. Thus, the sum of their teacher preparation may produce actants that reinforce and are reinforced by the conservative network of school science that prioritises knowledge consumption. In order to disrupt this hegemonic culture, orientations beneficial to STSE-activism need to be inculcated in new teachers preferably throughout teacher education programs.

Placing Student Teachers in Supportive Communities

The experiences by student teachers in our study was moderated to a great degree by their AT. Had the AT been more amenable to student-directed science approaches, participants would more likely have been able to develop and implement lessons based on STEPWISE. It, therefore, seems necessary that student teachers are matched to ATs that regularly employ student-directed science lessons, such as STSE-activism, or at least value these approaches enough to allow new teachers to attempt such activities. Matching student teachers with schools that are likely to support the implementation of STSE-activism may also be an important consideration. In recent years, we have partnered with science teachers in private schools, resulting in some highly successful STEPWISE projects. For example, student-directed science projects in a prominent private school in Montreal (Hoeg, Lemelin and Bencze, In Press), resulted in student activism that challenged international corporations to be more accountable for their practices. There are several structural factors that apparently advantage private schools in providing student-led activist education; for example, they appear to have a high degree of content and pedagogical freedom, uncommon in public schools (Moore 2007; Schulteis 2010). Moreover, because elite private schools, by their very nature, can establish and maintain particular collective norms, demonstrated by relatively uniform mission statements, common purposes can be effectively implemented amongst staff members. These norms frequently contain goals outlining requirements for community service, social justice, and altruism (Boerema 2006, 2009; Dill 2009), which justifies and supports inclusion of lessons and projects aligned with these goals. Some participants noted the potential of private schools in offering STSE-activist education during interviews; for example, Mark commented he felt private schools or public schools in wealthier neighborhoods might be more suitable to develop STSE-activism education than the urban school in which his practicum occurred. However, matching good candidates for teaching STSE-activism with supportive ATs and schools likely requires efficient integration and communication within the various components of teacher preparation programs, such as between course directors and administrators who organise the teaching practicum.

Initial experiences for new teachers in schools can have an enormous influence in orienting them towards a certain culture and in various networks. If we are to enable school science education that strives to affect sociocultural criticism and change, it is imperative that student teachers' experiences are supportive of their attempts to implement STSE-activism. The research we examined here points out the concerning effects of practicum experiences that restrict the new teachers' agency, and align their practice towards conservative, traditional, neoliberal structures and culture. Although this is not an absolute predictor of how teachers will enact science lessons in the future, it certainly is less desirable than having experiences causing new science teachers to believe implementing education such as STSE-activism is both possible and valued. Science teacher preparation must be better coordinated and aligned toward more common activist oriented goals and outcomes in order to provide for new teachers these experiences.

References

Barthes, R. (1972). *Mythologies*. New York: Hill and Wang.

Bencze, J. L. (2008). Private profit, science and science education: Critical problems and possibilities for action. *Canadian Journal of Science, Mathematics & Technology Education, 8*(4), 297–312.

Bencze, J. L. (2010). Exposing and deposing hyper-economized school science. *Cultural Studies of Science Education, 5*(2), 293–303.

Bencze, L., & Carter, L. (2011). Globalizing students acting for the common good. *Journal of Research in Science Teaching, 48*(6), 648–669.

Bencze, L., & Sperling, E. (2012). Student teachers as advocates for student-led research-informed socioscientific activism. *Canadian Journal of Science, Mathematics, and Technology Education, 12*(1), 62–85.

Bencze, L., Sperling, E., & Carter, L. (2012). Students' research-informed socioscientific activism: Re/visions for a sustainable future. *Research in Science Education, 42*(1), 129–148.

Beyer, L. E. (1998). Schooling for democracy: What kind? In L. E. Beyer & M. W. Apple (Eds.), *The curriculum: Problems, politics, and possibilities* (pp. 245–263). Albany: SUNY Press.

Boas, T. C., & Gans-Morse, J. (2009). Neoliberalism: From new liberal philosophy to anti-liberal slogan. *Studies in Comparative International Development, 44*(2), 137–161.

Boerema, A. J. (2006). An analysis of private school mission statements. *Peabody Journal of Education, 81*(1), 180–202.

Boerema, A. J. (2009). Does mission matter? An analysis of private school achievement differences. *Journal of School Choice, 3*(2), 112–137.

Bourdieu, P. (1990). *Reproduction in education, society and culture* (2nd ed.). London: Sage.

Dill, J. S. (2009). Preparing for public life: School sector and the educational context of lasting citizen formation. *Social Forces, 87*(3), 1265–1290.

dos Santos, W. L. P. (2009). Scientific literacy: A Freirean perspective as a radical view of humanistic science education. *Science Education, 93*(2), 361–382.

Ehrlich, P. R., & Ehrlich, A. H. (1972). *Population, resources, environment: Issues in human ecology*. San Francisco: Freeman.

Eisenhart, M., Finkel, E., & Marion, S. F. (1996). Creating the conditions for scientific literacy: A re-examination. *American Educational Research Journal, 33*(2), 261–295.

Engestrom, Y. (1999). Activity theory and individual and social transformation. In Y. Engestrom, R. Miettinen, & R.-L. Punamaki (Eds.), *Perspectives on activity theory* (pp. 19–38). New York: Cambridge University Press.

Foucault, M. (1980). Truth and power. In C. Gordon (Ed.), *Power/knowledge: Selected interviews and other writings, 1972–1977* (pp. 48–79). New York: Pantheon Books.

Freire, P. (1997). *Pedagogy of the oppressed* (New Rev. 20th-Anniversary ed.). New York: Continuum.

Gabbard, D. A. (Ed.). (2000). *Knowledge and power in the global economy: Politics and the rhetoric of school reform*. Mahwah: Lawrence Erlbaum.

Gall, M., Gall, J., & Borg, W. (Eds.). (2007). *Educational research: An introduction*. Boston: Pearson Education.

Gramsci, A. (1982). *Selections from the prison books*. New York: Lawrence and Wishart.

Hegel, G. W. F. (1975). *Lectures on the philosophy of world history: Introduction, reason in history* (Trans. by H. B. Nisbett, from the German edition of Hegel papers assembled by Eduard Gans, 1837). New York: Cambridge University Press.

Hodson, D. (2003). Time for action: Science education for an alternative future. *International Journal of Science Education, 25*(6), 645–670.

Hodson, D. (2009). *Teaching and learning about science: Language, theories, methods, history, traditions and values*. Rotterdam: Sense Publishers.

Hoeg, D., Lemelin, N., & Bencze, L. (in press). Socio-political development of private school children mobilising for disadvantaged others. *Cultural Studies of Science Education*.

Kane, R., Sandretto, S., & Heath, C. (2002). Telling half the story: A critical review of research on the teaching beliefs and practices of university academics. *Review of Educational Research, 72*(2), 177–228.

Kelly, G. (1955). *The psychology of personal constructs*. New York: W.W. Norton.

Kleinman, D. L. (2003). *Impure cultures: University biology and the world of commerce*. Madison: University of Wisconsin Press.

Krimsky, S. (2003). *Science in the private interest: Has the lure of profits corrupted biomedical research?* Lanham: Rowman & Littlefield.

Latour, B. (2004). *Politics of nature: How to bring the sciences into democracy*. Cambridge, MA: Harvard University Press.

Lester, B. T., Ma, L., Lee, O., & Lambert, J. (2006). Social activism in elementary science education: A science, technology, and society approach to teach global warming. *International Journal of Science Education, 28*(4), 315–339.

Likens, G. E., Wright, R. F., Galloway, J. N., & Butler, T. J. (1979). Acid rain. *Scientific American, 241*(11), 43–51.

Loving, C. C. (1991). The scientific theory profile: A philosophy of science model for science teachers. *Journal of Research in Science Teaching, 28*(9), 823–838.

Marbach-Ad, G., & McGinnis, J. R. (2008). To what extent do reform-prepared upper elementary and middle school science teachers maintain their beliefs and intended instructional actions as they are inducted into schools? *Journal of Science Teacher Education, 19*(2), 157–182.

McMurtry, J. (1999). *The cancer stage of capitalism*. London, England: Pluto.

Ministry of Education. (2008). *The Ontario curriculum, Grades 9 and 10: Science*. Toronto: Queen's Printer for Ontario.

Moore, R. (2007). What are students taught about evolution? *McGill Journal of Education, 42*(2), 177–187.

Nadeau, R., & Desautels, J. (1984). *Epistemology and the teaching of science*. Ottawa: Science Council of Canada.

NVivo (Version 10) [Computer software]. Burlington: QSR International.

Schulteis, M. W. (2010). Education's missing link: How private school teachers approach evolution. *American Biology Teacher, 72*(2), 91–94.

Sewell, W. H. (1999). The concept(s) of culture. In V. E. Bonnell, L. A. Hunt, & R. Biernacki (Eds.), *Beyond the cultural turn: New directions in the study of society and culture*. Los Angeles: University of California Press.

Stetsenko, A. (2012). Personhood: An activist project of historical becoming through collaborative pursuits of social transformation. *New Ideas in Psychology, 30*(1), 144–172.
Stobaugh, R., & Yergin, D. (Eds.). (1979). *Energy future: Report of the energy project at the Harvard Business School*. New York: Random House.
Thomas, D. R. (2006). A general inductive approach for analyzing qualitative evaluation data. *American Journal of Evaluation, 27*, 237–247.
Wasser, J. D., & Bresler, L. (1996). Working in the interpretive zone: Conceptualizing collaboration in qualitative research teams. *Educational Researcher, 25*(5), 5–15.
Watts, N., Williams, C., & Jagers, R. (2003). Sociopolitical development. *American Journal of Community Psychology, 31*(1), 185–194.
Wax, M. (1993). How culture misdirects multiculturalism. *Anthropology and Education Quarterly, 24*, 99–115.
Wood, G. H. (1998). Democracy and the curriculum. In L. E. Beyer & M. W. Apple (Eds.), *The curriculum: Problems, politics and possibilities* (pp. 177–198). Albany: SUNY Press.
Zeidler, D. L., Sadler, T. D., Simmons, M., & Howe, E. (2005). Beyond STS: A research-based framework for socioscientific issues education. *Science Education, 89*(3), 357–377.
Ziman, J. (2000). *Real science: What it is, and what it means*. Cambridge, UK: Cambridge University Press.

Chapter 33
Implementing Practical Pedagogical Strategies for the Widespread Adoption of Renewable Energy

Jose Etcheverry

Abstract The Sustainable Energy Initiative (SEI) of the Faculty of Environmental Studies of York University was established to develop practical solutions to the numerous social and environmental problems created by society's reliance on nuclear power and fossil fuels. The SEI's research and pedagogical approaches are based on knowledge mobilization, experiential learning, peer mentoring and internationalization strategies. This chapter provides an overview of those strategies and also an applied analysis of the pragmatic roles that learning institutions and educators can undertake to address social and environmental challenges.

Keywords Renewable energy • Conservation • Experiential education • Knowledge mobilization

> The essential content in any educational program – whether on syntax, biology, physics, mathematics, or the social sciences – is that which makes possible discussions of the mutable nature of natural reality, as well as of history, and which sees men and women as beings capable not only of adapting to the world but above all of changing it. (Paulo Freire 2004, p. x)

Introduction

The most vexing environmental challenges currently faced by society share a common denominator: local, regional and international overdependence on highly polluting forms of energy generation. Two of the clearest illustrations of that fact are climate change and radioactive waste. The bulk of the greenhouse gas emissions, and short-lived climate pollutants, which are now widely understood as the

J. Etcheverry (✉)
Faculty of Environmental Studies, York University, Toronto, ON, Canada
e-mail: rejose@yorku.ca

L. Bencze and S. Alsop (eds.), *Activist Science and Technology Education*,
Cultural Studies of Science Education 9, DOI 10.1007/978-94-007-4360-1_33,

main contributors to human-induced climate change, originate from the burning of fossil fuels which are used mainly to satisfy energy needs. Radioactive waste, a disturbing, yet compared to climate change, vastly ignored environmental problem is directly related to the use of nuclear technology. Although less omnipresent than fossil fuel use, nuclear technology is utilized by about 56 countries for electricity generation and also to power ships, submarines and research reactors.[1] In addition, many of those 56 nuclear nations already use, or covet, the technology for the production of nuclear weapons. All nuclear applications create an assorted variety of radioactive wastes that for the past 70 years have been stored in "temporary" facilities. That dangerous and abundant toxic legacy will remain in limbo until scientists, financiers, and policy-makers can find and secure safe permanent locations. The common English expression of "out of sight, out of mind" and the many decades that have passed since the introduction of the first commercial nuclear power stations (which started operation in the 1950s), have kept the challenge of radioactive waste away from the public eye. That situation is particularly prevalent in the few OECD countries that still enthusiastically support the use of the technology (i.e. Canada, England, and France). Recently, the Fukushima nuclear disaster is showing again to the world, just as Chernobyl once did, the massive economic costs and socio-environmental risks posed by the technology. Fukushima is also helping to re-ignite nuclear phase-out strategies in many advanced industrialized nuclear countries (e.g. Belgium, Germany, Italy, Sweden, and Switzerland).

Recognizing the multiple social, environmental, and security problems created by the use of fossil fuels and nuclear technology, researchers at York University's Faculty of Environmental Studies decided to start in 2007 the Sustainable Energy Initiative (SEI).[2] The SEI was formed by a group of tenured FES professors and was expanded to include leading energy experts from a variety of interdisciplinary disciplines that have become adjunct professors and SEI advisors. From its birth, the SEI was conceived as a collaborative effort aimed at moving beyond conventional academic approaches to concentrate not just on conducting research and teaching. In addition to those activities, a key SEI goal is helping the university and local communities in the adoption of a new energy system based on low-impact renewable energy options, conservation and efficiency.

This chapter describes some of the pedagogical strategies employed by the SEI to help advance the adoption of renewable energy and conservation solutions. The most innovative of these pedagogical strategies can be summarized under three inter-linked categories, which form the main subsections of this chapter, namely: Knowledge Mobilization, Peer Mentoring and Internationalization. Before discussing these three key SEI pedagogical strategies it is imperative to at least briefly describe the sustainability energy system that is becoming prevalent in a number of leading jurisdictions around the world. A system that many of us hope will soon become widespread around the world.

---

[1] See article titled "Nuclear Power in the World Today" (adapted April 2012) at the World Nuclear Association site at http://www.world-nuclear.org

[2] For details about the SEI see www.yorku.ca/sei

Rise of the Sustainable Energy Paradigm

As of 2011, renewable energy sources were supplying about 19 % of the global primary energy use and during 2012 modern renewable energy sources grew strongly in all energy end-use sectors i.e. electricity generation, heating and cooling, and transport.[3] In the electricity sector, almost half of thc total 280 gigawatts (GW) added globally during 2012 were renewable energy. Furthermore, wind and solar photovoltaics (PV) accounted for 39 and 26 % of all the new renewable capacity installed in 2012 for electricity generation. In terms of annual investment, renewable energy additions represented in 2012 $244 billion dollars, or an exponential increase of about six times the comparable investment figure for 2004. Although such figures provide a clear indication of a sector that is growing rapidly, a description of the global situation of renewable energy development requires emphasis on the most remarkable leading jurisdictions. Germany and Denmark are two nations that provide inspirational and practical examples of what can be achieved when policy is harnessed to achieve sustained change in the energy sector. Germany is a country that is internalizing the price of carbon through domestic carbon taxation and participation in the EU emissions trading system (the world's largest cap and trade system). The official targets of the German government are to reduce greenhouse gas emissions by 40 % by 2020 (compared with 1990 levels) and by 80 % by 2050. Germany also implemented in 2000 a comprehensive renewable energy law (known as EEG from its acronym in German), which prioritized the use of a system of feed-in tariffs, a policy mechanism that provides long-term premiums (20 years) for all the electricity generated by renewable energy projects. In addition, Germany has adopted a nuclear phase-out, for all its atomic generation facilities, which has to be completed by 2022. Germany has also implemented one of the world's most comprehensive R&D and training strategies focused on developing its renewable energy and conservation capacities. As a result of these policies Germany now has the largest installed capacity of renewable energy in Europe (with 33 GW of wind power and 36 GW of solar photovoltaics) and has also achieved a workforce of over 380,000 people in its renewable energy sector.[4] Denmark, although a much smaller nation, has also implemented many of the German policies described above but has also innovated with policies of its own. For example, Denmark avoided the implementation of nuclear energy and instead has focused for decades on developing a comprehensive district energy and combined heat and power (CHP) network that provides all the country with a reliable back-up system for its wind

---

[3] All the figures of this section are sourced from REN21 (2013a). Readers interested in learning more about the status of renewable energy and the future of the sector are encouraged to visit the REN 21 site and analyze the aforementioned report plus their forward looking report REN21 (2013b).

[4] For official information about Germany's renewable energy initiatives see http://www.erneuerbare-energien.de/en/ and for reliable statistics see http://www.ise.fraunhofer.de/en/renewable-energy-data To learn more about the German energy transition see http://energytransition.de

power sector. CHP enables the production of two products at the same time (electricity and heat) and for the same price than conventional thermal generators get only one (i.e. electricity). Instead of venting the heat to warm the atmosphere, as conventional thermal generators routinely do, Danish CHP plants store that heat as hot water, which is then sold at the cost of production to warm up homes and all kinds of buildings, which are easily connected to district energy networks (Danish policy prescribes that heating cannot be sold at a profit). The model is elegant and simple, when wind power is not available, CHP plants are quickly turned-on to provide electricity to Denmark and the heat is simultaneously stored to provide affordable heating when needed. District energy networks also provide a versatile option to use wind power in the many windy days with low electricity demand, which are increasingly common in energy efficient Denmark. By installing electric boilers the operators of the Danish district energy networks can heat water whenever wind energy is plentiful which is then stored to be used when needed. The Danish advances in district energy and CHP enable the country to have one of the world's highest levels of wind power in the grid (currently close to 30 %) and also has permitted the implementation of a government target of supplying 50 % of the total Danish electricity demand with wind power by the year 2020. Denmark today has developed unparalleled expertise on the highly complimentary wind power, CHP, and district energy sectors and is successfully exporting its products and know-how worldwide.[5]

These brief overviews, on the sustainable energy advances achieved by Germany and Denmark, are not comprehensive or detailed enough, but are provided here to show what can be achieved today when political commitment is matched with synergistic approaches and smart policies. These examples also illustrate the type of advances that the SEI is trying to help achieve in Ontario, Canada which is a jurisdiction that has implemented some of the policies that have propelled Denmark and Germany to their current leadership positions (e.g. Ontario has a new renewable energy law and feed-in tariffs based on the German EEG) but that is still highly dependent on fossil fuels and nuclear power. The pedagogical strategies summarized below aim at trying to materialize the quote by Paulo Freire, which opens this chapter, as we also see our educational activities as an essential strategy that can enable leaders capable not only of adapting to the world but above all of making it a better place for everyone.

Knowledge Mobilization Strategies

York University is located in Toronto and is amongst Canada's largest public universities. The university has developed and organized an approach to promote active knowledge mobilization (KM) and the Sustainable Energy Initiative (SEI)

[5] For official information about Denmark's renewable energy and district energy efforts see http://www.energinet.dk

Table 33.1 Sample of Knowledge Mobilization (KM) events organized by York University's Sustainable Energy Initiative (SEI)

2011/2012
Community Power Approaches
Combined Heat and Power and District Heating Strategies
Smart Grid Solutions for Sustainable Buildings: Advances in renewable energy and efficiency within the context of Buildings
Advances in Conservation, Efficiency and Smart Grids: Danish and Ontario Perspectives
New Research in Sustainable Energy
Ontario's Feed-in Tariff Program in Transition
Ontario and Baden-Württemberg: Leading Partners for a Sustainable Energy Economy
2012/2013
Storage Options for Renewable Energy: Developing RE to Commercialization
The State of Ontario's Green Energy Strategy
Green Energy Finance
Community Energy Planning I: Current state of provincial policy and municipal initiatives in Ontario
Community Energy Planning II: Operational context of community energy planning (CEP) in Ontario
From Northern Gateway to Line 9: The New Law and Politics of Energy Pipelines in Canada
California Experience: Energy Efficiency
Energy Technology Forum: International Overview of Energy Storage
History and Rise of Wind Power
2013 Energy Expert Update from Fukushima, Japan

uses that approach regularly to bring together, at monthly events, the best Canadian and international experts on conservation, efficiency and renewable energy together with key stakeholders of the energy sector (from academia, NGOs, the private and governmental sectors).[6] These monthly events, see Table 33.1, are used to share expert knowledge to help develop networks and nurture the development of a local expert community. The KM events are also used to systematically accumulate know-how that is translated into a variety of communication tools such as media interventions, technology road maps and policy briefs. SEI researchers carefully design the monthly SEI KM events, to help fill knowledge gaps that prevent the adoption of renewable energy and conservation in Ontario. The SEI KM events, summarized in Table 33.1, are also used as a pedagogical tool so undergraduate and graduate students from York University (and from other learning institutions) can access networks of local and international experts and to help them contextualize and integrate information discussed in the classroom. In addition, the topics analyzed in the SEI events are used to inform the ongoing development of the Faculty of Environmental Studies energy-related curriculum, which currently consists of several courses offered under a Sustainable Energy Certificate (offered at the

[6] For detailed information about York University's KM efforts see http://www.yorku.ca/research/innovation/knowledgemobilization/

undergraduate level) and to inform the development of a new graduate level Sustainable Energy Diploma.

In addition to its KM events the SEI developed, in 2011, a pedagogical partnership with Natural Resources Canada to host the RETScreen International Training Institute at the Faculty of Environmental Studies (FES) of York University. RETScreen is a world leading software tool for energy decision-making that is offered free of charge in 35 languages (which collectively cover about two thirds of the world's population).[7] As of May of 2013 RETScreen is used by over 386,000 people living in 222 countries.[8] The RETScreen Training Institute brings together international and Canadian students that come regularly to FES to learn directly from the software creators. The Institute also allows SEI to provide scholarships to develop the knowledge base of graduate and undergraduate students that can thereafter become qualified to teach others how to use RETScreen. This initiative is in essence a 'train the trainers' approach that represents another key pedagogical strategy used by the SEI, which is described below in more detail.

Peer Mentoring Strategies

By training trainers the SEI aims at empowering students to become leaders in their community and amongst their peers. In addition to the RETScreen Institute, which like most classroom experiences, provides a very abstract learning environment SEI students are part of experiential learning opportunities that aim at providing practical hands-on learning experiences to advance the use of renewable energy. A key SEI experiential peer mentoring initiative involves teaching graduate students how to use a versatile portable electronic device that permits to accurately estimate the solar potential of any building, parking lot, or parcel of land. This tool contains an advanced GPS system, a fish eye camera and specialized software that allows users to quickly determine the yearly solar insolation and shading of any area. The data generated can be then used to develop a very detailed report that enables people to easily understand how much solar energy can be harnessed to produce electricity (or heat), how much money can be made through electricity sales and how much carbon can be reduced through a solar project. Acquiring this sophisticated tool was a first step on this peer mentoring initiative and enabled the SEI to attract the help of a solar industry professional, well versed in the technology and in the development of solar assessment reports, to train SEI graduate students. After the graduate students became experts on how to use the device they were empowered to both conduct their own solar site assessments and producing accurate solar resource reports. After achieving fluency in the use of the technology the graduate students were actively encouraged to train undergraduate students, which following a

[7] For details about RETScreen see www.retscreen.net

[8] Op.cit. note 7.

similar learning curve were subsequently able to train high school students. This peer mentoring strategy was sponsored by the City of Markham and the York Region District School Board (the third largest public school board in Ontario) and has already enabled the mapping by students of valuable solar resources in several publicly-owned buildings in Markham and also the entire York University Keele campus. This solar mapping project is part of a broader initiative aimed at converting the York University Keele Campus into a smart mini-grid and a sustainable energy living laboratory. Through this initiative, York University could use solar PV to power buildings and electric vehicle charging stations, which can operate in tandem with electricity storage in buildings and a new CHP plant to produce electricity plus heating and cooling for district energy. The end result will be a very efficient energy system that can provide multiple opportunities for interdisciplinary research and training. The solar mapping project was achieved by enlisting the support of the university's director of parking and energy manager who helped identify parking areas and buildings that will not be affected by the construction boom that the university is currently experiencing. York University is expanding its building infrastructure at a very rapid pace due to the construction of a new school of engineering and two new subway stations that will connect the university with the rest of the Toronto subway system (and with other public transportation networks). In addition, the university is building several facilities due to the Pan American games that Ontario will host in 2015. The solar assessment facilitate understanding of the long-term solar energy potential of the Keele Campus and also provides a clear idea of the revenue generation potential of installing solar photovoltaic systems at the university.[9]

The success of that learning strategy inspired a related peer mentoring initiative, which involves a pragmatic experiential opportunity where graduate students learn all the details of how to install solar photovoltaic systems at Kortright Centre, which is Canada's largest renewable energy facility.[10] Through this 5-day 'hands-on' course students learn from an expert solar practitioner how to install their own solar PV systems. After completing the solar PV course at Kortright the graduate students provide practical training to undergraduate students that are part of a course on fundamentals of renewable energy. That practical training includes a "hands-on" introduction that teaches students how PV systems work by using PV equipment outdoors, which then prepares students for the more advanced practical courses that are offered by Kortright Center.[11]

---

[9] The province of Ontario currently provides long-term payment (i.e. 20 years) for all the electricity that is generated using renewable energy technologies such as photovoltaic systems as part of the Ontario Feed-in Tariff Program for details about this program see the official site http://fit.powerauthority.on.ca

[10] Kortright is a SEI partner located 12 km north of York University, for details see their internet site at www.kortright.org

[11] All the practical courses of Kortright are summarized at http://kortright.org/groups-and-education/energy-workshop/

A subsequent peer mentoring initiative was implemented in collaboration with the Nordic Folkecentre for Renewable Energy, Denmark's leading renewable energy educational facility.[12] The goal of this peer mentoring initiative was to provide a hands-on learning experience focused on the implementation of wind turbines. Through design assistance from Folkecentre a portable wind turbine system was built which permits quick installation and disassembly so students can safely learn key wind power principles through the implementation of a real wind turbine. The Director of Folkecenter, a pioneering expert with over four decades of practical experience in the Danish wind industry, visited Toronto in 2012 to provide theoretical and practical training to professors and students of York University, which are now able to train others using the portable wind system developed through this peer mentoring initiative. That portable wind structure is now regularly used to train university students and also permits to show high school and primary students how to implement wind turbines by bringing wind power technology to their own school backyards. The pedagogical approach is to first use the classroom to provide an overview of key wind principles and then move outdoors to raise a wind turbine in much the same manner that a sailboat raises and lowers its mast and sails.

Through all these SEI peer-mentoring initiatives students of different ages and levels of expertise learn by doing and experience first hand the transformational potential that experiential learning can bring to those that become closely involved.

Internationalization

A third pedagogical strategy of the SEI has been to develop collaborative links with leading renewable energy organizations that are able to provide training and research opportunities for students and professors. As part of that strategy the SEI has partnered with university and research organizations in Europe, Latin America and the Middle East. However, to achieve wider international impact the SEI partnered in 2011 with the International Renewable Energy Agency (IRENA) to develop and launch a new educational strategy based on new social media concepts that encourage online collaboration.[13] IRENA was created in 2009 with the sole goal of advancing the adoption of renewable energy worldwide. As of 2014 IRENA is supported by 167 nations representing all continents. A key area of work for IRENA has been international capacity development in recognition of the significant barrier that is posed to the implementation of renewable energy projects when local know-how is limited or absent. The IRENA Renewable Energy Learning Partnership (IRELP) was conceived as a capacity development effort that would

[12] Folkecenter is a SEI partner located in Northern Denmark, for details see their internet site http://www.folkecenter.net/gb/

[13] For information about IRENA see their site at www.irena.org

start by providing an internet-based meeting place for people looking for renewable energy training courses, university and college programmes, webinars, training guides and internship opportunities in the renewable energy sector. Although renewable energy education and training opportunities and materials existed prior to the launch of IRELP they tended to be difficult to find and widely dispersed throughout the Internet. IRELP was designed to help individuals, including students, professionals and decision makers, enter the renewable energy sector and to help them develop successful careers in the field. The IRELP portal currently features over 1,500 training courses, college and university programmes, training guides, educational webinars, and internship opportunities, and is being accessed by users in more than 140 countries worldwide. Also as of May of 2013 IRELP has over 27,000 followers on Facebook and Twitter. IRENA envisions that the next steps for IRELP will be to assist learning organizations on the development of standardized renewable energy curriculum.[14] The launch of IRELP represents a step in the right direction to ensure that pedagogical experience and opportunities can be shared in different cultural contexts. It is hoped that the IRELP partnership can grow to include more experiential learning and peer mentoring opportunities and that focus remains an area of further work for the SEI.

Conclusions

Based on the SEI pedagogical experiences summarized here it must be emphasized that innovative training strategies can be highly labour-intensive (particularly if they involve experiential learning and peer mentoring opportunities) but are nevertheless essential to help buttress the sustainability-inspired changes that are rapidly transforming the energy sector. That realization provided the original spark that helped give birth to the SEI and it still constantly motivates the students, professors and numerous partners actively involved in the initiative. As anyone involved deeply in transformative learning can testify it often can be a highly rewarding experience but it is also a hard approach to implement as the dominant systems in place are set-up to favour a very different educational model. From an energy perspective the educational contrast between two wealthy countries like Germany and Canada is stark. Germany invests heavily in developing its educational capabilities on renewable energy and conservation while Canada is currently almost solely focused on fossil fuel development and nuclear power. Even within Canada in the provinces that have focused on fostering their renewable energy sector (e.g. Ontario and Nova Scotia), capacity development and education is not yet a priority.

As understanding grows amongst citizens, policy and decision-makers about the essential roles that renewable energy, conservation, efficiency, electric mobility,

[14] For details about IRELP see www.irelp.org

district energy and other sustainable energy approaches have in helping achieve environmental protection and economic prosperity more attention needs to be paid to the implementation of effective educational strategies. Such strategies require not only support but also encouragement to share results, positive and negative, as widely as possible so different jurisdictions struggling with issues such as air pollution, climate change mitigation, price volatility, employment creation and energy security (to name a few energy-related challenges) can learn from existing experience and avoid reinventing initiatives and repeating mistakes.

References

Freire, P. (2004). *Pedagogy of indignation*. Boulder: Paradigm.
REN21. (2013a). *Renewables 2012: Global status report*. Paris: REN21 Secretariat.
REN21. (2013b). *Renewables 2012: Global futures report*. Paris: REN21 Secretariat.

Chapter 34
Afterword: Towards Technoscience Education for Healthier Networks of Being

Larry Bencze and Steve Alsop

Abstract In this final chapter, we weave together salient ideas and examples gleaned from other authors' writing in this edited collection with our perspectives on relevant literature. As we stated in the introductory chapter here, we believe that contributing authors have bravely engaged in critical scholarship regarding socioscientific issues and, related to that, offer very progressive perspectives and practices for science and technology education that we all hope will help contribute to increases in social justice and environmental wellbeing. We agree with many scholars here that the wellbeing of individuals, societies and environments are under considerable stress – very likely associated with the immense power held within a vast and complex network of actants (material and semiotic), largely controlled by relatively few individuals and groups whose main purpose appears to be personal enrichment, often at the expense of others and environments. Our world is a strange – and, we believe, highly problematic – mix of never-satiated, largely celebratory, consumerism and gut-wrenching, but largely submerged, poverty and environmental degradation. Enmeshed in the global capitalist network behind this scenario are many fields of science and technology and much of science education. Science education networks, particularly under the current 'STEM' movement, appear to focus on generation of the relatively few students who may supply capitalists with immaterial labour – professionals, such as engineers, scientists, accountants, etc., who may develop innovative designs for commodities and

L. Bencze (✉)
Ontario Institute for Studies in Education, University of Toronto, Bloor St W 252, Toronto, ON M5S 1V6, Canada
e-mail: larry.bencze@utoronto.ca

S. Alsop
Faculty of Education and Department of Science and Technology Studies, York University, 4700 Keele Street, Toronto, ON M3J 1P3, Canada
e-mail: salsop@edu.yorku.ca

L. Bencze and S. Alsop (eds.), *Activist Science and Technology Education*, Cultural Studies of Science Education 9, DOI 10.1007/978-94-007-4360-1_34,

their marketing. Complementing this function, appears to be generation of large numbers of citizens mostly prepared to serve as consumers (e.g., of labour instructions and commodities). Their consumerism, while enriching a few capitalists, appears to be largely-responsible for many potential problems associated with socioscientific issues discussed here. While several authors in this book lament difficulties challenging capitalist hegemony, rays of hope also are provided in several chapters regarding socio-political actions in the public sphere, elementary and secondary science education and in higher education contexts.

Keywords Neoliberalism • Capitalists-supporting network governance • Research-informed activism • Hyperreality • STEPWISE • STEM

Introduction

As Derek Hodson (2010) advised, a point emphasized in this book by Chantal Pouliot (p. 529), "[t]hose teachers who promote involvement and develop action skills are 'riding a tiger,' but it is a tiger that may well have to be ridden if we really mean what we say about education for [and through] civic participation" (2010, p. 205). Indeed, many authors in this book made explicit or implicit references to something 'wild' and dangerous needing to be tamed before activism can be a significant feature in primary, secondary and tertiary educational contexts. At the same time, while the tiger analogy/metaphor has merits in describing this inertial entity, other useful analogies/metaphors have emerged in view of recent research and publication. Forces apparently governing much of the world seem considerably less singular, animate and knowable than a 'tiger.' In this chapter, after a review of insights about this entity and its effects on fields of science and technology and their educational counterparts, individuals and societies and environments, orientations towards a 'healthier' world are discussed – particularly in light of the many wonderful chapters in this book.

Our Economized World

Although some of us have a sense of having *private* thoughts possibly leading to self-directed actions, it can be argued that our personal 'agency' (e.g., thoughts and actions) are reciprocally-related (within contextual limitations) to some general social 'structures' (e.g., discourse patterns) (Giddens 2006). In other words, our independence of thought may, to some extent, be illusionary. Instead, each of us could be considered to be a 'collective' – although likely a dynamic, constantly-changing, one – of influences from other people, groups, etc. circulating in our worlds. It may be, for instance, that people in specific countries share religious/spiritual thoughts/practices at least partly gleaned from such influences as their interactions with others and common iconography and media communications.

Among the many possible general social structures possibly-permeating individuals' minds, it seems that few can currently rival that of *neoliberal* capitalism – as many authors in this volume indicated. Arturo Rodriguez' chapter (pp. 55–66), for example, is a particularly-direct and forceful message about neoliberal influences. Other authors here also have attended, explicitly or implicitly, to this set of ideological perspectives. Albeit considered somewhat amorphous and, moreover, unpredictable in its applications in different contexts (Sovacool 2010), neoliberal capitalism generally seems to be an intensified and strategic renewal of traditional economic liberalism – which had prioritized liberating individuals (including corporations) from state intervention/regulations so that they might freely pursue their economic self-interests (profit) (Clarke 2008; McMurtry 2013; Shamir 2008). Under traditional economic liberalism, capitalists enjoyed considerable autonomy from government regulation to focus on employing workers to produce for-profit products and services that were to enrich everyone's lives. Neo-liberalism, which seems to have arisen in the early 1980s, after a period of increased social and infrastructure spending following the second 'world war' (McMurtry 2013), apparently goes well beyond traditional capitalists' focus on direct for-profit production-consumption cycles. The emphasis is not so much on production and consumption of physical commodities (e.g., products & services, like cars and car washes) as on *abstractions*, such as ideas, feelings, moods, etc. (Barber 2007; Baudrillard 1998; Bauman 2005; Sandlin and McLaren 2010). When we purchase a car, for instance, we often are led to focus on abstractions like its sleekness, its sense of social class and power and the extent to which we, as drivers of such a car, construct notions of 'cool' (e.g., imperturbable) about ourselves. Lazzarato (1996) referred to capitalist activities aimed at production of such for-profit abstractions as *immaterial labour*. A focus on such labour has many benefits for capitalists; often, for example, because frequently-revised abstractions associated with advertising can generate in consumers such strong desires for commodities that they can be convinced to relatively-quickly discard them in favour of the 'latest' replacements that supposedly provide users with current most-popular semiotic images (e.g., new forms of 'cool') (Leonard 2010). Marx (1867/1977) long ago recognized this as 'commodity fetishism,' an intense attachment to commodities distracting people from more critical relationships than those of seller and consumer, such as the labourer-capitalist relationship – which may not be particularly fair or environmentally sustainable. For capitalists, it is apparent that there is much more wealth-generation potential in highly-flexible abstractions – which can, however temporarily, become part of consumers' identities – than on less flexible material commodities (Latour 1987). This focus on the abstract has, moreover, apparently been extended to include *potential* development of ideas, identities, etc. Although capitalizing on uncertain futures seems risky, large fractions of investments appear to be committed to speculative idealized futures – in what Pierce (2012) calls "casino capitalism" (p. 25).

As Lyn Carter and Jesse Bazzul specifically emphasized in this volume (pp. 23–36 and 37–53, respectively), neoliberal capitalists appear to have much more influence over people than simply through capital exchanges and associated abstractions. Largely drawing on Foucault's (1977, 2008) conceptions of bio-power and, more specifically, bio-capitalism, they suggest that the neoliberal programme of wealth generation/concentration involves considerable influence over a broad range of thoughts/identities and actions ('subjectivities') of populations. Regarding the *materials economy* outlined by Leonard (2010), for example, pro-capitalist subjectivities may infiltrate the minds of people at every stage of the life cycle of products/services. Regarding cell phones, for instance, miners and manufacturing workers in poor countries may feel that, although they appreciate having a job, it is normal to work long hours in difficult conditions for wages most of us in 'wealthier' countries would reject. In 'wealthier' countries, meanwhile, some sales people may sell customers 'new' (perhaps only slightly-revised) cell phones, not questioning for a moment the possibility that the customer's previous, not-so-old, phone may end up in a landfill because s/he had been convinced of perceived obsolescence (e.g., slightly faster processer) communicated in advertising for the phone. Bio-power/capitalism seems, however, to also now extend to bio-technological formation of new life forms. In other words, pro-capitalist subjectivities may involve fundamental changes in individuals' acceptance of normality with regards to life forms – as in the case of genetically-modified organisms used in various new commodities; including, for example, foods and beverages and pharmaceuticals – currently and for speculative futures (Kleinman 2003; Pierce 2012).

Not only does neoliberalism appear highly-influential, affecting many living things, it also seems very *resilient* – resistant to change or replacement. While allowing periodic stresses and strains, like the global financial crisis (GFC) of c2008, which could have alerted many people to failings of deregulation, it seems to have more than survived this 'crisis' – given, for instance, that the richest people in the world continue to enjoy wealth increases while most others have experienced financial declines (Freeland 2012). To maintain influence over entities, such as workers, for wealth generation, economic and social elite have, in some contexts, as in Pinochet's Chile, used military might. A much more effective strategy, one that seems to account for capitalists' aforementioned resilience in many contexts appears to be its highly murky networked character. In light of actor network theory (Latour 2005), any one capitalist entity – such as an individual financier or corporation – may be considered to be a 'complex,' composed of influences (and influences on) many other relevant 'actants' (i.e., *material* (living & non-living) and *semiotic* (symbolic) entities). Ball (2012) suggests, for instance, that the *Atlas Economic Research Foundation* (atlasnetwork.org), which aims to establish and support free-market think tanks, is co-related with *at least* 38 supportive actants around the world – including the Fraser Institute in Canada, the Koch Family Foundations in the USA, James Tooley in the UK and Education for All Brazil (p. 20). Indeed, in Ball's (2012) analysis (see p. 83), it seems clear that James Tooley is connected to countless other actants supportive of free-market thinking, including Deutsche Bank, eBay, the Government of Ghana, and the Bill & Melinda Gates Foundation. The same can be said for each of the other networks he provides. Overall, therefore, it seems that there is a vast and complex network of actors supportive of neoliberalism. From Ball's (2012) analyses, it is apparent that the network consists of many human actants, including corporations, think tanks, advocacy organizations, charitable organizations and individual financiers. As Lyn Carter and Jesse Bazzul (Chaps. 2 and 3, respectively) point out, however, we must also consider, through biocapitalist analyses, numerous other living things – such as the sea scallops in Callon's (1986) analyses of fishermen in St. Brieuc Bay and various products of genetic engineering, like the AquaAdvantage™ salmon analyzed by Pierce (2013). Moreover, such actor networks must also include various non-living actants, such as the salt water growing pens in which genetically-modified salmon are contained (Pierce 2013).

The vast and complex pro-capitalist networks described above have been analyzed using metaphors/analogies of living things – such as rhizomes, as described by Deleuze and Guattari (1988) – particularly because of how such networks seem to change, evolve, and adapt like living things. Although we like the rhizome analogy, we think it also may be helpful to think of such networks as behaving like *slime moulds*, which seem to be able to extend their cell bodies in networks to encompass and devour food in ways suggesting they have intelligence (Davies 2012). We are not suggesting that there is an overall 'conspiracy' or 'central intelligence' governing these networks, although analyses of decisions made by about 43,000 trans-national corporations suggest that most (~80 %) economic control resides in a relatively small number of firms (~150 firms, mostly banks and corporations, the top three being Barclays Bank, Capital Group Companies and FMR Corporation) (Vitali et al. 2011). Regardless of which metaphor/analogy is best, it seems clear that neoliberal capitalism derives considerable durability from the sheer vastness and complexity of networks of actors supporting its causes.

In addition to the sheer vastness and complexity of their networks, capitalists also appear to derive considerable durability from their frequent enjoyment of veritable (and actual) *states of exception* described by Agamben (2005). Through a process of *destatization* (Jessop 2002), for instance, 'multi-national' corporations are allowed to operate parts of their businesses in many different countries; thus limiting their links to any one country, a status enhanced through invention of such 'extra-national' organizations as the World Bank, World Trade Organization and International Monetary Fund and their associated international trade agreements. In addition to this exceptional status, corporations – often treated as individuals in law – are legally allowed to *externalize* (lead others to pay) costs of, for example, production (e.g., lower labour standards) and uses of their commodities (e.g., environmental restoration). In a sense, neoliberal capitalists appear to live in a kind of alternate reality, still, of course globally linked to countless actants, but somewhat of a rarified world parallel to the rest of us – a 'republic of property,' as Wallerstein (2011/1974) described it. As a republic of exceptions, then, it seems that neoliberal capitalists have hegemonic influences enabling them to concentrate wealth into relatively few hands while inflicting damage of various sorts on other living and non-living things.

Technosciences, Science Education and Socioscientific Issues

Among the numerous and diverse actants integrated into the neoliberal capitalist network are professionals in fields of technoscience[1] and science education (at primary, secondary & tertiary levels) – which, to a great extent, co-affect each other. Science education, of course, supplies fields of technoscience with practitioners who, as knowledge workers, engage in immaterial production that is essential for development and marketing of a range of products and services and associated semiotic messages – for now and for a speculative future. Few can argue that many of the products of fields of technoscience, such as electronic communication devices, medical diagnostic and treatment innovations and agricultural and food production systems, have enriched lives and extended life spans of many people around the world. Consequently, there are many supporters of pro-capitalist technoscience innovations. Krimsky (2003) advises that advocates claim that "scientists who can turn ideas into profits are the ones who are contributing to a better world" (p. 2). Such praise is, however, controversial. There are many people and groups highly critical of practices and products of fields of technoscience where they are closely-aligned with powerful actants within neoliberal capitalist networks – such as corporations, individual financiers, certain (e.g., 'right wing') governments, private foundations, the World Trade Organization, etc. (Bencze 2008; Krimsky 2003; McMurtry 2013; Ziman 2000). Capitalists' critics suggest, for example, that the legal status of corporations, which are major capitalist actants, is highly problematic. McMurtry (2013), for one, claims that corporate board members' fiduciary duty to maximize profits for shareholders, regardless of negative side-effects of their actions on the wellbeing of other living things and non-living environments, is a fundamental concern – as is their right, as described above, to *externalize* many associated costs; including, for example, health costs linked to their products/services (Giroux 2008). About such ethics, McLaren and Jaramillo (2007) said:

> Capitalists accept collateral damage as part of the overall process, and whether it happens to be the deaths of thousands of human beings or eco-destructivity that leads to the elimination of large clusters of biospheres doesn't really seem to matter to the Masters of Capital – as long as this collateral damage has a minimal effect on the lives of the transnational capitalist class (p. 18).

With such an ethic, it may be difficult for technoscientists to maintain adherence to such claimed 'standards of practice' as conducting investigations without bias and sharing findings within communities of practitioners and with the public. Indeed, it seems such challenges exist. A major aspect of this issue seems to be pressure on many technoscientists to orient their work towards development of for-profit goods

[1] Although fields of science and technology sometimes work together, they often are integrated and individual practitioners conduct a combination of the two. Moreover, some suggest that the two fields share the same general epistemology. Consequently, they often are thought of as one merged field, perhaps called *technoscience* (Sismondo 2008).

and services (and semiotic messages). Such a focus has, unfortunately, often led to compromises to the integrity of technoscientists' work (Mirowski 2011; Ziman 2000). Pharmaceutical companies are among the richest in the world and it is common to find technoscience indiscretions in that field. To begin with, pharmaceutical companies sometimes create considerable desire for medications by, for example, promoting self-perceptions of illnesses (e.g., erectile dysfunction; ADHD [Attention Deficit/Hyperactivity Disorder]), leading people to consume drugs they may not need (Angell 2004). Such consumption can be problematic for many people. Technoscientists working for or with financial ties to drug companies sometimes, for example, have tested their medications using small sample sizes, younger, healthier subjects (who may be less susceptible to negative side-effects), lower doses than those to be prescribed, higher doses of the new drug tested against lower doses of the old drug (for which the patent period had expired), ineffective drug delivery techniques for tests of older drugs that companies want off the market, and short test periods – all aimed at maximizing the probability of drug approval by regulatory agencies (Angell 2004). Meanwhile, even if the integrity of the science may be sound, some drug companies have prevented technoscientists from releasing to the public negative findings about their drugs, as reportedly was the situation when Dr. Nancy Olivieri, while working at Sick Children's Hospital in Toronto and under contract with the drug company Apotex™, found negative side-effects of the drug deferiprone, which was being tested for treatment of the disease thalassemia major (Krimsky 2003). These and many other potential problems stemming from capitalist associations with technoscientists appear to represent significant threats to the wellbeing of individuals, societies and environments. As fundamentally associated with capitalists' wealth concentration, many of these potential problems appear to be a matter of *class*; that is, certain advantaged individuals and groups capture wealth and wellbeing from other living (including humans) and non-living things.

For at least the last 40 years, school science systems have given attention in curricula and, to some extent, in classroom instruction to controversies regarding relationships among fields of technoscience, members of societies and environments (Pedretti and Nazir 2011) – which many call *socioscientific issues* (SSIs).[2] Such an education could, in principle, involve considerations of potential problems for individuals, societies and environments stemming from influences of neoliberal capitalists on fields of technoscience. It could, as well, involve engagement of students in sociopolitical actions in response to SSIs – as all authors in this book recommend, including the two of us. Implementation of critical SSI education and sociopolitical actions within the context of formal science education has, however, been difficult. Often, SSI education appears to be limited to an emphasis on

[2] Depending on their jurisdictions, different people – including authors in this book – refer to such issues by different names (with acronyms like 'STSE,' 'SSI' & 'SAQ'). Although we acknowledge differences among these various terms and associated foci, we use the term *socio-scientific issues* – mainly because the site of its origin, the USA, continues to have significant influence on education around the world.

individualized logical decision making, although frequently within social contexts, after consideration of conflicting data-sets and claims (Hodson 2011; Levinson 2013). Zeidler et al. (2009), for example, who have had significant influences on the nature of SSI education, summarize the emphasis this way:

> Central to this approach is the concerted effort to provide opportunities for students to reflect on issues in order to evaluate claims, analyze evidence, and assess multiple viewpoints regarding ethical issues on scientific topics through social interaction and discourse. (p. 75)

On the one hand, such argumentation-based approaches seem to benefit some students, including in terms of: learning products of science (e.g., laws & theories) (Venville and Dawson 2010), development of socioscientific reasoning skills (Sadler et al. 2007); and, learning about the nature of science (Khishfe and Lederman 2006). On the other hand, an emphasis on personal decision-making suggests an orientation towards societal *individualism* – a priority of neoliberal capitalism. Although students often work in groups in arriving at socioscientific decisions, an over-emphasis on individual choice can be discriminatory, favouring students with advantages over others in terms of cultural and social capital (Bourdieu 1986). While this may be a concern regarding many SSI approaches, perhaps more serious may be a general de-emphasis on socio-political critique (Carter, pp. 23–35 here; Levinson 2013; Pedretti and Nazir 2011). Of particular concern is lack of reference to possible adverse effects of globalized neoliberal capitalism on practices and products of fields of technoscience (Bencze 2008; Carter 2005; Hodson 2008). Without insights into possible compromises to the integrity of technoscience because of influences from capitalists, students' SSI education may remain, in essence, *isolationist* – that is, detached from reference to a range of actants, such as trans-national corporations and free-market think tanks, that would give students more authentic bases for socioscientific decision-making (Pierce 2013) and, moreover, focused excessively on the self instead of communities.

Not only does socioscientific education appear to be isolationist in terms of students' awareness of various neoliberalism-influenced actants, but it also seems that educators struggle encouraging and enabling students to take sociopolitical actions to address issues (Hodson 2011; dos Santos 2009). As Derek Hodson said in his chapter here, "[t]he simple point is that it is almost always much easier to *proclaim* that one cares about an issue than to do something about it, and to do it consistently, coherently and effectively" (p. 68, emphasis added). Indeed, it seems to be easier for teachers – with support from many actants – to emphasize instruction in and assessment of highly decontextualized products, such as laws and theories, of fields of technoscience than to allow students to more independently explore controversial issues and create forms of action that may (or may not) bring about a healthier world – one, for instance, prioritizing social justice and environmental wellbeing. Encouraging and enabling activism seems particularly difficult in the current era, one in which governments in many places in the world claim that they need to, essentially, use science education as a mechanism for identifying and educating future workers in fields of 'science, mathematics, engineering and technology' (STEM) who may assist their nations to successfully compete in the global economy. This appears to be the major thrust of curriculum revision in the USA, for instance, which seems to be experiencing a crisis of confidence evident in the so-called *neo-Sputnik* science curriculum renewal project – which appears to be largely driven by concerns about losing out to countries like China and India in a global economic battle (NAS 2007; Pierce 2013). On the other hand, as we have discussed above, the economic battle may, actually, be *extra-national* (independent of nations). Therefore, a discourse of international economic competitiveness *among* nations may be – intentionally or otherwise – a *ruse*; used as a semiotic agent for structuring science education, among numerous actants, in ways conducive to *transnational* economic competitiveness among multi-national corporations and financiers.

Towards Healthier Networks of Being Through Science & Technology Education

On the one hand, since neoliberal capitalists appear to be successful through strong influences on a large/global and complex actor network and as advised by Lyn Carter in this volume (pp. 23–36), individual or groups of teachers may struggle to overturn capitalist hegemony by way of their efforts to encourage students to carry out actions to address socioscientific issues. Major shifts in pro-capitalist networks regarding education require re-alignment of numerous and varied types of actants – such as the mind-set of parents, school board officials, textbook publishers, etc., along with material resources such as videos depicting alternative ways of thinking and doing. Nevertheless, there appear to be several tacks teachers could take – many of which have been recommended by authors in this book – that may contribute to possibilities for transformations in students' attention to critical socioscientific issues that, in turn, could contribute to challenges to neoliberal hegemony.

Several authors suggest that teachers begin students' SSI education by having them first express their pre-instructional conceptions about issues. Mirjan Krstovic (pp. 339–417) showed students several cartoons – such as one showing two clouds, with one being a 'boss' advising a 'worker' that a report had indicated acid in the worker cloud's rain – and asked it to explain the meaning and implications of this shocking finding. Such a tack seems warranted in terms of classic constructivist learning theory (Osborne and Wittrock 1985), which suggests that personal reflections can help students to become more conscious of their (often) sub-conscious conceptions prior to being confronted with alternatives. It is also recommended, according to Carter, Castano and Jones (pp. 531–545), as the beginning stage of *transformative learning* – perhaps helping students to develop personal identities with issues through free expression of their initial ideas about them. Although it may occur later in students' learning about issues, they also can develop personal identities with them by being encouraged and enabled to conduct primary (e.g., correlational studies) and secondary (e.g., internet searches) research to learn more about them. The chapters in this book drawn from the 'STEPWISE[3]' project (Chaps. 20 [Sperling et al.], 22 [Krstovic], 24 [Zoras & Bencze], 32 [Hoeg & Bencze] (Bencze and Carter 2011) and others (e.g., Martinez & Alsop, pp. 477–484) encourage such student engagement in research, rationale being that increased personal student involvement in reciprocal relations between *phenomena* of the world (e.g., students' modes of transportation to and from school) and *representations* of them (e.g., graphs and posters promoting walking or bicycling) can increase their personal identification with issues (Wenger 1998). Promotion of student research also can help them to think of themselves as 'co-producers of knowledge' with scientists, rather than simply as consumers of scientists' conclusions – a shift Chantal Pouliot (Chap. 29) recommends as a contribution to greater democratization.

Students' pre-instructional conceptions about socioscientific issues are likely to vary considerably, given that, at the very least, it is common for students'/citizens' cultural and social capital to vary – depending, for example, somewhat on their socio-economic status (Bourdieu 1986). Consequently, teachers typically need to provide students with insights into SSIs. The chapters in the Public Sphere section in this volume provide documentary material for teaching about a range of SSIs – including those relating to manufactured foods (Chap. 13 [Mueller]), medical practices in states of emergency (Chap. 12 [Weinstein]), citizens' interactions with municipal authorities (e.g., Chap. 14 [Roth]) and climate change (e.g., Chaps. 16 [Bowen], 18 [Elshof] and 17 [Isopp]).

Given that 'issues,' by definition, involve controversies among people and organizations holding different points of view, one stance for teachers is to attempt to balance their presentations of perspectives, supporting data and theories about issues. An alternative, which many authors in this volume appear to support, is to present students with strong cases for positions that would challenge what they consider to be a dominant discourse in many societies, that is, neoliberalism. Using Laurence Simonneaux's terminology, they are, essentially, calling for "hot" attention to socioscientific issues (pp. 99–111). Leo Elshof's justification for this tack is that it is necessary to counter what he calls "corporate activism" (p. 324). In his chapter, he cites, for example, research reported by Oreskes and Conway (2010), who suggest that capitalists often have employed prominent scientists, journalists, and others to cast doubt on science findings – along with the general integrity of fields of science – that might damage the reputation of capitalist commodities. In this vein, Bernhard Isopp (Chap. 17) describes media opposition encountered by a climate scientist who wanted to enlighten the public about humans' contributions to global warming. Similarly, Michael Bowen (Chap. 16) and Derek Hodson (Chap. 5) advise that media's influences on public perceptions of fields of technoscience and

[3] You can learn more about this framework at: www.stepwiser.ca

business may need scrutiny. Gramsci (2007), who felt that power was enacted in democracies through tacit assumptions implanted into people's minds (e.g., doubt about climate science), advocated – as noted by Haluza-Delay and Carter in this volume (pp. 343–362) – for what he called an explicit 'war of position' against such assumptions. At the same time, while struggles with extreme capitalists may be warranted, we also must keep in mind Hardt and Negri's (2009) advice to not, in effect, engage in a reactionary conflict, in which we continue to play by opponents' rules; such as by engaging in a dialectic like this: Capitalism ↔ Anti-capitalism. Rather, they suggest, much more emphasis needs to be placed on promotion of – hopefully healthier – alternatives for the world.

Despite arguments for it, urging teachers to pose strong alternatives to neoliberal capitalism is controversial. By virtue of their roles as evaluators of their 'achievement,' teachers hold positions of power over students who, consequently, may defer to teachers' views. Indeed, as advised by Wolff-Michael Roth (Chap. 14), strong positions advocated by teachers can be viewed as *colonial* – given, for example, that it is difficult to know and appreciate views of all students and sub-groups (e.g., indigenous people). This may particularly be the case if teachers were to explicitly reveal to students their personal positions on issues, perhaps at least tacitly suggesting to students that the teacher holds the 'correct' position. Nevertheless, this is a tack strongly-advocated by Derek Hodson (Chap. 5). His rationale is complex, but a strong point he makes is that teachers need to not only reveal their positions, but be ready to have them critiqued and shown to be potentially problematic. In short, he advises that teachers also need to address the nature of science and the *nature* of socioscientific issues – including, for example, the role of commercial and non-commercial media in interpreting claims from fields of technoscience about issues. This recommendation appears to align well with that by Pierce (2013), who encourages educators to ask students to develop and explore actor networks relating to issues as a way to help them to realize the complex material-semiotic contexts surrounding every actant. Students could become aware, regarding televisions, for example, that the teacher (and others, including them) may be reciprocally connected in a web of interactions among actants like: electricity generators; materials used to make transmission lines; media companies; advertising companies; magazines discussing TV programs; electronics retailers; 'consumers' talking to each other about TV programs; actors' unions; workers in countries like China who manufacture TV components; labour laws in countries where TVs are manufactured; etc. In a similar vein regarding genetically-modified fish, Pierce (2013) said that: "scientific literacy needs to be radically rethought in an age where genes of an Ocean Pout (an eel fish) are spliced with those of a Chinook (king) salmon, implanted in Atlantic salmon eggs, and a corporation patents this process *and* the new species of the fish itself ..." (p. 113; emphasis in the original).

Again drawing on Wenger's (1998) claim that deeper, more committed, learning arises when students are engaged in reciprocal relations (e.g., inductive & deductive) between *phenomena* and *representations* of them, students not only need exposure to conceptions about SSIs, research and actions, they need to be given opportunities to increasingly self-determine (often in social settings) all decisions concerning them (Bencze and Carter 2011). In all of the chapters reporting results from the STEPWISE project (i.e., Chaps. 20 [Sperling et al.], 22 [Krstovic], 24 [Zoras & Bencze] and 32 [Hoeg & Bencze]), for instance, students are engaged in apprenticeship activities (in which control of decision-making is progressively ceded to students) aimed at helping them develop expertise and confidence for conducting research-informed action (RiA) projects to address SSIs of their interest. For example, as reported in Chap. 22 [Krstovic], after participating in a teacher-guided correlational study of peers' possible contributions to climate change, tenth-grade students drew from the class study, their secondary research and what they had been taught about climate change to develop various local social actions, including: posters and brochures placed in strategic places in the school; morning public address announcements to encourage recycling; and, a letter to the school principal asking for more prominent placement of recycling bins (p. 406). In yet-to-be-published findings (Bencze & Krstovic), students' engagement in issues, research and actions seems to be enhanced – as recommended by Pierce (2013) and others – by being guided through development and revision of actor network drawings. A particularly-successful strategy appears to be to introduce students to the idea that many commercial products and services act like Trojan horses.[4] To promote consumer desires, in other words, prominent positive semiotic messages (e.g., sex-appeal) often are associated with commodities; messages that may serve to occlude actants that may engender quite negative semiotic messages (e.g., pain & suffering) (Bencze 2013).

Although authors in this book have recommended that teachers take a strong stance in opposition or as alternatives to neoliberal capitalism, there also is in their writing an ultimate deference to the democratic principle of self-determination (although, often, in community contexts). Hodson advises, for instance, that, after a more teacher-guided apprenticeship in the nature of and ways of dealing with socioscientific issues, students should be left:

> responsible for the whole process, from initial problem identification to final evaluation. Students identify the issue or problem, collect, organize and analyse information, define the problem from a variety of perspectives, formulate and appraise alternative actions, choose which action to take, develop and carry out a plan of action, and evaluate the outcome and the entire undertaking (p. 88).

It may be that it is only through such autonomous (although social, with peers) project work that students are significantly-liberated from oppressive forces of powerful others (McLaren 2000).

Ways Forward

When we undertook this book project, we were somewhat unsure about the nature and extent of thinking and research that would be provided by the many scholars invited to contribute to our volume dealing with activism in/for/from science and

[4] http://en.wikipedia.org/wiki/Trojan_Horse

technology education. However, it is quite clear to us that this topic has generated considerable thought, ideas, strategies, emotions, etc. among our contributing authors. For this, we are very grateful. Among the more salient ideas emerging from many of the chapters pertains to humans' positions within complex, uncertain and power-infused networks. Although some might think of themselves as independent agents, largely in control of decisions, it seems clear that, as Ball (2012) suggests and many others concur, we are greatly influenced by capitalists-supporting network governance. This governance appears to orchestrate actors in networks in ways that primarily benefits relatively few economic elite, at the expense of the wellbeing of individuals, societies and environments. As a consequence, it has been extremely difficult for sub-cultures, like science and technology education, to change in dramatic ways. Indeed, several authors in this volume lament that shifts towards more critical views of socioscientific issues and socio-political actions by students, for instance, are fraught with difficulties. The broad and complex neoliberal capitalist network described in this chapter and supported by several others in this book seems to be at the root of such frustrations; and, it seems highly resistant to fundamental change.

We are left wondering how science and technology education might more purposefully aspire to help bring about a healthier alternative world than the one we believe is currently dominated by neoliberal capitalist networks. Although there are, undoubtedly, many possible answers to this question, one way forward seems to revolve around the concept of *the commons*. Hardt and Negri (2009) suggest that there is a tension in neoliberal capitalism involving common spaces, resources, etc. On the one hand, neoliberal capitalists often attempt to privatize aspects of the commons, such as living things (e.g., seeds) with commercial value (Shiva 1993). On the other hand, partly through their tendency to externalize costs, they often rely on public assets, such as road systems and energy supply networks. Perhaps our planet's healthier alternative future lies in recapturing part of the commons, as Hardt and Negri (2009) suggest. Indeed, in this book, several authors noted benefits of social media for engaging in actions for the common good. Much of this has, apparently, been central to successes of different 'Occupy' movements, like Occupy Wall Street (see Chap. 2 [Carter]). It also has been successfully incorporated by students in cases reported here (e.g., pp. 339–417 [Krstovic] & pp. 435–449 [Zoras & Bencze]). Perhaps ironically, activists have had some successes through uses of corporate electronic tools of social media, such as Facebook™, Twitter™ and YouTube™. This has, apparently, been possible because capitalists have, so far, kept many of these facilities open to the commons. But, as Kate Milberry (Chap. 15) advises, activists may not always be able to count on this openness. Indeed, she says that some activists have recognized this and begun to develop alternative communication ('disruptive') technologies with more anarchic characteristics – based on *pre-figurative* politics; that is, "modelling the change [activists] sought to bring about in the broader society" (p. 256). Such a politics may serve educators well in their ongoing efforts to bring about a healthier world through technoscience education and youth culture.

References

Agamben, G. (2005). *State of exception* (K. Attell, Trans.). Chicago: University of Chicago Press.

Angell, M. (2004). *The truth about the drug companies: How they deceive us and what to do about it*. New York: Random House.

Ball, S. J. (2012). *Global Education Inc.: New policy networks and the neo-liberal imaginary*. Abingdon: Routledge.

Barber, B. R. (2007). *Consumed: How markets corrupt children, infantilize adults, and swallow citizens whole*. New York: Norton.

Baudrillard, J. (1998). *The consumer society*. London: Sage.

Bauman, Z. (2005). *Liquid life*. Cambridge: Polity Press.

Bencze, J. L. (2008). Private profit, science and science education: Critical problems and possibilities for action. *Canadian Journal of Science, Mathematics & Technology Education, 8*(4), 297–312.

Bencze, J. L. (2013, April 6–9). Consumerism, science/education and gatedness. A presentation at the annual conference of the *National Association for Research in Science Teaching*, Rio Grande, Puerto Rico.

Bencze, L., & Carter, L. (2011). Globalizing students acting for the common good. *Journal of Research in Science Teaching, 48*(6), 648–669.

Bourdieu, P. (1986). The forms of capital. In J. G. Richardson (Ed.), *The handbook of theory: Research for the sociology of education* (pp. 241–258). New York: Greenwood Press.

Callon, M. (1986). Some elements of a sociology of translation: Domestication of the scallops and the fishermen of St Brieuc Bay. In J. Law (Ed.), *Power, action and belief: A new sociology of knowledge* (pp. 196–233). London: Routledge & Kegan Paul.

Carter, L. (2005). Globalisation and science education: Rethinking science education reforms. *Journal of Research in Science Teaching, 42*(5), 561–580.

Clarke, J. (2008). Living with/in and without neo-liberalism. *Focaal – European Journal of Anthropology, 51*(1), 135–147.

Davies, E. (2012, October 9). Brainless slime mould has an external memory. *BBC Nature*. Retrieved June 19, 2013, from http://www.bbc.co.uk/nature/19846365

Deleuze, G., & Guattari, F. (1988). *A thousand plateaus*. London: Athlone Press.

dos Santos, W. L. P. (2009). Scientific literacy: A Freirean perspective as a radical view of humanistic science education. *Science Education, 93*(2), 361–382.

Foucault, M. (1977). *Discipline and punish: The birth of the prison*. New York: Pantheon Books.

Foucault, M. (2008). *The birth of biopolitics: Lectures at the Collège de France 1978–1979*. (M. Sennelart (Ed.), F. Ewald & A. Fontana (General Editors), G. Burchell, Trans.). Basingstoke: Palgrave Macmillan.

Freeland, C. (2012). *Plutocrats: The rise of the new global super-rich and the fall of everyone else*. New York: Penguin.

Giddens, A. (2006). *Sociology* (5th ed.). Cambridge: Polity Press.

Giroux, H. (2008). *Against the terror of neoliberalism: Politics beyond the age of greed*. Boulder: Paradigm Publishers.

Gramsci, A. (2007). *Prison notebooks* (Vol. 3, J. A. Buttigieg, Trans.). New York: Columbia University Press.

Hardt, M., & Negri, A. (2009). *Commonwealth*. Cambridge: Belknap Press of Harvard University Press.

Hodson, D. (2008). *Towards scientific literacy: A teachers' guide to the history, philosophy and sociology of science*. Rotterdam: Sense.

Hodson, D. (2010). Science education as a call to action. *Canadian Journal of Science, Mathematics, and Technology Education, 10*(3), 197–206.

Hodson, D. (2011). *Looking to the future: Building a curriculum for social activism*. Rotterdam: Sense.

Jessop, B. (2002). *The future of the capitalist state*. Cambridge: Polity.

Khishfe, R., & Lederman, N. G. (2006). Teaching nature of science within a controversial topic: Integrated versus nonintegrated. *Journal of Research in Science Teaching, 43*(4), 395–418.
Kleinman, D. L. (2003). *Impure cultures: University biology and the world of commerce*. Madison: University of Wisconsin Press.
Krimsky, S. (2003). *Science in the private interest: Has the lure of profits corrupted biomedical research?* Lanham: Rowman & Littlefield.
Latour, B. (1987). *Science in action: How to follow scientists and engineers through society*. Cambridge: Harvard University Press.
Latour, B. (2005). *Reassembling the social: An introduction to actor-network-theory*. Oxford: Oxford University Press.
Lazzarato, M. (1996). Immaterial labor. In P. Virno & M. Hardt (Eds.), *Radical thought in Italy: A potential politics*. Minneapolis: University of Minnesota Press.
Leonard, A. (2010). *The story of stuff: How our obsession with stuff is trashing the planet, our communities, and our health – And a vision for change*. New York: Free Press.
Levinson, R. (2013). Practice and theory of socio-scientific issues: An authentic model? *Studies in Science Education, 49*(10), 99–116.
Marx, K. (1867). *Capital: A critique of political economy* (Vol. 1). New York: Vintage Books.
McLaren, P. (2000). *Che Guevara, Paulo Freire, and the pedagogy of the revolution*. Lanham: Rowman & Littlefield.
McLaren, P., & Jaramillo, N. (2007). *Pedagogy and praxis in the age of empire: Towards a new humanism*. Rotterdam: Sense.
McMurtry, J. (2013). *The cancer stage of capitalism: From crisis to cure*. London: Pluto.
Mirowski, P. (2011). *Science-mart: Privatizing American science*. Cambridge, MA: Harvard University Press.
National Academy of Sciences [NAS]. (2007). *Rising above the gathering storm: Energizing and employing America for a brighter economic future*. Washington, DC: National Academy Press.
Oreskes, N., & Conway, E. (2010). *Merchants of doubt*. London: Bloomsbury Press.
Osborne, R., & Wittrock, M. (1985). The generative learning model and its implications for science education. *Studies in Science Education, 12*, 59–87.
Pedretti, E., & Nazir, J. (2011). Currents in STSE education: Mapping a complex field, 40 years on. *Science Education, 95*(4), 601–626.
Pierce, C. (2012). The promissory future(s) of education: Rethinking scientific literacy in the era of biocapitalism. *Educational Philosophy and Theory, 44*(7), 721–745.
Pierce, C. (2013). *Education in the age of biocapitalism: Optimizing educational life for a flat world*. New York: Palgrave Macmillan.
Sadler, T. D., Barab, S. A., & Scott, B. (2007). What do students gain by engaging in socioscientific inquiry? *Research in Science Education, 37*(4), 371–391.
Sandlin, J. A., & McLaren, P. (Eds.). (2010). *Critical pedagogies of consumption: Living and learning in the shadow of the "Shopocalypse"*. New York: Routledge.
Shamir, R. (2008). The age of responsibilization: On market an embedded morality. *Economy and Society, 37*(1), 1–19.
Shiva, V. (1993). *Monocultures of the mind: Perspectives on biodiversity in biotechnology*. London: Zed Books.
Sismondo, S. (2008). Science and technology studies and an engaged program. In E. J. Hackett, O. Amsterdamska, M. Lynch, & J. Wajcman (Eds.), *The handbook of science and technology studies* (3rd ed., pp. 13–31). Cambridge, MA: The MIT Press.
Sovacool, B. K. (2010). Erasing knowledge: The discursive structure of globalization. *Social Epistemology, 24*(1), 15–28.
Venville, G. J., & Dawson, V. M. (2010). The impact of an argumentation intervention on Grade 10 students' conceptual understanding of genetics. *Journal of Research in Science Teaching, 48*(8), 952–977.
Vitali, S., Glattfelder, J. B., & Battison, S. (2011). The network of global corporate control. *PLoS ONE 6*(10), Available at: http://www.plosone.org/article/info%3Adoi%2F10.1371%2Fjournal.pone.0025995

Wallerstein, I. (2011/1974). *The modern world-system, I: Capitalist agriculture and the origins of the European world-economy in the sixteenth century* (Revised). San Francisco: University of California Press.
Wenger, E. (1998). *Communities of practice*. Cambridge: Cambridge University Press.
Zeidler, D. L., Sadler, T. D., Applebaum, S., & Callahan, B. E. (2009). Advancing reflective judgement through socioscientific issues. *Journal of Research in Science Teaching, 46*(1), 74–101.
Ziman, J. (2000). *Real science: What it is, and what it means*. Cambridge: Cambridge University Press.

Author Biographies

Steve Alsop is a Professor in the Faculty of Education and The Department of Science and Technology Studies at York University, Canada. Steve has held a series of academic positions including associate dean, academic director, coordinator, honorary professor, secondary and primary school teacher. He has a wide range of research interests exploring the personal, social and political organisation of scientific and technological knowledge, pedagogies of science and technology, and environmental sustainability. His latest research project is entitled PEASE (Policy Enactments and Sustainability Education). This focuses on environmental sustainability in universities and schools in Canada and is a collaboration with Lakehead University and a number of educational ENGOS including ECOSOURCE, ACER, The David Suzuki foundation and Learning for a Sustainable Future. Steve lives in Toronto.

Email: SAlsop@edu.yorku.ca

Barbara Bader holds the *Chaire de leadership en enseignement des sciences et dé veloppement durable* and is Professor of Environmental Education and Science Education at Université Laval in Québec City. She is member of an interdisciplinary research centre in Environmental Education, the "Centre de recherche en éducation et formation relatives à l'environnement et à l'écocitoyenneté (Centr'ERE – UQAM)" in Montréal and of the Environnement-Développement-Société Institut at Université Laval. Her latest book was co-edited with Lucie Sauvé and titled *Éducation, environnement et développement durable: vers une écocitoyenneté critique* (Les Presses de l'Université Laval, 2011). She just edited the 11 volume of the journal *Éducation relative à l'environnement, Regards, Recherches et Réflexions* with Angela Barthes and Alain Legardez. Barbara Bader wrote many articles in academic journals and contributed an entry on Education in the Encyclopedia of global warming & climate change (S. George Philander, Ed., 2nd ed., Thousand Oaks, CA: SAGE, pp. 491–495).

Email: Barbara.Bader@fse.ulaval.ca

Jesse Bazzul is an assistant professor in the department of STEM Education and Teacher Development at the University of Massachusetts, Dartmouth. He sees

L. Bencze and S. Alsop (eds.), *Activist Science and Technology Education*,
Cultural Studies of Science Education 9, DOI 10.1007/978-94-007-4360-1,

science education as a key site in confronting urgent problems such as social inequality and environmental destruction. He believes (science) educators must come together in community to transform the field into one that actively enables social transformation(s).

Email: jbazzul@gmail.com

Larry Bencze is an Associate Professor in Science Education at the Ontario Institute for Studies in Education, University of Toronto, where he teaches in the graduate studies and teacher education programmes. In addition to his Ph.D. (U. Toronto, 1995) and B.Ed. (Queen's U., 1977) in education, he holds B.Sc. (Queen's U., 1974) and M.Sc. (Queen's U., 1977) degrees in biology. Prior to his work as a professor, he worked as a teacher of science in elementary and secondary schools and as a science education consultant in Ontario, Canada. His teaching and research emphasize history, philosophy and sociology of science and technology, along with student-led, research-informed, socio-political actions to address personal, social and environmental issues associated with fields of science and technology.

Email: larry.bencze@utoronto.ca

David W. Blades is Professor of Science Education and Curriculum Studies at the University of Victoria. An award-winning teacher, David was also a Killam Scholar and recipient of the Ted Aoki Award for his work on curriculum change. David teaches courses on secondary and elementary school science education and curriculum theory. He has published and presented widely his research on the implications of curriculum change to the practice and theory of science education as well in the areas of ethics and global citizenship. In his spare time, which is never enough, David loves to explore with his family the bays and marinas of the Pacific Northwest aboard their boat, *Polar Star*.

Email: dblades@uvic.ca

Erica N. Blatt is an assistant professor of science education at the College of Staten Island, City University of New York, teaching courses in science teaching methods and curriculum at the elementary and secondary level. She earned her Ph.D. in Education from the University of New Hampshire, specializing in science education and curriculum development. Prior to this, Erica taught high school Biology, Earth Science, and Environmental Science in settings as diverse as inner city Philadelphia, Phillips Academy in Andover, MA, and in Costa Rica. Her current research focuses on pre-service teachers' youth experiences in the outdoors and their intentions and perceived obstacles to taking their own students outdoors, in addition to determining best practices for helping teachers effectively incorporate environmental topics into their science teaching. She is currently serving as co-chair of the Environmental Education strand of the National Association for Research in Science Teaching.

Email: Erica.Blatt@csi.cuny.edu

G. Michael Bowen is an Associate Professor at Mount Saint Vincent University in Halifax, Nova Scotia. He teaches in the Bachelor of Education program (including courses in science methods, science content, and media literacy) and in the graduate program (including courses in curriculum studies and research methodology). In addition to his Ph.D. (U. Victoria, 1999) and B.Ed. (Dalhousie U., 1989) in science education, he holds a B.Sc. in Marine Biology (U. Guelph, 1984) an M.Sc. in biology (U. Guelph, 1990), an MA in sociology (U. Guelph, 1995) and a Bachelor of Journalism (University of Kings College, 2013). He has taught science from grades 5 to 13 in Canada and the United States and he has also worked in informal science education working in the whale watching industry in several places in Atlantic Canada. His research includes topics such as science teacher education, learning science in authentic contexts (from elementary to post-doctoral levels), informal science education and news media portrayals of science and education.

Email: gmrwood@hotmail.com

Angela Calabrese Barton is a Professor in the Department of Teacher Education at Michigan State University. Her research focuses on equity and social justice in science education in urban contexts. Drawing from qualitative and critical methodologies, she conducts critical ethnography and longitudinal ethnographic case study research in both formal and informal settings on the cultural processes of learning and identity development.

Email: acb@msu.eduacb@msu.edu

Elena Camino gained a first degree in Physics from the University of Turin (Italy) and started doing research in electrophysiology. Since 1985 she became involved in research in science education and in 1991 she founded a Research Group in Didactics of the Natural Sciences.

In 2002, together with colleagues from other disciplines, Dr. Elena Camino set up IRIS, the Interuniversity Centre for Interdisciplinary Research on Sustainability (www.iris.unito.it).

Over the course of her career, Dr. Camino has been teaching undergraduate students enrolled in the degree of Natural Sciences (Natural Science Education, Science Communication, Sustainability Science), and post-graduate students enrolled in teacher education courses.

Dr. Camino's research interests include:

- the production and evaluation of educational activities on complex and controversial environmental issues, by integrating aspects of sustainability with the nonviolent approach to conflict
- the investigation of conceptual tools which can promote the understanding of complexity of the biosphere, the scale and intensity of human impact, the dependency of the living beings from the living planet.

Email: elena.camino@unito.it

Angela V. Carter is an Assistant Professor in the Department of Political Science at the University of Waterloo. Her degrees include a B.A. in Communication (University of Ottawa, 2001), an M.A. in Political Economy (Carleton University, 2003), and degrees in Government (M.A. Cornell University, 2006, and Ph.D. Cornell University 2011). Carter researches comparative environmental policy regimes surrounding oil developments in key oil-dependent cases. This research is supported by a SSHRC Strategic Research Grant entitled "The Environmental Assessment Processes of Canadian 'Frontier' Oil and Gas" with Dr. Gail Fraser and Dr. Anna Zalik at York University's Faculty of Environmental Studies. Carter's teaching focuses on political economy and environmental politics and policy.

Email: avcarter@uwaterloo.ca

Lyn Carter is an experienced teacher and university senior lecturer, currently lecturing in science and technology education to undergraduate primary and secondary teacher education students. She also lectures in postgraduate education particularly in the areas of postmodern research methodologies and contemporary issues in curriculum. Her Ph.D. thesis explored the ways in which globalisation shapes and influences science education. Her current research focuses on the use of cultural theory as a way of expanding research within science education. In particular, Dr. Carter is investigating how postcolonial theory can be used in science education. Dr. Carter is a well-known international conference speaker and is a member of a number of Editorial Boards, including at present Associate Editor of the international journal RISE. She has extensive experience with professional development programs having worked on the major SIMMER and ASISTM projects in the last number of years.

Email: Lyn.Carter@acu.edu.au

Carolina Castano Rodriguez is a lecturer in science and environmental education at the Australian Catholic University in Melbourne, Australia. She holds a Ph.D. from Deakin University, Australia, and a M.Ed. and B.Sc. from Los Andes University, Colombia. Her research interests include exploring the relevance of science within socio-economic disadvantaged populations, studying the role of science education in society and incorporating transformative learning pedagogies to a science education that foster care and respect for all species.

She is the leader of the Research group: Transformative Science, which focuses on Transformative Learning Theory. She is also a member of the research group "From Pedagogy to practice" from Universidad Nacional de Colombia. She has worked in research and consultancy projects with numerous international agencies including UNESCO, Iberoamerican organization (OEI), Department of Education in Ecuador and the World Society for the Protection of Animals (WSPA). She is a member of the editorial board of the journal Cultural Studies in Science Education and the professional group Coordinadora de Profesionales por la Prevencion de abusos (Coppa – Professionals for the prevention of abuse).

Email: carolina.rodriguez@acu.edu.au

Laura Colucci-Gray has a degree in Natural Sciences and a doctorate in science education. She is currently a lecturer in science education and sustainability education in the School of Education at the University of Aberdeen (United Kingdom). Since gaining her first degree Laura has been involved in science education research at the University of Turin looking at the development and evaluation of role-play simulations to deal with complex and controversial socio-environmental issues. After completing her doctorate in science education, Laura has been involved in a number of action-research projects looking at:

- the use of creative and participatory methodologies to deal with the controversial aspects of science and technology in sustainability studies; of particular importance in this work is the introduction of the perspective of nonviolence in current debates on sustainability;
- the philosophical dimension of learning outdoors and the integration of this strand with the epistemological reflection on science in the transition towards sustainability;

Laura teaches on a number of interdisciplinary courses in sustainability and science education for prospective teachers and university graduates (Science, Education and Society; Mind Body Nature; the Natural world; Sustainability: challenges and opportunities). Laura is currently the Director of the Postgraduate Research Programme at the University of Aberdeen and she is currently an executive member of Interuniversity Centre for Interdisciplinary Research on Sustainability (www.iris.unito.it).

Email: l.t.gray@abdn.ac.uk

Leo Elshof is an associate professor at Acadia University in Wolfville Nova Scotia Canada. He has worked in the fields of science, engineering and sustainability education for 30 years. He holds degrees from the University of Waterloo, McMaster University, the University of Western Ontario and a doctorate from the University of Toronto. At Acadia he teaches courses in environmental and sustainability studies and pre-service and graduate education programs in the areas of science and environmental education, critical media literacy and environmental technologies. Leo has published articles in a number of international journals, authored numerous book chapters and presented at over 40 national and international academic conferences on the themes of sustainability in technological and science education. His current research interests involve eco-technological literacy, climate change education, critical media literacy and environmental communications.

Email: leo.elshof@acadiau.ca

Jose Etcheverry is a professor at the Faculty of Environmental Studies of York University. Dr. Etcheverry researches topics such as climate change mitigation and renewable energy policies, focused on developing practical policy solutions through collaboration, and finding ways to communicate solutions effectively.

Dr. Etcheverry is Co-Chair of the Sustainable Energy Initiative (SEI) a new research and training effort that involves the Faculty of Environmental Studies, the Toronto Region and Conservation Authority, the Energy Institute of Centennial College and leading international organizations such as Aalborg University and the Nordic Folkecenter for Renewable Energy. SEI was established to advance renewable energy solutions in Canada and around the world.

Dr. Etcheverry is also working on the International Renewable Energy Agency's Learning Partnership (details at www.irelp.org) and the RETScreen International Training Institute. He is the current President of the Canadian Renewable Energy Association, a member of the Committee of Chairpersons of the World Council for

Renewable Energy and member of the Innovation Network of the Japan Renewable Energy Foundation.

Email: rejose@yorku.ca

Randolph Haluza-DeLay is an Associate Professor of Sociology at The King's University College in Edmonton Alberta, having received his Ph.D. in Education from the University of Western Ontario in 2007. He has considerable background and prior research projects on environmental education while current research focuses on environmental social movements and cultures of sustainability. He has co-edited two books – *Speaking for Ourselves: Environmental Justice in Canada* which was published by The University of British Columbia Press in 2009, and the recently released *How the World's Religions are Responding to Climate Change: Social Science Investigations* (Routledge, 2013). As an interdisciplinary scholar, his publications have appeared in the fields of sociology, education, leisure studies, religious studies, anthropology, and environmental ethics. As a citizen, he is active in local sustainability initiatives, peace and justice and anti-racism activities, and interfaith dialogue.

Email: Randy.Haluza-Delay@kingsu.ca

Derek Hodson, whose undergraduate and doctoral studies were in chemistry, is Emeritus Professor of Science Education at the Ontario Institute for Studies in Education (OISE) and Professor of Science Education at the University of Auckland. He has 45+ years experience in science education in schools and universities in UK, Canada, Hong Kong and New Zealand. His research interests include history, philosophy and sociology of science and science education, science curriculum history, multicultural and antiracist education, and action research approaches to science teacher education. Recently published books include *Towards Scientific Literacy: A Teachers' Guide to the History, Philosophy and Sociology of Science* (2008), *Teaching and Learning about Science: Language, Theories, Methods, History, Traditions and Values* (2009) and *Looking to the Future: Building a Curriculum for Social Activism* (2011) – all published by Sense Publishers.

Email: dandsehodson@xtra.co.nz

Darren Hoeg is a Ph.D. candidate in his fourth year at the Ontario Institute for Studies in Education, University of Toronto. He obtained his M.Ed. from York University in Toronto in 2010. Prior to this, he was a middle school general science and high school biology teacher for 9 years. Darren's Ph.D. research consists of an institutional ethnographic study on the practices of grade 9/10 science teachers, focusing on how these practices constitute relationships between humans and nature. Darren is interested more widely in how discourses, such as neoliberalism, biocapitalism, and anthropocentrism, are involved in the institution of school. He has conducted critical research on how these discourses influence science policy and science teacher practice. He has also done research on preservice teachers' orientations to and preparation for teaching about nature, environmental education, and STSE-activist education.

Email: darren.hoeg@utoronto.ca

Rebecca Houwer is the Research Coordinator for a critical community service learning initiative at York University called New Opportunities for Innovative Student Engagement ('NOISE') for Social Change. NOISE for Social Change engages university and high school students in local collaborative research and action projects. Ms. Houwer is a Ph.D. Candidate in the Faculty of Education at York University. Her doctoral research explores the relationships between post-secondary education institutions, eco-social sustainability and ethics, and community-based action-oriented pedagogies.

Email: rebecca_houwer@edu.yorku.ca

Bernhard Isopp is a Ph.D. candidate in the Science and Technology Studies Program at York University. He holds a BA in history (Queen's U., 2007) and a MA in humanities (York U., 2010). His research is located at the intersections between the public understanding of science and the sociology of science and technology. His dissertation explores how popular climate change discourse influences scientists' public engagement activities, and how this engagement reshapes scientific practices, identities, and norms.

Email: b_isopp@hotmail.com

Mellita Jones has been a lecturer in mathematics and science education for the past 10 years. Her teaching is centred on science teacher education, mathematics education and general foundation education principles. Her research is focused on science teacher pedagogy and reflective practice and this informs her work with practising and preservice teachers in these areas. Mellita's work with schools also gives her a range of experiences in teacher professional learning in science education, inquiry learning, curriculum design and reflective practice. Her teacher professional learning consultancies have been based in national and international settings including work in developing countries.

Email: Mellita.Jones@acu.edu.au

Ashley S. Kerckhoff is an elementary school teacher with the Rainbow District School Board. Her Master of Arts in Education thesis (U. Ottawa, 2011) focused on youth environmental activism. Also, she has a B.Ed. degree (U. Ottawa, 2009) as well as a B.A.Sc. in *Child, Youth, and Family* (U. Guelph, 2008). Ashley has had various opportunities in the environmental education field, such as her position as Editor of *Interactions: The Ontario Journal of Environmental Education,* teaching outdoor programs at a nature centre, and working as a Lake Water Quality Technician for the District of Muskoka.

Email: ashleykerckhoff@gmail.com

Mirjan Krstovic is a high school science teacher in the Peel District School Board in Mississauga, Ontario. In addition to being a fulltime teacher he is currently a par-time graduate student in the Masters of Education (M.Ed.) program at the Ontario Institute for Studies in Education, University of Toronto, where he also completed his Bachelor of Education degree in 2005. In addition, Mirjan holds a Hon. B.Sc. (University of Toronto Mississauga, 2002) degree in molecular biology. His teaching and research interests include innovation in science teaching, science, technology, society and environment (STSE) education, student success, instructional leadership and teacher learning in professional learning communities. Mirjan is also a strong advocate of an issues-based and action-oriented science education in his district. He has been promoting student-led, research-informed action projects to address issues affecting the well-being of individuals, societies and environments.

Email: mirjan.krstovic@peelsb.com

Yves Laberge is a sociologist and serves as the Senior Book Series Editor for the book series "L'espace public" and "Cinéma et société" at the Presses de l'Université Laval in Québec city (Canada). He holds a Ph.D. in sociology. He is member of the interdisciplinary research centre in Environmental Education, the "Centre de recherche en éducation et formation relatives à l'environnement et à l'écocitoyenneté (Centr'ERE – UQAM)" in Montréal. Yves Laberge has held visiting professorships at the University of Iceland, University of Ottawa, Université de Provence in France and at the Université de Haute-Bretagne in Rennes, where he held the Chair in American Studies for 2011.

Among various peer-reviewed publications in French and English, Yves Laberge contributed more than 160 entries and chapters in 20 encyclopedias (published by Oxford University Press; SAGE; Routledge; Facts On File, ABC-CLio, etc.). A member of the advisory board for six peer-reviewed journals (*The Electronic Green Journal; Journal of Religion and Popular Culture; Canadian Review of American Studies; Intersections: Gender and Sexuality in Asia and the Pacific; Empedocles: European Journal for the Philosophy of Communication; The European Legacy: Toward New Paradigms*).

Email: ylaberge@uottawa.ca

Ralph Levinson is Reader in Education at the Institute of Education University of London. After teaching science for 13 years in London comprehensive schools he worked for the Open University where he co-wrote a chemistry course for teachers and was responsible for the Open University's first distance-learning postgraduate training course for science teachers. He has taught at the Institute of Education since 1998 where he leads the Masters in Science Education as well as teaching on teacher education courses and supervising doctoral students. His main research interests are in socio-political issues in science education, the role of the arts and philosophy in science teaching and learning, and chemistry education.

Email: Ralph.Levinson@ioe.ac.uk

Ana Maria Martinez is a doctorate candidate in the Faculty of Education at York University in Toronto. She holds a B.Sc. in marine biology (Jorge Tadeo Lozano University, Colombia, 1998), a Master's in Environmental Studies (York University, 2010) and a specialization in Environmental Law (Externado de Colombia University, 2000). Her current research focuses on climate change policy and practice enactments in post secondary institutions. Her previous work focused on biodiversity conservation issues and the influence of social, political and economical arrangements on ecosystems and human diversity in the Tropics.

Email: Ana_Martinez@edu.yorku.ca

Kate Milberry is an instructor with the University of Alberta's Communication and Technology graduate program. Her research examines the democratization of internet technology as a tool and technique for progressive social change. In particular, she examines the relationship between free software development and activist web applications, considering how new values and unmet user needs are designed into technology. Recent work for the Privacy Commissioner of Canada has focused on the privacy implications of e-government and digital identity management schemes. Dr. Milberry completed a postdoctoral fellowship at the University of Toronto's iSchool and a Ph.D. in Communication at Simon Fraser University. She is a technology commentator for CBC radio and blogs at geeksandglobaljustice.com. She is @KateMilberry on Twitter.

Email: kate.milberry@gmail.com

Michael P. Mueller is a professor of secondary education in the Department of Teaching and Learning at the University of Alaska Anchorage. He explores trends in cultural and environmental studies, educational theory, and science education policy. His work with teachers and youth emphasizes how we share responsibility for cultural diversity, biodiversity, natural habitats, and nature's harmony.

Email: ak.mikem@gmail.com

Janet Newbury is currently engaged in post-doctoral research exploring the synergies between human service practices and community-based approaches to economic development. In addition to teaching in the School of Child and Youth Care at the University of Victoria, she lives in Powell River BC where she is an active member of boards and steering committees for: PR Voices, the Diversity Initiative, Sunshine Musicfest, and the Community Resource Centre. She is also a research associate with the Canadian Centre for Policy Alternatives, an associate with the Taos Institute, and a member of the Community Economies Collective, and was a co-founding member of the Post Growth Institute. In all of this work, her role involves pursuing sustainable paths to community wellbeing. She is particularly interested in activities that foster intergenerational civic engagement.

Chantal Pouliot is an Associate Professor in science education at the Faculty of Education of Université Laval, Québec City, and researcher at the Centre de recherche et d'intervention sur la réussite scolaire (CRIRES). She studied in science education (Ph.D., 2007) and in biology (B.Sc., 1997), postmodern literature (M.A., 2001) and science teaching (CEC., 2001). Prior to her work as a professor, she worked as research professional in entomology (1997–2002) and as a teacher of biology in a postsecondary school in Québec City (Cégep Limoilou, 2000–2007). Her main research focuses include the ways students and prospective science teachers picture the management of socio-scientific issues and conceive of the roles, capacities and legitimacy of the social actors concerned. Her research draws on the analyses and findings of science studies, public understanding of science and political philosophy.

Email: Chantal.Pouliot@fse.ulaval.ca

Giuliano Reis is an Associate Professor of Science Education at the University of Ottawa, where he teaches science and environmental education courses in both the Teacher Education Program and the Graduate Program in Education. He is the former director of the research unit Eco-Science 21 (2010–11) and is currently a member of the editorial boards of *Cultural Studies of Science Education* (CSSE) and the *Journal of Research in Science Teaching* (JRST). Giuliano is the lead guest editor of a thematic issue of the Brazilian Journal of Research in Science Education (RBPEC) entitled "Environmental Discourses in Science Education: Contributions to Democracy, Citizenship and Social Justice."

Email: Giuliano.Reis@uottawa.ca

Pedro Reis is an Associate Professor in Science Education at the Institute of Education, University of Lisbon, where he coordinates the Ph.D. programme in Science Education.

In addition to his Ph.D. (U. Lisbon, 2004) and M.Ed. (U. Lisbon, 1997) in education, he holds a B.Sc. (U. Lisbon, 1988) degree in biology. Prior to his work as a professor, he worked as a teacher of science in elementary and secondary schools.

He has published and been involved in research about: (a) teaching the nature of science and technology and assessing its understanding; (b) the discussion of socio-scientific issues in science education; (c) teachers' and students' informed and negotiated actions to address social and environmental issues associated with the fields of science and technology; and (d) ICT integration in science education. He has been involved in research, teacher training and curriculum development projects in Portugal, other European countries, Africa and Latin America.

Email: preis@ie.ul.pt

Arturo Rodriguez is an Associate Professor in Bilingual Education, College of Education, Boise State University, where he teaches in the graduate, undergraduate, and teacher education programmes. Arturo holds the Ph.D. and M.A. in Language, Literacy and Culture (New Mexico State University, 2006) and B.A. in English and American Literature (The University of Texas, El Paso, 1996). Prior to his work as a professor, he worked as a Spanish and English teacher in Texas and California and Dean of Students in East Harlem, New York. His teaching and research interests include critical theory/pedagogy, theoretical and applied linguistics, philosophy of mind and science and technology education.

Email: arturorodriguez@boisestate.edu

Wolff-Michael Roth is Lansdowne Professor of applied cognitive science at the University of Victoria. His research concerns the learning of science and mathematics across the life span. He uses a broad range of research methods and theoretical frameworks from across multiple disciplines. His most recent works include What more in/for Science Education: An Ethnomethodological Perspective (Sense, 2013), First person methods: For a rigorous approach to the study of lived experience (Sense Publishers, 2012), Passibility: At the limits of the constructivist metaphor (Springer, 2011), Geometry as objective science in elementary classrooms: Mathematics in the flesh (Routledge, 2011), and, with L. Radford, A cultural-historical perspective on mathematics teaching and learning (Sense Publishers, 2011).

Email: mroth@uvic.ca

David E. Selby is Founding Director of Sustainability Frontiers, a not-for-profit international organization with offices in Canada and the United Kingdom. He is also Adjunct Professor in the Faculty of Education, Mount St Vincent University, Halifax, Nova Scotia. He was previously (2003–9) Professor of Education for Sustainability and Director of the Centre for Sustainable Futures, University of Plymouth, United Kingdom, and, before that (1992–2003), Professor of Education and Director of the International Institute for Global Education at the Ontario Institute for Studies in Education of the University of Toronto, Canada. He regularly acts as consultant in the areas of disaster risk reduction education, climate change education and peace building education to international organizations such as UNESCO, UNICEF, Plan International and Save the Children. His recent publications include: *Climate Change in the Classroom* (UNESCO, 2013), *Disaster Risk Reduction in School Curricula: Case Studies from Thirty Countries* (UNESCO/UNICEF, 2012) and *Education and Climate Change: Living and Learning in Interesting Times* (Routledge, 2010). His co-edited book, *Sustainability Frontiers: Transformative Voices from the Borderlands of Sustainability Education* (Barbara Budrich, 2014), is published to coincide with the close of the UN Decade of Sustainable Development.

Email: dselby@sustainabilityfrontiers.org

Laurence Simonneaux has been a professor of agronomy and science education since 2001 at Ecole Nationale de Formation Agronomique. She has been a farmer. She has participated in three European research projects on biotechnology education 'European Initiative for Biotechnology Education' (about cloning, transgenesis, genetic screening, stem cells, GMO and biodiversity). She has led French projects on Socially Acute Questions education: for example for the Agence Nationale pour la Recherche (ANR) a for years project on Education for Sustainability and for the Ministry of Agriculture projects on the reduction of the use of pesticides and on the welfare of animals. She participates to two European projects: CASSIS (Communication About SocioScientific Issues) and PARRISE (Promoting Attainment of Responsible Research and Innovation in Science Education).

She coordinates a research team on didactics and is the director of two Masters degree programmes. Laurence has been a member of the Academic Committee of the European Researchers in Didactics of Biology (ERIDOB). She has coordinated five books dealing with discussions, debate, argumentation and the teaching of socially acute questions.

Email: laurence.simonneaux@educagri.fr

Erin Sperling is a doctoral candidate at the Ontario Institute for Studies in Education at the University of Toronto, in Ontario, Canada. She has been a key collaborator for the Environmental and Sustainability Education Initiative at OISE. Her research focus is food justice education and activism for educators and youth. She is a certified teacher and has taught in Ontario, the UK and in Tanzania, including teaching French as a fourth language to 3 year olds. Along with her research work with Dr. Larry Bencze, she has collaborated with educators and researchers in schools and community organizations to consider the impact we have as humans on the well-being of our society and the environment, in order to make positive changes. She is an urban gardener, a step-mother, an avid cyclist and an oenophile. She has a B.Sc. from Dalhousie University and a B.Ed. from the University of Ottawa.

Email: erin.sperling@utoronto.ca

Edna Tan is an assistant professor of science education at UNC-Greensboro. She takes a critical, sociocultural approach in her work with youth and science teachers. Her work focuses on how youth from non-dominant backgrounds can be empowered to work with their teachers in creating hybrid spaces for meaningful science engagement, thereby authoring positive science identities and identity trajectories.

Email: e_tan@uncg.edu

Kenneth Tobin came to the Urban Education doctoral program at the Graduate Center of CUNY in the fall semester of 2003. Prior to that he had positions as tenured full professor at Florida State University (1987–1997) and the University of Pennsylvania (1997–2003). Also, he held university appointments at the Western Australian Institute of Technology (now Curtin University), Mount Lawley College and Graylands College (now Edith Cowan University).

Prior to becoming a university science educator in Australia in 1974, Tobin taught high school physics, chemistry, biology general science, and mathematics for 10 years. He began a program of research in 1973 that continues to the present day – teaching and learning of science and learning to teach science.

Email: KTobin@gc.cuny.edu

Matthew Weinstein directs the Secondary Science Education Program at the University of Washington-Tacoma. He is the author of *Robot World: Education, Popular Culture, Science* and more recently (with Dr. Nidaa Makki) of *Bodies Out of Control: Rethinking Science Texts*. His work draws on anthropology and cultural studies to analyze science education in and out of schools as forms of contested public culture. His recent work has analyzed the politics of human subjects, the nationalist discourse of the anthrax attacks of 2001, and the framing of disease and illness in school textbooks. His current research explores the history and culture of the street medics movement.

Email: matthewW@u.washington.edu

Terry Wilkinson is an elementary classroom teacher for the Toronto District School Board. She began her career as a youth worker and educator in the fields of mental health and special education, and has taught in a variety of middle school settings. Terry's specialist background in visual arts and design and technology education, along with her interests in visual literacy, creativity and problem-solving have also informed her practice as an instructor and sessional lecturer for the teacher pre-service and continuing education programs at the Ontario Institute for Studies in Education, University of Toronto. She is currently enrolled in the Faculty of Education doctoral studies program at York University.

Email: terry_wilkinson@edu.yorku.ca

Brandon Zoras is a science teacher with the Toronto District School Board (TDSB). He has completed his Hon. B.Sc. in biochemistry (U. Toronto, 2008), B.Ed. in inner city education (OISE, U. Toronto, 2009) and M.Ed. in urban education (OISE, U. Toronto 2012). He is interested in social justice and environmental issues and how students can use science to overcome those issues. Technology and social media is a major part of his classrooms, which is used to engage students and teach science. Through problem based learning (PBL) and research information action projects (RiA) his students use inquiry to examine science, technology, society and environment (STSE) problems and propose solutions.

Email: Brandon.Zoras@tdsb.on.ca

Index

L. Bencze and S. Alsop (eds.), *Activist Science and Technology Education*, Cultural Studies of Science Education 9, DOI 10.1007/978-94-007-4360-1,

C

P

T

U

V

W

The manufacturer's authorised representative in the EU is Springer
Nature Customer Service Centre GmbH, Europaplatz 3, 69115 Heidelberg,
Germany. If you have any concerns regarding our products, please
contact ProductSafety@springernature.com

Printed and bound by CPI Group (UK) Ltd, Croydon, CR0 4YY
25/09/2026
02222885-0002